Astrophysical Techniques

Astrophysical Techniques

Seventh Edition

C. R. Kitchin

CRC Press is an imprint of the
Taylor & Francis Group, an **informa** business

Seventh edition published 2020
by CRC Press
6000 Broken Sound Parkway NW, Suite 300, Boca Raton, FL 33487-2742

and by CRC Press
2 Park Square, Milton Park, Abingdon, Oxon, OX14 4RN

Sixth edition published by CRC Press 2013

CRC Press is an imprint of Taylor & Francis Group, LLC

Library of Congress Cataloging-in-Publication Data

Names: Kitchin, C. R. (Christopher R.), author.
Title: Astrophysical techniques / C.R. Kitchin.
Description: Seventh edition. | Boca Raton : CRC Press, 2020. | Includes
bibliographical references and index.
Identifiers: LCCN 2020017234 | ISBN 9781138590168 (hardback) | ISBN
9780429491139 (ebook)
Subjects: LCSH: Astrophysics--Technique. | Astronomy--Technique. |
Astronomical instruments. | Imaging systems in astronomy.
Classification: LCC QB461 .K57 2020 | DDC 522--dc23
LC record available at https://lccn.loc.gov/2020017234

ISBN: 978-1-138-59016-8 (hbk)
ISBN: 978-1-138-59120-2 (pbk)
ISBN: 978-0-429-49113-9 (ebk)

Typeset in Times
by Lumina Datamatics Limited

For

Christine,

Rowan and Willow,

Arthur and Lottie,

and

Bob.

Contents

Preface

The aim of this book is to provide a coherent state-of-the-art account of the instruments and techniques used in astronomy and astrophysics today. Whilst every effort has been made to make it as complete and up to date as possible, the author is only too aware of the many omissions and skimpily treated subjects throughout the work. For some types of instrumentation, it is possible to give full details of the instrument in its finally developed form. However, for the 'new astronomies' and even some aspects of established fields, development is occurring at a rapid pace, and the details will change between the writing and publishing of this edition. For those areas of astronomy, therefore, a fairly general guide to the principles behind the techniques is given, and this should enable the reader to follow the detailed designs in the scientific literature.

The coverage of this book is restricted, as it has been since its first edition in 1984, largely to the techniques used to study objects beyond the solar system. The reason for this is that 35 years ago, while the study of solar system objects had expanded enormously since 1957 (Sputnik 1) through the use of spacecraft, the individual instruments and techniques onboard those spacecraft were still being invented, designed, and developed very rapidly. Then, therefore, a book covering these subjects, would have been outdated almost before it could be published. By 2020 though, the science of the study of the solar system using spacecraft has matured considerably, with many of the experiments and instruments being third- or fourth-generation versions of those used in the 1970s and 1980s. The author's companion book to this one; *Remote and Robotic Investigations of the Solar System* (Taylor & Francis Group, 2018), therefore, provides a comparable review to that given here (for the techniques used to study objects beyond the solar system), for those instruments, experiments, and techniques now used to investigate objects inside the solar system. That book and this one, though, are free-standing and should not be considered as 'Volumes 1 and 2'; however, some readers may well find it helpful, useful, or interesting to study both.

With this seventh edition of *Astrophysical Techniques*, many new instruments and techniques are included for the first time, and some topics have been eliminated on the grounds that they have not been used by either professional or amateur astronomers[1] for many years. Other topics, although no longer employed by professional observers for current work, are included because archive material that is still in use was obtained using them or because amateur astronomers use the techniques. Insofar as may be possible, the material in this edition is up to date to the end of January 2020.

The section on gravitational wave detection, at long last, no longer begins with the statement that gravitational waves have 'yet to be detected' – the first such wave, resulting from the merger of a 31 $M_\odot$[2] black hole with a 36 $M_\odot$ black hole, was picked up by the AdvLIGO detectors on 14th September 2015 – and at the time of writing, a further 55 similar events have been detected. Major efforts are now being made to detect dark matter and dark energy, though without success to date. The developments, new instruments, and techniques for dark matter and energy investigations are covered in a much-expanded Section 1.7. Multi-messenger astrophysics is now becoming a reality, albeit this does not much affect the individual techniques and instruments used for each of the 'messengers'. Additionally, the number of known exoplanets is now several thousand, though the techniques used for their detection have not changed much in the last seven years.

A few references to Internet sites have been included, but not many, because the sites change so frequently and search engines are now so good. However, this resource now is usually the point of first call for most scientists when they have a query and much material, such as large sky surveys, is

[1] The distinction between amateur and professional astronomers can be blurred. Often both work to similar standards and the professionals have just found someone to pay their salary whilst pursuing their interests. Amateurs, though, need their day job to be able to pursue their *real* vocation.

[2] A widely used symbol for the mass of the Sun = $\sim 2 \times 10^{30}$ kg.

only available over the Internet. Furthermore, the Internet is used for the operation of some remote telescopes, and it forms the basis of 'virtual' observatories discussed in the last chapter.

As in previous editions, another aim has always been to try and reduce the trend towards fragmentation of astronomical studies, and this is retained in this edition. The new techniques that are required for observing in exotic regions of the spectrum bring their own concepts and terminology with them. This can lead to the impression that the underlying processes are quite different, when in fact they are identical but are merely being discussed from differing points of view. Thus, for example, the Airy disc and its rings and the polar diagram of a radio dish do not at first sight look related, but they are simply two different ways of presenting the same data. As far as possible, therefore, broad regions of the spectrum are dealt with as a single area, rather than as many smaller disciplines. The underlying unity of all of astronomical observation is also emphasised by the layout of the book; the pattern of detection ➔ imaging ➔ ancillary techniques has been adopted, so that one stage of an observation is encountered together with the similar stages required for related information carriers. This is not an absolutely hard-and-fast rule, however and in some places, it seemed appropriate to deal with topics out of this sequence – either to prevent a multiplicity of very small sections or to keep the continuity of the argument going.

The treatment of the topics is at a level appropriate to a science-based undergraduate degree. As far as possible, the mathematics or physics background which may be needed for a topic is developed or given within that section. In some places, it was felt that some astronomy background might be needed as well, so that the significance of the technique under discussion could be properly realised. Although aimed at an undergraduate level, most of the mathematics should be understandable by anyone who has attended a competently taught mathematics course in their final years at school and some sections are non-mathematical. Thus, many amateur astronomers will find the aspects of the book to be of great use and interest. The fragmentation of astronomy, which has already been mentioned, means that there is a third group of people who may find the book useful and that is professional astronomers themselves. The treatment of the topics in general is at a sufficiently high level, yet in a convenient and accessible form, to be of some use to those professionals seeking information on techniques in areas of astronomy with which they might not be totally familiar.

Lastly, I must pay a tribute to the many astronomers and other scientists whose work is summarised here. It is not possible to list them by name and to study the text with detailed references would have ruined the intention of making the book readable. I would, however, like to take the opportunity afforded by this preface to express my deepest gratitude to them all.

Clear Skies and Good Observing to you all!

C. R. Kitchin

Author

C. R. Kitchin is currently professor emeritus at the University of Hertfordshire and a freelance writer of astrophysics textbooks. From 1987 to 2001, he was the director of the University's Observatory at Bayfordbury and from 1996 to 2001, he served as the head of the division of physics and astronomy. He took early retirement in 2001 to concentrate on his writing interests. Kitchin has written 16 books as sole author and contributed to another dozen or so, as well as writing hundreds of articles covering interests ranging from popular to specialist research.

When not writing books, Kitchin enjoys gardening and being taken for walks by his two border collies.

1 Detectors

1.1 OPTICAL DETECTION

1.1.1 INTRODUCTION

In this and the immediately succeeding sections, the emphasis is upon the *detection* of the radiation or other information carrier and upon the instruments and techniques used to facilitate and optimise that detection. There is inevitably some overlap with other sections and chapters, and some material might arguably be more logically discussed in a different order from the one chosen. In particular in this section, telescopes are included as a necessary adjunct to the detectors themselves. The theory of the formation of an image of a point source by a telescope, which is all that is required for simple detection, takes us most of the way through the theory of imaging extended sources. Both theories are, therefore, discussed together, even though the latter should perhaps be in Chapter 2. There are many other examples such as X-ray spectroscopy and polarimetry that appear in Section 1.3 instead of Sections 4.2 and 5.2. In general, the author has tried to follow the route *detection→imaging→ancillary techniques* throughout the book but has dealt with items out of this order when it seemed more natural to do so.

The optical region is taken to include the mid/near infrared, the visible and the long-wave ultraviolet (UV) regions (sometimes called the Ultraviolet, Optical and Infrared [UVOIR] region) and, thus, roughly covers the wavelength range from 100 μm to 10 nm (3 THz to 30 PHz, 10 meV to 100 eV).[1] The techniques and physical processes employed for investigations over this region bear at least a generic resemblance to each other and so may conveniently be discussed together.

1.1.2 DETECTOR TYPES

In the optical region, detectors fall into two main groups: thermal and quantum (or photon) detectors. Both these types are incoherent; that is to say, only the amplitude of the electromagnetic wave is detected, and the phase information is lost. Coherent detectors are common at long wavelengths (Section 1.2), where the signal is mixed with that from a local oscillator (heterodyne principle). But only recently have heterodyne techniques been developed for wavelengths as short as the infrared (IR) and applied to astronomy (see Mid Infrared Laser Heterodyne Instrument [MILAHI], for example, Section 1.1.15.4). We may, therefore, safely regard almost all current astronomical optical detectors as still being incoherent in practice. With optical aperture synthesis (Section 2.5), some phase information may be obtained providing that three or more telescopes are used.

In quantum detectors, the individual photons of the optical signal interact directly with the electrons of the detector. Sometimes individual detections are monitored (photon counting), and at other times, the detections are integrated to give an analogue output like that of the thermal detectors. Examples of quantum detectors include the eye, photographic emulsion, photomultiplier, photodiode, charge-coupled devices (CCDs) and many other solid-state detectors.

[1] At longer wavelengths, the frequency equivalent of the wavelength is listed (using $c = 300{,}000$ km/s for the conversion) because many astronomers researching in this region are accustomed to working with frequency. Usually this applies for wavelengths of 10 μm (30 THz) or longer, but the conversion is given at shorter wavelengths when (as here) it seems appropriate to do so. Similarly, at short wavelengths the energy in electron-volts (eV) is listed as well as the wavelength. One eV, equals 1.6×10^{-19} J and is a convenient unit for use in this spectral region (and also when discussing cosmic rays in Section 1.4). Since the values of the corresponding frequencies are 10^{17} Hz and above, they are not commonly encountered in the literature.

Thermal detectors, by contrast, detect radiation through the increase in temperature that its absorption causes in the sensing element. They are generally less sensitive and slower in response than quantum detectors, but they have a much broader spectral response. Examples include thermocouples, thermistors, pyroelectric detectors, and bolometers.

1.1.3 The Eye

This is undoubtedly the most fundamental of detectors to a human astronomer, although it is a way from being the simplest. It is now rarely used for primary detection except, of course, for the billions of people who gaze into the skies for pleasure – and that includes most professional astronomers.

If you live in a city, only a few stars may be visible, even on the clearest of nights. From a city's suburbs you might see a few tens of stars, but throughout most urban or semi-rural parts of most countries, you will be lucky to see more than a hundred stars with the unaided eye. Because around 5,000 to 6,000 stars *should* be visible (2,500 to 3,000 at any given moment, of course, because half the sky is below the horizon), thousands of stars are 'missing'. The 'missing' stars, though, have not truly vanished; they have just been swamped by the lights from the city.

In the last three decades or so, many high-publicity campaigns have been initiated to try and reduce light pollution and restore the glories of really dark skies to millions of city dwellers, but regrettably such efforts have had little effect. However, there are still some areas of most countries where light pollution is much reduced. Some of those areas now advertise their dark skies as tourist attractions and so can be found via an internet search. At the time of writing, eleven sites around the world have been identified as 'International Dark Sky Reserves'. These are areas of at least 700 km^2 that have exceptional night sky quality and within which further human developments which might increase levels of light pollution are (at least somewhat) controlled. An internet search will quickly list these reserves and any more that may have been recognised recently. If you think you know your constellations but have learnt to identify them only from a 'normal' darkish site, then a visit to one of the reserves will amaze you; the constellations will no longer be recognisable because of the hundreds of additional stars that have suddenly become visible (of course, it *will* need to be a clear night, but *that* problem remains to be solved!).

Much more detailed coverage of the ways in which eyes and vision work, and how their idiosyncrasies may affect astronomical observations have been included in previous editions of this book. However, most astronomers, both amateur and professional, now use their eyes mainly for monitoring their instruments and for looking at computer VDUs or printouts, so such detailed coverage is no longer appropriate. If needed, further information of how the nature of the human eye affects astronomical observation may be found in the author's book *Telescopes and Techniques* (2013), previous editions of *Astrophysical Techniques*, other introductory and observational astronomy sources and biological and medical sources dealing with the eye and ophthalmology.

One aspect of the eye's behaviour though, will still be discussed briefly here because it is to be encountered in many aspects of observational astronomy, even today – and that is the use of *magnitudes* to measure the brightness or luminosity of objects in the sky, especially over the visual region (see also Section 3.1). The magnitude scale is used to measure the brightness of stars and other celestial objects as seen in the sky (i.e., the light energy arriving from the star at the surface of the Earth) and has its roots in eye estimates of stellar brightness made more than 2,000 years ago for Hipparchus' star catalogue. In that catalogue, the brightest stars were first class, the next brightest were second class, then third class and so on down to the sixth-class stars which could only just be seen.

These classes became roughly our modern stellar magnitudes and so retain the (now unusual) custom of denoting the *brighter* of two stars by a numerically *smaller* value of the magnitude i.e., a magnitude 2 star is brighter than a magnitude 3 star, etc.

Additionally, the eye's response to changes in illumination is approximately logarithmic (Fechner's law[2]), so the magnitude scale varies *logarithmically* with the visual brightness of the stars and *not linearly*. That is to say; if two sources, A and C are observed to differ in brightness by a certain amount and a third source, B, appears *to the eye* as being midway in brightness between them, then the energy from B will differ from that from A by the same *factor* as that of C differs from B. Thus, if we use L to denote the perceived luminosity and E to denote the actual energy at the Earth's surface of the sources, then for

$$L_B = \frac{1}{2}\left(L_A + L_C\right) \tag{1.1}$$

we have

$$\frac{E_A}{E_B} = \frac{E_B}{E_C} \quad \left[\text{or } \log E_B = \frac{1}{2}\left(\log E_A + \log E_C\right)\right] \tag{1.2}$$

In other words, a magnitude 2 star is not only brighter than a magnitude 3 star, it is 2.51 times brighter in energy terms. Similarly, a magnitude 3 star is not only brighter than a magnitude 4 star, it is 2.51 times brighter. The magnitude 2 star is thus 6.30 times (2.51 × 2.51) brighter than the magnitude 4 star.[3]

Fechner's law and the magnitude scale's ancient historic background are the reasons for the (awkward) nature of the magnitude scale used by astronomers to measure stellar brightness (Section 3.1).

The faintest stars visible to the dark-adapted naked eye from a good observing site on a clear moonless night are (still) of about magnitude six. This corresponds to the detection of about 3×10^{-15} W or about 8,000 visible light photons per second. Special circumstances or especially sensitive vision may enable this limit to be improved upon by one to one and a half stellar magnitudes (×2 to ×4). Conversely, the normal ageing processes in the eye, such as a decreasing ability to dilate the eye's pupil and increasing numbers of 'floaters', etc., mean that the retina of a 60-year old person receives only about 30% of the amount of light seen by a person half that age.[4] Eye diseases and problems, such as cataracts, may reduce this much further. Observers should thus expect a reduction in their ability to perceive very faint objects as time goes by.

1.1.4 Semiconductors

The photomultiplier, CCD and several of the other detectors considered later derive their properties from the behaviour of semiconductors. Thus, some discussion of the relevant physics of these materials is a necessary prerequisite to a full understanding of detectors of this type.

Conduction in a solid may be understood by considering the electron energy levels. For a single isolated atom, they are unperturbed and are the normal atomic energy levels. As two such isolated atoms approach each other, though, their interaction causes the levels to split (Figure 1.1). If further atoms approach, then the levels develop further splits. So that for N atoms in close proximity to each other, each original level is split into N sublevels (Figure 1.2). Within a solid, therefore, each level becomes a pseudo-continuous band of permitted energies because the individual sublevels overlap

[2] Sometimes called the Weber-Fechner law, although strictly speaking the Weber law is a different formulation of the same physical phenomenon.

[3] The factor is actually 2.511886 …, or the fifth root of 100. Further details on the magnitude scale, including how it is used to measure the intrinsic brightness of objects in the sky, not just the brightness as we see them, may be found in Section 3.1.

[4] The eye is fully formed by the time a person is about age 13 and almost so from the age of about 3 onwards. There are thus no equivalent age-related changes amongst young people.

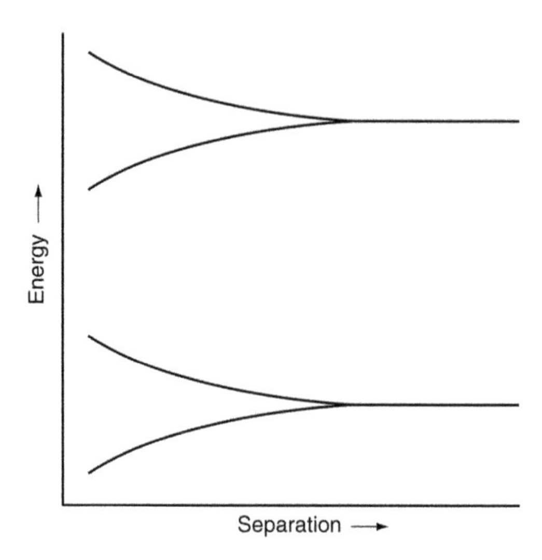

FIGURE 1.1 Schematic diagram of the splitting of two of the energy levels of an atom resulting from its proximity to another similar atom.

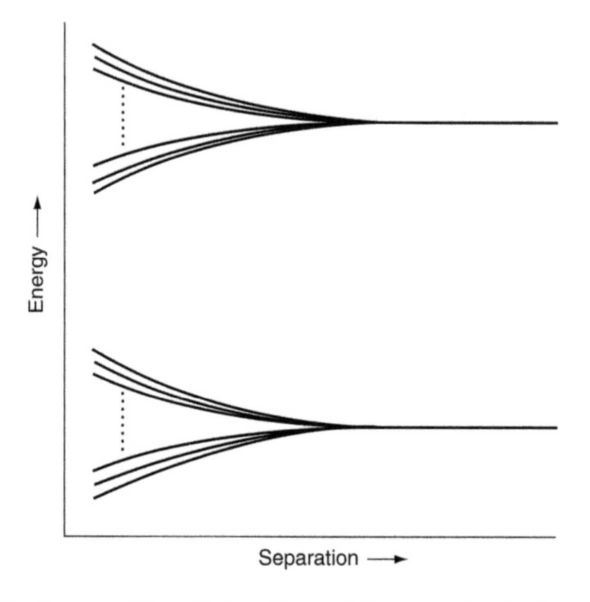

FIGURE 1.2 Schematic diagram of the splitting of two of the energy levels of an atom resulting from its proximity to many similar atoms.

each other. The energy level diagram for a solid, thus, has the appearance shown in Figure 1.3. The innermost electrons remain bound to their nuclei, while the outermost electrons interact to bind the atoms together. They occupy an energy band called the 'valence band'.

To conduct electricity through such a solid, the electrons must be able to move within the solid. From Figure 1.3 we may see that free movement could occur for electrons within the valence and higher bands. However, if the original atomic level that became the valence band upon the formation of the solid was *fully occupied* by electrons, then all the sublevels within the valence band will *still* be fully occupied. If any given electron is to move under the influence of an electric potential, then its energy must increase. This it cannot do because all the sublevels are occupied and so there is

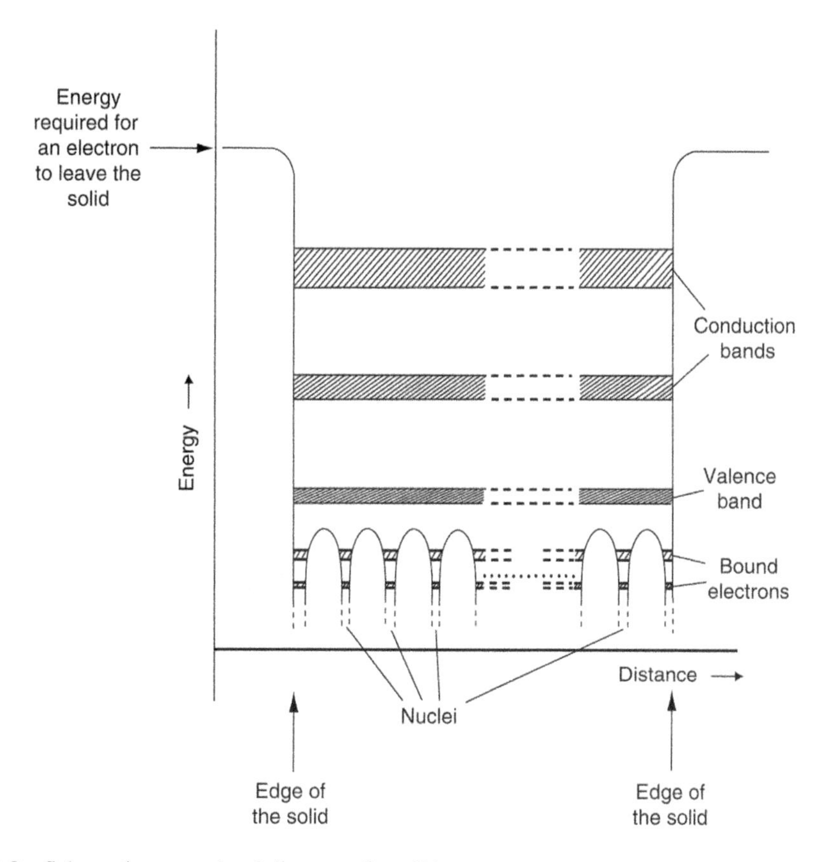

FIGURE 1.3 Schematic energy level diagram of a solid.

no vacant level available for it at this higher energy. Thus, the electron cannot move after all. Under these conditions we have an electrical insulator.

If the material is to be a conductor, we can now see that there must be vacant sublevels which the conduction electron may enter. There are two ways in which such empty sublevels may become available. Either the valence band is unfilled, for example when the original atomic level had only a single electron in an *s* sub-shell,[5] or one of the higher-energy bands becomes sufficiently broadened to overlap the valence band. In the latter case at a temperature of absolute zero, all the sublevels of both bands will be filled up to some energy that is called the 'Fermi level'. Higher sublevels will be unoccupied. As the temperature rises, some electrons will be excited to some of these higher sublevels but will still leave room for conduction electrons.

A third type of behaviour occurs when the valence and higher bands do not actually overlap but have only a small energy separation.[6] Thermal excitation may then be sufficient to push a few valence electrons into some of the higher bands. An electric potential can now cause the electrons in either the valence or the higher band to move. The material is known as a 'semiconductor' because its conductivity is generally better than that of an insulator but considerably poorer than that of a true conductor. The higher-energy bands are usually known as the 'conduction bands'.

[5] See, for example, the author's book *Optical Astronomical Spectroscopy* (1995), or other sources dealing with atomic structure and/or spectroscopy for further details of this topic and its notation.

[6] The energy gaps (band gaps) in the widely used semiconductors, silicon and germanium, are 1.09 and 0.72 eV, respectively. These energies correspond to infrared photons with wavelengths of 1.14 and 1.72 μm, respectively, so these wavelengths are the nominal lower-energy limits for detectors based upon silicon or germanium.

A pure substance will have equal numbers of electrons in its conduction bands and of spaces in its valence band. However, an imbalance can be induced by the presence of atoms of different elements from that forming the main semiconductor. If the valence band is full and one atom is replaced by another that has a larger number of valence electrons (a donor atom), then the excess electron(s) usually occupy new levels in the gap between the valence and conduction bands close to the bottom of the conduction band. From there they may more easily be excited into the conduction band by thermal motions or other energy sources. The semiconductor is then an *n*-type because any current within it will largely be carried by the (*n*egative) electrons in the conduction band.

In the other case, when an atom is replaced with the one that has few valence electrons (an acceptor atom), then new empty levels will be formed just above the top of the valence band. Electrons from the valence band can easily be excited into these new levels leaving energy gaps in the valence band. Physically this means that an electron bound to an atom within the crystal is now at a higher, but still bound, energy level. However, the excited electron no longer has an energy placing it within the valence band. Other electrons in the valence band can now take up the energy 'abandoned' by the excited electron. Physically this means that a valence electron is now able to move within the crystal. Without an externally applied voltage, the electron movement will be random, but if there is an externally applied voltage then the electron will move from the negative potential side towards the positive potential side. The movement will comprise that of an electron from a neighbouring atom hopping over to the atom which has the gap in its electron structure. The moving atom will then be bound to its new atom but will have left a gap in the electron structure of its originating atom. An electron from the next atom over can now therefore hop into that gap. A third electron from the next atom over can then hop into this new gap – and so on. Since the hops occur rapidly, the appearance to an external observer would be that of the *absence* of an electron within atoms' electron structures moving continuously from the positive side towards the negative side. The *absence* of a (negatively charged) electron however is equivalent to the *presence* of an equivalent positive charge. The process is thus usually visualised, not as comprising a series of electron hops, but as the continuous movement of a positively charged particle, called a 'hole' in the opposite direction. This type of semiconductor is called *p*-type (from *p*ositive charge carrier), and its electric currents are thought of as mainly being transported by the movement of the positive holes in the valence band.

1.1.4.1 The Photoelectric Effect

The principle of the photoelectric effect is well known; the material absorbs a photon with a wavelength less than the limit for the material and an electron is then emitted from the surface of the material. The *energy* of that electron is a function of the energy of the photon, while the *number* of electrons depends upon the intensity of the illumination. In practice, the situation is somewhat more complex, particularly when what we are interested in is specifying the properties of a *good* photoemitter.

The main requirements are that the material should absorb the required radiation efficiently and that the mean free paths of the released electrons within the material should be greater than that of the photons. The relevance of these two requirements may be most simply understood from looking at the behaviour of metals in which neither condition is fulfilled. Since metals are conductors there are many vacant sublevels near their Fermi levels (see the previous discussion). After absorption of a photon, an electron is moving rapidly and energetically within the metal. Collisions will occur with other electrons and because these electrons have other nearby sublevels that they may occupy, they can absorb some of the energy from our photoelectron. Thus, the photoelectron is slowed by collisions until it may no longer have sufficient energy to escape from the metal even if it does reach the surface. For most metals, the mean free path of a photon is about 10 nm and that of the released electron is less than 1 nm; thus, the eventually emitted number of electrons is considerably reduced by collisional energy losses. Furthermore, the high reflectivity of metals results in only a small fraction of the suitable photons being absorbed, and so the actual number of electrons emitted is only a small proportion of those potentially available.

A good photoemitter must thus have low-energy loss mechanisms for its released electrons whilst they are within its confines. The loss mechanism in metals (collision) can be eliminated by the use of semiconductors or insulators. Then the photoelectron cannot lose significant amounts of energy to the valence electrons because there are no vacant levels for the latter to occupy. Neither can it lose much energy to the conduction electrons because there are few of these around. In insulators and semiconductors, the two important energy-loss mechanisms are pair production and sound production. If the photoelectron is energetic enough, then it may collisionally excite other valence electrons into the conduction band thus producing *pairs* of electrons and holes. This process may be eliminated by requiring that $E_{1,}$ the minimum energy to excite a valence electron into the conduction band of the material (Figure 1.4), is larger than E_2, the excess energy available to the photoelectron. Sound waves or phonons are the vibrations of the atoms in the material and can be produced by collisions between the photoelectrons and the atoms, especially at discontinuities in the crystal lattice, etc. Only 1% or so of the electron's energy will be lost at each collision because the atom is so much more massive than the electron. However, the mean free path between such collisions is only 1 or 2 nm, so that this energy loss mechanism becomes significant when the photoelectron originates deep in the material. The losses may be reduced by cooling the material because this then reduces the available number of quantised vibration levels in the crystal and also increases the electron's mean free path.

The minimum energy of a photon, if it is to be able to produce photoemission, is known as the 'work function' and is the difference between the ionisation level and the top of the valence band (Figure 1.4). Its value may, however, be increased by some or all of the energy-loss mechanisms mentioned previously. In practical photoemitters pair production is particularly important at photon energies above the minimum and may reduce the expected flux considerably as the wavelength decreases. The work function is also strongly dependent upon the surface properties of the material; surface defects, oxidation products and impurities can cause it to vary widely even amongst samples of the same substance. The work function may be reduced if the material is strongly *p*-type and is at an elevated temperature. Vacant levels in the valence band may then be populated by thermally excited electrons, bringing them nearer to the ionisation level and so reducing the energy required to let them escape. Since most practical photoemitters are strongly *p*-type, this is an important

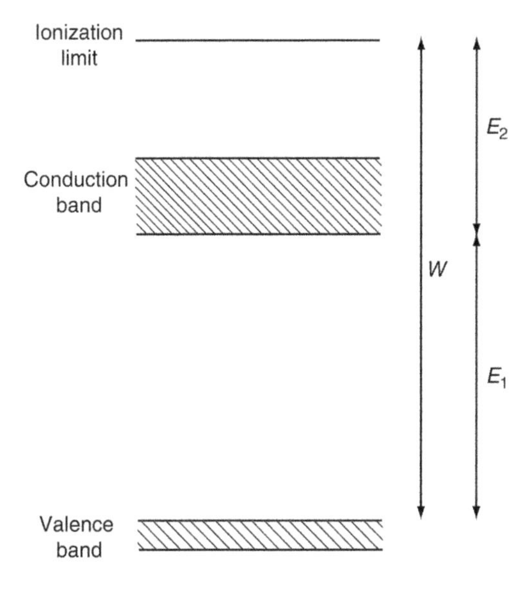

FIGURE 1.4 Schematic partial Grotrian diagram of a good photoemitter.

process and confers sensitivity at longer wavelengths than the nominal cut-off point. The long-wave sensitivity, however, is variable because the degree to which the substance is *p*-type is strongly dependent upon the presence of impurities and so is sensitive to small changes in the composition of the material.

1.1.5 A Detector Index

After the 'natural' detector formed by the eye, there are numerous types of 'artificial' optical detectors. Before going to look at some of them in more detail, however, it is necessary to place them in some sort of logical framework or the reader is likely to become confused rather than enlightened by this section. We may idealise any detector as simply a device wherein some measurable property changes in response to the effects of electromagnetic radiation. We may then classify the types of detector according to the property that is changing and this is shown in Table 1.1.

Other processes may be sensitive to radiation but fail to form the basis of a useful detector. For example, we can feel the (largely IR) radiation from a fire directly on our skins, but this process is not sensitive enough to form an IR detector for any celestial objects other than the Sun. Conversely as yet unutilised properties of matter, such as the initiation of stimulated radiation from excited atomic states as in the laser or maser, may become the basis of detectors in the future.

1.1.6 Detector Parameters

Before resuming discussion of the detector types listed, we need to establish the definitions of some of the criteria used to assess and compare detectors. The most important of these are listed in Table 1.2.

For purposes of comparison, D^* is generally the most useful parameter. For photomultipliers in the visible region, it is around 10^{15}–10^{16}. Figures for the eye and for photographic emulsion are not directly obtainable, but values of 10^{12}–10^{14} can perhaps be used for both to give an idea of their relative performances.

TABLE 1.1
Classification Scheme for Types of Detector

Sensitive Parameter	Detector Names	Class
Voltage	Photovoltaic cells	Quantum
	Thermocouples	Thermal
	Pyroelectric detectors	Thermal
Resistance	Blocked impurity band device (BIB)	Quantum
	Bolometer/Calorimeter	Thermal
	Photoconductive cell	Quantum
	Phototransistor/Photodiode	Quantum
	Transition Edge Sensor (TES)	Thermal
Charge	Charge-coupled device (CCD)	Quantum
	Charge injection device (CID)	Quantum
Current	Superconducting tunnel junction (STJ)	Quantum
Electron excitation	Photographic emulsion	Quantum
Electron emission	Photomultiplier	Quantum
	Television	Quantum
	Image intensifier	Quantum
Chemical composition	Eye	Quantum

TABLE 1.2
Some Criteria for Assessment and Comparison of Detectors

QE (quantum efficiency)	Ratio of the actual number of photons that are detected to the number of incident photons.
DQE (detective quantum efficiency)	Square of the ratio of the output signal-to-noise ratio to the input signal-to-noise ratio.
τ (time constant)	This has various precise definitions. Probably the most widespread is that τ is the time required for the detector output to approach to within $(1 - e^{-1})$ of its final value after a change in the illumination; i.e., the time required for about 63% of the final change to have occurred.
Dark noise	The output from the detector when it is unilluminated. It is usually measured as a root-mean-square voltage or current.
NEP (noise equivalent power or minimum detectable power)	The radiative flux as an input that gives an output signal-to-noise ratio of unity. It can be defined for monochromatic or black body radiation and is usually measured in watts.
D (detectivity)	Reciprocal of NEP. The signal-to-noise ratio for incident radiation of unit intensity.
*D** (normalised detectivity)	The detectivity normalised by multiplying by the square root of the detector area and by the electrical bandwidth. It is usually pronounced 'dee star'. $$D^* = \frac{(a\,\Delta f)^{1/2}}{NEP} \quad (1.3)$$ The units, cm $Hz^{1/2}$ W^{-1}, are commonly used, and it then represents the signal-to-noise ratio when 1 W of radiation is incident on a detector with an area of 1 cm^2 and the electrical bandwidth is 1 Hz.
R (responsivity)	Detector output for unit intensity input. Units are usually volts per watt or amps per watt. For the human eye, it is called the 'luminosity function' and its unit is lumens per watt.
Dynamic range	Ratio of the saturation output to the dark signal. Sometimes only defined over the region of linear response.
Spectral response	The change in output signal as a function of changes in the wavelength of the input signal. Usually given as the range of wavelengths over which the detector is useful.
λ_m (peak wavelength)	The wavelength for which the detectivity is a maximum.
λ_c (cut-off wavelength)	There are various definitions. Amongst the commonest are: wavelength(s) at which the detectivity falls to zero; wavelength(s) at which the detectivity falls to 1% of its peak value; wavelength(s) at which *D** has fallen to half its peak value.

1.1.7 Cryostats

The noise level in many detectors can be reduced by cooling them to below ambient temperature. Indeed, for some detectors, such as superconducting tunnel junctions (STJs; see Section 1.1.12) and transition edge sensors (TESs; see Section 1.1.15.3) cooling to very low temperatures is essential for their operation. Small CCDs produced for the amateur market are usually chilled by Peltier-effect coolers, but almost all other approaches require the use of liquid nitrogen or liquid helium as the coolant. Since these materials are both liquids, they must be contained in such a way that the liquid does not spill out as the telescope moves. The container is called a 'cryostat', and while there are many different detailed designs, the basic requirements are much the same for all detectors. In addition to acting as a container for the coolant, the cryostat must ensure that the detector and sometimes pre-amplifiers, filters, and optical components are cooled whatsoever the position of the telescope, that the coolant is retained for as long as possible, that the evaporated coolant can escape and that the detector and other cooled items do not ice up.

These requirements invariably mean that the container is a Dewar (vacuum flask) and that the detector is behind a window within a vacuum or dry atmosphere. Sometimes the window is heated to avoid ice forming on *its* surface. Most cryostats are of the 'bath' design and are essentially tanks containing the coolant with the detector attached to the outside of the tank or linked to it by a thermally conducting rod. Such devices can only be (roughly) half-filled with the coolant if none is to overflow as they are tilted by the telescope's motion. If the movement of the telescope is restricted though, higher levels of filling may be possible. Hold times between refilling with coolant of a few days can currently be achieved. When operating at Nasmyth or Coudé foci or if the instrument is in a separate laboratory fed by fibre-optics from the telescope, so that the detector is not tilted, continuous flow cryostats can be used, in which the coolant is supplied from a large external reservoir; therefore, hold times of weeks, months or longer are then possible.

Closed-cycle cryostats have the coolant contained within a completely sealed enclosure. The exhausted (warm) coolant is collected, re-cooled and reused. This type of cryostat design is mostly needed when (expensive) liquid or gaseous helium is the coolant. Open cycle cryostats allow the used coolant to escape from the system. It is then simply vented to the atmosphere, or may, especially for helium, be collected and stored for future use, in a separate operation.

The detector, its immediate electronics and sometimes filters or other nearby optical components are usually cooled to the lowest temperature. Other parts of the instrument such as heat shields, windows and more distant optical components may not need to be quite so cold. Hence, especially when helium is the main coolant, cheaper cryogens may be used for these less critical components. Thus, a Stirling engine used in reverse can routinely cool down to 70 K and in exceptional circumstances down to 40 K (a multi-stage variant on the Stirling system called a 'Pulse Tube Refrigerator' and utilising helium as its working fluid has recently achieved 1.7 K), liquid nitrogen is usually around 77 K, dry ice (solid carbon dioxide – not now used much by astronomers) reaches 195 K and Peltier coolers can go down to around 220 K, depending upon the ambient temperature.

Bolometers, STJs, and TESs require cooling to temperatures well below 1 K. Temperatures down to about 250 mK can be reached using liquid helium-3. The helium-3 itself has to be cooled to below 2 K before it liquifies, and this is achieved by using helium-4 under reduced pressure. Temperatures down to a few mK require a dilution refrigerator. This uses a mix of helium-3 and helium-4 at a temperature lower than 900 mK. The two isotopes partially separate out under gravity, but the lower helium-4 layer still contains some helium-3. The helium-3 is removed from the lower layer distilling it off at 600 mK in a separate chamber. This forces some helium-3 from the upper layer to cross the boundary between the two layers to maintain the equilibrium concentration. However, crossing the boundary requires energy and this is drawn from the surroundings, so cooling them. The Submillimetre Common User Bolometer Array (SCUBA-2)[7], for example (see bolometers in Section 1.1.15.2), uses a dilution refrigeration system to operate at 100 mK. An alternative route to mK temperatures is the adiabatic demagnetisation refrigerator. The ions in a paramagnetic salt are first aligned by a strong magnetic field. The heat generated in this process is transferred to liquid helium via a thermally conducting rod. The rod is then moved from contact with the salt, the magnetic field is reduced and the salt cools adiabatically.

1.1.8 Charge-Coupled Devices (CCDs)

1.1.8.1 CCDs

Willard Boyle and George Smith invented CCDs in 1969 at the Bell telephone labs for use as a computer memory. The first application of CCDs within astronomy as optical detectors occurred in the late 1970s. Since then they have come to dominate the detection of optical radiation at professional observatories and to be widely used amongst amateur astronomers. Their popularity arises from their ability to integrate the detection over long intervals, their dynamic range (>10^5), linear response, direct digital output, high quantum efficiency, robustness, wide spectral range and from

[7] On the James Clerk Maxwell telescope.

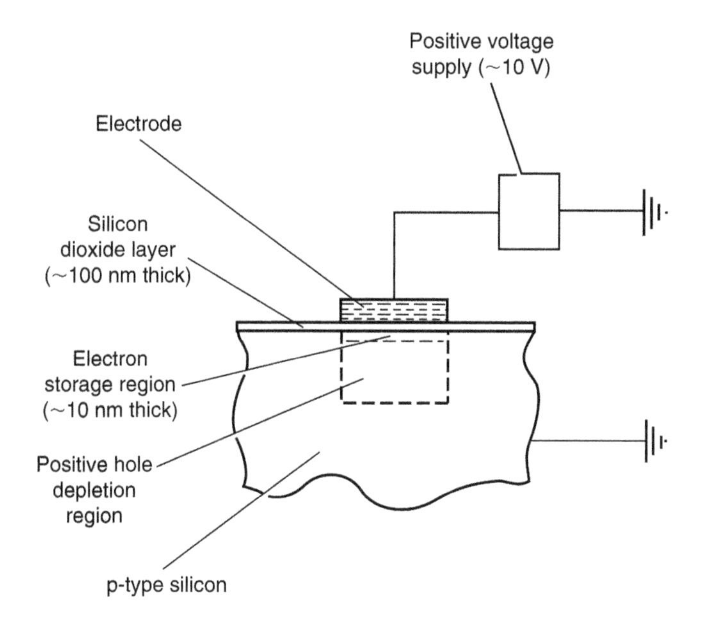

FIGURE 1.5 Basic unit of a charge-coupled device (CCD).

the ease with which arrays can be formed to give two-dimensional (2D) imaging. In fact, CCDs can only be formed as an array; a single unit is of little use by itself.

The basic detection mechanism is related to the photoelectric effect. Light incident on a semiconductor (usually silicon) produces electron-hole pairs as we have already seen. These electrons are then trapped in potential wells produced by numerous small electrodes. There, they accumulate until their total number is read out by charge coupling the detecting electrodes to a single read-out electrode.

An individual unit of a CCD is shown in Figure 1.5. The electrode is insulated from the semiconductor by a thin oxide layer. In other words, the device is related to metal-oxide-silicon (MOS) transistors.[8] The electrode is held at a small positive voltage that is sufficient to drive the positive holes in the *p*-type[9] silicon away from its vicinity and to attract the electrons into a thin layer immediately beneath it. The electron-hole pairs produced in this depletion region by the incident radiation are thereby separated and the electrons accumulate in the storage region. Thus, an electron charge is formed whose magnitude is a function of the intensity of the illuminating radiation. In effect, the unit is a radiation-driven capacitor.

Now if several such electrodes are formed on a single silicon chip and zones of very high *p*-type doping[10] insulate the depletion regions from each other, then each will develop a charge that is proportional to its illuminating intensity.[11] Thus, we have a spatially and electrically digitised reproduction of the original optical image (Figure 1.6).

All that remains is to retrieve this electron image in some usable form. This is accomplished by the charge coupling. Imagine an array of electrodes such as those we have already seen in Figure 1.6 but without their insulating separating layers. Then if one such electrode acquired a charge, it would

[8] Also known as Metal Oxide Field Effects Transistors (MOSFETs), these are widely used in electronics, where devices based upon diamond (band gap 5.47 eV), silicon carbide (2.3 to 3.3 eV), gallium nitride (3.4 eV) and other wide band gap semiconductors are now under development. Possibly, improved UV CCD-type detectors may follow from this work in due course.

[9] Usually boron is the doping element.

[10] Near atomic-scale precision in the levels of doping can now be achieved using molecular beam epitaxy. The technique is known as 'delta-doping' because it can (almost) produce a spike change in the doping levels mimicking the mathematical delta function. Such precise control of doping levels helps to increase detection efficiency and reduce noise levels in CCDs.

[11] Unless anti-blooming is used (see below), the electron charge is linearly related to the optical intensity, at least until the number of electrons approaches the maximum number that the electrode can hold (known as its 'well capacity').

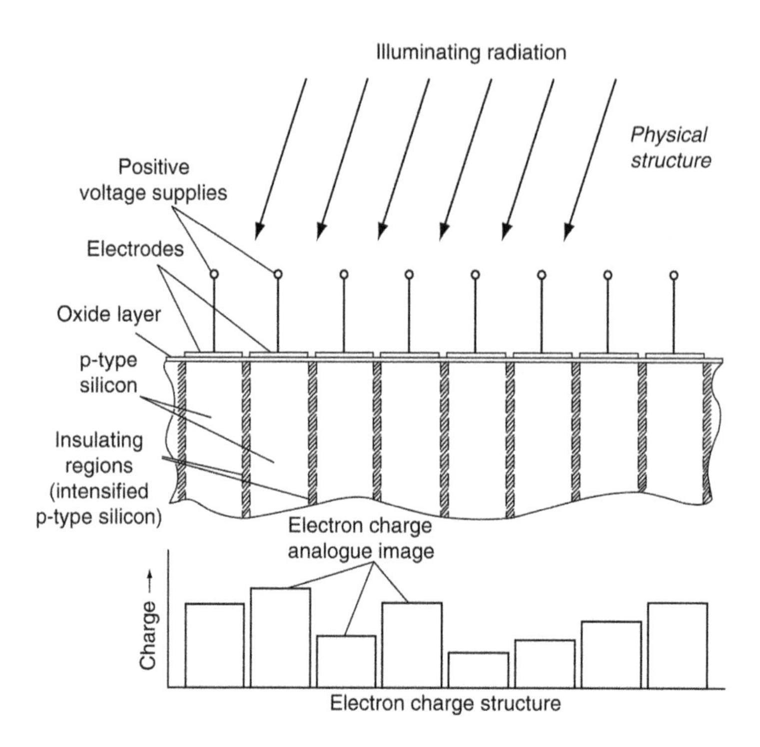

FIGURE 1.6 Array of charge-coupled device (CCD) basic units.

diffuse across to the nearby electrodes. However, if the voltage of the electrodes on either side of the one containing the charge were to be reduced, then their hole-depletion regions would disappear and the charge would once again be contained within two *p*-type insulating regions (Figure 1.7). This time, however, the insulating regions are not permanent but may be changed by varying the electrode voltage. Thus, the stored electric charge may be moved physically through the structure of the device by sequentially changing the voltages of the electrodes. Hence, in Figure 1.7, if the voltage on electrode C is changed to about +10 V, then a second hole-depletion zone will form adjacent to the first. The stored charge will diffuse across between the two regions until it is shared equally. Now if the voltage on electrode B is gradually reduced to +2 V, its hole-depletion zone will gradually disappear and the remaining electrons will transfer across to be stored under electrode C. Thus, by cycling the voltages of the electrodes as shown in Figure 1.8, the electron charge is moved from electrode B to electrode C.

With careful design, the efficiency of this charge transfer (or coupling) may be made as high as 99.9999%. Furthermore, we may obviously continue to move the charge through the structure to electrodes D, E, F and so on by continuing to cycle the voltages in an appropriate fashion. Eventually the charge may be brought to an output electrode from whence its value may be determined by discharging it through an integrating current meter or some similar device. In the scheme outlined here, the system requires three separate voltage cycles to the electrodes to move the charge. Hence, it is known as a three-phase CCD (variants on the basic CCD are discussed below). Three separate circuits are formed, with each electrode connected to those three before and three after it (Figure 1.9). The voltage supplies, α, β and γ (Figure 1.9), follow the cycles shown in Figure 1.10 to move charge packets from the left towards the right (Figure 1.11). Since only every third electrode holds a charge in this scheme, the output follows the pattern shown schematically at the bottom of Figure 1.10. The order of appearance of an output pulse at the output electrode is directly related to the position of its originating electrode in the array. Thus, the original spatial charge pattern and, hence, the original optical image may easily be inferred from the time-varying output.

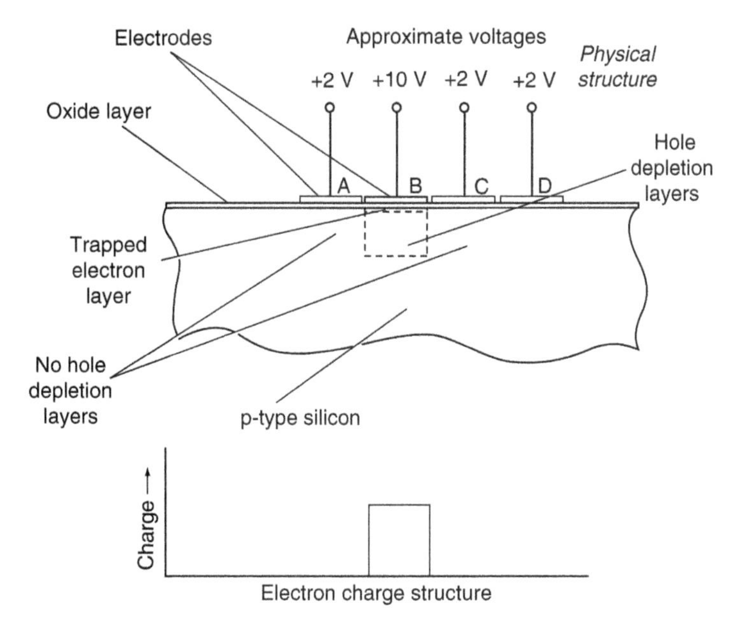

FIGURE 1.7 Active electron charge trapping in charge-coupled device (CCD).

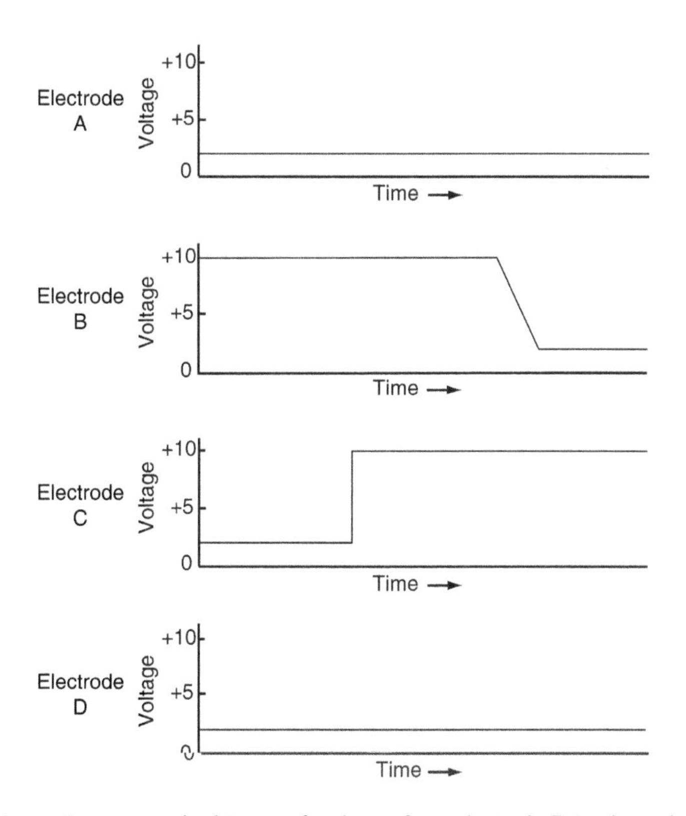

FIGURE 1.8 Voltage changes required to transfer charge from electrode B to electrode C in the array as shown in Figure 1.7.

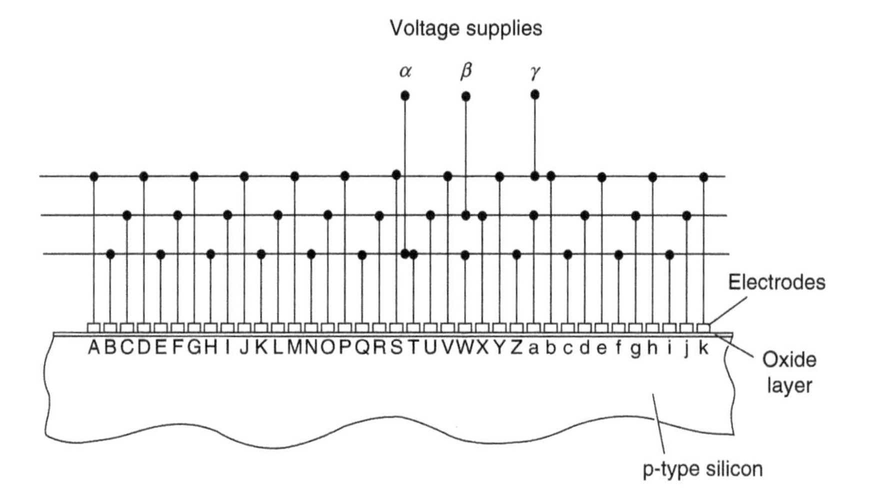

FIGURE 1.9 Connection diagram for a three-phase charge-coupled device (CCD).

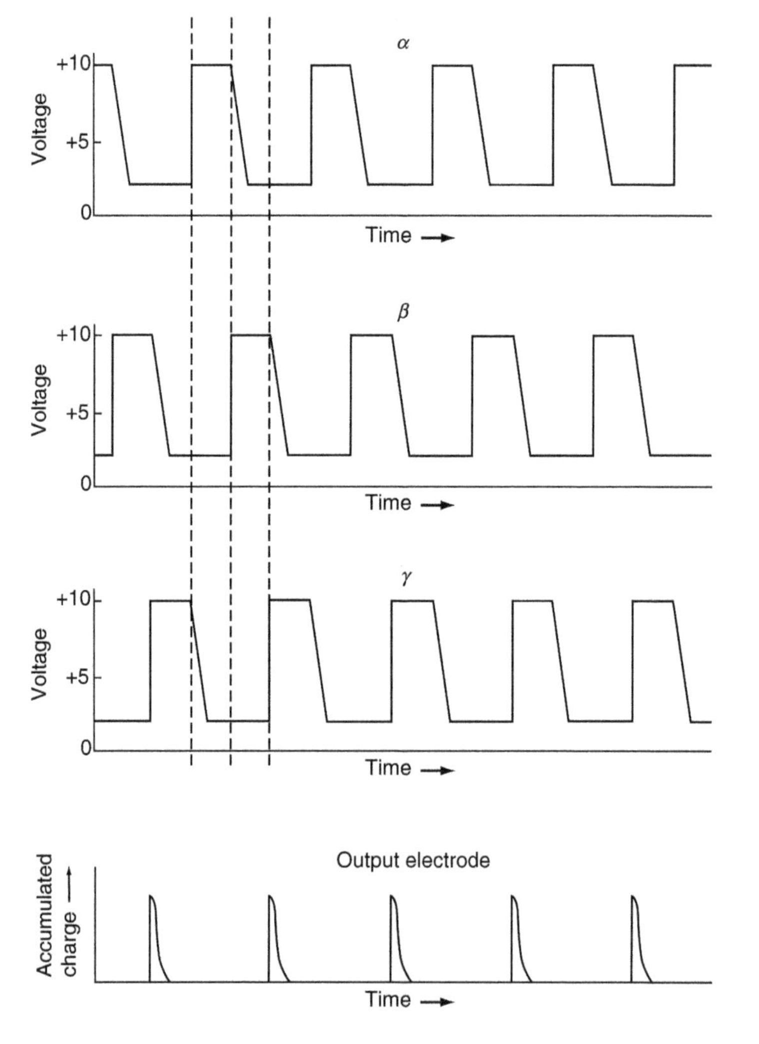

FIGURE 1.10 Voltage and output cycles for a three-phase charge-coupled device (CCD).

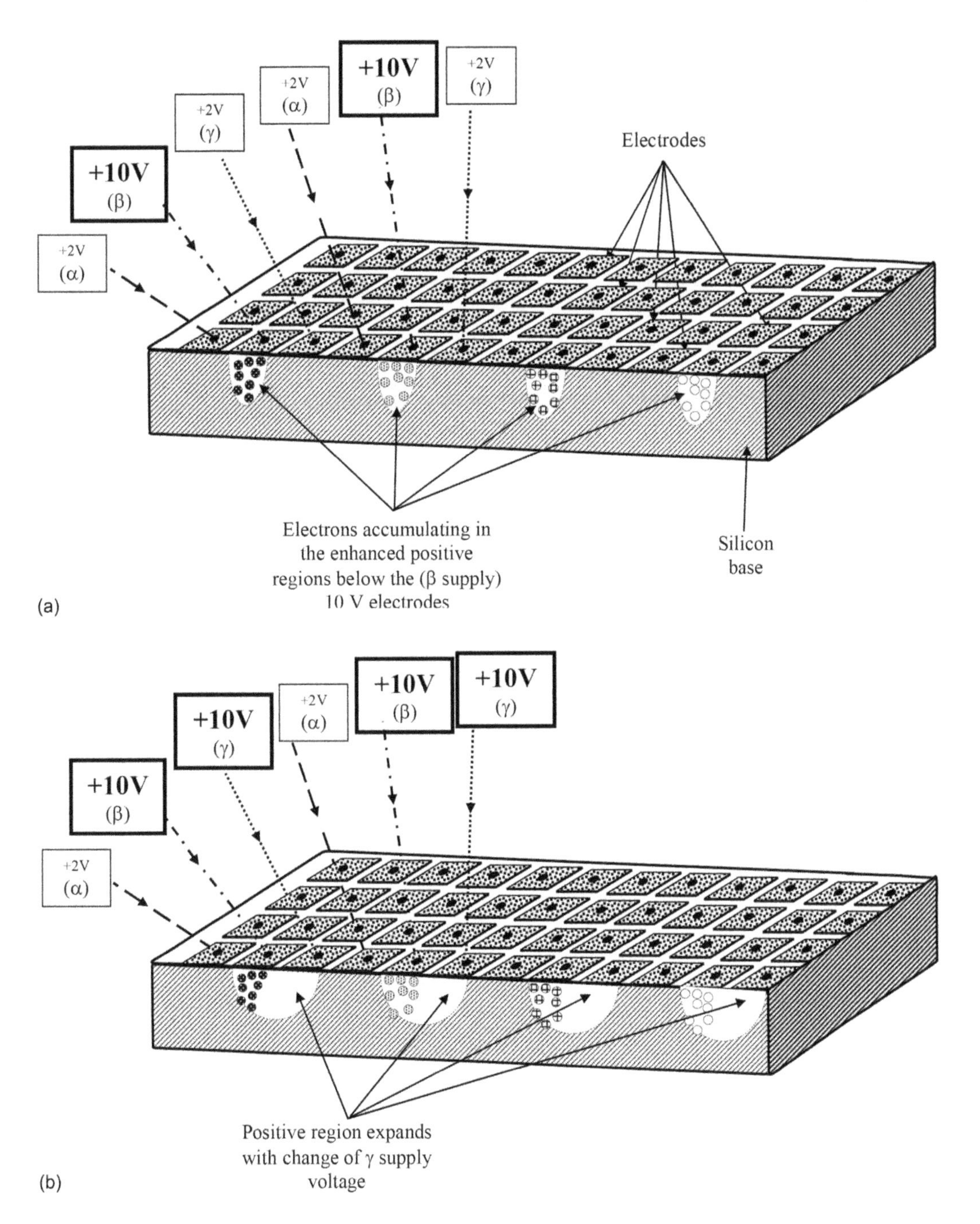

FIGURE 1.11 The movements of several packets of electrons through the physical structure of a three-phase charge-coupled device (CCD) array. N.B. For clarity only six electrical connections are shown, but all the electrodes are connected to one of the three power supplies (α, β or γ) also, the full CCD would comprise upwards of a million such electrodes. (a) Power supply β is at +10 V and power supplies α and γ at +2 V. Every third electrode thus has an enhanced positive region (hole-depletion zone) below it into which the photoelectrons accumulate. (b) Power supplies β and γ are at +10 V and power supply α is at +2 V. The enhanced positive regions have expanded to cover the original electrodes and their right-hand neighbours. *(Continued)*

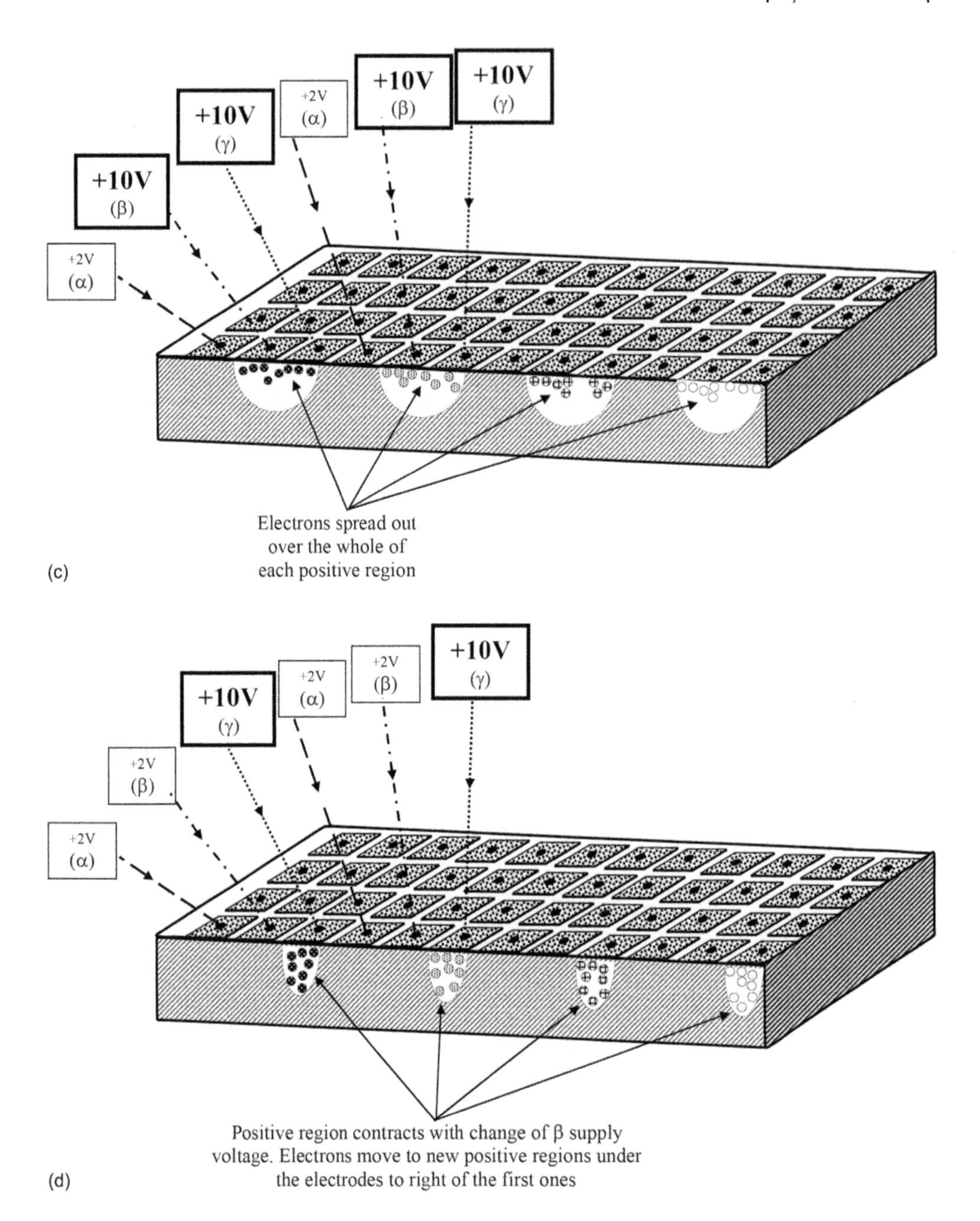

FIGURE 1.11 (Continued) (c) Power supplies β and γ are at +10 V and power supply α is at +2 V. The electrons move physically through the silicon substrate until they are shared between each pair of +10 V electrodes. (d) Power supply γ is at +10 V and power supplies α and β at +2 V. The enhanced positive regions under the original electrodes disappear and the electrons continue to move through the silicon substrate until they are all accumulated below the right-hand neighbours of the original electrodes. Every third electrode thus again has an enhanced positive region below it containing the photoelectrons. We are back to the situation shown in Figure 1.11(A) except that all the electron packages have been moved one electrode to the right of their previous positions. Further, similar, voltage cycles will continue to move the electron packages towards the right (movement to the left would simply require a slightly different phasing of the power supply changes).

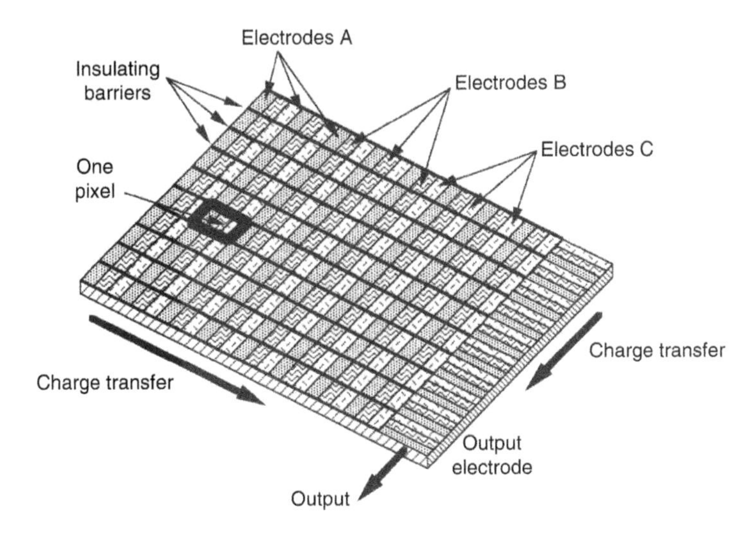

FIGURE 1.12 Schematic structure of a three-phase charge-coupled device (CCD).

The complete three-phase CCD is a combination of the detecting and charge transfer systems. Each pixel[12] has three electrodes and is isolated from pixels in adjacent columns by insulating barriers (Figure 1.12). During an exposure, electrodes B are at their full voltage and the electrons from the whole area of the pixel accumulate beneath them. Electrodes A and C meanwhile are at a reduced voltage and so act to isolate each pixel from its neighbours along the column. When the exposure is completed, the voltages in the three electrode groups are cycled as shown in Figure 1.10 until the first set of charges reaches the end of the column. At the end of the column, a second set of electrodes running orthogonally to the columns (Figure 1.12) receives the charges into the middle electrode for each column. That electrode is at the full voltage and its neighbours are at reduced voltages, so that each charge package retains its identity. The voltages in the read-out row of electrodes are then cycled to move the charges to the output electrode where they appear as a series of pulses. When the first row of charges has been read out, the voltages on the column electrodes are cycled to bring the second row of charges to the read-out electrodes and so on until the whole image has been retrieved.

In the early days, a small number of basic CCD units were simply strung together to form a linear array. However, whilst these have uses (as in bar code readers), they can only produce 2D images if scanned orthogonally to their long axes. Two-dimensional arrays can easily be made though by stacking linear arrays side by side. The operating principle is unchanged and the correlation of the output with position within the image is only slightly more complex than before. The largest single CCD arrays currently produced are 10,560 × 10,560 pixels (10k × 10k), giving a physical size for the whole device of about 95 mm to a side. The US Naval Observatory uses four such arrays for its Robotic Astrometric Telescope providing it with a 28-square degree field of view.

Most astronomical CCD arrays though are 2k × 4k[13] or 4k × 4k. For an adaptive optics telescope (see later in this section) operating at 0.2″ resolution, a 2k × 4k CCD covers only 400″ × 800″ of the

[12] Pixel is a commonly used term for an individual detecting unit within an array detector of any type. It is derived from 'picture element'.

[13] In this context, the number of pixels in a device is frequently a power of 2. The habit has thus developed of using 'k' as shorthand. Thus 2^{11} = 2048 and 2^{12} = 4096. These numbers would then be written as 2k or 4k, etc. A 2k × 4k array is actually 2048 pixels by 4096 pixels – a total of 8,388,608 pixels – it is *not* 2,000 × 4,000 pixels (a total of 8,000,000 pixels). Unfortunately, 'k' is also used more loosely meaning '1,000' or some rough approximation to '1,000', as in the previous paragraph where '10k' is used for 10,560. Care is therefore needed when encountering numbers expressed in this way to ensure that the true meaning is understood.

sky if the resolution is to be preserved. Many applications require larger areas of the sky than this to be imaged, so several such CCD arrays must then be formed into a mosaic. However, the normal construction of a CCD with electrical connections on all four edges means that there will then be a gap of up to 10 mm between each device, resulting in large dead-space areas in the eventual image. To minimise the dead space, 'three-edge-buttable' CCDs are used. These have all the connections brought to one of the short edges, reducing the gaps between adjacent devices to about 0.2 mm on the other three edges. It is then possible to form mosaics with two rows and as many columns as desired with a minimum of dead space.

The Vera C. Rubin Observatory (VRO),[14] thus, will have a 3.2-gigapixel camera made up from a mosaic of one hundred and eighty-nine 4k × 4k arrays based upon 10 μm-sized pixels, giving it nominal resolution of 0.2″, although the atmosphere will limit this to 0.7″ most of the time.[15] The instrument's aim is to obtain 15-s exposure images of areas of the sky, each 3.5° × 3.5° across, every 20 seconds, thus covering the whole visible night sky three times every week. First light is expected at the time of writing during 2020, with regular surveys starting from 2022 onwards. Similarly, the European Space Agency's (ESA's) Gaia[16] spacecraft (launched 2013 and positioned near the Earth's L2 Lagrangian point) has a 106 CCD arrays each with 1,966 × 4,500 pixels, making a total of 1.41 gigapixels (Figure 1.13). However, only 62 of the arrays are used for direct imaging, resulting in a 0.7° × 0.7° field of view for the instrument.

With the larger format CCD arrays, the read-out process can take some time (typically several hundred milliseconds).[17] To improve observing efficiency, some devices therefore have a storage

FIGURE 1.13 Part of the Gaia charge-coupled device (CCD) mosaic during its assembly. (Courtesy of Astrium.)

[14] Previously known as the LSST (Large Synoptic Survey Telescope).

[15] Although multi-conjugate adaptive optics (see later in this section) has recently enabled fields of view of up to 4′ × 4′ to be imaged at near diffraction-limited resolution, it seems unlikely that this will have improved to 3.5° × 3.5° by 2020.

[16] The name originated as an acronym for Global Astrometric Interferometer for Astronomy. It is not now planned to use interferometry for the project, but the name is retained as 'just' a name.

[17] The delay, of course, is insignificant when an exposure has a duration of 10 seconds or more. However, for some applications, such as adaptive optics (see later in this section), images have to be obtained, processed and the correcting optics adjusted on a timescale of a millisecond or so. It is then vital to have a detector with the shortest possible read-out time. Rapidly varying objects and transients must also be observed quickly. The rapidity with which observations may be repeated is now often called the 'cadence' of the instrument (high cadence = very rapid).

area between the imaging area and the read-out electrodes. This is simply a second CCD that is not exposed to radiation or half of the CCD is covered by a mask (frame transfer CCD). The image is transferred to the storage area in less than 0.1 ms, and while it is being read-out from there, the next exposure can commence on the detecting part of the CCD. Even without a separate storage area, reading the top half of the pixels in a column upwards and the other half downwards will halve read-out times. Column parallel CCDs (CPCCD) have independent outputs for each column of pixels, thus allowing read-out times as short as 50 μs. More rapid read-out of a CCD can also be achieved by binning. In this process, the electron charges from two or more pixels are added together before they are read-out. There is, of course, a consequent reduction in the spatial resolution of the CCD, and so the procedure will rarely if ever be encountered in astronomical applications. If binning is needed for other reasons, such as noise reduction, it is easily undertaken as a part of subsequent image processing (Section 2.9).

A two-phase CCD requires only a single clock, but needs double electrodes to provide directionality to the charge transfer (Figure 1.14). The second electrode, buried in the oxide layer, provides a deeper well than that under the surface, electrode and so the charge accumulates under the former. When voltages cycle between 2 and 10 V (Figure 1.15), the stored charge is attracted over to the nearer of the two neighbouring surface electrodes and then accumulates again under the buried electrode. Thus, cycling the electrode voltages, which may be done from a single clock, causes the charge packets to move through the structure of the CCD, just as for the three-phase device.

A virtual phase CCD requires just one set of electrodes. Additional wells with a fixed potential are produced by p and n implants directly into the silicon substrate. The active electrode can then be at

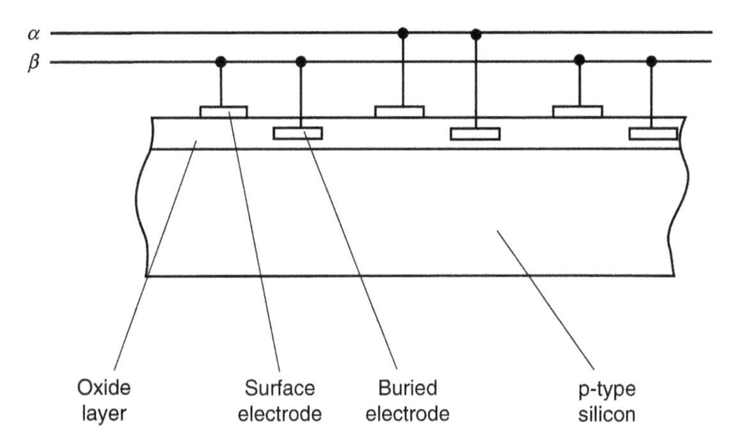

FIGURE 1.14 Physical structure of a two-phase charge-coupled device (CCD).

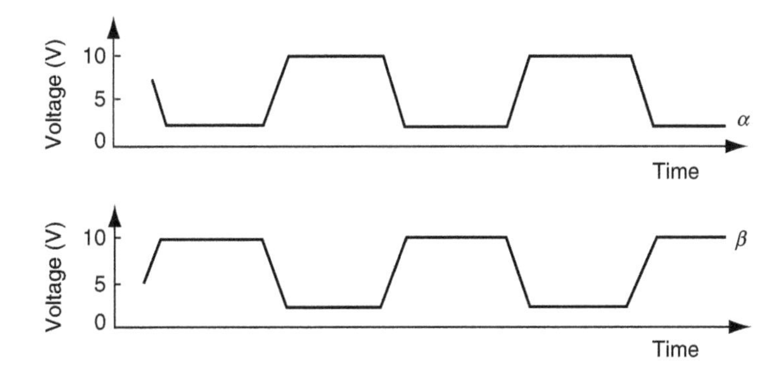

FIGURE 1.15 Voltage cycles for a two-phase charge-coupled device (CCD).

a higher or lower potential as required to move the charge through the device. The active electrodes in a virtual phase CCD are physically separate from each other, leaving parts of the substrate directly exposed to the incoming radiation. This enhances their sensitivity, especially at short wavelengths.

Non-tracking instruments such as the Carlsberg Meridian Telescope (Figure 5.4), liquid mirror, the 11-m Hobby-Eberly and South African Large Telescope (SALT) telescopes (see below) can follow the motion of objects in the sky by transferring the charges in their CCD detectors at the same speed as the image drifts across their focal planes (time-delayed integration [TDI]). To facilitate this, orthogonal transfer CCDs (OTCCDs) are used. These can transfer the charge in up to eight directions (up/down, left/right and at 45° between these directions). OTCCDs can also be used for active image motion compensation arising from scintillation, wind shake, etc. Other telescopes use OTCCDs for the 'Nod and Shuffle' technique. This permits the almost simultaneous observation of the object and the sky background through the same light path by a combination of moving the telescope slightly (the nod) and the charge packets within the CCD (the shuffle). The recently commissioned one-degree imager on the WIYN[18] 3.5-m telescope uses four thousand and ninety-six 512 × 512-pixel OTCCDs to cover its its 1-square degree field of view. Sixty-four of the individual arrays at a time are grouped into 8 × 8 mosaics called 'array packages' and 64 array packages in an 8 × 8 mosaic then form the complete detector comprising 1.07 gigapixels in total.

For faint objects, a combination of photomultiplier/image intensifier (Sections 2.1 and 2.3) and CCD, known as an electron bombarded CCD (EBCCD),[19] may be used. This places a negatively charged photocathode before the CCD. The photoelectron from the photocathode is accelerated by the voltage difference between the photocathode and the CCD and hits the CCD at high energy, producing many electron-hole pairs in the CCD for each incident photon. This might appear to give the device a quantum efficiency of more than 100%, but it is in effect merely another type of amplifier; the signal-to-noise ratio remains that of the basic device (or worse) and so no additional information is obtained.

Confusingly, devices sometimes called low-light level CCDs (L3CCDs)[20] are also known as EMCCDs because they have a high on-chip amplification, and so their basic output is many electrons for a single photon input. Unlike the EMCCDs, where the detection of the photon occurs at the photo-cathode, the photon detection in an L3CCD is within the CCD as usual and the amplification occurs after the detection. In a similar manner to the frame-transfer CCD, the output from the detector is moved to a storage area called the 'extended output register'. The extended output register is operated at up to 40 V, and this voltage is high enough for electrons entering it to gain sufficient energy that they may collide with silicon atoms and occasionally liberate a second electron-hole pair – therefore, effectively giving an amplification of the original detection by a factor of two (cf. avalanche photodiodes, Section 1.1.9). The voltage is adjusted so that the probability of such an electron-hole-producing-collision is only 1% or 2%. The average amplification is thus reduced to about ×1.01 or ×1.02. The extended output register however has up to 600 stages, each of which has this average intrinsic amplification. The output from the extended output register is thus hundreds or thousands of times[21] the number of electrons produced by the incoming photons directly in the CCD.

The output from the extended output register then goes through the remaining output stages required for a 'normal' CCD. The read noise (see below) of an L3CCD is thus the same as that for any other variety of CCD, but when superimposed upon a basic output which is hundreds or

[18] The name derives from the operating institutes: University of Wisconsin–Madison, Indiana University, Yale University and the National Optical Astronomy Observatory.

[19] Also known as an Electron Multiplying CCD (EMCCD, though this term is also used for a quite different type of CCD as discussed in the next paragraph) or an Intensified CCD (ICCD – though this term may also be used to designate a CCD being fed by a normal image intensifier).

[20] LLLCCD; hence, L3CCD.

[21] $1.01^{600} \approx 400$, while $1.02^{600} \approx 150{,}000$.

thousands of times larger than that of the normal CCD, it becomes almost negligible. The main noise source in an L3CCD is the variation in the amplification factor arising from the stochastic nature of the amplification process. The detections of several otherwise identical single photons will thus be a number of pulses of varying amplitudes. As with the photomultiplier tube (below), this will not be a problem if those individual pulses can be individually counted. If the number of photons however increases to the point where the pulses merge into each other, then this pulse strength variation will reduce the signal-to-noise ratio by about a factor of 1.4 ($=\sqrt{2}$) – effectively halving the device's quantum efficiency. The L3CCDs are thus best suited to counting individual photons and so must either be used for observing faint sources or must be operated at very high frame rates.[22] The high amplification of L3CCDs also means that other noise sources may become important. In particular, electrons produced by induction from the clock-driven power supplies can be amplified and appear as bright pixels anywhere within the image. This clock-induced noise is intrinsic to all CCDs but only becomes obtrusive in L3CCDs. The L3CCD arrays are currently produced up to about 1k × 1k in size.

Interline transfer CCDs have an opaque column adjacent to each detecting column. The charge can be rapidly transferred into the opaque columns and read-out from there more slowly while the next exposure is obtained using the detecting columns. This enables rapid exposures to be made, but half the detector is dead space. They are mainly used for digital video cameras and rarely find astronomical applications.

To astronomers working 50 years ago and struggling to hypersensitize photographic emulsions using noxious chemicals, dry ice and acetone mixtures or potentially dangerous hydrogen-air combinations, a detector with all the advantages of a CCD would have seemed to be beyond their wildest dreams. Yet despite their superiority, CCDs are not quite the ultimate perfect detector. Their outputs are affected by the general sources of noise (see later in this section) that bedevil all forms of measurement, but they also have some problems that are unique to themselves and those we consider here.

The electrodes on or near the surface of a CCD can reflect some of the incident light, thereby reducing the quantum efficiency and changing the spectral response. To overcome this problem, several approaches have been adopted. Firstly, transparent polysilicon electrodes replace the metallic electrodes used in early devices. Secondly, the CCD may be illuminated from the back so that the radiation does not have to pass through the electrode structure at all. This, however, requires that the silicon substrate forming the CCD be very thin so that the electrons produced by the incident radiation are collected efficiently. Nonetheless, even with thicknesses of only 10–20 μm, some additional cross talk (see below) may occur. More importantly, the process of thinning the CCD chips (by etching away the material with an acid) is risky and many devices may be damaged during the operation. Successfully thinned CCDs are therefore expensive to cover the cost of the failures. They are also fragile and can become warped, but they do have the advantage of being less affected by cosmic rays than the thicker CCDs (see below). With a suitable anti-reflection coating, a back-illuminated CCD can now reach a quantum efficiency of 90% in the red and near infrared (NIR). Other methods of reducing reflection losses include using extremely thin electrodes, or an open electrode structure (as in a virtual phase CCD) which leaves some of the silicon exposed directly to the radiation. At longer wavelengths, the thinned CCD may become semi-transparent. Not only does this reduce the efficiency because not all the photons are absorbed, but interference fringes may occur between the two faces of the chip.[23] These can become obtrusive and have to be removed as a part of the data-reduction process.

[22] CCDs used as the detectors within instruments on board spacecraft may be subject to intense ionising radiation from the Earth's Van Allen belts, solar flares, etc. One way of radiation-hardening such detectors is to use very fast read-out rates so that detections of individual photons do not have time to be affected by the radiation.

[23] The effect is sometimes called 'etaloning' because similar internal reflections are utilised productively in etalons (Section 4.1.4.1).

The CCD as so far described suffers from loss of charge during transfer because of imperfections at the interface between the substrate and the insulating oxide layer. This affects the faintest images most badly because the few electrons that have been accumulated are physically close to the interface. Artificially illuminating the CCD with a low-level uniform background during an exposure will ensure that even the faintest parts of the image have the surface states filled. This offset or 'fat zero' can then be removed later in signal processing. Alternatively, a positively charged layer of *n*-type[24] silicon between the substrate and the insulating layer may be added to force the charge packets away from the interface, producing a buried-channel CCD. A variant on the buried channel CCD, termed a 'peristaltic CCD' (from a rather fanciful analogy between the way electrons move in the device and the way food is moved down the oesophagus) intensifies the transfer speed of the electrons by using additional implanted electrodes. This allows the devices to operate at up to frequencies of 100 MHz and perhaps in the future after further development, at frequencies of 1 GHz or more.

Charge transfer efficiencies for CCDs now approach 99.9999%, leading to typical read-out noise levels of one to two electrons per pixel. Using a non-destructive read-out mechanism and repeating and averaging many read-out sequences can reduce the read-out noise further. To save time, only the lowest intensity parts of the image are repeatedly read out, with the high intensity parts being skipped. The devices have therefore become known as 'skipper CCDs' (cf. charge injection devices [CIDs]; see Section 1.1.8.2).

At some point and usually within the CCD integrated circuit itself, the electron charge packages are converted to digital voltage signals. The analogue-to-digital converters (ADCs) used for this do not normally respond to single electrons but only groups of them. Typically, ten electrons might be needed to give a unit output from the ADC. The group of electrons required for a unit output is termed the analogue-to-digital unit (ADU), and the number of electrons is termed 'the gain'.[25] The gain (and the ADU) is usually determined by matching the maximum value that can be output by the ADC to the well capacity of a single pixel. Thus, a 16-bit ADC has an output range from 0 to 65,535. If such an ADC is used for a CCD with a well capacity of 500,000 electrons, then the gain must be around 7 or 8 (500,000/65,536 = 7.6). The rounding (up or down) of the number of electrons to the number of ADUs is one component of the read noise from the CCD. The others are the errors in the conversion of electron numbers to ADUs and noise from the ADCs' electronics and other electronic components, such as amplifiers, incorporated into the CCD chip. Typically, in a science-grade CCD, the read noise corresponds to a few ($\leq$10) electrons. The read noise is effectively the dark signal of the CCD. The dynamic range (Table 1.2) of a CCD with a well capacity of 500,000 electrons, but a read noise of 10 electrons, is thus $\times$50,000 (= 500,000/10). The use of a 16-bit ADC for such a CCD, as cited previously, is thus not quite as contrary as it might seem. The read noise would need to reduce to two electrons before 18-bit conversion would be justified. With L3CCDs, the read noise is similar to that of other CCDs, but the amplification that occurs before read-out means that the read noise is superimposed upon a much larger signal. In proportion to the number of originally detected photons, the read noise in L3CCDs can thus be down to 0.01 electrons per pixel.

The spectral sensitivity of CCDs ranges from 400 to 1100 nm, with a peak near 750 nm where the quantum efficiency can approach 90%. The quantum efficiency drops off further into the IR as the silicon becomes more transparent. Short-wave sensitivity can be conferred by coating the device with an appropriate phosphor to convert the radiation to longer wavelengths, and the device becomes directly sensitive again in the X-ray region (Section 1.3).

---

[24] Usually phosphorous is the doping element.

[25] An odd terminology because the number of ADUs is *lower* than the number of photons originally detected by the factor given by the 'gain' (except in L3CCDs and EBCCDs).

Personal digital video and still cameras automatically obtain their images in colour. To do this, small filters are placed over each CCD element. In the commonly used Bayer arrangement, for each set of four pixels, two have green-transmission filters, one a blue transmission filter and one a red transmission filter (allowing, roughly, for the eye's intrinsic spectral response). This means that the spatial resolution of the image is degraded – a 10 megapixel colour camera only has the spatial resolution of a 2.5 megapixel monochromatic camera. More importantly for rigorous scientific analysis of the images, the three colours do not come from exactly the same locality within the original object. If the properties of the original object are varying significantly on a size scale equivalent to an individual pixel, then the colour image will give false results. When images at two or more wavelengths are needed for astronomical purposes (including the beautiful colour representations of nebulae and galaxies, etc., ornamenting many astronomy books), it is usual to obtain individual images at each wavelength through appropriate filters and then combine them into a colour (or false-colour) final image. The full spatial resolution of the detector is thus retained and, if wanted, a truer visual representation of the appearance of the object may be obtained by better relative weightings of the different images than that given by the crude Bayer system.[26]

The CCD regains sensitivity at short wavelengths because X-rays are able to penetrate the device's surface electrode structure. For both X-ray and IR detection (and possibly also for the direct detection of low energy dark matter particles – see Section 1.7), a deep-depletion CCD[27] may provide higher sensitivity. These devices have a thick silicon substrate which has a high resistivity and uses a bias voltage. The hole-depletion zones are deeper than normal so that there is more opportunity for penetrating photons to interact with the material and so produce electron-hole pairs. As an example, the Soft X-ray Imager camera on the Japanese Astro-H spacecraft (Hitomi – lost after launch in 2016) would have used four 640 × 640-pixel deep depletion CCDs to observe an area of the sky 38′ square to a resolution of 1.74″ in the soft X-ray region (energy < 10 keV, wavelength > 0.1 nm). The eROSITA X-ray instrument on board the Spektr-RG[28] spacecraft uses 384 × 384-pixel frame store deep depletion CCDs for its seven X-ray telescopes and covers the spectral region from 300 eV to 10 keV.

For integration times longer than a fraction of a second,[29] it is usually necessary to cool the device to reduce its dark signal. Using liquid nitrogen for this purpose and operating at around 170 K, integration times of minutes to hours are easily attained. Small commercial CCDs produced for the amateur astronomy market (which often also find applications at professional observatories for guiding, etc.) usually use Peltier coolers to get down to about 50° below the ambient temperature. Subtracting a dark frame from the image can further reduce the dark signal. The dark frame is in all respects (exposure length, temperature, etc.) identical to the main image, excepting that the camera shutter remains closed whilst it is obtained. It therefore just comprises the noise elements present within the detector and its electronics. Because the dark noise itself is noisy, it may be better to use the average of several dark frames obtained under identical conditions.

For typical noise levels and pixel capacities, astronomical CCDs have dynamic ranges of 50,000 to 500,000 (see previous discussion of read noise), which results in a usable magnitude range in accurate brightness determination of up to 14.5^{m}. This compares favourably with the dynamic range of less than 1,000 (brightness range of less than 7.5^{m}) available from a typical photographic image.

A major problem with CCDs used as astronomical detectors is the noise introduced by cosmic rays. A single cosmic ray particle passing through one of the pixels of the detector can cause a large

[26] Of course, if the object is varying on a *time* scale faster than the interval between the separate images, then the standard astronomical approach will also give false results. When an object is changing rapidly on both the spatial and the time scales, such as with a solar flare, then special techniques have to be devised.

[27] Also known as *pn*-CCDs or *p*-channel CCDs.

[28] Spectrum–Roentgen-Gamma. Note that it is not the same as the Spektr-R spacecraft (Section 1.2.3).

[29] For astronomical purposes even CCDs used for short exposures will be cooled.

number of ionisations. The resulting electrons accumulate in the storage region along with those produced by the photons. It is usually possible for the observer to recognise such events in the final image because of the intense 'spike' that is produced. Replacing the signal from the affected pixel by the average of the eight surrounding pixels improves the *appearance* of the image but does *not* retrieve the original information. This correction often has to be done 'by hand' and is a time-consuming process. Though, when two or more images of the same area are available, automatic removal of the cosmic ray spikes is possible with reasonable success rates.

Another serious defect of CCDs is the variation in background noise between pixels. This takes two forms. There may be a large-scale variation of 10%–20% over the whole sensitive area, and there may be individual pixels with permanent high background levels (hot spots). The first problem has been much reduced by improved production techniques and may largely be eliminated by flat fielding during subsequent signal processing, if the effect can be determined by observing a uniform source. Commonly used sources for the flat field include the twilit sky and a white screen inside the telescope dome illuminated by a single distant light source. The flat field image is divided into the main image after dark frame subtraction from both images to reduce the large-scale sensitivity variations. The effect of a single hot spot may also be reduced in signal processing by replacing its value with the mean of the four or eight surrounding pixels. However, the hot spots are additionally often poor transferors of charges, so all preceding pixels are then affected as their charge packets pass through the hot spot or bad pixel, introducing a spurious line into the image. There is little that can be done to correct this last problem, other than to buy a new CCD. Even the 'good' pixels do not have 100% charge transfer efficiency, so that images containing bright stars show a 'tail' to the star caused by the electrons lost to other pixels as the star's image is read-out.

Yet another problem is that of cross talk or blooming. This occurs when an electron strays from its intended pixel to one nearby. It affects rear-illuminated CCDs because the electrons are produced at some distance from the electrodes and is the reason why such CCDs have to be thinned. It can also occur for any CCD when the accumulating charge approaches the maximum capacity of the pixel. Then the mutual repulsion of the negatively charged electrons may force some over into adjacent pixels. The well capacity of each pixel depends upon its physical size and is around half a million electrons for 25 μm-sized pixels. Some CCDs have extra electrodes to enable excess charges to be bled away before they spread into nearby pixels. Such electrodes are known as 'drains', and their effect is often adjustable by means of an anti-blooming circuit. Anti-blooming should not be used if you intend making photometric measurements on the final image, but otherwise it can be effective in improving the appearance of images containing both bright and faint objects.

The size of the pixels in a CCD can be too large to allow a telescope to operate at its limiting resolution.[30] In such a situation, the images can be 'dithered' to provide improved resolution. Dithering consists simply of obtaining multiple images of the same field of view, with a shift of the position of the image on the CCD between each exposure by a fraction of the size of a pixel. The images are then combined to give a single image with sub-pixel resolution. A further improvement is given by 'drizzling' (also known as 'variable pixel linear reconstruction') in which the pixel size is first shrunk leaving gaps between the pixels. The images are then rotated before mapping onto a finer scale output grid.

One of the major advantages of a CCD over a photographic emulsion is the improvement in the sensitivity. However, widely different estimates of the degree of that improvement may be found in the literature, based upon different ways of assessing performance. At one extreme, there is the ability of a CCD to provide a recognisable image at short exposures because of its low noise. Based upon this measure, the CCD is perhaps 1000 times faster than photographic emulsion. At the other extreme, one may use the time taken to reach the mid-point of the dynamic range. Because of the much greater dynamic range of the CCD, this measure suggests CCDs are about 5–10 times faster

[30] The well capacity of a CCD pixel varies with its size. For a given noise level, therefore, a larger dynamic range requires physically larger pixels. Most CCDs have pixel sizes in the range 10–50 μm.

than photography. The most sensible measure of the increase in speed, however, is based upon the time required to reach the same signal-to-noise ratio in the images (i.e., to obtain the same information). This latter measure suggests, that in their most sensitive wavelength ranges (around 750 nm for CCDs and around 450 nm for photographic emulsion), CCDs are 20–50 times faster than photographic emulsions.

1.1.8.2 Charge Injection Devices (CIDs)

CIDs were covered in previous editions of this book but were dropped because their astrophysical usage was rare compared with that of the all-conquering CCD. They have been reinstated because they may find future applications in stellar coronagraphs and studies of exoplanets (i.e., in very high contrast imaging).

The detection mechanism of these devices is identical to that of the CCD. The difference between them occurs in their read-out system. With these devices two electrodes are used for each sensitive element and the charge can be switched between them by alternating their voltages. If the voltages of both elements are simultaneously reduced to zero, then the accumulated charge is deposited (injected) into the substrate. The resulting pulse can be detected and its magnitude is a measure of the original charge. Connecting all the first electrodes of each sensitive element serially along rows, while all the second electrodes are connected serially along columns provides a 2D read-out. Switching a given row and column to zero volts then causes the charge in the sensitive element at their intersection to be read out, whilst all other charges are either unaffected or jump to their alternative electrode.

The connections for a CID device are more complex than those for a CCD, but its data are easier to retrieve separately and can be read-out non-destructively. In high-contrast imaging, therefore, those pixels with high levels of illumination may be read-out quickly, whilst the pixels with low illumination levels are allowed to continue accumulating their signals.

1.1.8.3 CCDs – The Future

CCDs are close to being ideal detectors (see above), but they fail to be ideal in one important respect: they have no intrinsic sensitivity to the energy (wavelength, frequency) of an incoming photon. A red photon (1.8 eV) produces the same signal as a violet photon (3.3 eV). Of course, we can separate two such photons using filters or spectrographs, but ideally the detector should do that job itself – as currently happens with some detectors in the x- and γ-ray regions. As we have seen, the band gap for silicon is 1.09 eV and this corresponds to the energies of photons with a wavelength 1.14 μm. A photon with half this wavelength will have twice that energy. A 570 nm photon (yellow-green) being absorbed by a silicon-based CCD should, therefore, in principle, be able to release two electrons and a 380 nm photon (extreme violet – energy 3×1.09 eV) be able to release three electrons. In principle then, a CCD used with short exposures, so that none of its elements received more than one photon before it was read-out, could distinguish intrinsically between violet photons, those in the blue-green region and those in the yellow to NIR region. This in spectroscopic terms (Section 4.1) would be a spectral resolution of one or two – and 'real' spectrographs use spectral resolutions ranging from 50 to 1,000,000!

It is easy to see from the discussion in the previous paragraph that to enable CCDs to have useful intrinsic spectral resolutions within the optical part of the spectrum they would need to be formed from semiconductors with band gaps of 20 meV for spectral resolutions of 50 and 1 meV for spectral resolutions of 1,000. They would also need to operate in a photo-counting mode. Now materials with suitable such band gaps do exist (see STJs and Microwave Kinetic Inductance Detectors [MKIDs][31]; see Sections 1.1.12 and 1.1.13, respectively) and are generally based upon the band gaps between different energy levels in superconducting materials. A CCD-type device could therefore be produced

[31] Also known as kinetic inductance detector (KID).

with intrinsic spectral resolution but would probably be have few or no advantages over STJs or MKIDs or other future super-conductor-based detectors. On the whole, therefore, it now seems unlikely that CCDs will go down the road of being developed to have intrinsic spectral resolution.

For non-astronomical applications, CCDs are being replaced by complementary metal oxide superconductor (CMOS; see Section 1.1.16) detectors because of manufacturing advantages and their greater read-out speeds. Perhaps the future of astronomical CCD detectors also lies in this direction.

1.1.9 Avalanche Photodiodes (APDs)

The APD[32] is finding increasing use as an optical detector within astronomy. It is a variant on the basic photodiode (which can also be an optical detector but which is not now employed by astronomers as a primary detector). However, to understand APDs, we need to know something about the basic photodiode's operating principles and so we look at those first.

1.1.9.1 Photodiodes

These are also known as 'photovoltaic cells', 'photoconductors', and 'barrier junction detectors'. They rely upon the properties of a *p-n* junction[33] in a semiconductor. The energy-level diagram of such a junction is shown in Figure 1.16. Electrons in the conduction band of the *n*-type material are at a higher energy than the holes in the valence band of the *p*-type material. Electrons, therefore, diffuse across the junction and produce a potential difference across it. Equilibrium occurs when the potential difference is sufficient to halt the electron flow. The two Fermi levels are then

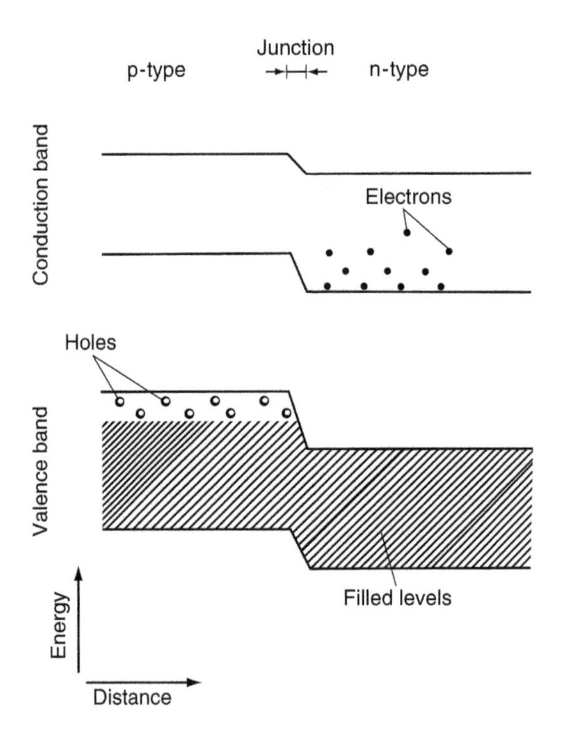

FIGURE 1.16 Schematic energy level diagram of a *p-n* junction at the instant of its formation. The (free) electrons and holes have yet to diffuse across the junction.

[32] A type of image intensifier using APDs is also discussed in Section 2.3.3.

[33] This is the union between a *p*-doped semiconductor and an *n*-doped semiconductor, usually with the same bulk semiconductor on both sides of the join.

coincident, the potential across the junction is equal to their original difference and there is a depletion zone containing neither type of charge carrier across the junction. The *n*-type material is positive, the *p*-type negative, and we have a simple *p-n* diode.

Now, if light of sufficiently short wavelength falls onto such a junction then it can generate electron-hole pairs in both the *p*- and the *n*-type materials. The electrons in the conduction band in the *p* region will be attracted towards the *n* region by the intrinsic potential difference across the junction and they will be quite free to flow in that direction. The holes in the valence band of the *p*-type material will be opposed by the potential across the junction and so will not move. In the *n*-type region, the electrons will be similarly trapped while the holes will be pushed across the junction. Thus, a current is generated by the illuminating radiation and this may simply be monitored and used as a measure of the light intensity. For use as a radiation detector the *p-n* junction often has a region of undoped (or intrinsic) material between the *p* and *n* regions to increase the size of the detecting area. These devices are then known as *p-i-n* junctions. Their operating principles do not differ from those of the simple *p-n* junction.

The response of a *p-n* junction to radiation is shown in Figure 1.17. It can be operated under three different regimes. The simplest, which is labelled B in Figure 1.17, has the junction short-circuited through a low-impedance meter. The current through the circuit is the measure of the illumination. In regime C, the junction is connected to a high impedance so that the current is close to zero and it is the change in voltage that is the measure of the illumination.

Finally, in regime A, the junction is back (or reverse) biased – the *p*-type material is made more negative and the *n*-type material made more positive. When an incident photon produces an electron-hole pair, electrons in the conduction band of the *p*-type material are strongly attracted over to the *n*-type material. Similarly holes in the valence band of the *n*-type material are strongly attracted towards the *p*-type material. The movement of holes in the valence band of the *p*-type

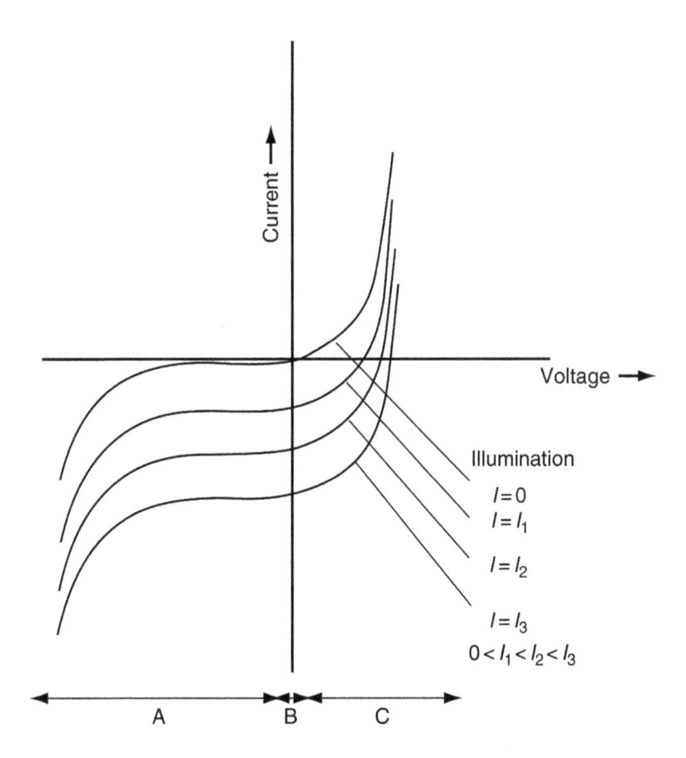

FIGURE 1.17 Schematic Voltage/Illumination (V/I) curves for a *p-n* junction under different levels of illumination.

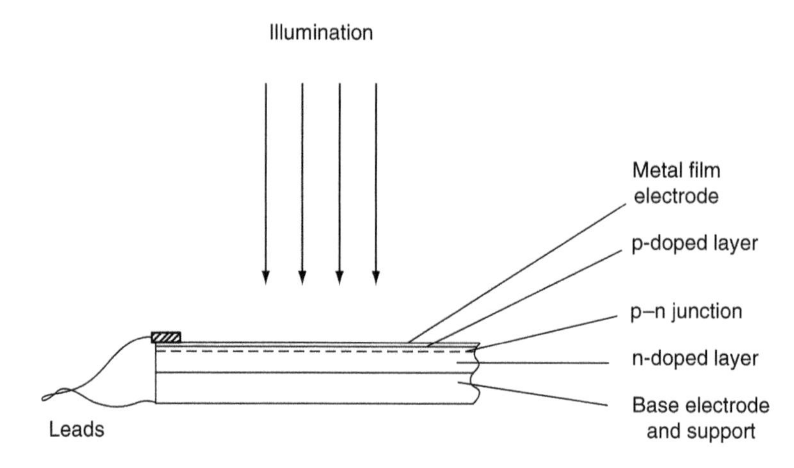

FIGURE 1.18 Cross section through a *p-n* junction photovoltaic detector.

material and of electrons in the conduction band of the *n*-type material is prevented by the biasing. In regime A, the voltage across a load resistor in series with the junction measures the radiation intensity. In this mode, the device is known as a 'photoconductor' (see infrared detectors in Section 1.1.15).

The construction of a typical photovoltaic cell is shown in Figure 1.18. The material in most widespread use for the *p* and *n* semiconductors is silicon which has been doped with appropriate impurities. Solar power cells are of this type but often use cadmium telluride (band gap ~1.5 eV) or copper indium gallium selenide (band gap ~1.0 to ~1.7 eV) for their semiconductors in place of silicon. The silicon-based cells have a peak sensitivity near 900 nm and cut-off wavelengths near 400 and 1,100 nm. Their quantum efficiency can be up to 50% near their peak sensitivity and D^* can be up to 10^{12}.

Indium arsenide, indium selenide, indium antimonide, gallium arsenide and indium gallium arsenide can also all be fabricated into photodiodes. They are particularly useful in the IR where germanium doped with gallium out-performs bolometers (see Section 1.1.15.2) for wavelengths up to 100 μm.

1.1.9.2 Avalanche Photodiode

If a *p-n* junction is reverse-biased to more than half its breakdown voltage then an APD[34] results. The original electron-hole pair produced by the absorption of a photon will be accelerated by the applied field sufficiently to cause further pair production through inelastic collisions. These secondary electrons and holes can in their turn produce further ionisations and so on (cf. previous discussion of L3CCDs and see Sections 1.3 and 1.4 for Geiger and proportional counters). Eventually, an avalanche of carriers is created, leading to an intrinsic gain in the signal by a factor of 100 or more. The typical voltage used is 100–200 V and the response of the device is linear.

The basic APD has a physical structure which has the radiation entering through a thin layer of *p*-doped semiconductor and then being absorbed in a thicker intrinsic semiconductor layer. The *n*-doped semiconductor forms the bottom layer of the device. The electrons accelerate towards the *n*-doped layer, producing the avalanche along the way. The semiconductor material that is used varies with the wavelength of the radiation to be detected. Silicon is used for the 200-nm to 1.1-μm region, gallium nitride operates down to around 250 nm in the UV region, indium-gallium-arsenide

[34] Sometimes called a 'solid photomultiplier' or 'silicon photomultiplier' or SiPMT (ss Section 1.1.11).

is used for the region from 1.0 to 2.6 μm with high gains, germanium for 800 nm to 1.7 μm and mercury-cadmium-telluride alloy (HgCdTe) based APDs can reach out to 14 μm. Quantum efficiencies approach 50% in the optimum regions of the spectrum for each material. Response times to the detection of an incoming photon are generally in the region of tens to hundreds of picoseconds. A major drawback though for APDs is that the gain depends sensitively upon the bias. The power supply thus typically has to be stable to a factor of ten better than the desired accuracy of the output. The devices also have to be cooled and maintained at a stable temperature to reduce noise. Arrays of APDs are coming into production but are still small (256 × 256 pixels) at the time of writing. Hybrid APDs are just APD versions that have amplifiers and other electronic circuits integrated onto their chips.

Once an incident photon has produced an electron-hole pair, both are accelerated by the back-biasing voltage. However, materials are generally chosen for construction of APDs that make collisions by the electrons more likely to produce more electrons than the collisions by the holes are to produce more holes.[35] The ratio of the likelihood of hole production to that of electron production within the avalanche is called the '*k*-factor'. For some materials such as HgCdTe, the value of k can reach zero (i.e., the avalanche contains only electrons). For silicon, $k \approx 0.02$ while for germanium, $k \approx 0.9$. The APDs with low or zero values of k are termed 'electron-APDs' or 'E-APDs', and they can have high gains at low bias voltages. The E-APDs have yet to find direct astronomical applications, but at the time of writing their possible use is being investigated at several observatories. The E-APDs are also being investigated as wavefront sensors for active optics systems (Section 1.1.22) and have been trialled on the 8.2 m Subaru telescope for use in lucky imaging (Section 1.1.22). They are also being investigated for use within instruments being considered for the 30-m telescope (see below).

APDs can also be used for direct X-ray detection, although have not been so used for studying astronomical sources yet.[36] The physical structure of these devices differs somewhat from that of the basic APD and allows much higher bias voltages to be employed. The difference lies in the presence of a large drift region between the layer in which the electron-hole pairs are produced and the layer in which the avalanche occurs. The highest bias voltages are to be found in the bevelled-edge APDs. The bevelled edge acts to reduce the electrical field along the edges thus preventing unwanted electrical breakdowns. Bias voltages of up to 2,000 V and gains of up to ×10,000 are possible. Reach-through APDs have the avalanche layer at the back of the structure, are operated at up to 500 V and offer gains of up to ×200. Reverse or buried junction APDs have the avalanche layer immediately behind the X-ray entry window, are operated at less than 500 V and offer gains that are generally less than ×200.

Once started, the avalanche is quenched (because otherwise the current would continue to flow) by connecting a resistor of several hundred kΩ in series with the diode. Then as the current starts to flow the voltage across, the diode is reduced and the breakdown ended. Quenching the avalanche using a series resistor, though, is a relatively slow process and leads to long dead times before the APD recovers and is ready to detect another photon. Most modern APDs, therefore, employ active quenching which reduces the dead time to a few tens of nanoseconds. In active quenching, there is a circuit that detects the onset of the avalanche and which rapidly reduces and then restores the bias voltage.

[35] If this seems counter-intuitive then remember that the motions of the positive 'holes' in one direction are actually due to bound electrons jumping from one atom within the crystal to another in the opposite direction. Their motions and interactions may thus be expected to differ in some ways from those of the free electrons.

[36] APDs formed part of the active shield (Section 1.3) for the soft γ-ray detector launched on board the Hitomi (Astro-H) spacecraft in 2016. They were not intended to detect the γ-rays directly but to pick up the visible light flashes produced in the bismuth germanate (BGO) scintillator that surrounds the main Compton detector. Hitomi unfortunately disintegrated five weeks after launch. A possible replacement for the mission is under consideration at the time of writing, but the details of its possible instrumentation are still unknown.

1.1.9.3 Single Photon Avalanche Photodiodes

Single-photon avalanche photodiodes (SPADs)[37] are versions of the basic APD in which the bias voltage is well above the breakdown voltage of the semiconductor in use (typically >30,000 V/mm). Like the Geiger counter (Section 1.3), the electron avalanche saturates; however many photons may have entered the device to initiate it. The gain is thus in the region of $\times 10^8$ to 10^{12} or more, but the response of the detector is non-linear. This, though, is unimportant because the devices are used to detect single photons and so only operate in circumstances (low light levels and/or high frame rates) where average the arrival time between individual photons is at least ten times longer than the dead time of the device. Because electron-hole pairs produced by any means, including thermal motions, will result in avalanches, these form a major noise source in SPADs. The SPAD arrays up to 8×8 pixels can currently be fabricated.

1.1.10 Photography

Photography has a long and honourable history as a detecting process in astronomy, but it has now vanished entirely from professional observatories and almost entirely from being used by amateur astronomers. Its detection mechanism is the production of electron-hole pairs in (usually) small grains of silver bromide suspended in a thin film of gelatine. The free electrons chemically convert molecules of silver bromide into atoms of silver and bromine. After obtaining an image, the photographic film has to be developed and fixed before becoming visible as a negative version of the original. Developing and fixing are chemical processes which convert silver bromide grains that have absorbed photons into silver and dissolve away the silver bromide grains which have not absorbed photons. A positive image, if needed, has to be obtained by re-photographing the negative image – a process usually called 'printing'.

Details of film photography and its astronomical applications may be found in previous editions of this book and on the internet, but its astronomical usage is now so minimal that further details are not included in this edition.

1.1.11 Photomultipliers (PMTs)

The photomultipliers (PMTs)[38] were once considered as the workhorses of optical photometry (Chapter 3). For this purpose, they have largely been superseded by CCDs. They continue though, to be used for UV measurements in the 10–300 nm regions where CCDs are insensitive. One of their primary astronomy-related uses now lies within cosmic ray detectors such as High-Energy Stereoscopic System (HESS) and the Telescope Array Project (Section 1.4) and neutrino detectors like Super-Kamiokande (Section 1.5) where, because PMTs can be made physically large, they can cover extensive areas relatively inexpensively. A brief discussion of their principles of operation and properties is included here to cover these applications.

The PMTs detects photons through the photoelectric effect. The basic components and construction of the device are shown in Figure 1.19. The photoemitter is coated onto the cathode, and this is at a negative potential of some 1000 V. Once a photoelectron has escaped from the photoemitter, it is accelerated by an electric potential until it strikes a second electron emitter. The primary electron's energy then goes into pair production and secondary electrons are emitted from the substance in a manner analogous to photoelectron emission. Typically, 1 eV of energy is required for pair production, and the primary's energy can reach 100 eV or more by the time of impact; thus several secondary electron emissions result from a single primary electron.

---

[37] Also known as Geiger-mode APDs.

[38] They should more accurately be called *electron* multiplier phototubes.

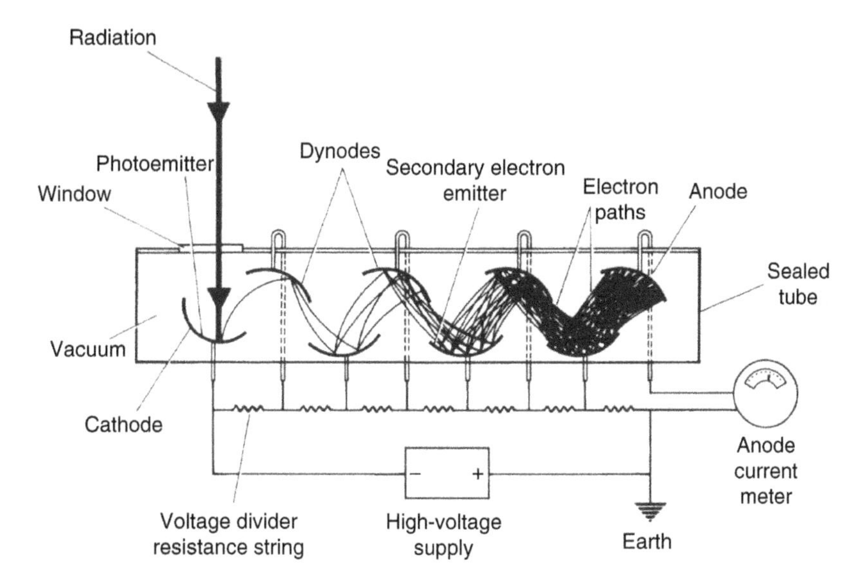

FIGURE 1.19 Schematic arrangement for a photomultiplier.

The secondary emitter is coated onto dynodes, which are successively more positive than the cathode by 100 V or so for each stage. The various electrodes are shaped and positioned so that the electrons are channelled towards the correct next electrode in the sequence after each interaction. The final signal pulse may contain 10^6 electrons for each incoming photon, and after arriving at the anode, it may be further amplified and detected in any of the usual ways. This large intrinsic amplification of the PMT is one of the major advantages of the device.

Noise in the signal from a PMT arises from many sources. The amplification of each pulse can vary by as much as a factor of ten through the sensitivity variations and changes in the number of secondary electrons lost between each stage. The final registration of the signal can be by analogue means and then the pulse strength variation is an important noise source. Alternatively, individual pulses may be counted and then it is less significant. Indeed, when using pulse counting even the effects of other noise sources may be reduced by using a discriminator to eliminate the large and small pulses which do not originate in primary photon interactions. Unfortunately, however, pulse counting is limited in its usefulness to faint sources otherwise the individual pulses start to overlap.

Electrons can be emitted from either the primary or secondary electron emitters through processes other than the required ones, and these electrons then contribute to the noise of the system. The most important such processes are thermionic emission and radioactivity. Cosmic rays and other sources of energetic particles and gamma rays contribute to the noise in several ways. Their direct interactions with the cathode or dynodes early in the chain can expel electrons that are then amplified in the normal manner. Alternatively, electrons or ions may be produced from the residual gas or from other components of the structure of the PMT. The most important interaction, however, is generally Ĉerenkov radiation produced in the window or directly in the cathode. Such Ĉerenkov pulses can be up to a hundred times more intense than a normal pulse. They can thus be easily discounted when using the PMT in a pulse counting mode but make a significant contribution to the signal when its overall strength is simply measured.

PMTs are essentially non-imaging detectors, but recently versions with some spatial resolution have been produced and are called 'Multiple-anode PMTs'. These are essentially several tens of individual PMTs housed in a single vacuum tube. The electrodes are designed so that the cascades of electrons preserve their individual natures during their travel through the dynodes. Multiple-anode PMTs are currently being assessed for possible use on the proposed Japanese

Experiment Module-Extreme Universe Space Observatory (JEM-EUSO) experiment for launch to the International Space Station (ISS) sometime in the 2020s.

The microchannel plate (MCP) (Section 1.3 and Figure 1.85) is closely related to the PMT tube in terms of its operating principles and can also be used at optical wavelengths. With an array of anodes to collect the clouds of electrons emerging from the plate, it provides an imaging detector with a high degree of intrinsic amplification. Such devices are often referred to as 'multi-anode microchannel array' (MAMA) detectors (Section 1.3.2.8). The APD (Section 1.1.9) is sometimes called a silicon photomultiplier (SiPMT), but although an avalanche of many electrons results from the first single photoelectron, the physical structure of APDs is quite different from that of PMTs.

1.1.12 Superconducting Tunnel Junction (STJ) Detectors

A possible replacement for the CCD in a few years, at least at well-equipped major observatories, is the STJ. The STJ can operate from the UV to long-wave IR and also in the X-ray and radio regions, can detect individual photons, has a very rapid response, and perhaps most importantly, provides an intrinsic spectral resolution (Sections 1.1.8 and 4.1) of around 500 or 1000 in the visible. Its operating principle is based upon a Josephson junction. This has two superconducting layers separated by a thin insulating layer. Electrons are able to tunnel across the junction because they have a wave-like behaviour as well as a particle-like behaviour and so a current may flow across the junction despite the presence of the insulating layer. Within the superconductor, the lowest energy state for the electrons occurs when they link together to form Cooper pairs. The current flowing across the junction because of the paired electrons can be suppressed by a magnetic field.

The STJ detector therefore comprises a Josephson junction based upon tantalum, hafnium, niobium, aluminium, etc., and placed within a magnetic field to suppress the current and having an electric field applied across it. It is cooled to about a tenth of the critical temperature of the superconductor – normally less than 1 K. A photon absorbed in the superconductor may split one of the Cooper pairs. This requires an energy of a milli-electron-volt or so compared with about an electron-volt for pair production in a CCD. Potentially, therefore, the STJ can detect photons with wavelengths up to a millimetre (300 GHz). Shorter wavelength photons will split many Cooper pairs, with the number split being dependent upon the energy of the photon, hence, the device's intrinsic spectral resolution. The free electrons are able to tunnel across the junction under the influence of the electric field and produce a detectable burst of current.

The STJ detectors and the arrays made from them are still very much under development at the time of writing but have been tried on the 4.2-m William Herschel Telescope (WHT) around 2004. ESA/ESTEC's S-Cam 3 used a 10×12 array of 35 μm^2 STJs and, although somewhat damaged, made successful observations while on the WHT and also with the 1-m Optical Ground Telescope.

The STJs are also used as heterodyne receivers in the 100 GHz to 1 THz frequency range (Section 1.2) and in some dark matter detectors (Section 1.7). They will also detect X-rays directly up to energies of several keV and arrays of up to 48×48 basic units have been considered as possibilities for some future X-ray spacecraft observatories. Outside astronomy, they find applications within quantum computing, as voltage standards and as magnetometers.

1.1.13 Microwave Kinetic Inductance Detectors (MKIDs) or Kinetic Inductance Detectors (KIDs)

A second device currently showing great future potential as a wide-band detector and which is also based upon photon interactions within a superconductor is the MKID or KID. These detectors potentially operate from the submillimetre region to X-rays. The use of 'microwave' in one version of the name for the devices refers to the way in which they operate and not to the e-m radiation region which they detect.

A superconducting circuit has a non-zero impedance ('resistance' to an alternating current) which arises from the energy that is required to accelerate the electrons (in the form of Cooper pairs) in one direction during the 0°–90° phase part of the AC cycle, to decelerate them to zero velocity during the 90°–180° part of the cycle, accelerate them in the opposite direction during the 180°–270° part and then to decelerate them again to zero velocity during the 270°–360° (0°) part. This kinetic inductance increases as the density of the Cooper pairs of electrons decreases. A photon interacting within the superconductor will disrupt one or more Cooper pairs, thus decreasing their density and so increasing the kinetic inductance. By combining the device with a capacitor, a resonant circuit is formed whose frequency lies in the microwave region (10^8–10^{11} Hz or so, hence, the 'M' in MKIDs). The change in kinetic inductance lowers the resonant frequency, and it is this change that is registered externally and provides the detection signal.

Like STJs, MKIDs can detect photons into the far infrared (FIR) and into the microwave and even shorter wavelength regions because the energy required to split the Cooper pairs of electrons is so small. The number of Cooper pairs split by an incoming photon will also be proportional to that photon's energy, giving the devices an intrinsic spectral resolution. Their quantum efficiencies are currently up to around 75%. Another major advantage of MKIDs is their relatively easy fabrication into large arrays. This arises because several hundred or more of the basic units can be constructed so that each has a slightly different resonant frequency. They can then all be read using a single output and fed to a single amplifier. Aluminium is commonly used as the superconductor material together with an absorber appropriate to the spectral region of interest (tin, for example, at long wavelengths). Also, like STJs, MKIDs have to be operated at temperatures below 100 mK so that their use is likely to be restricted to major observatories where suitable cryogenic facilities can be provided. Recently the performance of MKIDs at THz frequencies has been improved by using separate discrete elements separated from the capacitor (a lumped-element KID, or LEKID).

The first visual-region astronomical use of MKIDs was in ARCONS (Array Camera for Optical to Near Infrared Spectrophotometry (ARCONS), an Integral Field Spectrograph (IFS; see Section 4.2.4) designed for the 5-m Palomar and the 3-m Lick telescopes. A 44 × 46-pixel array of titanium nitride-based MKIDS was the instrument's detector and its intrinsic spectral resolution was ~8 and its time resolution was better than 2 μs.

The Dark Speckle Near Infrared Energy Resolved Superconducting Spectrophotometer (DARKNESS) saw its first light in 2016 and operates with the 5-m Palomar telescope's coronagraphs. It uses a 10,000-pixel array of MKIDs operating over the 800-nm to 1.4-μm waveband. The first arrays were based upon titanium nitride, but more recently platinum silicide has been used because of its higher uniformity and better intrinsic spectral resolution.

A 20,160-pixel array (140 × 144) of MKIDs is used for the MKID Exoplanet Camera (MEC) This was deployed in 2017 and also operates over the 800-nm to 1.4-μm waveband. Its intrinsic spectral resolution is ~8 to ~10, its time resolution about 1 μs and its angular resolution on the sky about 0.01″/pixel. Initially used on the 5-m Palomar telescope, it now operates on the 8.2- Subaru telescope with Subaru Coronographic Extreme Adaptive Optics (SCExAO).[39]

Even larger MKID arrays are planned in the near future for the Keck 1 telescope's Keck Radiometer Array using KID Energy Sensors (KRAKENS) IFS. This is expected to use arrays with up to 30,000 pixels and to cover from 380 nm to 1.35 μm at an intrinsic spectral resolution of better than 20.

1.1.14 Future Possibilities

One technique for improving detectors has recently started to be used and seems likely soon to become much more widely applied and that is improving the physical structure of the detector.

[39] Extreme adaptive optics are generally regarded as systems that operate at kHz frequencies and use 1,000s of actuators for their correcting optics.

Reducing reflection from the surface of the detector is one approach which has been in use for decades, using interference coatings on the detector's surface (Sections 1.1.15.1 and 1.1.18.1). The eye of a moth however, improves its absorption by a physical structure of fine, closely spaced, protuberances on its surface. The protuberances are smaller than the wavelength of the incident radiation and, in effect, mean that the refractive index of the outer layer of the moth's eye *gradually* varies between that of the atmosphere and that of the material forming the eye, thus, substantially reducing reflection. Similar structures may be formed by micro-machining the surfaces of man-made detectors. Thus, the High resolution Airborne Wide-band Camera (HAWC+) FIR camera on board the Stratospheric Observatory for Infrared Astronomy (SOFIA) (Section 1.1.15.4 and Figure 1.59) uses the technique on its two 2,560-pixel bolometer arrays. A somewhat related technique that has yet to be applied to astronomy is to recycle the unabsorbed photons, that is, reflect the unabsorbed photons back into the detector and give them a second chance to be absorbed (cf. power recycling in gravitational wave detectors; see Section 1.6). In this technique, a thin layer of silver is used as the reflector, and this is separated from the thin layer comprising the detector by a layer of tiny cavities.

As a general rule, though, we may expect those detectors and techniques currently in use to be developed and improved (better quantum efficiency, wider spectral coverage, faster, bigger (or smaller if that is more desirable), cheaper, larger dynamic range, lower noise, etc.) on timescales of a few years or so – and (without underplaying the hard and brilliant work that is needed for such developments) we may regard this process as being routine, or at least as normal. Equally and clearly, detectors and techniques that rely upon the invention or discovery of some radically new technical process or scientific principle cannot currently be speculated about. We are left therefore to look at the in-between possibilities.

Given that we already have detectors such as STJs and MKIDs that are capable of detecting individual photons with good quantum efficiencies over a wide spectral range, with reasonable spectral and time resolutions and with MKIDs now reaching array sizes comparable with those of present-day CCDs, our optical detectors are getting pretty close to the ideal detector conceivable for astronomy (if only they operated at room temperature and were a lot cheaper!), What, therefore, apart from (relatively) minor improvements in these various present-day attributes of optical detectors, could be demanded of an ideal optical astronomical detector? The answer is intrinsic sensitivities to:

- the phase of the incoming radiation,
- its state of polarization, and
- radiation from 1,000-km radio wavelengths to TeV γ rays.

The first two of these desirable attributes of these may not be too far distant; heterodyne detection and hence, information about the phase of the radiation at frequencies up to a few tens of terahertz (sub-10 μm wavelengths) is starting to be implemented in astronomical instrumentation (Sections 1.1.15.4 and 1.2.2).

Many crystals absorb or reflect light that is polarised in one fashion better than light polarised in a different fashion (see, e.g., Section 1.3). Incorporating such materials into CCDs, STJs, TESs (see Section 1.1.15.3) and others could endow them with an intrinsic sensitivity to polarisation and perhaps this might be possible in years rather than decades.

A bit more speculatively carbon nanotubes might be developable to act as antennas and waveguides for visible light (they are at least of roughly the right size – cf. moth's eye anti-reflection coating, above). Then all the techniques, including the direct detection of phase and polarisation, which are currently the tools of radio astronomers would become available to optical astronomy.

Extreme wide-band sensitivity, however, is probably a long way off if, indeed, it is ever possible. The reason for this pessimism is that although electromagnetic radiation has the same basic nature whether we consider the highest energy γ rays or the longest wavelength radio waves, the ways in which that radiation interacts with matter *does* vary from one part of the spectrum to another. Thus (to a first approximation):

- radio waves interact with matter by direct induction of electric currents in conductors
- FIR and millimetre waves interact with the vibrations and rotations of molecules
- NIR, visible and UV light interacts with the electrons in the outer levels of atoms and molecules
- far UV light and soft x-rays interact with the inner electrons of atoms.
- hard x-rays and γ rays interact with the particles within atoms' nuclei and
- the highest energy γ rays interact with themselves (producing particle/anti-particle pairs of various subatomic particles).

It is, thus, most unlikely that high-quality[40] detectors capable of picking-up (say) hard x-rays and NIR photons simultaneously and via the same physical process will be found, simply because the two varieties of e-m radiation interact with the material forming the detector in different ways.

1.1.15 Infrared Detectors

Many of the detectors just considered have some IR sensitivity, especially out to 1 μm. However, at longer wavelengths, other types of detectors are needed, although if the STJ or MKID fulfils their promise, one or other may replace some of these devices in the future.

In astronomy, the IR region is conventionally divided into three sections:[41] the NIR, 0.7–5 μm (4.3 PHz–600 THz), the Mid Infrared (MIR), 5–30 μm (600–100 THz) and the FIR, 30–1000 μm (100 THz–300 GHz). At the long-wavelength end of the FIR region, there is overlap with the submillimetre region, or as it is now quite frequently labelled, the terahertz region (radiation of frequency 1 THz has a wavelength of 300 μm), and so there is some duplication with the techniques considered in Section 1.2.

All IR detectors need to be cooled,[42] with the longer the operating wavelength, the lower the required temperature. Thus, in the NIR, liquid nitrogen (77 K) generally suffices, in the MIR, liquid helium (4 K) is needed and in the FIR, temperatures down to 100 mK are used. Currently, there are two main types of IR detector; the photoconductor (see Section 1.1.19.1) for the NIR and MIR and somewhat into the FIR and the bolometer for the FIR. As just discussed, various varieties of superconductor-based detectors seem likely to add a third strand to this list in the future.

---

[40] Thermal detectors, such as thermocouples (see previous editions of this book), can detect over wide ranges of the spectrum provided only that they can absorb and so be heated by, the e-m radiation that is involved. They are of low sensitivity but have in the past been used to cross-calibrate more sensitive detectors operating in different parts of the spectrum. Future developments in the superconducting detectors that operate by detecting photons via the disruption of Cooper pairs of electrons (STJs, MKIDs, Superconducting Nanowire Single-Photon Detectors [SNSPDs] and Quantum Capacitance Detectors [QCDs]) may give the lie to this statement. However, it still seems likely that different absorbers would be needed for the devices when operating in different spectral regions even though the detecting mechanism may be the same in each case. So effectively, different detectors will still be being used.

[41] Much narrower subdivisions are used within photometry (Section 3.1), especially for the NIR. There are, thus, seven defined photometric bands (I, Z, J, H, K, L and M) within the 780-nm to 4.75-μm region for the Johnson, Cousins and Glass (JCG) photometric system alone. Also, there are many other conventions in use – for example in military and surveillance applications, the definitions; Short Wavelength IR (SWIR: 1–3 μm), Middle Wavelength IR (MWIR: 3–5 μm), Long Wavelength IR (LWIR: 8–14 μm) and Very Long Wavelength IR (VLWIR:14–30 μm) are likely to be encountered.

[42] NASA's Spitzer infrared space observatory was launched in August 2003 and for five and three-quarter years its instruments were cooled to below 3 K using liquid helium, thus enabling observations to be made over the wavelength range from 3 to 180 μm. The helium coolant was exhausted in May 2009. The spacecraft then entered the 'warm' phase of its mission, observing just at 3.6 and 4.5 μm. This terminology however is misleading; the instruments are operating at 31 K – so 'warm' in this context is still decidedly chilly.

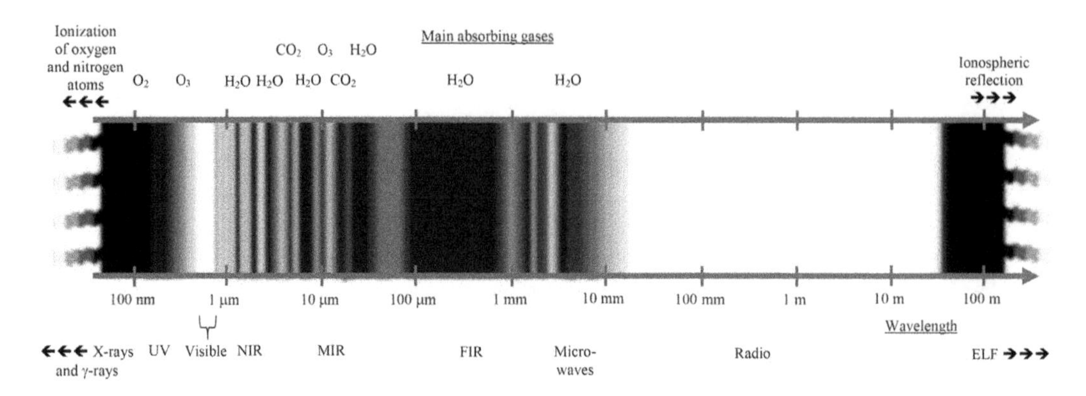

FIGURE 1.20 The transparent and partially transparent spectral regions (windows) of the Earth's atmosphere. The main, but not the only, causes of the opaque regions are listed at the top of the diagram. The spectrum goes off to both the left (X-rays and γ rays) and right (extremely low frequencies [ELF]), theoretically to infinity, but there are no further windows. The abbreviations stand for; CO_2, carbon dioxide; FIR, Far Infrared; H_2O, water; MIR, Mid Infrared; NIR, Near Infrared; O_2, oxygen; O_3, ozone (or trioxygen) and UV, ultraviolet.

The Earth's atmosphere is opaque over much of the IR region, although there are narrow wavelength ranges (windows – Figure 1.20) where it becomes transparent to a greater or lesser degree. The windows can be enhanced by observing from high altitude, dry sites or by flying telescopes on balloons or aircraft. Nonetheless, the sky background can still be sufficiently high that images have to be read-out several hundred times a second so that they do not saturate. Much of the observing, therefore, has to be done from spacecraft. Conventional reflecting optics can be used for the telescope and the instruments, though the longer wavelength means that lower surface accuracies are adequate. Refractive optics, including achromatic lenses, can be used in the NIR, using materials such as barium, lithium and strontium fluoride, zinc sulphate or selenide and IR-transmitting glasses.

1.1.15.1 Photoconductive Cells

Photoconductive cells exhibit a change in conductivity with the intensity of their illumination. The mechanism for that change is the absorption of the radiation by the electrons in the valence band of a semiconductor and their consequent elevation to the conduction band. The conductivity, therefore, increases with increasing illumination and is monitored by a small bias current. There is a cut-off point determined by the minimum energy required to excite a valence electron over the band gap. A wide variety of materials may be used, with widely differing sensitivities, cut-off wavelengths, operating temperatures, and so on. The semiconductor may be intrinsic, such as silicon, germanium, lead sulphide, indium antimonide or HgCdTe. For the NIR and the short wavelength end of the MIR, the HgCdTe alloy approaches being the ideal material for detectors. It is a mix of cadmium and mercury tellurides. Its sensitivity can be tailored to requirements over the range from 1 to 30 μm (300–10 THz) by adjusting the relative proportion of cadmium in the mix and it has a quantum efficiency, with anti-reflection coatings, of up to 90%. By removing the substrate after its manufacture, the detector can be made simultaneously sensitive to visible and IR radiation.

The band gaps in intrinsic semiconductors however tend to be large, restricting their use to the NIR. Doping of an intrinsic semiconductor produces an extrinsic semiconductor with the electrons or gaps from the doping atom occupying isolated levels within the band gap. These levels can be just above the top of the valence band or close to the bottom of the conduction band so that much less energy is needed to excite electrons to or from them. Extrinsic semiconductors can therefore be made that are sensitive across most of the IR. Doping is normally carried out during the melt stage of the formation of the material; however, this can lead to variable concentrations of the dopant and so to variable responses for the detectors. For germanium doped with gallium, Ge(Ga), the most widely used detector material at wavelengths longer than 50 μm, extremely uniform doping has been achieved by exposing pure germanium to a flux of thermal neutrons in a nuclear reactor. Some of the germanium nuclei absorb a neutron and become radioactive. The germanium-70 nucleus transmutes to germanium-71 which in turn decays to gallium-71 via β decay. Arsenic, an *n*-type dopant, is also produced from germanium-74 during the process; however, only at 20% of the rate of production of the gallium. The process is known as neutron transmutation doping (NTD). The response of Ge(Ga) detectors may additionally be changed by applying pressure or by stressing the material along one of its crystal axes. The pressure is applied by a spring (Figure 1.23) and can change the detectivity range of the material (which is normally from ~40 to ~115 μm: 7.5–2.6 THz) to ~80 to ~240 μm (3.8–1.3 THz).

Ge(Ga) along with silicon doped with arsenic, Si(As), or antimony, Si(Sb), is also one of the materials used in the relatively recently developed blocked impurity band (BIB)[43] detectors (see also Section 1.1.15.4). These use a thin layer of heavily doped semiconductor to absorb the radiation. Such heavy doping would normally lead to high dark currents, but a layer of undoped semiconductor blocks these. Ge(Ga) BIB detectors are sensitive out to about 180 μm (1.7 THz) and are twice as sensitive as normal stressed Ge(Ga) photoconductors around the 140-μm (2.1-THz) region. Further extension of the cut-off wavelength limit of these devices to 300 μm (1 THz) may soon be achievable using GaAs(Te), Ge(Sb) and/or Ge(Ga).

Photoconductive detectors that do not require the electrons to be excited all the way to the conduction band have been made using alternating layers of gallium arsenide (GaAs) and indium gallium arsenide phosphide (InGaAsP) or aluminium gallium arsenide (AlGaAs), each layer being only ten or so atoms thick. The detectors are known as quantum well infrared photodetectors (QWIPs). The lower energy required to excite the electron gives the devices a wavelength sensitivity ranging from 1 to 12 μm. The sensitivity region is quite narrow and can be tuned by changing the proportions of the elements. Recently, the National Aeronautics and Space Administration (NASA) has produced a broadband 1k × 1k QWIP with sensitivity from 8 to 12 μm by combining more than a hundred different layers ranging from 10 to 700 atoms thick. Quantum dot infrared photodetectors (QDIPs) have recently been developed wherein the 'well' is replaced by a 'dot', i.e., a region that is confined in all spatial directions. It remains to be seen if QDIPs have any advantages for astronomy over QWIPs.

The superlattice has a similar structure to that of QWIPs with alternating layers of semiconductor materials each a few nanometres thick. In the type II superlattice, the conduction and valence bands do not overlap so that the electrons and holes are trapped. These structures may be developed into IR detectors suitable for astronomical applications in the future because they can have quantum efficiencies of up to 30%.

Details of some of the materials used for IR photoconductors are listed in Table 1.3, and those in current widespread use are underlined.

1.1.15.2 Bolometers

A bolometer is simply a device that changes its electrical resistivity in response to heating by illuminating radiation. At its simplest, two strips of the material are used as arms of a Wheatstone bridge. When one is heated by the radiation, its resistance changes and so the balance of the bridge alters. Two strips of the material are used to counteract the effect of slower environmental temperature changes because they will both vary in the same manner under that influence.

Bolometers are simply a type of thermometer, and they respond to temperature changes from any cause. If the bolometer material itself (see below) does not absorb the desired radiation frequencies efficiently, then a separate absorber may be used and the bolometer thermally connected to

[43] Also known as Impurity Band Conduction (IBC) detectors.

TABLE 1.3
Some Materials Used for Infrared Photoconductors

Material	Cut-off Wavelength/Frequency or Wavelength/Frequency Range	
	μm	THz
Silicon (Si)	1.11	270
Germanium (Ge)	1.8	167
Gold-doped germanium (Ge[Au])	1–9	300–33
Mercury-cadmium-telluride (HgCdTe)	1–30	300–10
Gallium arsenide QWIPs (GaAs + InGaAsP or AlGaAs)	1–12	300–25
Lead selenide (PbSe)	1.5–5.2	200–60
Lead sulphide (PbS)	3.5	86
Mercury-doped germanium (Ge[Hg])	4	75
Indium antimonide (InSb)	6.5	46
Copper-doped germanium (Ge[Cu])	6–30	50–10
Gallium-doped silicon (Si[Ga])	17	18
Arsenic-doped silicon BIB (Si[As])	23	13
Gallium-doped germanium (Ge[Ga])	~40–~115	~7.5–~2.6
Gallium-doped germanium stressed (Ge[Ga])	~80–~240	~3.8–~1.3
Boron-doped germanium (Ge[B])	120	2.5
Gallium-doped germanium BIB (Ge[Ga])	~180	~1.7
Antimony-doped germanium (Ge[Sb])	130	2.3

the absorber. The current state of the art uses an absorber that is a mesh of metallised silicon nitride like a spider's web, with a much smaller bolometer bonded to its centre. This minimises the noise produced by cosmic rays, but because the mesh size is much smaller than the operating wavelength, it still absorbs all the radiation. Arrays of bolometers up to 800 × 1,000 pixels can now be produced (see also HAWC+, Section 1.1.14) and their pixel sizes can be down to 12 μm. Bolometers and/or their absorbers can be found to cover a wide range of frequencies and so they are found as microwave detectors (Section 1.2.2) as well as IR detectors.

Cooled semiconductor bolometers were once used as astronomical detectors throughout most of the IR region. Photoconductive cells have now replaced them for NIR and MIR work, but they are still used for the FIR (~100 μm to a few mm, 3 THz to 100 GHz or so). Germanium doped with gallium (a *p*-type dopant) is widely used for the bolometer material with a metal-coated dielectric as the absorber. The bolometer is cooled to around 100 mK when in operation to reduce the thermal noise. Germanium doped with beryllium, silicon, silicon nitride, and several vanadium oxides are other possible bolometer materials.

Thermistors are a type of bolometer which can be fabricated so that their resistance either decreases as the temperature rises (negative temperature coefficient [NTC] thermistors) or increases as the temperature rises (positive temperature coefficient [PTC]). They are widely used as temperature sensors in every-day life, but have found fewer applications within astronomy. However, the high frequency instrument carried by the Planck spacecraft (2009 to 2013) to study the cosmic microwave background (CMB) radiation, used NTD germanium thermistors to observe up to 857 GHz. Whilst silicon-based thermistors with silicon wire absorbers are currently planned[44] to be used as detectors for the Primordial Inflation Explorer (PIXIE) spacecraft's spectrograph to study the CMB up to a frequency of 6 THz (50 μm).

[44] Possible launch around 2024, if the proposal continues to go ahead.

Hot electron bolometers (HEBs) and cold electron bolometers (CEBs) and the closely related TES (Section 1.1.15.3) are all based upon superconductors. The terms 'hot' and 'cold' are therefore somewhat misleading; all these devices operate at very close to zero Kelvin temperatures. In HEBs though, the electron temperature is slightly higher (hotter) than that of their substrate, whilst in CEBs the electron temperature is slightly lower (cooler). Both bolometers rely upon superconductor-insulator-normal (SIN)[45] metal tunnel junctions (cf. STJs, Section 1.1.12) wherein the superconductor's resistance changes with electron temperature, whereas that of the normal metal at such low temperatures does not. The HEBs are based upon materials such as magnesium bromide, niobium nitride, niobium titanium nitride, and gallium arsenide/aluminium arsenide alloy. The CEBs are based upon materials such as silicon and silicon nitride.

Uncooled micro-bolometer arrays have been used for some time for civilian and military applications. They are based upon amorphous silicon or vanadium oxides and can detect over the 8–14-μm region. They are able to operate uncooled because the IR absorbing layer is thermally isolated from the rest of the instrument. A commercial 512 × 640-pixel micro-bolometer operating at 10 μm has recently been trialled on the 2.0-m Liverpool telescope and successfully detected some of the brighter known MIR sources. Further development of such arrays could clearly make astronomical IR instruments cheaper and/or larger in the future and/or make them available to smaller or even to portable telescopes.

1.1.15.3 Other Types of Detectors

A recent development that promises to result in much larger arrays is the Transition Edge Sensor (TES). These detectors are thin films of a superconductor, such as tungsten, held at their transition temperature from the superconducting to the normally conducting state (cf. HEBs and CEBs, Section 1.1.15.2). There is thus a very strong dependence of the resistivity upon temperature in this region. The absorption of a photon increases the temperature slightly and so increases the resistance. The resistance is monitored through a bias voltage. Their quantum efficiency can reach 95%. MKIDs (see Section 1.1.13), also based upon thin superconducting films, are already in use as astronomical IR detectors and promise to find much wider application as their array sizes increase.

The Q and U Bolometric Interferometer for Cosmology (QUBIC), which is under construction at the time of writing, is a millimetre-wave interferometer (Section 2.5) to be sited in Argentina. The radiation will be collected by 400 horn feeds covering a 0.4-m diameter aperture and then split into two beams at frequencies of 150 and 220 GHz by a dichroic beam splitter (Section 1.1.22.1). The two beams will then each detected by separate 32 × 32-pixel InSi back-biased TES arrays operating at 320 mK and the read-outs will be via Superconducting Quantum Interference Devices (SQUIDs). The KIDs Interference Spectrum Survey (KISS) instrument, recently installed on one of the two 2.25-m Q-U-I Joint Tenerife (QUIJOTE) telescopes on Tenerife, has two 316-pixel MKID arrays for its detectors. Ti-Al and Al absorbers are used to enable frequencies from 80 to 300 GHz to be observed and the MKID arrays are cooled to 150 mK.

The Astro2020's concept project, Origins (see also Large UV Optical Infrared Surveyor [LUVOIR] and Habitable Exoplanet Imaging Mission [HabEx]– Section 1.1.23 and Lynx – Section 1.3.8), envisages using either TES or MKID arrays for the detectors of a 5.9-m cryogenic (4.5 K) telescope observing across the FIR and MIR from the Sun-Earth Lagrange L2 point and with a thousand times the sensitivity of the Herschel and Spitzer[46] missions.

Two more superconducting detectors (in addition to STJs and MKIDs) whose operating principle is based upon the disruption of Cooper pairs are the superconducting nanowire single-photo detector

[45] The term 'NIS' may also be encountered.

[46] Mission ended January 2020.

(SNSPD) and the quantum capacitance detector (QCD). These devices have yet to be applied to astronomy but may find such application in the future.

The basic element of the SNSPD is a tightly concertinaed niobium nitride wire whose dimensions are typically 10 nm × 100 nm × 100 μm. The wire is cooled to well below its critical temperature and has a DC bias current running through it which is just below the critical current. Absorption of a photon breaks some of the Cooper pairs of electrons, reducing the critical current to below the bias current. The resulting transition to a non-superconducting state is used to shunt the bias current to an amplifier and the ensuing voltage pulse indicates the absorption of the photon. Its speed of operation is higher than that of TES detectors, and it has most of the other advantages of superconducting detectors.

In the QCD, the electrons from Cooper pairs that have been disrupted by an absorbed photon, tunnel through to a microwave resonator. They change the capacitance of the resonator and so also its resonant frequency. The read-out of the detection is similar to that in the MKID. The device's advantage over the MKID may be that of improved sensitivity.

For detection from the visual out to 30 μm (10 THz) or so, an SiPMT may be used. This is closely related to the APD (Section 1.1.11). It uses a layer of arsenic-doped silicon on a layer of undoped silicon. A potential difference is applied across the device. An incident photon produces an electron-hole pair in the doped silicon layer. The electron drifts under the potential difference.

Platinum silicide acting as a Schottky diode (a metal-semiconductor junction diode) can operate out to 5.6 μm. It is easy to fabricate into large arrays but is of low sensitivity compared with photoconductors. Its main application is for terrestrial surveillance cameras.

Large[47] (2k × 2k pixel) arrays can now be produced for some of the NIR detectors and 1k × 1k pixel arrays for some MIR detectors, though at the long wave end of the MIR, arrays are still at a 256 × 256 maximum. In the FIR, array sizes are limited to a maximum of around 100 × 100 pixels at the short-wave end and to a few tens of pixels (not always in rectangular arrangements) in total at the long wave end. Unlike CCDs, infrared arrays are read-out pixel by pixel. Although this complicates the connecting circuits, there are advantages; the pixels are read-out non-destructively and so can be read-out several times and the results averaged to reduce the noise, there is no cross-talk (blooming) between the pixels and one bad pixel does not affect any of the others. The sensitive material is usually bonded to a silicon substrate which contains the read-out electronics.

A recent technological development may however lead to cheaper and larger arrays. Typically, a 2k × 2k IR array currently costs up to $0.5 million. One of the major constraints on building larger sizes is the mismatch between the properties of the silicon substrate and that of the IR detector materials – particularly the inter-atomic spacings in the crystals and their thermal expansion coefficients. Molecular beam epitaxy,[48] already a technique widely used in the semiconductor industry, is starting to be used to apply HgCdTe to a silicon wafer substrate. Because silicon wafers can come in sizes up to 300 mm, potentially arrays up to 14k × 14k could be being produced within a few years and at much lower costs than the smaller arrays available today.

In the NIR, large format arrays have led to the abandonment of the once common practice of alternately observing the source and the background (chopping), but at longer wavelengths the sky is sufficiently bright that chopping is still needed. Chopping may be via a rotating 'windmill' whose blades reflect the background (say) onto the detector, while the gaps between the blades allow the

[47] For two or more decades the sizes of infrared arrays have been doubling every 18 to 20 months. This is similar to the Moore law growth in the capabilities of computers (named after Gordon Moore). However, the sizes of individual CCD arrays have not followed such a law in recent years and have perhaps reached a naturally useful size limit at around 4k × 4k (the maximum single CCD array currently available is about 10k × 10k – see Section 1.1.1. Increasingly large mosaics, of course, *do* carry on with the Moore-law trend). Because some IR arrays are now approaching the 4k × 4k size, their growth may also soon slow down or come to a halt.

[48] A process whereby a layer of a different material is deposited onto a substrate within a vacuum chamber at a rate slow enough that the new layer adopts the crystalline structure of the substrate.

radiation from the source to fall onto the detector directly. Alternatively, some telescopes designed specifically for IR work have secondary mirrors that can be oscillated to achieve this switching.

It might seem that the larger the array, the better. However, with limited resources (i.e., money!) this may not always be the case. The relative noise within the combined output of n detectors is reduced by a factor of $n^{1/2}$ compared with the relative noise from a single detector. If the noise level from a single detector however can be reduced, the improvement in the performance is directly proportional to the degree of that noise reduction. Thus, it may be better to spend the available money on a few good detectors rather than on lots of poorer detectors. Of course, lots of good detectors are even better. In the early 1990s, J. N. Bahcall attempted to quantify this concept with the astronomical capability of a system. This he defined as:

$$\text{Astronomical capability} = \frac{\text{Lifetime} \times \text{Efficiency} \times \text{ Number of pixels}}{(\text{Sensitivity})^2} \quad (1.4)$$

If used with some caution, this is a useful parameter for deciding between the different options that may be available during the early planning stages of a new instrument. For example, using Equation 1.4, the potential astronomical capability of the mid-infrared instrument (MIRI) (see Section 1.1.15.4.5) on the James Webb Space Telescope (JWST) is 10^4–10^5 better than the capability of the Spitzer[49] spacecraft's instruments.

1.1.15.4 Astronomical Applications

There are many recent, current, and possible future examples of the use of IR detectors within astronomy. Some selected examples are briefly discussed next (see also Section 1.2).

1.1.15.4.1 MKIDS (Section 1.1.13, See also Section 1.2.2)

The currently under construction, 6-m Cerro Chajnantor Atacama Telescope Prime (CCAT-p; first light expected for 2021), is planned to use 18,216 KIDs in a circular array for 860 kHz detection (and also TES detectors; see below). The telescope will be of the Dragone design and be remotely controlled. A 24 × 24 array of MKIDs (Section 1.1.13) formed the heart of Multiband Submillimetre Inductance Camera (MUSIC) used on the 10.4-m telescope of the Caltech Submillimeter Observatory (CCAT-p's predecessor, decommissioned in 2015) to observe simultaneously at 850 μm, 1.03 mm, 1.36 mm, and 2 mm (350, 290, 220, and 150 GHz).

The APEX-MKID (A-MKID) is an MKID-based camera used on Atacama Pathfinder Experiment (APEX). Its two detector arrays, which operate concurrently and which have 3,520 and 21,600 pixels, respectively, observe at 347 GHz (864 μm) and 850 GHz (353 μm). Nèel IRAM KID Arrays (NIKA-2) on the 30-m Institute de Radioastronomie Millimétrique (IRAM) telescope similarly observes at 150 and 260 GHz (2.0 and 1.15 mm) with one 616-pixel and two 1,140-pixel lumped MKID arrays, respectively.

1.1.15.4.2 HgCdTe (Section 1.1.15.1)

The JWST (currently due for launch in March 2021) will carry two main instruments for NIR observations (and one for the MIR – see below). Near Infrared Camera (NIRCAM) will use four 2k × 2k HgCdTe arrays for the range from 600 nm to 2.3 μm and one 2k × 2k HgCdTe array for the range from 2.4- to 5.0-μm region. Both sets of detectors and their optics are doubled-up to provide redundancy. They are mounted side-by-side, can operate concurrently and each will view a separate 2.2′ × 2.2′ area of the sky. While the JWST's multi-object Near Infrared

[49] Initially known as SIRTF (Space Infrared Telescope Facility). Mission ended 2020.

FIGURE 1.21 European Southern Observatory's (ESO's) 4.1-m Visible and Infrared Survey Telescope for Astronomy (VISTA) telescope. The large metal cylinder mounted at and projecting beyond the top ring of the telescope is the infrared camera containing the cryostat, baffles, filters, and of course, the 67.1-megapixel detector array. (Courtesy of ESO.)

Spectrograph (NIRSpec; see Section 4.2) will use two 2k × 2k HgCdTe arrays for the 600-nm to 5.0-μm region.

Visible and Infrared Survey Telescope for Astronomy (VISTA), which had first light in 2009 (Figure 1.21), uses sixteen 2k × 2k HgCdTe arrays that are sensitive to the range 800 nm to 2.5 μm and has a field of view 1.65° across. The individual arrays are separated by large dead spaces so that six exposures with slight shifts between each of them are needed to give complete coverage of one area of the sky.

A second-generation instrument for European Southern Observatory's (ESO's) Very Large Telescope (VLT), K-band Multi Object Spectrograph (KMOS), is a NIR multi-object, integral field unit (IFU) spectrograph (Section 4.1) covering the 1- to 2.5-μm region. The instrument uses three separate spectrographs, each of which employs a single 2k × 2k HgCdTe detector. While the PRAXIS[50] (Section 1.1.7) OH line suppression unit for the 3.9-m Anglo-Australian Telescope (AAT) uses a 2048 × 2048-pixel HAWAII[51] array detector.

MILAHI uses 2 × 2-pixel HgCdTe devices as its final detectors but operates as a heterodyne spectrometer (Section 1.2.2.2) using quantum cascade lasers as the local oscillators. The intermediate frequency is in the radio region and the source and local oscillators' signals are mixed at a beam splitter and then fed to the detectors. It is mounted on a 0.6-m dedicated telescope sited on Hawaii and has a spectral resolution of $>10^6$ over the 7- to 12-μm region. It is used remotely to monitor the atmospheres of Venus and Mars and is due to be moved to a dedicated 1.8-m telescope. The development of related devices small and light enough to be carried by spacecraft is currently underway.

Small HgCdTe arrays are similarly proposed as some of the detectors for Time Domain Spectroscopic Observatory (TSO), a spectroscopic space mission for the late 2020s proposed as a concept for the Astro2020 Decadal survey. The TSO's aim would be to acquire transients and other targets of opportunity within about 5 minutes of time and to cover the 300-nm to 5-μm region using a 1.3-m telescope. It might be placed either in a geosynchronous orbit or at the Lagrange L2 point.

---

[50] A name, also Greek for 'practice'.

[51] Given the large number of astronomical instruments sited on the island of Hawaii, it can cause confusion that some HgCdTe-based detectors are named HAWAII detectors. In fact, the name is an acronym derived from HgCdTe Astronomical Wide Area Infrared Imager.

1.1.15.4.3 Indium Antimonide (Section 1.1.15.1)

The first-generation instrument, First Light Infrared Test Camera (FLITECAM, now retired), for the 2.5-m airborne SOFIA telescope, used a 1k × 1k indium antimonide array to observe over the 1- to 5-μm region. The Spitzer spacecraft (2003–2020) carried an NIR camera (Infrared Array Camera [IRAC]) using 256 × 256 indium antimonide arrays to observe at 3.6 and 4.5 μm.

1.1.15.4.4 QWIPs

QWIPs have been used for terrestrial and atmospheric remote sensing and for commercial and military purposes. Landsat 8, for example, carries a thermal infrared sensor based upon three 512 × 640 gallium arsenide QWIPs. However, while a 256 × 256 QWIP array has been used on the 5-m Hale telescope and others in the Search for Extraterrestrial Intelligence (SETI), they have found few other astronomical applications so far, perhaps because of their low quantum efficiencies (around 10%).

1.1.15.4.5 BIB

In the MIR, the JWST will carry the MIRI imager which will cover the 5 to 28 μm (60–11 THz) region using a 1k × 1k Si(As) BIB array. While the MIRI spectrograph will cover the same spectral region, with low and high spectral resolution options, using two 1k × 1k Si(As) BIB arrays.

The VLT Mid-Infrared Imager and Spectrometer (VISIR), which is used on ESO's VLT for imaging and spectroscopy between 8 to 13 μm (38–23 THz) and 16.5 to 24.5 μm (18–12 THz), originally used two 256 × 256 BIB detectors. These were replaced in 2016 by a 1024 × 1024-pixel Si(As) BIB array constructed as two 512 × 1024 adjacent arrays. The venerable 3.0-m Infrared Telescope Facility's (IRTF's) 2 to 29 μm imager, Mid-Infrared Imager and Spectrometer (MIRSI), uses a 240 × 320-pixel Si(As) BIB detector. The Spitzer spacecraft, before running out of coolant, used 256 × 256 Si(As) BIB detectors for 5.8 and 8.0 μm. While the Faint Object Infrared Camera for the SOFIA Telescope (FORCAST) MIR detector uses 256 × 256 Si(As) and Si(Sb) BIB arrays to observe over the 5- to 40-μm (60–7.5 THz) range.

1.1.15.4.6 Si and Ge Bolometers

In the FIR, ESA's Herschel Space Observatory was launched in May 2009 to occupy the second Lagrangian[52] point (L2) of the Earth-Sun system, some 1.5 million km away from the Earth. It was decommissioned in 2013. It carried a 3.5-m infrared telescope to observe in the FIR (60–670 μm, 5 THz to 450 GHz) and was cooled to below 2 K using liquid helium (which evaporated around March 2013). Three instruments were carried on board the spacecraft; Photodetector Array Camera and Spectrometer (PACS), Spectral and Photometric Imaging Receiver (SPIRE), and Heterodyne Instrument for the Far Infrared (HIFI; see Section 1.2).

The PACS comprised two sub-instruments. The first was an imaging photometer operating over three bands from 60 to 210 μm (5–1.4 THz) and using 32 × 64 pixel and 16 × 32-pixel monolithic silicon bolometer arrays. The second was an IFU spectrograph (Section 4.2) covering the 51- to 220-μm (5.9- to 1.4-THz) region with two 16 × 25-pixel Ge(Ga) photodetector arrays. An image slicer (Section 4.2) was used to reformat the 5 × 5-pixel field of view to 1 × 25 pixels to match the arrays' shapes. The SPIRE also comprised two sub-instruments. The first of these was again an imaging

[52] L1, L2, L3, L4 and L5 – the Lagrangian points in a binary system. Three of the points are to be found on the line joining the binary components – L1 between the components, L2 and L3 outside the components. While L4 and L5 are the third points of equilateral triangles in the plane of the binary's orbit, where the other two points of the triangles are occupied by the binary components. When the main binary components are quite different in mass, small objects placed at one of these points will be in (relatively) stable orbits with respect to the main components. Those points will be approximately in the same orbit as the less massive binary component and 60° ahead and behind that component. The Greek-Trojan and Trojan-Trojan asteroids respectively occupy regions of space near the F4 and F5 points in the Sun-Jupiter binary system, for example.

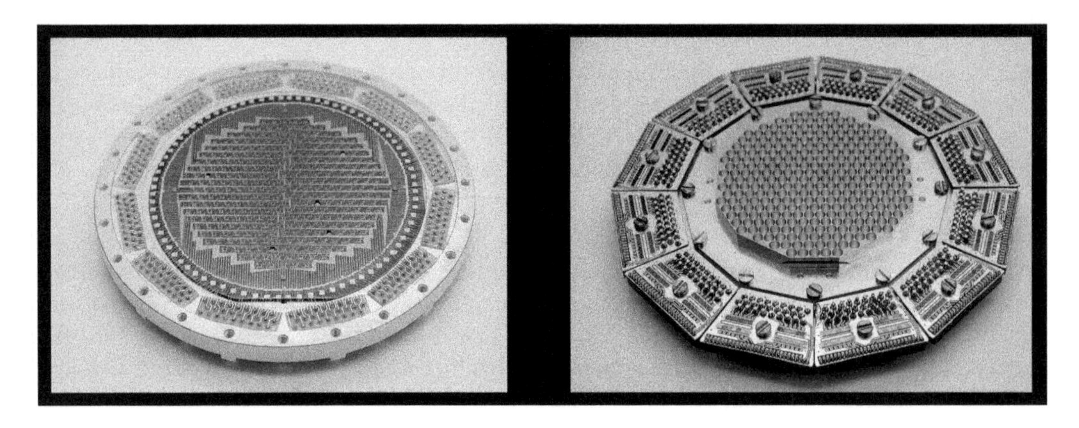

FIGURE 1.22 The bolometer array for the Large APEX Bolometer Camera (LABOCA) instrument on the Atacama Pathfinder Experiment (APEX) telescope. Left, the backs of the bolometers and the wiring assembly. Right, the feed horn array on the other side. (Courtesy of Giorgio Siringo.)

photometer. It observed bands centred at 250, 350 and 500 μm (1.2 THz, 860 and 600 GHz) using 9 × 15/16, 7 × 12/13 and 5 × 8/9 pixel hexagonally packed NTD spider web Germanium bolometer arrays with feed horns for each pixel. The second instrument was a Fourier transform spectrometer (Sections 2.1.1 and 4.1) covering the range from 194 to 671 μm (1.5 THz to 450 GHz) and with 19-pixel and 37-pixel hexagonally shaped arrays of NTD spider web germanium bolometers.

Large APEX Bolometer Camera (LABOCA, Figure 1.22) on the 12-m APEX telescope situated at the Atacama Large Millimetre Array (ALMA, see Section 2.5) site on the Chajnantor plateau in Chile comprises 295 NTD germanium bolometers mounted on silicon-nitride membranes to observe around 870 μm (340 GHz).

The Spitzer spacecraft (see Section 1.1.15.4.5), before running out of coolant, used a 32 × 32 array of unstressed Ge(Ga) detectors to observe at 70 μm (4.3 THz) and a 2 × 20 array of stressed Ge(Ga) detectors to observe at 160 μm (1.9 THz; Figure 1.23).

1.1.15.4.7 TES (Section 1.1.15.3)

The SOFIA uses HAWC+ for imaging and polarimetry over the 50- to 240-μm region (see also Section 1.1.4) with the capacity for up to two 40 × 64-pixel arrays of Si TES bolometers. Three 1024-pixel arrays of TES detectors are used by the 6-m Atacama Cosmology Telescope (ACT) to scan 5°-wide strips of the sky at frequencies of 145, 215 and 280 GHz.

The CCAT-p telescope's Prime-Cam is expected to use 8,640 TES detectors for observing a 12,000-deg^2 area of the sky over the 220- to 405-kHz region and 4,320 TES detectors for observing the same frequencies spectroscopically over a 16-square degree area with Epoch of Reionization Spectrometer (EoR-Spec). The detector arrays will be hexagonal in shape. It will also use KIDs detectors (see Section 1.1.13).

One of the possible instruments for the planned Space Infrared Telescope for Cosmology and Astrophysics (SPICA) space mission – with, perhaps, a 2.5-m telescope and a possible launch date in the late 2020s, is SPICA Far Infrared Instrument (SAFARI), a spectrometer designed for the 34- to 210-μm region. It is currently expected to use large arrays of TES bolometers as its detectors.

Bolometers may make a comeback at shorter wavelengths in the future as detectors for signals from spacecraft. Many spacecraft, particularly planetary probes, generate large quantities of data, which the current radio transmitters and receivers can send back to Earth slowly. Faster communications[53] require shorter wavelengths to be used and transmission speeds thousands of times those

[53] Faster data rates, not, of course, faster speeds for the radio eaves.

FIGURE 1.23 The 160-μm stressed Ge(Ga) detector array for Spitzer's Multiband Imaging Photometer (MIPS) instrument. Leaf springs apply 500 N to each of the 1-mm^3 detectors to extend their response out to the required wavelength. (Courtesy of G. Rieke.)

possible in the radio region could be attained if the 1.55-μm IR radiation, commonly used to transmit broadband signals along optical fibres on the Earth, were to be utilised. A recent development has a superconductor bolometer using a coil of extremely thin wire placed within a mirrored cavity (photon trap) that bounces unabsorbed photons back to the coil of wire to increase their chances of being absorbed. Just such a nanowire detector, based upon niobium nitride has achieved 67% detection efficiency (at 1.064 μm), perhaps enabling lasers with power requirements low enough for them to be used on spacecraft to be employed for communications in the future. A recently constructed tungsten/silicon nanowire array has 1,024 sensors with the individual wires being 3 nm × 180 nm × 3.5 mm in size. It may be used for the Origins spacecraft (see Section 1.1.15.3).

1.1.16 Ultraviolet Detectors

The UV spectral region is conventionally defined as extending from 10 to 400 nm (124 to 3 eV). The normal human eye can thus see slightly into the UV region since its short-wave limit is about 380 nm. Persons who have been born without a lens in the eye (aphakia) or who have lost it (e.g., through surgery for the treatment of cataracts) can see, reportedly, down to about 350 nm. The new 'colour' appears as white tinged with blue or violet.

There are a number of subdivisions of the UV which may be encountered. While usage does vary, the definitions given in Table 1.4 are widely employed.

Ground-based telescopes can observe down to a wavelength of about 300–340 nm depending upon altitude, state of the ozone layer and other atmospheric conditions), i.e., the UV-A and UV-B regions. Wavelengths shorter than this can only be detected from high-flying aircraft or balloons or by using rockets and spacecraft.

As with detectors for IR radiation, some of the detectors that we have already reviewed are intrinsically sensitive to short-wave radiation, although modification from their standard optical or IR forms may be required. For example, photomultipliers will require suitable photoemitters and

TABLE 1.4
Regions Within the UV

Name	Abbreviation	Wavelength Range (nm)	Energy Range (eV)
Ultraviolet-A	UV-A	315–400	3.94–3.10
Near ultraviolet	NUV	300–400	4.14–3.10
Ultraviolet-B	UV-B	280–315	4.43–3.94
Middle ultraviolet	MUV	200–300	6.20–4.14
Ultraviolet-C	UV-C	100–280	12.40–4.43
Far ultraviolet	FUV	122–200	10.16–6.20
Vacuum ultraviolet	VUV	10–200	123.98–6.20
Extreme ultraviolet	EUV or X-ray ultraviolet (XUV)	10–121	123.98–10.25

windows that are transparent to the required wavelengths. Lithium fluoride and sapphire are commonly used materials for such windows. Thinned, rear-illuminated CCDs have a moderate intrinsic sensitivity into the long wave UV region. The EBCCDs with an appropriate UV photocathode can also be used. At shorter wavelengths microchannel plates (Section 1.3) take over. Detectors sensitive to the visible region as well as to the UV need filters to exclude the usually much more intense longer wavelengths. Unfortunately, the filters also absorb some of the UV radiation and can bring the overall quantum efficiency down to a few per cent. The term 'solar blind' is used for detectors or detector/filter combinations that are *only* UV sensitive.

Another common method of short-wave detection is to use a standard detector for the visual region and to add a fluorescent or short glow phosphorescent material to convert the radiation to longer wavelengths. Sodium salicylate and tetraphenyl butadiene are the most popular such substances because their emission spectra are well matched to standard photocathodes. Sensitivity down to 60 nm can be achieved in this manner. Additional conversions to longer wavelengths can be added for matching to CCD and other solid-state detectors whose sensitivities peak in the red and IR. Ruby ($Al_2 O_3$) is suitable for this and its emission peaks near 700 nm. In this way, the sensitivity of CCDs can be extended down to about 200 nm.

Recently a variant on the CCD, much used in cell phone and web cameras, has found application as an astronomical UV detector and also for visual cameras aimed at the amateur astronomy market. It is called the Complementary Metal Oxide Semiconductor-Active Pixel Sensor (CMOS-APS).[54] It has the same detection mechanism as the CCD, but the pixels are read-out individually. The silicon detector part of the device is separate from the CMOS read-out section and the two sections are electrically connected together by small indium linkages (usually called 'bumps'). The CMOS-APS detectors are less power-hungry than CCDs, do not require mechanical shutters during the read-out process and are easier to harden against radiation. The dead spaces on CMOS-APS's though are larger than for CCDs because of the areas occupied by the read-out electronics. In other respects, including sensitising to the UV via fluorescent coatings, CMOS-APSs are similar to CCDs.

MCPs are discussed more fully in Section 1.3. Here we just note that as well as being X-ray detectors, they can be used as UV detectors for wavelengths shorter than about 200 nm (6.2 eV).

Many of the super-conducting detectors discussed previously (TESs, STJs, MKIDs, SNSPSs, and QCDs) potentially have the ability to detect into the UV and X-ray regions and some of these are currently starting to be so used in the laboratory.

[54] Also called hybrid CMOS devices.

1.1.16.1 Applications

During the Hubble Space Telescope's (HST's) 2009 servicing mission, the Cosmic Origins Spectrograph (COS) was installed to undertake spectroscopy over the 115- to 320-nm region (10.8–3.9 eV), and it is still in frequent use at the time of writing. It uses two windowless 1k × 16 k MCPs as the detectors for its far UV channel (115–205 nm, 10.8–6.0 eV) and a single 1k × 1k caesium telluride MCP for its near UV (170–320 nm, 7.3–3.9 eV) channel. Also installed in 2009, Wide Field Camera (WFC3) has two channels and covers the region from 200 nm to 1.7 μm. The UV and visible channel (200 nm to 1.0 μm) has two thinned and UV-optimised 2k × 4k CCDs as its detectors.

The NASA's Solar Dynamics Observatory (SDO) was launched in 2010 and is still active at the time of writing. It carries two UV instruments – Atmospheric Imaging Assembly (AIA) and Extreme Ultra-Violet Variability Experiment+ (EVE).

AIA comprises four separate telescopes (Figure 1.24) and observes the Sun at nine UV wavelengths ranging from 9.4 to 170 nm (132 to 7.3 eV) as well as in white light. The telescopes are of Ritchey-Chrétien design with 0.2-m diameter, normal incidence, primary mirrors coated with multiple alternating layers of silicon and molybdenum. The mirror surfaces are accurate to 0.3 nm (about 200 times better than a typical visual telescope). The primary mirrors are coated in two halves with the reflecting layers in each sector optimised for a different spectral region – effectively giving the instrument eight telescopes. There are filters at the front of each telescope to reject the visible and IR radiation which are also in two halves. Five of these sectors have thin metallic aluminium filters, two have zirconium filters and one magnesium fluoride filter. Each telescope has a 4k × 4k pixel CCD array as its detector.

The EVE observes the Sun from 0.1 to 105 nm (1,240 to 11.8 eV). The UV component of this range is measured by the two channels of the Multiple EUV Grating Spectrograph (MEGS) instrument. The two channels both use 1k × 2k split-frame transfer CCDs as their detectors.

ESA/NASA's planned Solar Orbiter (SolO) spacecraft (scheduled for launch in 2020), will carry the Spectral Imaging of the Coronal Environment (SPICE) EUV spectrograph which will use two MCP-intensified 1k × 1k APS arrays to cover the 70.2- to 79.2-nm (17.7–15.7 eV) and 97.2- to 105-nm (12.8–11.8 eV) spectral regions.

India's Astrosat, launched in 2016 has (amongst other instruments) two 0.375-m Ritchey-Chrétien telescopes for UV and visual observations. The Ultra-Violet Imaging Telescope (UVIT) instrument has three channels covering the range from 130 to 530 nm, and all three of these have 512 × 512-pixel MCP-intensified CMOS arrays as their detectors.

FIGURE 1.24 The Atmospheric Imaging Assembly (AIA) instrument for the Solar Dynamic Observatory (SDO) spacecraft during its assembly. The image shows the four 0.2-m telescopes with their protective front covers in place. The charge-coupled device (CCD) array detectors are at the far ends of the telescopes. (Courtesy of National Aeronautics and Space Administration [NASA].)

For ground-based observers, ESO's second-generation X-shooter spectrograph, covers the 300-nm to 2.48-μm spectra region using three channels. The UV-B channel (300–560 nm) employs a 2,048 × 4,102-pixel CCD as its detector.

1.1.17 Noise, Uncertainties, Errors, Precision and Accuracy

Noise is also often called the 'measurement uncertainty' or 'error'. The latter term though implies that the investigator is somehow to 'blame' for the problem. Unless the investigator however has made some sort of mistake (e.g., running a cooled detector at the wrong temperature) then the noise is usually intrinsic to the source and/or to the transmission path and/or to the measuring or detecting instrument. It may additionally be introduced or worsened during data reduction and analysis, but in that case, it *is* an error and the reduction/analysis process should be improved.

In the absence of noise any detector would be capable of detecting any source, howsoever faint. Noise, though, is *never* absent and generally provides the major limitation on detector performance. A minimum signal-to-noise ratio of unity is required for reliable detection. However, most research work (see also later in this section) requires signal-to-noise ratios of at least 10 and preferably 100 or 1,000.

Noise sources in PMTs and CCDs have already been mentioned and the noise for an un-illuminated detector (dark signal) is a part of the definitions of DQE, NEP, D* and dynamic range (see previous discussion). Now we must look at the nature of detector noise in more detail.

We may usefully separate noise sources into four main types: *intrinsic noise* (i.e., noise originating in the detector), *signal noise* (i.e., noise arising from the character of the incoming signal, particularly its quantum nature), *external noise* such as spurious signals from cosmic rays and so on and *processing noise*, arising from amplifiers and so on used to convert the signal from the detector into a usable form.

We may generally assume processing noise to be negligible in any good detection system.

Likewise, external noise sources should be reduced as far as possible by external measures. Thus, an IR detector should be in a cooled enclosure and be allied to a cooled telescope to reduce thermal emission from its surroundings. Similarly, a PMT used in a photon-counting mode should employ a discriminator to eliminate the large pulses arising from Ĉerenkov radiation from cosmic rays and so on.

In the NIR, recent developments in fibre-optics hold out the hope of almost completely suppressing the sky background noise. Atmospheric emission from the OH molecule forms about 98% of background noise in the NIR, and between the lines, the sky is very clear. The lines, though, are numerous and closely spaced, so that existing filters cannot separate the clear regions from the emission regions. However, by manufacturing a fibre whose refractive index varies rapidly along its length in a sinusoidal manner, a Bragg grating (Section 1.3) can be formed. Such a grating produces a narrow absorption band at a wavelength dependent upon the spacing of the refractive index variations. The absorption band can thus be tuned to centre on one of the OH emission lines' wavelengths. By making the refractive index variations aperiodic many absorption bands can be produced, each aligned with an OH emission line. In this way, Gemini Near infrared OH Suppression IFU System (GNOSIS)[55] was developed to feed the Infrared Imager and Spectrograph (IRIS2) spectrograph on the 3.9-m AAT with the partial suppression of 103 OH lines between 1.43 and 1.7 μm. The follow-on instrument, PRAXIS, which had first light in 2018 feeding a purpose-built seven-field IFU spectrograph (Section 4.2.4), improves on GNOSIS' performance, with the OH lines being suppressed by a factor of ~8. The significant improvement in the resulting spectra may clearly be seen in Figure 1.25. The planned Maryland OH Suppression IFU System (MOHSIS) for the 4.3-m Discovery Channel Telescope should suppress some 300 OH lines between 1.0 and 1.7 μm.

Thus, we are left with intrinsic and signal noise, which need to be considered further.

[55] Greek for 'knowledge'. The instrument was not, in the end, used on the Gemini telescope.

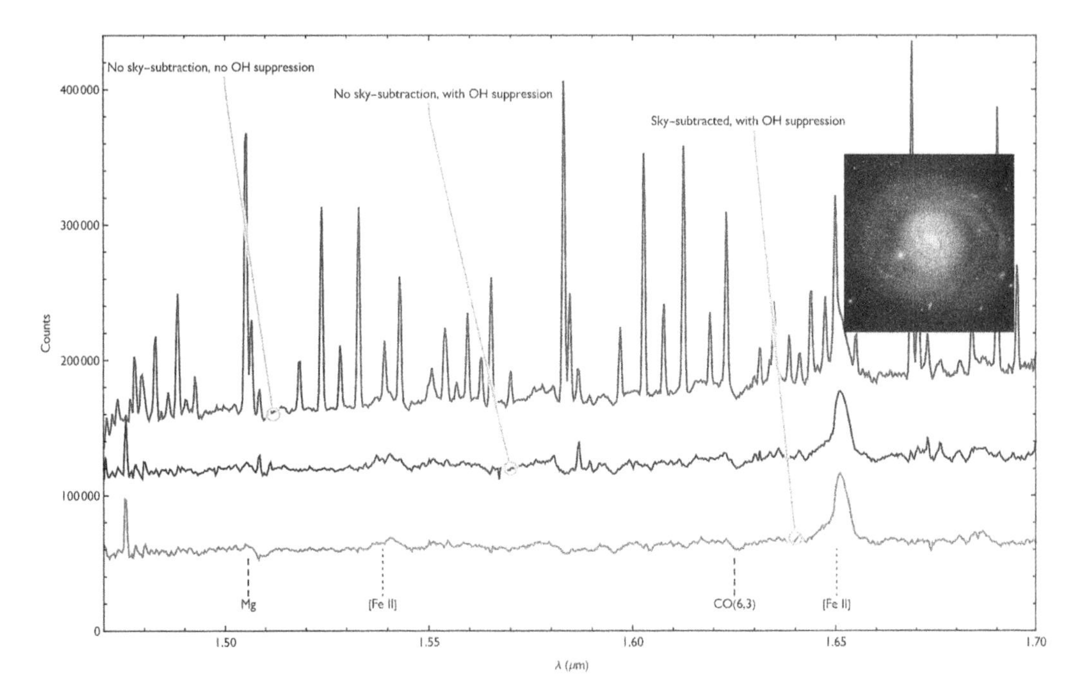

FIGURE 1.25 PRAXIS spectra for the Seyfert Galaxy M77 (NGC1068 – see inset image). Top graph, The raw spectrum dominated and largely obscured by the strong terrestrial atmospheric hydroxide (OH) emission lines. Centre graph, The same spectrum with PRAXIS operating to suppress the OH lines. The forbidden emission lines of ionised iron (marked 'Fe II'), which are characteristics of Seyfert galaxies are now much more clearly seen. Bottom graph, The same spectrum as the centre graph with the non-OH line background noise removed. (Courtesy of Simon Ellis, Macquarie University for PRAXIS graphs and ESO for inset galaxy image.)

1.1.17.1 Intrinsic Noise

Intrinsic noise in PMTs has already been discussed and arises from sources such as variation in photoemission efficiency over the photocathode, thermal electron emission from the photocathode and dynode chain, etc. In solid-state devices, intrinsic noise comes from four sources.

Thermal noise, also known as Johnson or Nyquist noise (see Section 1.2), arises in any resistive material. It is due to the thermal motion of the charge carriers. These motions give rise to a current, whose mean value is zero, but which may have non-zero instantaneous values. The resulting fluctuating voltage is given by Equation 1.65.

Shot noise[56] arises from the quantum nature of the signal (usually the signal is either a flow of electrons or of photons – for the latter see signal noise, see Section 1.1.17.2). It occurs, for example, in junction devices and is due to variation in the diffusion rates in the neutral zone of the junction because of random thermal motions. The general form of the shot noise current is

$$i = \left(2eI\Delta f + 4eI_0\Delta f\right) \tag{1.5}$$

where e is the charge on the electron, Δf is the measurement frequency bandwidth, I is the diode current, and I_0 is the reverse bias or leakage current. When the detector is reverse-biased, this equation simplifies to

$$i = \left(2eI_0\Delta f\right) \tag{1.6}$$

[56] The name arises from the sounds made by an uncoordinated barrage of shooting. Also called 'quantum noise' or the 'quantum limit'.

g-r noise (generation-recombination) is caused by fluctuations in the rate of generation and recombination of thermal charge carriers, which in turn leads to fluctuations in the device's resistivity. The g-r noise has a flat spectrum up to the inverse of the mean carrier lifetime and then decreases roughly with the square of the frequency.

Flicker noise, or $1/f$ noise, occurs when the signal is modulated in time, either because of its intrinsic variations or because it is being 'chopped' or 'nodded' (i.e., the source and background or comparison standard are alternately observed). The mechanism of flicker noise is unclear but its amplitude follows an f^{-n} power spectrum where f is the chopping frequency and n lies typically between 0.75 and 2.0. This noise source may obviously be minimised by increasing f. With large arrays of IR detectors now generally available, rapidly chopping between the detector alternately observing an astronomical source and an artificial calibration source is becoming less important, but slower speed nodding between the source and the nearby sky background is still widely needed.

The relative contributions of these noise sources are shown in Figure 1.26.

1.1.17.2 Signal Noise

Noise can be present in the signal for a variety of reasons. One obvious example is background noise. The source under observation will generally be superimposed upon a signal from the sky as a result of scattered terrestrial light sources, scattered starlight, diffuse galactic emission, zodiacal light, microwave background radiation, etc. The usual practice is to reduce the importance of this noise by measuring the background and subtracting it from the main signal. Often the source and its background are observed in quick succession (chopping or nodding, see *flicker noise* and Section 3.2). Alternatively, there may only be measurements of the background at the beginning and end of an observing run. In either case, some noise will remain because of fluctuations of the background signal about its mean level. This noise source reduces as the resolution of the telescope improves. If the resolution is 1″, then a point source has to have an energy equal to that coming from an arc-second squared of the background sky to have a signal-to-noise ratio of unity. But if the resolution were to be 0.1″, then the same point source would have a signal-to-noise ratio of 100 because it is only 'competing' with 0.01 square arc-seconds of the background. Since the light grasp of a telescope increases as the diameter squared, for *diffraction-limited* telescopes, the signal-to-noise ratio for point sources thus improves as D^4 (see discussion below on real-time atmospheric compensation).

Noise also arises from the quantum nature of light. At low signal levels, photons arrive at the detector sporadically. A Poisson distribution gives the probability of arrival, and this has a standard

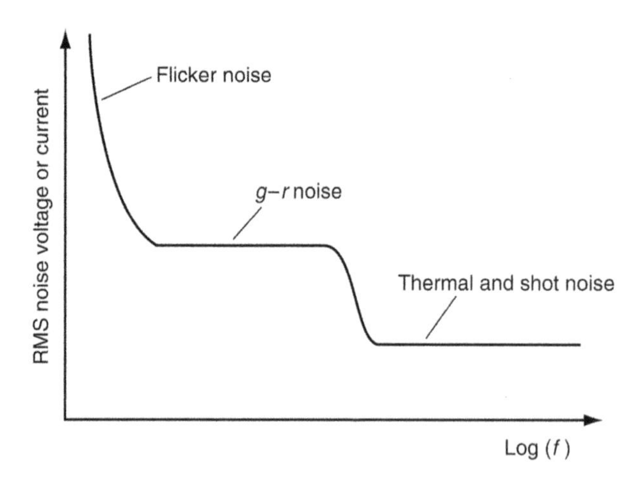

FIGURE 1.26 Relative contributions of various noise sources (schematic). g-r, generation-recombination; RMS, Root-Mean-Square.

deviation (see below) of $\sqrt{n}$ (where n is the mean number of photons per unit time). Thus, the signal will fluctuate about its mean value. To reduce the fluctuations to less than $x\%$, the signal must be integrated for $10^4/(nx^2)$ times the unit time. At high photon densities, photons tend to cluster more than a Poisson distribution would suggest because they are subject to Bose-Einstein statistics. This latter noise source may dominate at radio wavelengths (Section 1.2) but is not normally of importance over the optical region.

1.1.17.3 Digitisation

Signals are digitised in two ways, signal strength and time. The effect of the first is obvious; there is an uncertainty (i.e., noise) in the measurement corresponding to plus or minus half the measurement resolution. The effect of sampling a time varying signal is more complex. The well-known sampling theorem (Section 2.1) states that the highest frequency in a signal that can be determined is half the measurement frequency. Thus, if a signal is bandwidth-limited to some frequency, f, then it may be completely determined by sampling at $2f$ or higher frequencies. In fact, sampling at higher than twice the limiting (or Nyquist) frequency is a waste of effort; no further information will be gained. However, in a non-bandwidth-limited signal, or a signal containing components above the Nyquist frequency, errors or noise will be introduced into the measurements through the effect of those higher frequencies. In this latter effect, or 'aliasing' (see Section 1.2) as it is known, the beat frequencies between the sampling rate and the higher frequency components of the signal appear as spurious low frequency components of the signal.

1.1.17.4 Errors and Uncertainties in Data Reduction, Analysis, and Presentation

One source of error (not noise) arises if the equipment is in some way faulty, so that the results obtained from it are inaccurate. Clearly, equipment that is known to be faulty will not normally be used unless there is some overriding reason to do so. Thus, instruments on board a spacecraft may degrade over time through the effects of radiation but continue to be used because there is no alternative. In such circumstances it may be possible to compensate for the fault, perhaps by calibrating the instrument's later results against data obtained previously or comparing the faulty data with data from another information channel. An example of another type of common fault is a prism-based spectrograph. The dispersion of such a spectrograph is non-linear (Section 4.1), but the non-linearity is anticipated in the design of the instrument through the provision of the comparison spectrum, which enables the correct wavelengths of lines in the spectrum still to be determined.

Faults of this type are called 'systematic errors'. If the error is well-known and understood (as with the prism spectrograph), then its effects may be corrected during data reduction. The systematic error does not then form a part of the uncertainty in a measurement. However, the correction of the systematic error itself contains uncertainties, and these *will* contribute to the uncertainty in the actual measurement. Of more importance are systematic errors that are not understood well or maybe not even known to exist. These errors will then contribute to the uncertainty in the measurement. The possibility of the presence of this latter type of error is the main reason why scientists seek to confirm measurements by a second approach that is independent of the manner in which the initial data were obtained.

The presence of unknown systematic errors in an instrument is one of many reasons why its measurements may be inaccurate, even though they may also be precise. The difference between accuracy and precision is illustrated in Figure 1.27. An inaccurate but precise (or consistent) piece of apparatus always produces the same wrong result with little scatter in the measurements because the problem always affects the measurements in the same way. An accurate but imprecise instrument simply has large uncertainties in its results arising from random processes.

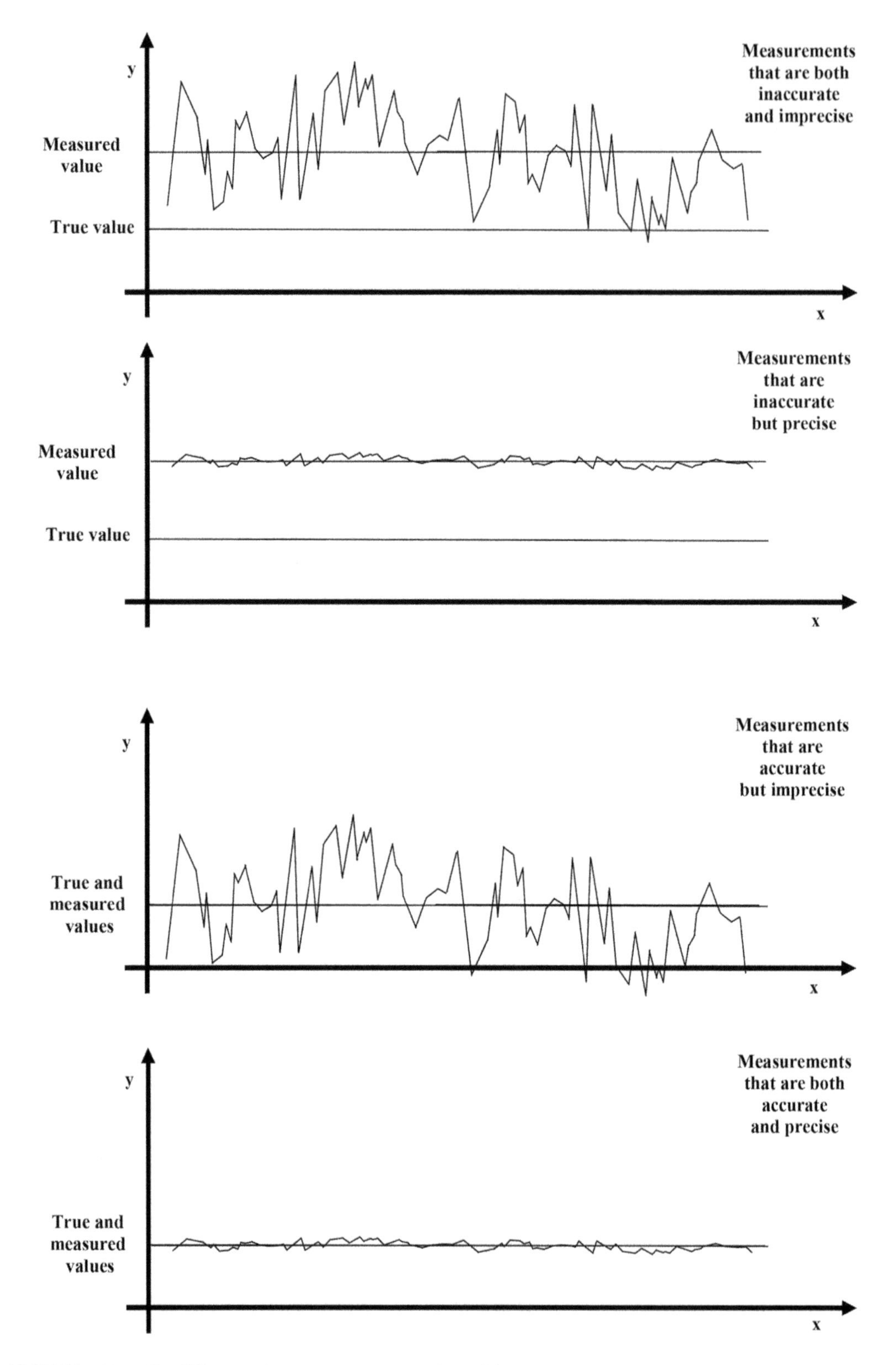

FIGURE 1.27 The difference between accuracy and precision.

It is also possible for errors to arise through a faulty theoretical understanding of the situation. Ptolemy's geocentric model of the solar system for example, with its epicycles, deferents, eccentrics and equants, was the most successful scientific theory ever devised, in that it continued to be accepted as correct and to make useful predictions of planetary positions for around one and a half millennia. Yet we now know that physically, the idea was completely incorrect. This book however is (mostly) concerned with practical astronomy, and so further consideration of theoretical misunderstandings will be left for the reader to pursue elsewhere.

In presenting results – whether they are simply raw data or the outcome from a complex and sophisticated piece of analysis and synthesis – the effects of noise, uncertainty and/or error on those results should always be included. Sometimes, especially in unique or unexpected situations such as the arrival of neutrinos from the 1987 Large Magellanic Cloud (LMC) supernova, it may be difficult to quantify some or all the sources of uncertainty, but they should always be estimated or discussed. More usually, the uncertainties will be reasonably well understood and the measurement is then given in the form

$$x \pm \Delta x \tag{1.7}$$

where x is the result and Δx is its uncertainty.

When repeated measurements of the same, unchanging, phenomenon can be obtained, then their grouping often approximates towards a Gaussian (or Normal) distribution. The Gaussian takes the form of a bell-shaped curve (Figure 1.28), which when normalised to have unit total area under the curve, is given mathematically by

$$P(x) = -\frac{1}{\sigma\sqrt{2\pi}} e^{(x-\mu)^2 / 2\sigma^2} \tag{1.8}$$

where $P(x)$ is the probability of value, x, μ is the value of x at the peak of the distribution and σ is a constant discussed below.

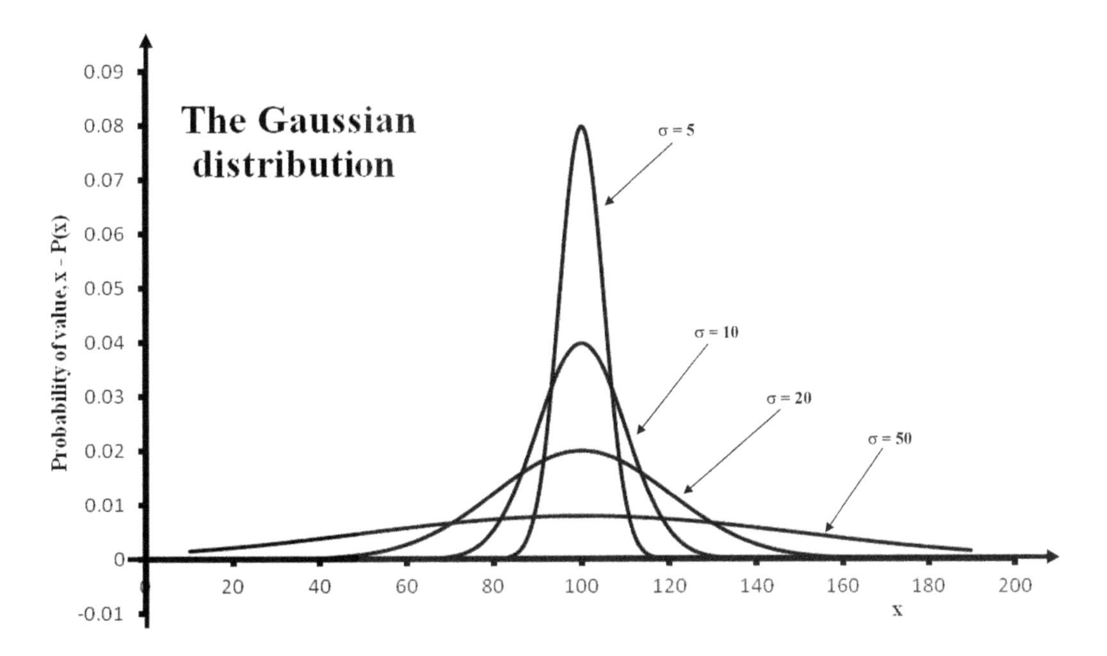

FIGURE 1.28 Gaussian distributions for a value of μ of 100 and for values of σ of 5, 10, 20, and 50.

TABLE 1.5
Areas Under the Gaussian Curve

Lying Between Values of x	Relative Area Under the Curve (%) Between the Values of x	Relative Area Under the Curve (%) Outside the Values of x
$\mu \pm 0.5\sigma$	38.293	61.707
$\mu \pm 0.67449\,\sigma$	50.000	50.000
$\mu \pm \sigma$	68.269	31.731
$\mu \pm 1.5\sigma$	86.639	13.361
$\mu \pm 2\sigma$	95.450	4.5500
$\mu \pm 2.5\sigma$	98.758	1.2419
$\mu \pm 3\sigma$	99.730	2.6998×10^{-1}
$\mu \pm 4\sigma$	99.994	6.3343×10^{-3}
$\mu \pm 5\sigma$	100.000	5.7330×10^{-5}
$\mu \pm 6\sigma$	100.000	1.9731×10^{-7}
$\pm \infty$	100 (exact)	0 (exact)

The constant, σ, is called the 'standard deviation of the curve'. From the mathematical definition of the Gaussian distribution, σ's value can be seen to equal the half-width of the curve between the points where the height of the curve is $e^{-0.5}$($\approx$60.7%) of its peak value. The areas under the curve between points equidistant from the curve's peak ($x = \mu$) on each side of that peak are listed in Table 1.5.

In a real scientific investigation, the number of independent measurements of a quantity will be limited, and their actual distribution will only approximate towards the shape of a Gaussian. The standard deviation in these circumstances is given by[57]

$$\sigma = \sqrt{\frac{\sum_{i=1}^{n} (\bar{x} - x_i)^2}{n - 1}} \tag{1.9}$$

where n is the number of measurements, $\bar{x}$ is the mean value of the measurements and x_i is the ith measurement.

The interpretation of the standard deviation is also now approximate, but is likely to be based upon the values given in Table 1.5. The area under the curve between a given range of values of x gives the probability that x will be found between those values. Thus, we will usually find that about 68%[58] of the individual measurements will lie between $\bar{x} \pm \sigma$, about 95% between $\bar{x} \pm 2\sigma$ and about 99% between $\bar{x} \pm 2.5\sigma$. Only about one measurement in two million might be expected to lie outside the range, $\bar{x} \pm 5\sigma$. These figures form the basis by which results of experimental data are usually presented in refereed journals. The value of Δx in Equation 1.8 is usually taken as 2.5σ[59] If this is not the case, then the definition of the uncertainty will need to be specified (and it's no bad practice *always* to specify the definition being used).

[57] An approximation to this equation that is useful when the number of measurements is large replaces the '$n - 1$' with 'n'. Many calculators and computers offer both versions. If working with small numbers of measurements, then always ensure that you select the correct equation.

[58] The reader should always bear in mind that statistics is a house built on shifting sand. Thus, not only is the interpretation of the values in Table 1.5 approximate, but the value of σ obtained from Equation 1.9 also has its own uncertainties. These are typically ±10% when $n \approx 100$, ±1% when $n \approx 10{,}000$ and ±0.1% when $n \approx 1{,}000{,}000$. Rarely, therefore, can the value of σ, when based upon experimental measurements, justifiably be given to more than *one* significant figure.

[59] Another measure of uncertainty is called the 'Probable Error' and is given by 0.67449σ. It gives equal chances of a result lying inside or outside the range $\mu \pm 0.67449\sigma$. Its use may be encountered, but it is not generally a preferred way of presenting results.

Recently and especially with regard to the detection of gravitational wave events (Section 1.6), measurement uncertainties have been specified as the false alarm rate (FAR).[60] The FAR is the length of time over which the phenomenon, event and /or measurement might be expected to occur because of a false detection (i.e., to arise from processes other than those leading to the 'real thing'). Gravitational wave events have to be detected by searching for the signature variations in the frequency and/or intensity output of the detector arising from specific types of events, such as neutron star–black hole mergers or black hole-black hole mergers. Such patterns could also arise from purely random variations in the output, so giving rise to a false event. To date, events picked up by the Advanced Laser Interferometer Gravitational-wave Observatory (AdvLIGO or aLIGO) and Advanced Virgo (AdvVirgo or aVirgo)[61] instruments and accepted as being real have had FARs ranging from 1 year (marginal) to more than 10^{25} years (you can bet the farm on this one).

When a number of measurements have already been made and an additional one is obtained, then the significance of that new result may be examined by using the mean and standard deviation obtained from the previous set of results. If, for example, a variable star is being observed and a single new measurement of its brightness differs from the previously obtained mean value, then the question will arise 'has the star really changed in brightness or is the difference just the result of the measurement uncertainties?' If the new measurement differs from the mean of the old measurements by more than $\pm 2\sigma$, then the difference is conventionally termed to be 'significant', if by more than $\pm 2.5\sigma$, then the difference is termed to be 'highly significant'. The chances that the star has *not* changed in brightness are then around 5% and 1%, respectively. A result at the 5σ level (a chance that the star has *not* changed of around 0.00006%) is likely to be as close to being certain as practising scientists can hope to achieve during their lifetimes!

If a number of measurements of a phenomenon have been obtained, then it is usually the mean value of those measurements that is used in subsequent analysis or published as the result. The uncertainty in the mean value of a set of individual measurements is called the 'standard error of the mean' and commonly symbolised as, S, with

$$S = \frac{\sigma}{\sqrt{n}} \tag{1.10}$$

and the result should be given in the form

$$\bar{x} \pm S \tag{1.11}$$

Equation 1.10 is the basis of the common assertion that the uncertainty in a result improves as the inverse square root of the number of measurements. However, this rule only applies when then uncertainties are random and approximate to a Gaussian distribution. There are other circumstances where the rule will be misleading, in particular, systematic errors are unlikely to improve much, if at all, with repeated measurements.

The practising researcher will need to understand far more about handling uncertainties in measurements then it is possible to include in this brief introduction. For example, even in the early years of an undergraduate science degree, the student will need to know how uncertainties propagate through formulae, how to compare two average values (Student's t-test), how to find the best (linear) formula to fit a set of measurements containing uncertainties (linear regression), whether the variations in two sets of data correlate with each other (correlation coefficient) and so on. However, that is left to be found in other sources (see Bibliography – including the author's book *Telescopes and Techniques*).

[60] Not to be confused with the False Alarm Ratio, which has the same initials but is the ratio of the numbers of false alarms to real events.

[61] Named for the Virgo cluster of galaxies and sited about 13 km southeast of the town of Pisa (of leaning tower fame).

1.1.18 Telescopes

The *apologia* for the inclusion of telescopes in the chapter titled detectors has already been given, but it is perhaps just worth repeating that for astronomers the use of a detector in isolation is almost unknown. Some device to restrict the angular acceptance zone and possibly to increase the radiation flux as well is almost invariably a necessity. In the optical region, this generally implies the use of a telescope and so they are discussed here as a required ancillary to detectors.

1.1.18.1 Telescopes from the Beginning

The telescope may justly be regarded as the symbol of the astronomer, for notwithstanding the heroic work done with the naked eye in the past and the more recent advances in 'exotic' regions of the spectrum; our picture of the Universe is still to a large extent based upon optical observations made through telescopes.

The development of the optical telescope has been reviewed in some detail in fifth and sixth editions of this book. New developments such as gravitational wave and dark matter detectors (Sections 1.6 and 1.7), however, are of great significance and deserve their much-increased coverage in this edition. Unfortunately, this means that some other things have to disappear to keep the size of this book within bounds. Regretfully, the developmental history of optical telescopes is one of those things that has had to go. A timeline is now included (Table 1.6, Figures 1.29 and 1.30), which may help a new reader, but anyone wishing to know more details on the subject will need to consult other sources, including the previous editions of this book.

TABLE 1.6
Timeline of Optical Telescope Development

Date	Event
640 BCE	First known example of a lens – Rock crystal – found at Nineveh
424 BCE	First known written account of a lens – Burning glasses mentioned in Aristophanes' play '*The Clouds*'.
c. 10th century	Rock crystal lenses found in Viking graves in Sweden.
c. 970	Ibn Dahl – First known statement of the law of refraction
c. 1000	Ibn al-Haitham – First known account of the optics of the eye
c. 1230	Bishop Robert Grosseteste gives an account of an optical device which might be a telescope.
1266	Roger Bacon gives an account of an optical device which might be a telescope.
c. 13th century	Convex lenses used as magnifiers
c. 1350	First reading spectacles (converging lenses) produced
c. 1450	First distance spectacles (diverging lenses) produced
c. Early 16th century	Domestic mirrors produced using a reflective coating of metal (tin-mercury amalgam) on a glass support.
1571	Thomas Digges gives an account of an optical device which might be a telescope and which had been invented by his father (Leonard).
c. 1575	Giambattista della Porta claims (dubiously) to have invented an optical device which might be a telescope.
1608	Hans Lipperhey applies for a patent on an optical device that is undoubtedly a telescope.
1608	Jacob Metius applies for a patent on an optical device that is probably a telescope, a few weeks after Lipperhey's application.
1608/1609	Evidence that Zacharias Janssen designed and perhaps built a telescope.
1609	Galileo Galilei invents from first principles and constructs a telescope which used a convex lens as the objective and a concave lens as the eyepiece. Now known as the Galilean telescope design, it is probably the same optical design as that used by Lipperhey.

(*Continued*)

TABLE 1.6 (*Continued*)
Timeline of Optical Telescope Development

Date	Event
1611	Johannes Kepler invents the astronomical telescope (a refractor with two convex lenses, producing inverted images).
1632	Bonaventura Cavalieri gives an account of an optical device which might be a reflecting telescope.
1637	Father Marin Mersenne invents optical beam compressors – whose optical designs are essentially identical to those of the Cassegrain and Gregorian reflecting telescopes.
1663	James Gregory invents the Gregorian design of reflecting telescope, but his prototype never worked.
1668	Sir Isaac Newton invents and builds a working model of the Newtonian design of reflecting telescope (Figure 1.29). The mirrors are fabricated from speculum metal.
c. 1670	Johannes Hevelius observes using a telescope 43-m long (known as an 'aerial telescope') to reduce the effects of chromatic aberration.
1672	Guillaume (or Laurent, or Jacques, or Nicolas) Cassegrain invents the Cassegrain reflecting telescope design – the basic design used for most major astronomical telescopes for the last three centuries.
1729	Chester Moor Hall invents the achromatic lens design.
1758	John Dollond produces the first commercially available achromatic lenses.
1789	Sir William Herschel builds his 1.2-m reflecting telescope using his own off-axis (Herschellian) design.
1835	Justus von Leibig invents a chemical process for depositing a thin reflecting layer of silver onto a glass substrate.
1845	Sir William Parsons builds his 1.8-m reflecting telescope – the last major telescope to utilise speculum metal mirrors.
1857	Léon Foucault builds the first telescope mirrors which use Leibig's metal-on-glass techniques for the reflecting layer.
1897	1-m Yerkes refractor built
1917	Hooker 2.5-m reflecting telescope built
1948	The Palomar Hale Telescope built. With its single monolithic 5-m mirror it would be the world's largest (effectively working) optical telescope for the next five decades.
1979 (and onwards)	The Multi-Mirror Telescope (MMT) built. It used six 1.8-m mirrors to emulate the performance of a single 4.7-m telescope. Although later replaced by a single 6.5-m mirror, the original six-mirror instrument foreshadowed the later development of multi-segment mirror telescopes.
1990	The 2.4-m Hubble Space Telescope (HST) launched.
1993	Keck 1 telescope built. This is the first major telescope to use a segmented primary mirror. It has 36 hexagonal segments making up the equivalent of a single 10-m mirror.
1999	The four 8.2-m reflector telescope units comprising the Very Large Telescope (VLT) built (Figure 1.30).
2004	The Large Binocular Telescope (LBT) built. Its two 8.4-m mirrors are the largest monolithic mirrors fabricated to date. It is also the world's largest optical telescope at the time of writing, with the two mirrors being equivalent to a single 11.9-m mirror.
2021	Launch of the 6.5-m James Webb Space Telescope (JWST) due.
2025	First light for the 39.3-m Extremely Large Telescope (ELT) planned. The mirror will comprise 798 hexagonal segments (Figure 1.30).
2027	First light for the Giant Magellan Telescope (GMT) planned. It will use seven 8.4-m monolithic mirrors giving a collecting area equivalent to a single 21.6-m mirror.
2027	First light for the 30-m Telescope (Thirty Metre Telescope, TMT) planned. The mirror will comprise 492 hexagonal segments.
2030s	One or more of the Habitable Exoplanet Observatory (HabEx), Large UV Optical IR (LUVOIR), Lynx and Origins space telescope concepts could be operational.

FIGURE 1.29 A full-sized replica of Newton's first telescope. It was to be found facing the visitors' gallery of the 2.5-m Isaac Newton telescope during the 1970s when that instrument was still in the United Kingdom. At the bottom of the case (and not visible to the visitors) can be seen the instruction 'When main telescope fails – Break glass'. (Courtesy of C. R. Kitchin, Copyright © 1977.)

FIGURE 1.30 Left, The 8.2-m monolithic primary mirror of 'Yepun' (Mapuche language word for the Evening Star, Venus) – one of the four main instruments comprising European Southern Observatory's (ESO's) Very Large Telescope (VLT). (Courtesy of ESO/G. Huedepohl). Right, An artist's concept of the 39.3-m main mirror of the future Extremely Large Telescope (ELT) clearly showing the hexagonal segments from which it is made. (Courtesy of ESO/L. Calçada/Ace Consortium.)

1.1.18.2 Optical Theory

Before looking further at the designs of telescopes and their properties, it is necessary to cover some of the optical principles upon which they are based. It is assumed that the reader is familiar with the basics of the optics of lenses and mirrors; such terminology as focal ratio, magnification, optical axis, and so on with laws such as Snell's law, the law of reflection and so on and with equations such as the lens and mirror formulae and the lens maker's formula. If it is required, however, any basic optics book or reasonably comprehensive physics book will be suffice to provide this background.

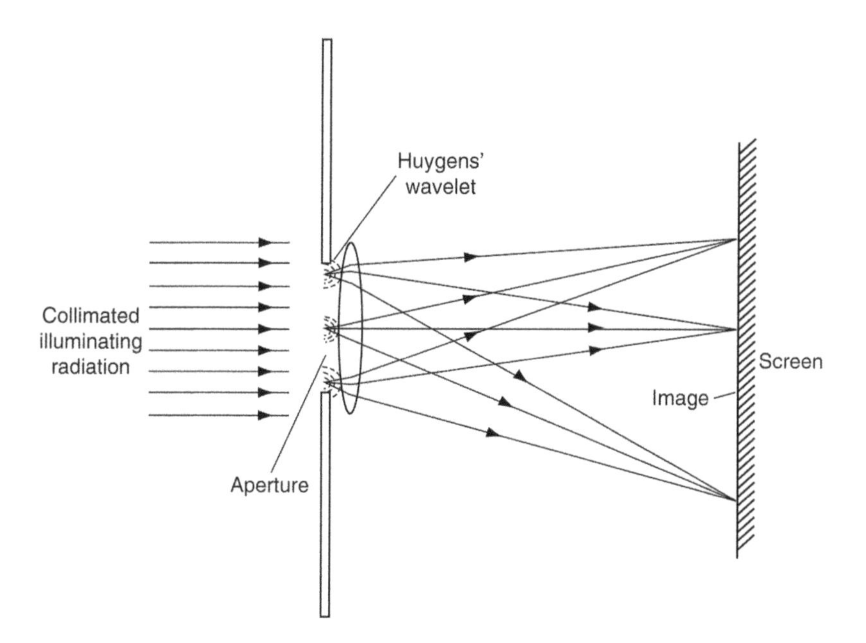

FIGURE 1.31 Fraunhofer diffraction at an aperture.

We start by considering the resolution of a lens (the same considerations apply to mirrors, but the light paths and so on are more easily pictured for lenses). That this is limited at all is due to the wave nature of light. As light passes through any opening it is diffracted and the wavefronts spread out in a shape given by the envelope of the Huygens' secondary wavelets (Figure 1.31). Huygens' secondary wavelets radiate spherically outwards from all points on a wavefront with the velocity of light in the medium concerned. Three examples are shown in Figure 1.31. Imaging the wavefront after passage through a slit-shaped aperture produces an image whose structure is shown in Figure 1.32.

The fringes shown in Figure 1.32 are due to interference between waves originating from different parts of the aperture. The paths taken by such waves to arrive at a given point will differ, and so the distances that they have travelled will also differ. The waves will be out of step with each other to a greater or lesser extent depending upon the magnitude of this path difference. When the path difference is a half wavelength or an integer number of wavelengths plus a half wavelength, then the waves will be 180° out of phase with each other and so will cancel out. When the path difference is

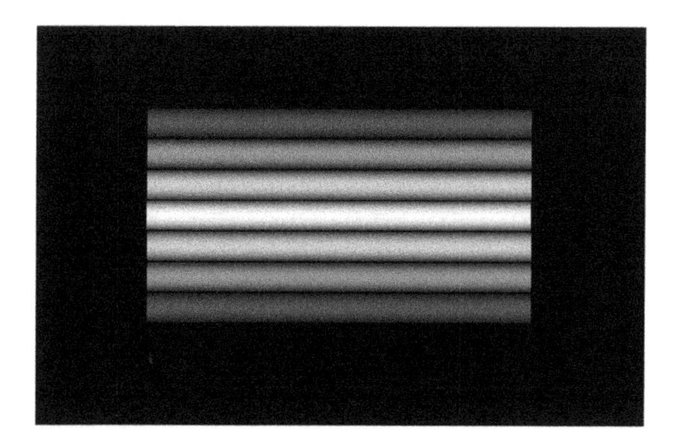

FIGURE 1.32 Image of a narrow slit.

a whole number of wavelengths, the waves will be in step and will reinforce each other. Other path differences will cause intermediate degrees of cancellation or reinforcement.

The central maximum arises from the reinforcement of many waves. Whilst the first minimum occurs when the path difference for waves originating at opposite edges of the aperture is one wavelength, then for every point in one half of the aperture there is a point in the other half such that their path difference is half a wavelength and all the waves cancel out completely. The intensity at a point within the image of a narrow slit may be obtained from the result that the diffraction pattern of an aperture is the power spectrum of the Fourier transform (Section 2.1.1) of its shape and is given by

$$I_\theta = I_0 \frac{\sin^2(\pi d \sin\theta / \lambda)}{(\pi d \sin\theta / \lambda)^2} \tag{1.12}$$

where θ is the angle to the normal from the slit, d is the slit width and I_0 and I_θ are the intensities within the image on the normal and at an angle θ to the normal from the slit, respectively. With the image focused onto a screen at distance F from the lens, the image structure is as shown in Figure 1.33, where d is assumed to be large when compared with the wavelength λ. For a rectangular aperture with dimensions $d \times l$, the image intensity is similarly

$$I(\theta,\varphi) = I_0 \frac{\sin^2(\pi d \sin\theta / \lambda)}{(\pi d \sin\theta / \lambda)^2} \frac{\sin^2(\pi l \sin\varphi / \lambda)}{(\pi l \sin\varphi / \lambda)^2} \tag{1.13}$$

where φ is the angle to the normal from the slit measured in the plane containing the side of length, l. To obtain the image structure for a circular aperture, which is the case normally of interest in astronomy, we must integrate over the surface of the aperture. The image is then circular with concentric light and dark fringes. The central maximum is known as 'Airy's disc' after George

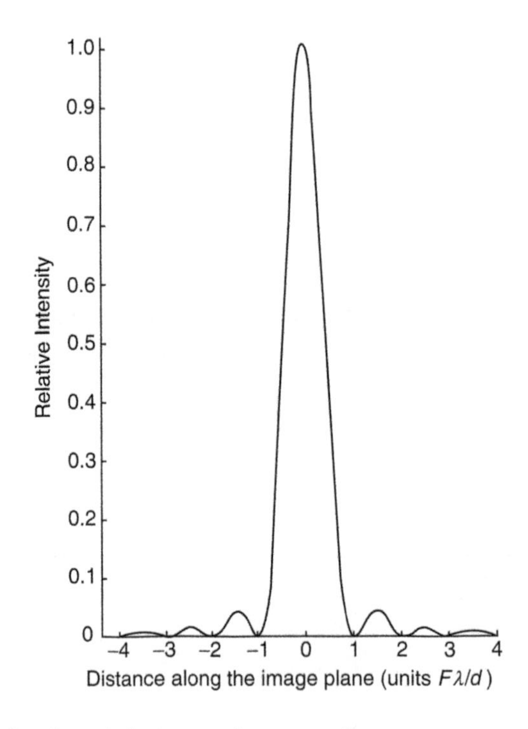

FIGURE 1.33 Cross section through the image of a narrow slit.

Biddell Airy, the Astronomer Royal who first succeeded in completing the integration.[62] The major difference from the previous case occurs in the position of the fringes and, of course, their circular shapes. The computer-generated image of two stars shown in Figure 1.35 will give the reader an idea of the structure of the image through a circular aperture. The dark fringes occur at angles from the centres of the Airy disks of

$$\theta \approx \frac{1.220\lambda}{d}, \quad \frac{2.233\lambda}{d}, \quad \frac{3.238\lambda}{d}, \ldots \tag{1.14}$$

where d is the diameter of the aperture.

Many of the recently constructed large telescopes have primary mirrors made up of numerous hexagonal segments (Figure 1.30) or of several circular monolithic mirrors (e.g., Large Binocular Telescope [LBT]; two 8.4-m mirrors, Giant Magellan Telescope [GMT]; seven 8.4-m mirrors, etc.). The matrix of hexagonal segments may approximate to the shape of a circle, or an oval or a hexagon but will still have a 'saw-tooth' edge. The diffraction patterns from point sources in these cases will differ slightly from those for simple circular apertures, but not by much. There will also be diffraction effects arising from the edges of the hexagonal segments themselves and for all reflecting telescopes from the secondary and tertiary mirrors and their supports and/or from equipment such cameras, photometers or spectroscopes that may obstruct parts of the light beam.

Most of these diffraction effects will show up as minor additional features superimposed upon the basic Airy diffraction pattern (in terms of the Fourier transform, they are the high frequency components of the pattern). The 'spikes' that often appear on images of bright stars are probably the best-known example of these effects, and those spikes arise from diffraction at the edges of the supports for the secondary mirror (Figure 1.34). In further discussions in this book and without much loss of generality, the diffraction pattern in these cases will be taken to be the appropriate Airy disk for a simple circular aperture with the same collecting area as the telescope concerned.

FIGURE 1.34 Diffraction spikes on images of bright stars arising from the secondary mirror supports. (Detail from an image of the globular cluster NGC 2257 obtained with the Wide Field Imager instrument on the 2.2-m Max Planck Gesellschaft/European Southern Observatory (MPG/ESO) telescope at La Silla. (Courtesy of ESO.)

[62] This calculation is non-trivial but should be within the grasp of a second-year physical sciences undergraduate. It has been given in full in previous editions of this book but is now omitted due to pressure on the book's length.

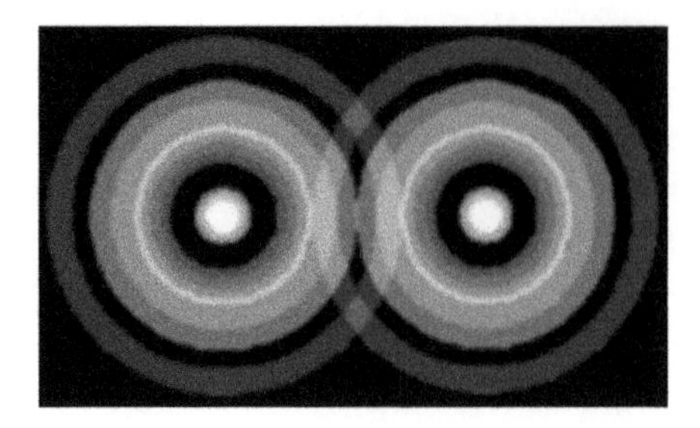

FIGURE 1.35 Image of two distant point sources through a circular aperture.

If we now consider two distant point sources separated by an angle α, then two circular diffraction patterns will be produced and will be superimposed. There will not be any interference effects between the two images because the sources are mutually incoherent and so their intensities will simply add together. The combined image will have an appearance akin to that shown in Figure 1.35. When the centre of the Airy disc of one image is superimposed upon the first minimum of the other image (and vice versa), then we have Rayleigh's criterion for the resolution of a lens (or mirror). This is the normally accepted measure of the theoretical resolution of a lens. It is given in radians by (from Equation 1.33)

$$\alpha = \frac{1.220\lambda}{d} \tag{1.15}$$

It is a convenient measure, but it is quite arbitrary. For sources of equal brightness, the image will appear non-circular for separations of about one third of α, while for sources differing appreciably in brightness, the separation may need to be an order of magnitude larger than α for them to be resolved. For visual observing, the effective wavelength is 510 nm for faint images so that the resolution of a telescope used visually is given by

$$R = \frac{0.128}{d} \tag{1.16}$$

where d is the objective's diameter in metres and R is the angular resolution in arc-seconds. An empirical expression for the resolution of a telescope of $0.116''/d$, known as Dawes' limit, is often used by amateur astronomers because the slightly better value it gives for the resolution takes some account of the abilities of a skilled observer. To achieve either resolution in practice, the magnification must be sufficient for the angular separation of the images in the eyepiece to exceed the resolution of the eye. Taking this to be an angle, β, we have the minimum magnification required to realise the Rayleigh limit of a telescope, M_m

$$M_m = \frac{\beta d}{1.220\lambda} \tag{1.17}$$

so that for β equal to three arc-minutes, which is about its average value,

$$M_m = 1300d \tag{1.18}$$

where d is again measured in metres.

Of course, most astronomical telescopes are actually limited in their resolution by the atmosphere. A 1-m telescope might reach its Rayleigh resolution on one night a year from a good observing site. On an average night, scintillation will spread stellar images to about 2″ so that only telescopes smaller than about 0.07 m can regularly attain their diffraction limit. Since telescopes are rarely used visually for serious work, such high magnifications as are implied by Equation 1.18 are hardly ever encountered today. However, some of William Herschel's eyepieces still exist and these, if used on his 1.2-m telescope, would have given magnifications of up to 8000 times.

The theoretical considerations of resolution that we have just seen are only applicable if the lens or mirror is of sufficient optical quality that the image is not already degraded beyond this limit. There are many effects that will blur the image and these are known as 'aberrations'. With one exception, they can all affect the images produced by either lenses or mirrors. The universal or monochromatic aberrations are known as the 'Seidel aberrations' after Ludwig von Seidel who first analysed them. The exception is chromatic aberration (Figure 1.36) and the related second-order effects of transverse chromatic aberration and secondary colour and these affect only lenses.

Chromatic aberration arises through the change in the refractive index of glass or other optical material with the wavelength of the illuminating radiation. Some typical values of the refractive index of some commonly used optical glasses are tabulated below (Table 1.7).

The degree to which the refractive index varies with wavelength is called the 'dispersion' and is measured by the constringence, ν

$$\nu = \frac{\mu_{589} - 1}{\mu_{486} - \mu_{656}} \tag{1.19}$$

where μ_λ is the refractive index at wavelength, λ. The three wavelengths that are chosen for the definition of ν are those of strong Fraunhofer lines: 486 nm – the F line (Hβ); 589 nm – the D lines (Na);

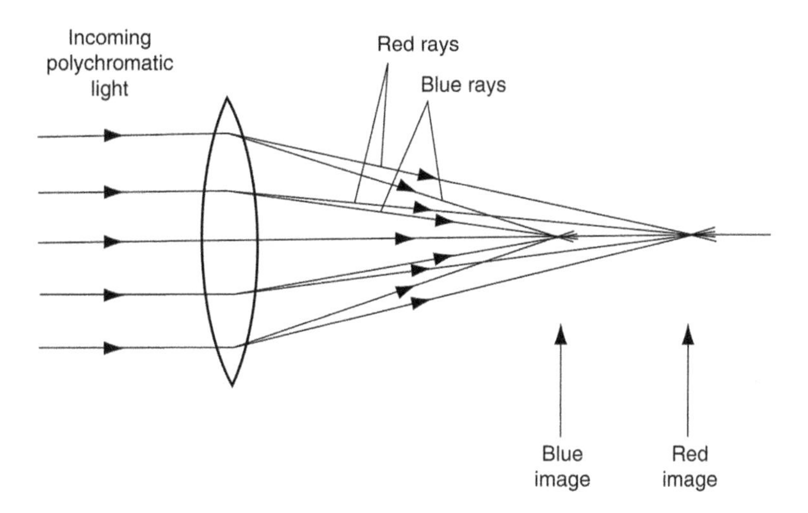

FIGURE 1.36 Chromatic aberration.

TABLE 1.7
Some Typical Values of the Refractive Index of Some Commonly Used Optical Glasses

	Refractive Index at the Specified Wavelengths				
Glass Type	**361 nm**	**486 nm**	**589 nm**	**656 nm**	**768 nm**
Crown	1.539	1.523	1.517	1.514	1.511
High dispersion crown	1.546	1.527	1.520	1.517	1.514
Light flint	1.614	1.585	1.575	1.571	1.567
Dense flint	1.705	1.664	1.650	1.644	1.638

and 656 nm – the C line (Hα). Thus, for the glasses listed previously, the constringence varies from 57 for the crown glass to 33 for the dense flint (note that the *higher* the value of the constringence, the *less* that rays of different wavelengths diverge from each other).

The effect of the dispersion upon an image is to string it out into a series of different coloured images along the optical axis (Figure 1.36). Looking at this sequence of images with an eyepiece, then at a particular point along the optical axis, the observed image will consist of a sharp image in the light of one wavelength surrounded by blurred images of varying sizes in the light of all the remaining wavelengths. To the eye, the best image occurs when yellow light is focussed because the retina is less sensitive to the red and blue light. The image size at this point is called the 'circle of least confusion'. The spread of colours along the optical axis is called the 'longitudinal chromatic aberration', while that along the image plane containing the circle of least confusion is called the 'transverse chromatic aberration'.

Two lenses of different glasses may be combined to reduce the effect of chromatic aberration. Commonly in astronomical refractors, a biconvex crown glass lens is allied to a plano-concave flint glass lens to produce an achromatic doublet. In the IR, achromats can be formed using barium, lithium and strontium fluorides, zinc sulphate or selenide and IR-transmitting glasses. In the submillimetre region (i.e., wavelengths of several hundred microns), crystal quartz and germanium can be used. The lenses are either cemented together or separated by only a small distance (Figure 1.37). Despite its name, there is still some chromatic aberration remaining in this design of lens because it can only bring two wavelengths to a common focus. If the radii of the curved surfaces are all equal, then the condition for two given wavelengths, λ_1 and λ_2, to have coincident images is

$$2\Delta\mu_C = \Delta\mu_F \tag{1.20}$$

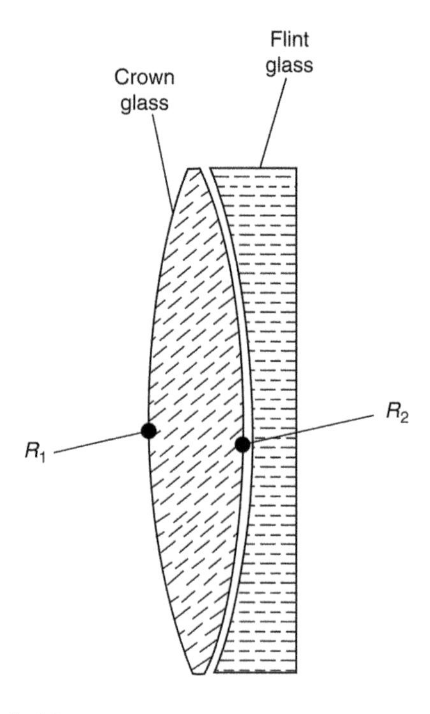

FIGURE 1.37 An achromatic doublet.

where $\Delta\mu_C$ and $\Delta\mu_F$ are the differences between the refractive indices at λ_1 and λ_2 for the crown glass and the flint glass, respectively. More flexibility in design can be attained if the two surfaces of the converging lens have differing radii. The condition for achromatism is then

$$\frac{|R_1|+|R_2|}{|R_1|}\Delta\mu_C = \Delta\mu_F \tag{1.21}$$

where R_2 is the radius of the surface of the crown glass lens that is in contact with the flint lens (NB: the radius of the flint lens surface is almost invariably R_2 as well, to facilitate alignment and cementing) and R_1 is the radius of the other surface of the crown glass lens. By a careful selection of λ_1 and λ_2 an achromatic doublet can be constructed to give tolerable images. For example, by achromatising at 486 and 656 nm, the longitudinal chromatic aberration is reduced when compared with a simple lens of the same focal length by a factor of about 30.

Nevertheless, since chromatic aberration varies as the square of the diameter of the objective and inversely with its focal length, achromatic refractors larger than about 0.25-m still have obtrusively coloured images. More seriously, if filters are used then the focal position will vary with the filter. Similarly, the image scale on array detectors will alter with the wavelengths of their sensitive regions. Further lenses may be added to produce apochromats that have three corrected wavelengths and super-apochromats with four corrected wavelengths. But such designs are impossibly expensive for telescope objectives of any size, although eyepieces and camera lenses may have eight or ten components and achieve very high levels of correction.

A common and severe aberration of both lenses and mirrors is spherical aberration. In this effect, annuli of the lens or mirror that are of different radii have different focal lengths. It is illustrated in Figure 1.38 for a spherical mirror. For rays parallel to the optical axis, it can be eliminated completely by deepening the sphere to a paraboloidal surface for the mirror. It cannot be eliminated from a simple lens without using aspheric surfaces, but for a given focal length it may be minimised. The shape of a simple lens is measured by the shape factor, q

$$q = \frac{R_2 + R_1}{R_2 - R_1} \tag{1.22}$$

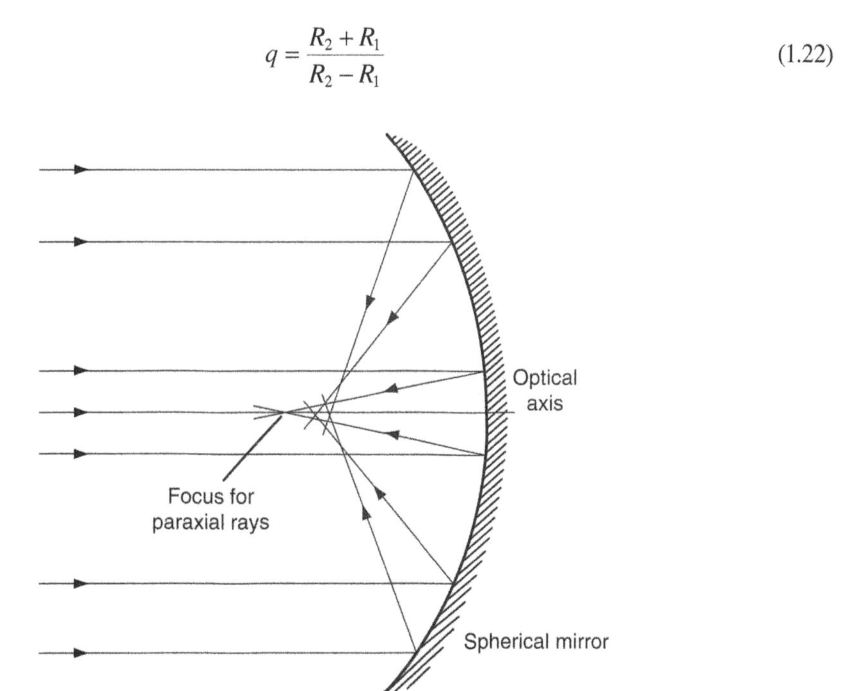

FIGURE 1.38 Spherical aberration.

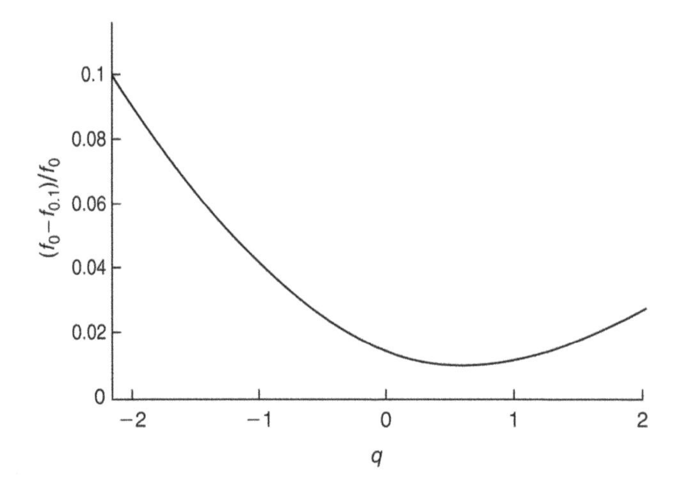

FIGURE 1.39 Spherical aberration in thin lenses. f_x is the focal length for rays parallel to the optical axis, and at a distance x times the paraxial focal length away from it.

where R_1 is the radius of the first surface of the lens and R_2 is the radius of the second surface of the lens. The spherical aberration of a thin lens then varies with q as shown in Figure 1.39, with a minimum at $q = +0.6$. The lens is then biconvex with the radius of the surface nearer to the image three times the radius of the other surface. Judicious choice of surface radii in an achromatic doublet can lead to some correction of spherical aberration while still retaining the colour correction. Spherical aberration in lenses may also be reduced by using high refractive index glass because the curvatures required for the lenses' surfaces are lessened, but this is likely to increase chromatic aberration. Spherical aberration increases as the cube of the aperture.

The deepening of a spherical mirror to a paraboloidal one to correct for spherical aberration unfortunately introduces a new aberration called 'coma'. This also afflicts mirrors of other shapes and lenses. It causes the images for objects away from the optical axis to consist of a series of circles that correspond to the various annular zones of the lens or mirror and which are progressively shifted towards or away from the optical axis (Figure 1.40). The severity of coma is proportional to the square of the aperture. It is zero in a system that obeys Abbe's sine condition

$$\frac{\sin\theta}{\sin\varphi} = \frac{\theta_p}{\varphi_p} = \text{constant} \tag{1.23}$$

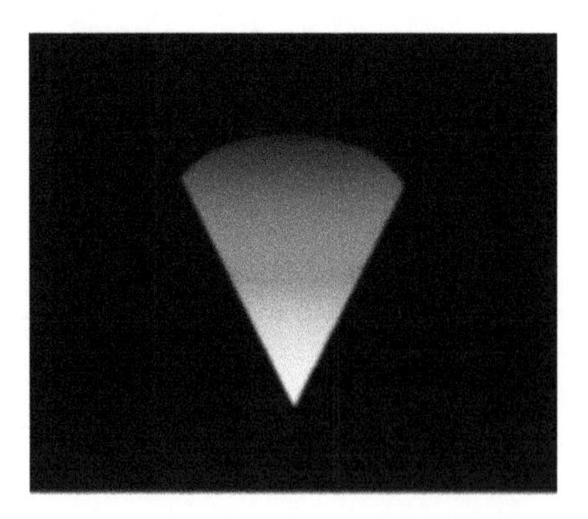

FIGURE 1.40 Shape of the image of a point source resulting from coma.

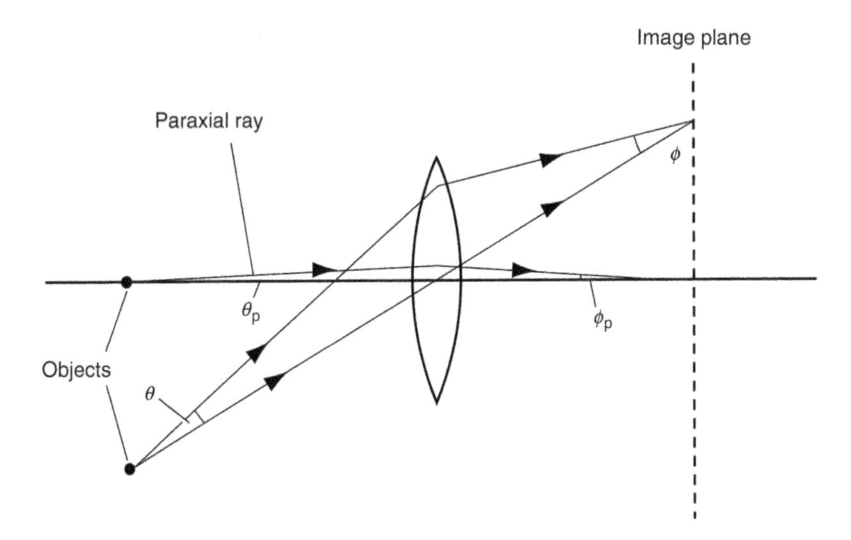

FIGURE 1.41 Parameters for Abbe's sine condition.

where the angles are defined in Figure 1.41. A doublet lens can be simultaneously corrected for chromatic and spherical aberrations and for coma within acceptable limits, if the two lenses can be separated. Such a system is called an 'aplanatic lens'. A parabolic mirror can be corrected for coma by adding thin correcting lenses before or after the mirror (as discussed in more detail later in this section under the heading 'telescope designs'). The severity of the coma at a given angular distance from the optical axis is inversely proportional to the square of the focal ratio. Hence, using as large a focal ratio as possible can also reduce its effect. In Newtonian reflectors, a focal ratio of *f*8 or larger gives acceptable coma for most purposes. At *f*3, coma will limit the useful field of view to about one minute of arc so that prime focus imaging almost always requires the use of a correcting lens to give reasonable fields of view.

Astigmatism (Figure 1.42) is an effect whereby the focal length differs for rays in the plane containing an off-axis object and the optical axis (the tangential plane), in comparison with rays in the plane at right angles to this (the sagittal plane). It decreases more slowly with focal ratio than coma so that it may become the dominant effect for large focal ratios. There is a point between the tangential and sagittal 'foci', known as the 'point of least confusion', where the image is at its sharpest. It is possible to correct astigmatism but only at the expense of introducing yet another aberration called field curvature. This is simply that the surface containing the sharply focused images is no longer a flat plane but is curved. A system in which a flat image plane is retained and astigmatism is corrected for at least two radii is termed an 'anastigmatic system'.

The final aberration is distortion, and this is a variation in the magnification over the image plane. An optical system will be free of distortion only if the condition

$$\frac{\tan\theta}{\tan\varphi} = \text{constant} \tag{1.24}$$

holds for all values of theta (see Figure 1.43 for the definition of the angles). Failure of this condition to hold results in pincushion or barrel distortion (Figure 1.44) accordingly as the magnification increases or decreases with distance from the optical axis. A simple lens is little affected by distortion, and it can frequently be reduced in more complex systems by the judicious placing of stops within the system.

There are higher-order aberrations than the ones just discussed which the reader may encounter, especially if venturing to study advanced optical design. These have names such as Trefoil,

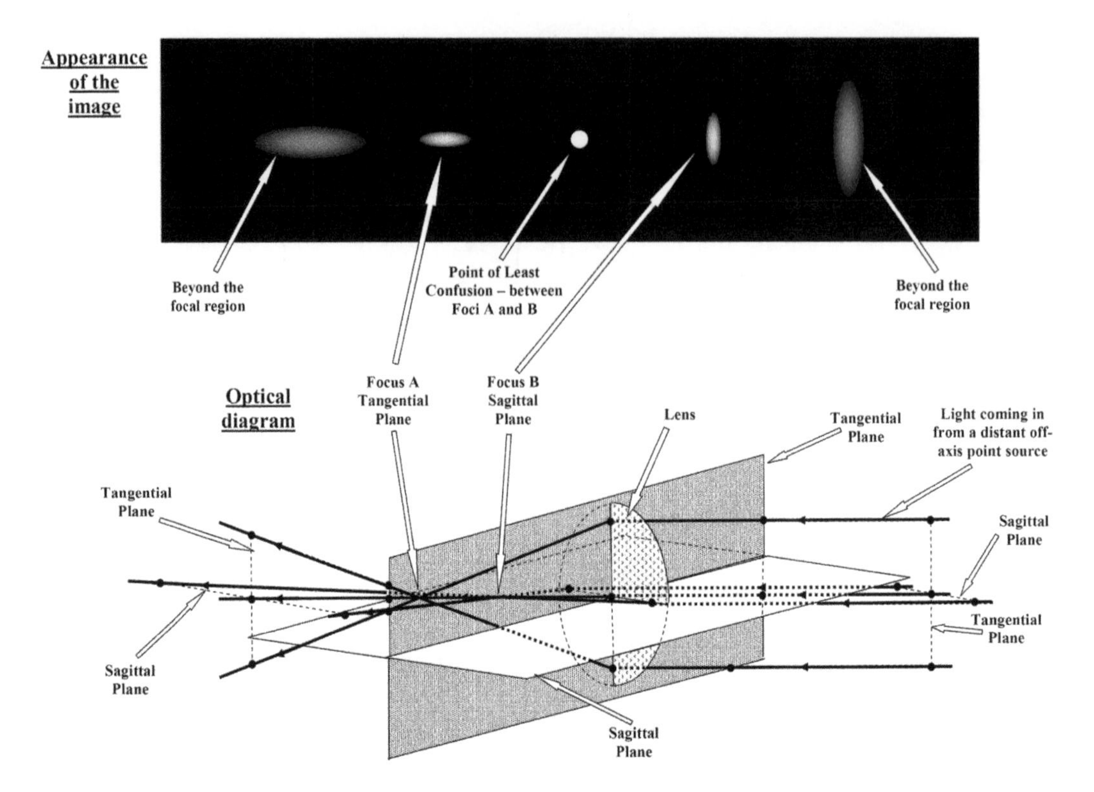

FIGURE 1.42 Shape of the image of a point source resulting from astigmatism, at different points along the optical axis.

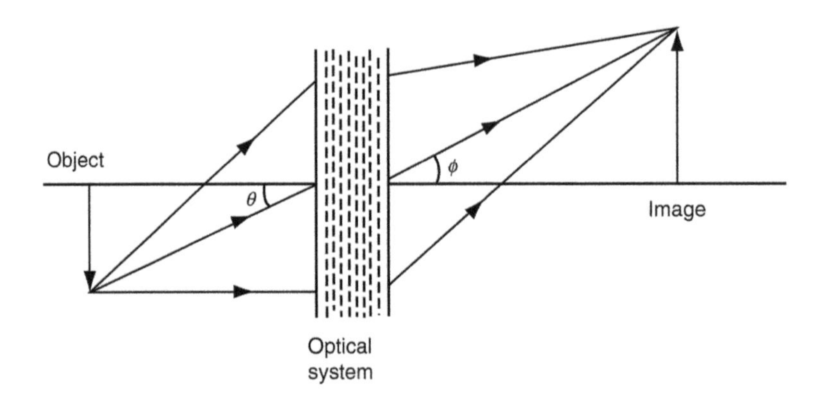

FIGURE 1.43 Terminology for distortion.

Quadrafoil, Pentafoil, and so on. Trefoil has effects upon images that are symmetrical about a three-fold axis (quadrafoil – four-fold axis, pentafoil – five-fold axis, and so on). However, interested readers will have to seek explanations of these from more specialised texts than this one (although the effects of trefoil aberration at least will be familiar to any amateur astronomer possessing a telescope with a mirror that is held in place by three supporting bolts and who has overtightened those bolts).

A fault of optical instruments as a whole is 'vignetting'. This is not an attribute of a lens or mirror and so is not included amongst the aberrations. It arises from the uneven illumination of the image plane, usually resulting from the obstruction of the light path by parts of the instrument. Normally,

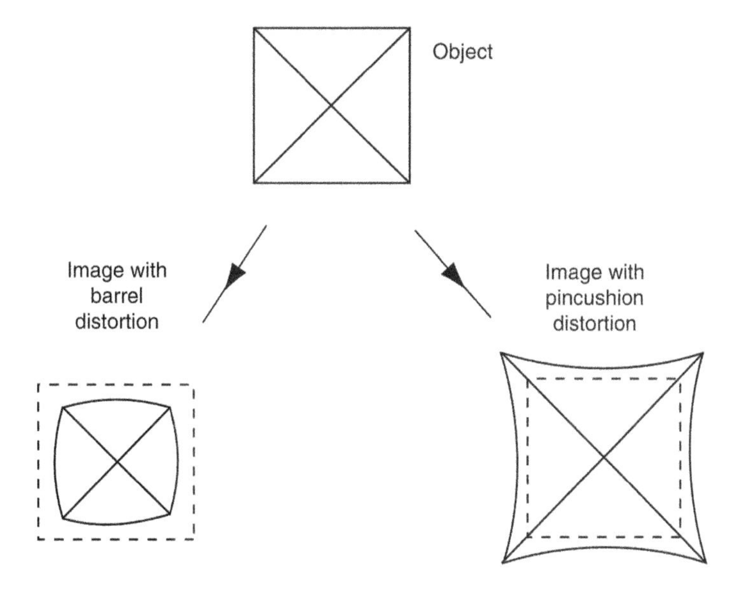

FIGURE 1.44 Distortion.

it can be avoided by careful design, but it may become important if stops are used in a system to reduce other aberrations.

This long catalogue of faults of optical systems may well have led the reader to wonder if the Rayleigh limit can ever be reached in practice. However, optical designers have a wide variety of variables to play with: refractive indices, dispersion, focal length, mirror surfaces in the form of various conic sections, spacings of the elements, number of elements and so on, so that it is usually possible to produce a system which will give an adequate image for a particular purpose. Other criteria such as cost, weight, production difficulties and others may well prevent the realisation of the system in practice, even though it is theoretically possible. Multipurpose systems can usually only be designed to give lower quality results than single-purpose systems. Thus, the practical telescope designs that are discussed later in this section are optimised for objects at infinity and if they are used for nearby objects their image quality will deteriorate.

The method of designing an optical system is still largely an empirical one. The older approach requires an analytical expression for each of the aberrations. For example, the third-order approximation for spherical aberration of a lens is

$$\frac{1}{f_x}-\frac{1}{f_p}=\frac{x^2}{8f^3\mu(\mu-1)}\left[\frac{\mu+2}{\mu-1}q^2+4(\mu+1)\left(\frac{2f}{v}-1\right)q+(3\mu+2)(\mu-1)\left(\frac{2f}{v}-1\right)^2+\frac{\mu^2}{\mu-1}\right] \quad (1.25)$$

where f_x is the focal distance for rays passing through the lens at a distance x from the optical axis, f_p is the focal distance for paraxial rays from the object (paraxial rays are rays which are always close to the optical axis and which are inclined to it by only small angles), f is the focal length for paraxial rays which are initially parallel to the optical axis and v is the object distance. For precise work it may be necessary to involve fifth-order approximations of the aberrations and so the calculations rapidly become unwieldy.

The alternative and now far more widely used method is via ray tracing. The concepts involved are much simpler because the method consists simply of accurately following the path of a selected ray from the object through the system and finding its arrival point on the image plane. Only the basic formulae are required; Snell's law for lenses

$$\sin i = \frac{\mu_1}{\mu_2}\sin r \tag{1.26}$$

and the law of reflection for mirrors

$$i = r \tag{1.27}$$

where i is the angle of incidence, r is the angle of refraction or reflection as appropriate, and μ_1 and μ_2 are the refractive indices of the materials on either side of the interface.

The calculation of i and r for a general ray requires knowledge of the ray's position and direction in space, and the specification of this is rather more cumbersome. Consider first of all a ray passing through a point P, which is within an optical system (Figure 1.45). We may completely describe the ray by the spatial coordinates of the point P, together with the angles that the ray makes with the coordinate system axes (Figure 1.46). We may without any loss of generality set the length of the ray, l, equal to unity and therefore write

$$\gamma = \cos\theta \tag{1.28}$$

$$\delta = \cos\varphi \tag{1.29}$$

$$\varepsilon = \cos\psi \tag{1.30}$$

The angular direction of the ray is thus specified by the vector

$$\mathrm{v} = (\gamma, \delta, \varepsilon) \tag{1.31}$$

The quantities γ, Δ, and ε are commonly referred to as the direction cosines of the ray. If we now consider the point P as a part of an optical surface, then we require the angle of the ray with the normal to that surface to obtain i and thereafter, r. Now we may similarly specify the normal at P by its direction cosines, forming the vector, $\mathbf{v}'$,

$$\mathrm{v}' = (\gamma', \delta', \varepsilon') \tag{1.32}$$

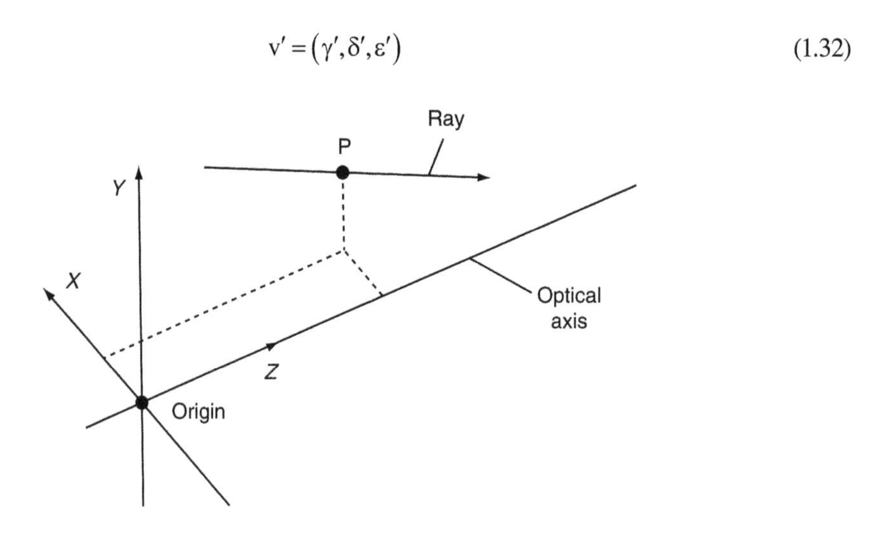

FIGURE 1.45 Ray tracing coordinate system.

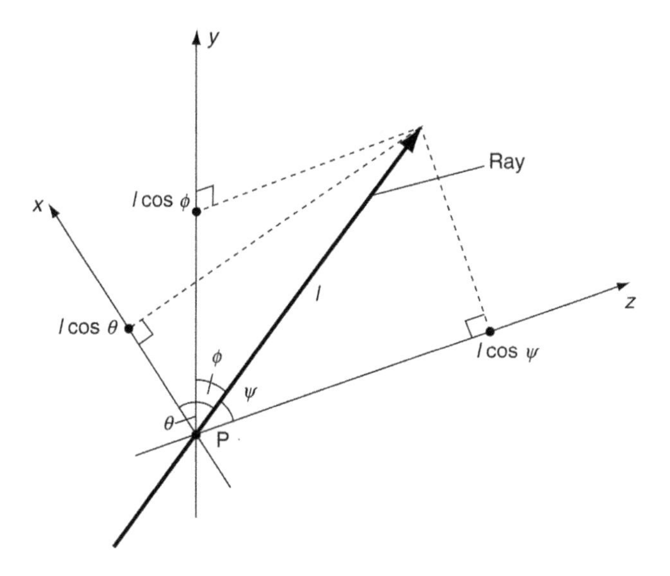

FIGURE 1.46 Ray tracing angular coordinate system.

and we then have

$$\cos i = \frac{\mathrm{v.v'}}{\mathrm{v}\,\|\mathrm{v'}\|} \tag{1.33}$$

or

$$i = \cos^{-1}(\gamma\gamma' + \delta\delta' + \varepsilon\varepsilon') \tag{1.34}$$

The value of r can now be obtained from Equations 1.26 or 1.27. We may again specify its track by direction cosines (the vector v″)

$$\mathrm{v}'' = (\gamma'', \delta'', \varepsilon'') \tag{1.35}$$

and we may find the values of the components as follows. Three simultaneous equations can be obtained; the first from the angle between the incident ray and the refracted or reflected ray, the second from the angle between the normal and the reflected or refracted ray and the third by requiring the incident ray, the normal and the refracted or reflected ray to be coplanar and these are

$$\gamma\gamma'' + \delta\delta'' + \varepsilon\varepsilon'' = \cos 2r \quad (\text{reflection}) \tag{1.36}$$

$$= \cos(i - r) \quad (\text{refraction}) \tag{1.37}$$

$$\gamma'\gamma'' + \delta'\delta'' + \varepsilon'\varepsilon'' = \cos r \tag{1.38}$$

$$(\varepsilon\delta' - \varepsilon'\delta)\gamma'' + (\gamma\varepsilon' - \gamma'\varepsilon)\delta'' + (\delta\gamma' - \delta'\gamma)\varepsilon'' = \cos r \tag{1.39}$$

After a considerable amount of manipulation, one obtains

$$\varepsilon'' = \frac{\{[(\varepsilon\delta' - \varepsilon'\delta)\delta' - (\gamma\varepsilon' - \gamma'\varepsilon)\gamma'](\gamma'\cos\alpha - \gamma\cos r) - (\gamma'\delta - \gamma\delta')(\varepsilon\delta' - \varepsilon'\delta)\cos r\}}{\{[(\varepsilon\delta' - \varepsilon'\delta)\delta' - (\gamma\varepsilon' - \gamma'\varepsilon)\gamma'](\gamma'\varepsilon - \gamma\varepsilon') - [(\varepsilon\delta' - \varepsilon'\delta)\varepsilon' - (\delta\gamma' - \delta'\gamma)\gamma'](\gamma'\delta - \gamma\delta')\}} \tag{1.40}$$

$$\delta'' = \frac{\gamma'\cos\alpha - \gamma\cos r - (\gamma'\varepsilon - \gamma\varepsilon')\varepsilon''}{(\gamma'\delta - \gamma\delta')} \tag{1.41}$$

$$\gamma'' = \frac{\cos\alpha - \delta\delta'' - \varepsilon\varepsilon''}{\gamma} \tag{1.42}$$

where α is equal to $2r$ for reflection and $(i - r)$ for refraction.

The direction cosines of the normal to the surface are easy to obtain when its centre of curvature is on the optical axis

$$\gamma' = \frac{x_1}{R} \tag{1.43}$$

$$\delta' = \frac{y_1}{R} \tag{1.44}$$

$$\varepsilon' = \frac{(z_1 - z_R)}{R} \tag{1.45}$$

where (x_1, y_1, z_1) is the position of P and $(0, 0, z_R)$ is the position of the centre of curvature. If the next surface is at a distance s from the surface under consideration, then the ray will arrive on it at a point (x_2, y_2, z_2) still with the direction cosines γ'', $\delta''\Delta''$ and ε'', where

$$x_2 = \frac{\gamma'' s}{\varepsilon''} + x_1 \tag{1.46}$$

$$y_2 = \frac{\delta'' s}{\varepsilon''} + y_1 \tag{1.47}$$

$$z_2 = s + z_1 \tag{1.48}$$

We may now repeat the calculation again for this surface and so on. Ray tracing has the advantage that all the aberrations are automatically included by it, but it has the disadvantage that many rays have to be followed to build up the structure of the image of a point source at any given place on the image plane, and many images have to be calculated to assess the overall performance of the system. Ray tracing, however, is eminently suitable for programming onto a computer, while the analytical approach is not because it requires frequent value judgements.

The approach of a designer, however, to an optical design problem is similar whichever method is used to assess the system. The initial prototype is set up purely on the basis of what the designer's experience suggests may fulfil the major specifications such as cost, size, weight, resolution, and so on. A general rule of thumb is that the number of optical surfaces needed will be at least as many as the number of aberrations to be corrected. The performance is then assessed either analytically or by ray tracing. In most cases, it will not be good enough, so a slight alteration is made with the intention of improving the performance and it is reassessed. This process continues until the original prototype has been optimised for its purpose. If the optimum solution is within the specifications, then there is no further problem. If it is outside the specifications, even after optimisation, then the whole procedure is repeated starting from a different prototype. The performances of some of the optical systems favoured by astronomers are considered in the next subsection.

Even after a design has been perfected, there remains the not inconsiderable task of physically producing the optical components to the accuracies specified by the design. The manufacturing steps for both lenses and mirrors are broadly similar, although the details may vary. The surface is roughly shaped by moulding or by diamond milling. It is then matched to another surface formed in the same material whose shape is its inverse, called the 'tool'. The two surfaces are ground together with coarse carborundum or other grinding powder between them until the required surface begins to approach its specifications. The pits left behind by this coarse grinding stage are removed by a second grinding stage in which finer powder is used. The pits left by this stage are then removed in turn by a third stage using still finer powder and so on. As many as eight or ten such stages may be necessary.

When the grinding pits are reduced to a micron or so in size, the surface may be polished. This process employs a softer powder such as iron oxide or cerium oxide which is embedded in a soft matrix such as pitch. Once the surface has been polished, it can be tested for the accuracy of its fit to its specifications. Since in general it will not be within the specifications after the initial polishing, a final stage, which is termed 'figuring', is necessary. This is simply additional polishing to adjust the surface's shape until it is correct. The magnitude of the changes involved during this stage is only about a micron or two, so that if the alteration that is needed is larger than this, it may be necessary to return to a previous stage in the grinding to obtain a better approximation. There are a number of tests that can determine the shape of the mirrors surface to a precision of ± 50 nm or better, such as the Foucault, Ronchi, Hartmann and Null tests. The details of these tests are beyond the scope of this book but may be found in books on optics and telescope making.

Small mirrors can be produced in this way by hand and many amateur astronomers have acquired their telescope at very low cost by making their own optics. Larger mirrors require machines that move the tool (now smaller than the mirror) in an epicyclic fashion. The motion of the tool is similar to that of the planets under the Ptolemaic model of the solar system and such machines have become known as 'planetary polishers'. The epicyclic motion can be produced by a mechanical arrangement, but commercial production of large mirrors now relies on computer-controlled planetary polishers. The deepening of the surface to a parabolic, hyperbolic or other shape is accomplished by preferential polishing and sometimes by the use of a flexible tool whose shape can be adjusted. The mirror segments for large instruments like the 10-m Keck telescopes are clearly small, off-axis parts of the total hyperbolic shape. These and also the complex shapes required for Schmidt telescope corrector lenses (see later discussion) have been produced using stressed polishing. The blank is deformed by carefully designed forces from a warping harness and then polished to a spherical shape (or flat for Schmidt corrector plates). When the deforming forces are released, the blank springs into the required non-spherical asymmetric shape. For the Keck and Hobby-Eberly mirror segments, the final polishing was undertaken using ion beams. The ion beam is an accelerated stream of ions, such as argon, that wears away the surface an atom at a time. Because it applies almost no force to the mirror, the process can be used to correct defects left by normal polishing techniques such as deformations at the edge and print-through of the honeycomb back on lightweight mirrors.

Recently, the requirements for non-axisymmetric mirrors for segmented mirror telescopes (see later discussion) and for glancing incidence X-ray telescopes (Section 1.3) have led to the development of numerically controlled diamond milling machines which can produce the required shaped and polished surface directly to an accuracy of 10 nm or better.

The defects in an image that are the result of surface imperfections on a mirror will not exceed the Rayleigh limit if those imperfections are less than about one-eighth of the wavelength of the radiation for which the mirror is intended. Thus, we have the commonly quoted $\lambda/8$ requirement for the maximum allowable deviation of a surface from its specifications. Sometimes the resulting deviations of the wavefront are specified rather than those of the optical surface. This is twice the value for the surface (i.e., a limit of $\lambda/4$). The restriction on lens surfaces is about twice as large as those of mirror surfaces because the ray deflection is distributed over the two lens faces. However, as we have seen, the Rayleigh limit is an arbitrary one and for some purposes the fit must be several times better than this limit. This is particularly important when viewing extended objects such as planets or for high-contrast images and improvements in the signal-to-noise ratio can continue to be obtained by figuring surfaces to $\lambda/20$ or better. The surfaces of the JWST segments (Figure 1.30) deviate from their correct form, for example, by generally less than 20 nm. To put this in perspective: if the Earth's surface were as smooth as the JWST's mirror, then the largest hills or valleys would be only 40 mm high (or deep).

The surface must normally receive its reflecting coating after its production. The vast majority of astronomical mirror surfaces have a thin layer of aluminium evaporated onto them by heating aluminium wires suspended over the mirror inside a vacuum chamber. The initial reflectivity of an aluminium coating in the visual region is around 90%. This however can fall to 75% or less within a few months, as the coating ages. Mirrors, therefore, have to be re-aluminised at regular intervals. The intervals between re-aluminising can be lengthened by gently cleaning the mirror every month or so. The currently favoured methods of cleaning are rinsing with deionised water and/or drifting carbon dioxide snow across the mirror surface. Mirrors coated with a suitably protected silver layer can achieve 99.5% reflectivity in the visible and are becoming more and more required because some modern telescope designs can have four or five reflections. With ancillary instrumentation the total number of reflections can then reach ten or more. A silver coating also lowers the IR emission from the mirror itself from around 4% at a wavelength of 10 μm for an aluminium coating to around 2%. For this reason, the 8.1-m Gemini telescopes use protected silver coatings that are applied to the surfaces of the mirrors by evaporation inside a vacuum chamber. Other materials, such as silicon carbide, are occasionally used as the reflective layer, especially for UV work because the reflectivity of aluminium falls off below 300 nm. For IR telescopes, such as the JWST (Figure 1.47), the mirror coating can be gold.

Lens surfaces also normally receive a coating after their manufacture, but in this case the purpose is to reduce reflection. Uncoated lenses reflect about 5% of the incident light from each surface so that a system containing, say, five uncoated lenses could lose 40% of its available light through this process. To reduce the reflection losses a thin layer of material covers the lens, for which

$$\mu' = \sqrt{\mu} \tag{1.49}$$

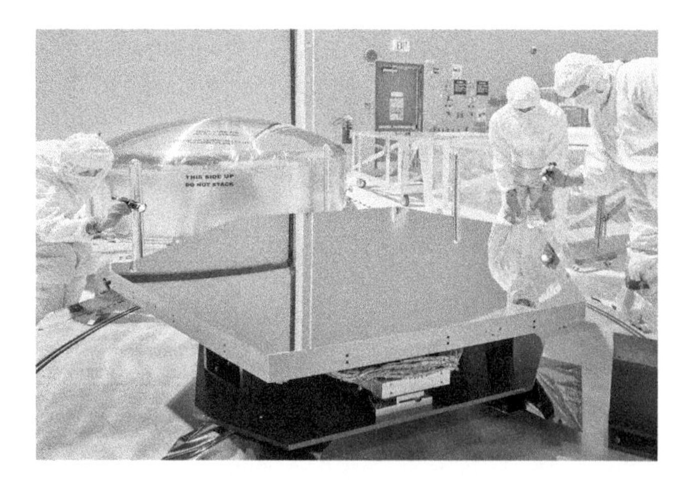

FIGURE 1.47 One of the 18 mirror segments for the James Webb Space Telescope (JWST). The segment is 1.32 m in diameter (between the flat edges) and weighs around 20 kg. The gold reflective coating can be clearly seen. (Courtesy of National Aeronautics and Space Administration [NASA]/Chris Gunn.)

where μ' is the refractive index of the coating and μ is the refractive index of the lens material. The thickness, t, should be

$$t = \frac{\lambda}{4\mu'} \tag{1.50}$$

This gives almost total elimination of reflection at the selected wavelength, but some will still remain at other wavelengths. Lithium fluoride and silicon dioxide are commonly used materials for the coatings. Recently, it has become possible to produce anti-reflection coatings that are effective simultaneously over a number of different wavelength regions through the use of several interfering layers.

Mirrors only require the glass as a support for their reflecting film. Thus, there is no requirement for it to be transparent; on the other hand, it is essential for its thermal expansion coefficient to be low. Many of today's large telescope mirrors are, therefore, made from materials other than glass. The coefficient of thermal expansion of ordinary glass is about 9×10^{-6} K^{-1}, that of Pyrex™ about 3×10^{-6} K^{-1} and for fused quartz it is about 4×10^{-7} K^{-1}.

Amongst the largest telescopes almost the only ordinary glass mirror is the one for the 2.5-m Hooker telescope[63] on Mount Wilson. After its construction, borosilicate glasses[64] such as Pyrex became the favoured material, until the last few decades, when quartz or artificial materials with a similar low coefficient of expansion such as 'CerVit', 'Clearceram', and 'Zerodur' but which are easier to manufacture, have been used. These low-expansion materials have a ceramic glass structure that is partially crystallised. The crystals are around 50 nm across and contract as the temperature rises, while the surrounding matrix of amorphous material expands. An appropriate ratio of crystals to matrix provides small expansion (10^{-7} K^{-1}) over a wide range of temperatures. Ultra-low expansion (ULE) fused silica has an even smaller coefficient of expansion (3×10^{-8} K^{-1}).

Borosilicate glasses however have returned to favour in the last ten years with monolithic mirrors up to 8.4 m in diameter being produced for the LBT, VRO, Synoptic All-Sky Infrared Imaging Survey (SASIR)[65] and the Magellan I and II[66] instruments amongst others. The GMT, due for completion around 2025 will also use seven 8.4-m monolithic borosilicate glass mirrors on a single mounting. The reasons for the material's current popularity include easy workability at (relatively) low temperatures thus facilitating the construction of honeycombed mirror blanks by the spin-casting process (see below), relatively low cost and resistance to the chemicals used to remove the reflective coating when it has to be replaced. The Extremely Large Telescope's (ELT's) mirror segments though, will be fabricated using Zerodur and the Thirty Metre Telescope's (TMT's) from Clearceram.

A radically different alternative to that of forming mirrors from low-expansion materials is to use a material with a high thermal conductivity, such as silicon carbide, graphite epoxy, steel, beryllium or aluminium. The mirror is then always at a uniform temperature and its surface has no stress distortion. The ESA's infrared Herschel Space Observatory (2003–2013), for example, carried a 3.5-m silicon carbide mirror, which, at the time of its manufacture, was the largest silicon carbide

[63] The Hooker telescope was completed in 1917 and the first borosilicate glass, Duran, was invented in 1893. Pyrex was produced from about 1915 onwards. The telescope is still in use today.

[64] These glasses are typically composed of around 70% silica with the remaining 30% made up by boron, sodium and potassium oxides in varying proportions.

[65] The survey will use the 6.5-m San Pedro Mártir telescope that is currently under construction for the San Pedro Mártir Observatory in Baja California, Mexico.

[66] The Magellan I and II telescopes are twin 6.5-m instruments at the Las Campanas Observatory which had first light early this millennium. They should not be confused with the Giant Magellan Telescope (GMT) which is currently under construction and which will also to be based at Las Campanas.

structure in the world. Most metallic mirrors, however, have a relatively coarse crystalline surface that cannot be polished adequately. They must, therefore, be coated with a thin layer of nickel before polishing, and there is always some risk then of this being worn through if the polishing process extends for too long.

Provided that they are mounted properly, small solid mirrors can be made sufficiently thick that once ground and polished, their shape is maintained simply by their mechanical rigidity. However, the thickness required for such rigidity scales as the cube of the size so that the weight of a solid rigid mirror scales as D^5. Increasing solid mirror blanks to larger sizes thus rapidly becomes expensive and impractical. Most mirrors larger than about 0.5–1 m, therefore, should have their weight reduced in some way. There are two main approaches to reducing the mirror weight, thin mirrors and honeycomb mirrors. The thin mirrors are also subdivided into monolithic and segmented mirrors. In both cases, active support of the mirror is required to retain its correct optical shape. The 8.2-m mirrors of ESO's VLT (Figure 1.30) are an example of thin monolithic mirrors, they are 178-mm thick Zerodur, with a weight of 23 tonnes each, but need 150 actuators to maintain their shape. The 10-m Keck telescopes use thin, segmented Zerodur mirrors. There are 36 individual hexagonal segments in each main mirror, with each segment 1.8-m across and about 70-mm thick, giving a total mirror weight of just 14.4 tonnes (compared with 14.8 tonnes for the 5-m Hale telescope mirror).

Honeycomb mirrors are essentially thick solid blanks that have had a lot of the material behind the reflecting surface removed, leaving only thin struts to support that surface. The struts often surround hexagonal cells of removed material giving the appearance of honeycomb, but other cell shapes such as square, circular or triangular can be used. The mould for a honeycomb mirror blank has shaped cores protruding from its base which produce the empty cells, with the channels between the cores filling with the molten material to produce the supporting ribs. Once the blank has solidified and been annealed it may be ground and polished as though it were solid. However, if the original surface of the blank is flat, then considerable amounts of material will remain near the edge after grinding, adding to the final weight of the mirror. Roger Angel of the University of Arizona has, therefore, pioneered the technique of spin-casting honeycomb mirror blanks. The whole furnace containing the mould and molten material is rotated so that the surface forms a parabola (see discussion of liquid mirrors) whose shape is close to that which is finally required. The surface shape is preserved as the furnace cools and the material solidifies. Honeycomb mirrors up to nearly 8 1/2-m diameter may be produced by this method. Howsoever, it may have been produced, thinning the ribs and the underside of the surface may further reduce the weight of the blank. The thinning is accomplished by milling and/or etching with hydrogen fluoride. In this way, the 1-m diameter Zerodur secondary mirrors produced for the 8.1-m Gemini telescopes have supporting ribs just 3-mm wide and weigh less than 50 kg.

Whatever approach is taken to reducing the mirror's weight, it will be aided if the material used for the blank has an intrinsic highly stiffness (resistance to bending). Beryllium has already been used to produce the 1.1-m secondary mirrors for the VLT, whose weights are only 42 kg. Silicon carbide (Herschel – see previous discussion) and graphite epoxy are other high-stiffness materials that may be expected to be used more in the future.

A quite different approach to producing mirrors that, perhaps surprisingly gives low weights, is to use a rotating bath of mercury. Isaac Newton was the first to realise that the surface of a steadily rotating liquid would take up a paraboloidal shape under the combined influences of gravity and centrifugal acceleration. If that liquid reflects light, like mercury, gallium, gallium-indium alloy or oil suffused with reflecting particles, then it can act as a telescope's primary mirror. Of course, the mirror has to remain accurately horizontal, so that it always points towards the zenith, but with suitable, perhaps active, correcting secondary optics the detector can be moved around the image plane to give a field of view currently tens of arc-minutes wide and conceivably in the future up to 8° across. Moving the electrons across the CCD detector at the same speed as the image movement enables short time exposures to be obtained (TDI, see Section 1.1.8.1). The Earth's rotation enables such a zenith telescope to sample up to 7% of the sky for less than 7% of the cost of an equivalent fully steerable telescope. The 'bath' containing the mercury is in fact a lightweight structure whose surface is parabolic to within 0.1 mm. Only a thin layer of mercury is thus needed to produce the accurate mirror surface, and so the mirror overall is a lightweight one despite the high density of mercury. The bath is rotated smoothly around an accurately vertical axis at a constant angular velocity using a large air bearing. Mercury is toxic, so that suitable precautions have to be taken to protect the operators. Also, its reflectivity is less than 80% in the visible, but because all the other optics in the instrument can be conventional mirrors or lenses, this penalty is not too serious.

The Large Zenith Telescope (LZT) in British Columbia saw first light in 2003 and was decommissioned in 2016. It used a 6-m diameter liquid mirror with 3 tonnes of mercury and rotated at 8.5 rpm. The rather ambitious follow-on proposal, Large Aperture Mirror Array (LAMA), which might have comprised sixty-six 6.15-m liquid mirror telescopes, so gathering the same amount of light as a single 50-m dish and with an angular resolution of a 70-m telescope, however, never proceeded beyond the concept stage.

Currently, actually under construction with first light expected for mid-2020, is the International Liquid Mirror Telescope (ILMT), which is to be a 4-m instrument located at Devasthal on the Himalayan foothills (Northern India) imaging a 0.5 deg^2 of the sky using a 4k × 4k TDI CCD array.

In space, a liquid mirror of potentially almost any size could be formed from a ferromagnetic liquid confined by electromagnetic forces. Another possible concept places a 20-m to 100-m liquid mirror telescope on the Moon. Since mercury evaporates in a vacuum, the proposal suggests using liquid salts with a thin surface layer of silver to act as the reflector.

An approach to producing mirrors whose shape can be rapidly adjusted is to use a thin membrane with a reflecting coating. The membrane forms one side of a pressure chamber and its shape can be altered by changing the gas pressure inside the chamber. $\lambda/2$ surface quality or better can currently be achieved over mirror diameters up to 5 mm and such mirrors have found application in optical aperture synthesis systems (Section 2.5).

1.1.19 Telescope Designs

1.1.19.1 Background

Most serious work with telescopes uses equipment placed directly at the focus of the telescope. But for visual work such as finding and guiding on objects, an eyepiece is necessary. Often it matters little whether the image produced by the eyepiece is of a high quality. Ideally, however, the eyepiece should not degrade the image noticeably more than the main optical system. There are an extremely large number of eyepiece designs, whose individual properties can vary widely. For example, one of the earliest eyepiece designs of reasonable quality is the Kellner. This combines an achromat and a simple lens and typically has a field of view of 40°–50°. The Plössl uses two achromats and has a slightly wider field of view. More recently, the Erfle design employs six or seven components and gives fields of view of 60°–70°, while the current state-of-the-art is represented by designs such as the Nagler with eight or more components and fields of view[67] up to 85°. Details of these and other designs may generally be found in books on general astronomy or on optics or from the manufacturers. For small telescopes used visually, a single low magnification wide angle eyepiece may be worth purchasing for the magnificent views of large objects like the Orion Nebula that it will provide. There is little point in the higher power eyepieces being wide angle ones (which are expensive – $200–$500 at the time of writing) because these will normally be used to look at angularly small objects. The use of a Barlow lens provides an adjustable

[67] The individual healthy human eye has a field of view that is about 150° across in the horizontal plane and about 130° in the vertical plane. The periphery of this region though only serves to detect something, especially if it is moving and so acts mainly as a defence mechanism – alerting observers to the need to move their head and/or eyes to look more directly at that object. The field of sharp vision is only about 5° across.

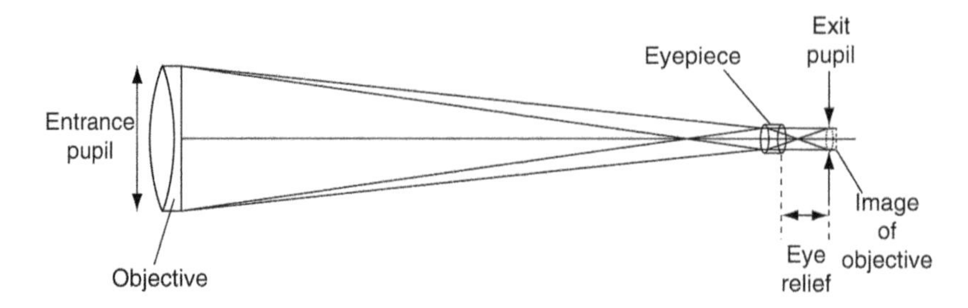

FIGURE 1.48 Exit pupil and eye relief.

magnification for any eyepiece. The Barlow lens is an achromatic negative lens placed just before the telescope's focus. It increases the telescope's effective focal length and so the magnification. More convenient eyepieces with an integral zoom are now available but with some loss of image quality over the single-focal length designs.

For our purposes, only four aspects of eyepieces are of concern: light losses, eye relief, exit pupil and angular field of view. Light loss occurs through vignetting when parts of the light beam fail to be intercepted by the eyepiece optics or are obstructed by a part of the structure of the eyepiece and also through reflection, scattering and absorption by and within the optical components. The first of these can generally be avoided by careful eyepiece design and selection, while the latter effects can be minimised by anti-reflection coatings and by keeping the eyepieces clean.

The exit pupil is the image of the objective produced by the eyepiece (Figure 1.48). All the rays from the object pass through the exit pupil, so that it must be smaller than the pupil of the human eye if all of the light gathered by the objective is to be utilised. Its diameter, E, is given by

$$E = \frac{F_e D}{F_o} \tag{1.51}$$

where D is the objective's diameter, F_e, is the focal length of the eyepiece, and F_o is the focal length of the objective. Since magnification is given by

$$M = \frac{F_o}{F_e} \tag{1.52}$$

and the diameter of the pupil of the dark-adapted eye is 6 or 7 mm, we must, therefore, have

$$M \geq \sim 170D \tag{1.53}$$

where D is in metres, if the whole of the light from the telescope is to pass into the eye.

The maximum magnification is not so precisely definable, but in practice, for small telescopes being used visually, the maximum useable magnification is around $\times 2000D$ (i.e., $\times 300$ for a 0.15-m telescope, etc.).

The eye relief is the distance from the final lens of the eyepiece to the exit pupil. It should be about 6–10 mm for comfortable viewing. If you wear spectacles however, then the eye relief may need to be up to 20 mm.

The angular field of view is defined by the acceptance angle of the eyepiece, θ'. Usually, this is about 40°–60°, but it may be up to 90° or 100° for wide-angle eyepieces. The angular diameter of the area of sky which is visible when the eye is positioned at the exit pupil and which is known as the angular field of view, θ, is then just

$$\theta = \frac{\theta'}{M} \tag{1.54}$$

The brightness of an image viewed through a telescope is generally expected to be greater than when it is viewed directly. However, this is not the case for *extended* objects. The naked eye brightness of a source is proportional to the eye's pupil diameter squared, while that of the image in a telescope is proportional to the objective diameter squared. If the eye looks at that image, then its angular size is increased by the telescope's magnification. Hence, the increased brightness of the image is spread over a greater area. Thus, we have

$$R = \frac{\text{Brightness through a telescope}}{\text{Brightness to the naked eye}} = \frac{D^2}{M^2 P^2} \tag{1.55}$$

where D is the objective's diameter, P is the diameter of the pupil of the eye, and M is the magnification. But from Equations 1.51 and 1.52, we have

$$R = 1 \tag{1.56}$$

when the exit pupil diameter is equal to the diameter of the eye's pupil and

$$R < 1 \tag{1.57}$$

when it is smaller than the eye's pupil.

If the magnification is less than $\times 170D$, then the exit pupil is larger than the pupil of the eye and some of the light gathered by the telescope will be lost. Since the telescope will generally have other light losses due to scattering, imperfect reflection and absorption, the surface brightness of an extended source is always fainter when viewed through a telescope than when viewed directly with the naked eye.

This result is in fact a simple consequence of the second law of thermodynamics because if it were not the case, then one could have a net energy flow from a cooler to a hotter object. The *apparent* increase in image brightness when using a telescope arises partly from the increased angular size of the object so that the image on the retina must fall to some extent onto the regions of the retina containing the more sensitive rod cells, even when looked at directly and partly from the increased contrast resulting from the exclusion of extraneous light by the optical system. Even with the naked eye, looking through a long cardboard tube enables faint extended sources such as M31 to be seen more easily.

The analysis that we have just seen does not apply to images that are physically smaller than the detecting element. For this situation, the image brightness is proportional to D^2. Again, however, there is an upper limit to the increase in brightness that is imposed when the energy density at the image is equal to that at the source. This limit is never approached in practice because it would require 4π steradians for the angular field of view. Thus, stars may be seen through a telescope which are fainter than those visible to the naked eye by a factor called the 'light grasp'. The light grasp is simply given by D^2/P^2. Taking $+6^m$ as the magnitude of the faintest star visible to the naked eye (see Section 3.1); the faintest star that may be seen through a telescope has a magnitude, m_l, which is the limiting magnitude for that telescope

$$m_l = 17 + 5\log_{10} D \tag{1.58}$$

where D is in metres. If the stellar image is magnified to the point where it spreads over more than one detecting element, then we must return to the analysis for the extended sources. For an average eye, this upper limit to the magnification is given by

$$M \approx 850D \tag{1.59}$$

where D is again in metres.

1.1.19.2 Designs

Probably, the commonest format for large telescopes is the Cassegrain system, although most large telescopes can usually be used in several alternative different modes by interchanging their secondary mirrors. The Cassegrain system is based upon a paraboloidal primary mirror and a convex hyperboloidal secondary mirror (Figure 1.49). The nearer focus of the conic section which forms the surface of the secondary is coincident with the focus of the primary and the Cassegrain focus is then at the more distant focus of the secondary mirror's surface. The major advantage of the Cassegrain system lies in its telephoto characteristic; the secondary mirror serves to expand the beam from the primary mirror so that the effective focal length of the whole system is several times that of the primary mirror. A compact and hence, rigid and relatively cheap mounting can thus be used to hold the optical components while retaining the advantages of long focal length and large image scale. The Cassegrain design is afflicted with coma and spherical aberration to about the same degree as an equivalent Newtonian telescope or indeed to just a single parabolic mirror with a focal length equal to the effective focal length of the Cassegrain. The beam expanding effect of the secondary mirror means that Cassegrain telescopes normally work at focal ratios between 12 and 30, although their primary mirror may be *f*3 or *f*4 (or even *f*1 as in the 4-m VISTA telescope). Thus, the images remain tolerable over a field of view that may be several tenths of a degree across (Figure 1.50). Astigmatism and field curvature are stronger than in an equivalent Newtonian system, however.

Focusing of the final image in a Cassegrain system is customarily accomplished by moving the secondary mirror along the optical axis. The amplification due to the secondary means that it only has to be moved a short distance away from its optimum position to move the focal plane considerably. The movement away from the optimum position, however, introduces severe spherical aberration. For the 0.25-m *f*4/*f*16 system whose images are shown in Figure 1.50, the secondary mirror can only be moved by 6 mm either side of the optimum position before even the on-axis images without diffraction broadening become comparable in size with the Airy disc. Critical work with Cassegrain telescopes should, therefore, always be undertaken with the secondary mirror at or near to its optimum position.

A great improvement to the quality of the images may be obtained if the Cassegrain design is altered slightly to the Ritchey-Chrétien system. The optical arrangement is identical with that shown in Figure 1.47 except that the primary mirror is deepened to an hyperboloid and a stronger hyperboloid is used for the secondary. With such a design both coma and spherical aberration can be corrected and we have an aplanatic system. The improvement in the images can be seen by

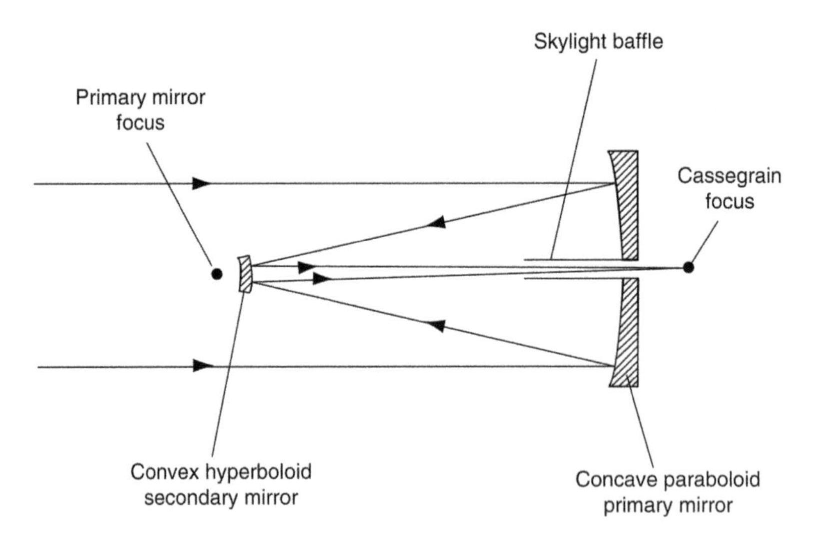

FIGURE 1.49 Cassegrain telescope optical system.

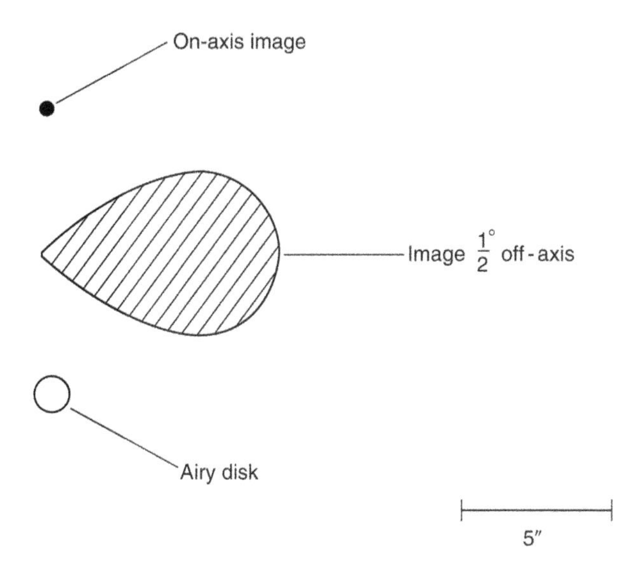

FIGURE 1.50 Images in a 0.25-m *f*4/*f*16 Cassegrain telescope. (The images were obtained by ray tracing. Since this does not allow for the effects of diffraction, the on-axis image in practice will be the same as the Airy disc and the 0.5° off-axis image will be blurred even further.)

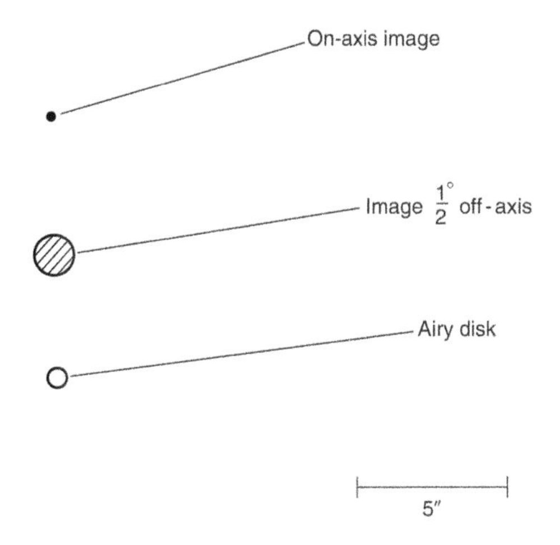

FIGURE 1.51 Ray tracing images in a 0.5-m *f*3/*f*8 Ritchey-Chrétien telescope.

comparing Figure 1.51 with Figure 1.50. It should be noted, however, that the improvement is in fact considerably more spectacular because we are comparing a 0.5-m *f*/8 Ritchey-Chrétien with a 0.25-m *f*/16 Cassegrain. A 0.5-m Cassegrain telescope would have its off-axis image twice the size of that shown in Figure 1.50 and its Airy disc half the size shown there.

A Cassegrain or Ritchey-Chrétien system (indeed, almost any design of telescope) can be improved by the addition of correctors just before the focus. The correctors are low or zero-power lenses whose aberrations oppose those of the main system. There are numerous successful designs for correctors although many of them require aspheric surfaces and/or the use of exotic materials such as fused quartz. Images can be reduced to less than the size of the seeing disk over fields of view of up to one degree or sometimes over even larger angles. The AAT, for example, uses correcting lenses to provide an unvignetted 2° field of view (two-degree field or 2dF), over which 400 optical fibres can be positioned to feed a spectroscope. The VISTA telescope uses a modified

Ritchey-Chrétien design that leaves some significant aberrations in images produced by the mirrors alone. However, by using three correcting lenses a fully corrected image over 2° across in the visible and 1.6° across in the infrared is produced. The 8.4-m VRO will use three correcting lenses. The front lens will be the largest lens ever manufactured, with a diameter of 1.55 m, while the third lens will additionally form the entrance window to the detectors' vacuum chamber. Reflective corrective optics are also possible. The ELT, with its all-aspheric, three-mirror design will effectively combine the correcting optics with the main imaging optics.

Corrective optics may be combined with a focal reducer to enable the increased field of view to be covered by the detector array. A focal reducer is the inverse of a Barlow lens and is a positive lens, usually an apochromatic triplet, placed just before the focal point of the telescope that decreases the effective focal length and so gives a smaller image scale.

Another telescope design that is again closely related to the Cassegrain is termed the 'Coudé system'. It is in effect a long focal length Cassegrain or Ritchey-Chrétien whose light beam is folded and guided by additional flat mirrors to give a focus whose position is fixed in space irrespective of the telescope position. One way of accomplishing this (there are many others) is shown in Figure 1.52. After reflection from the secondary, the light is reflected down the hollow declination axis by a diagonal flat mirror and then down the hollow polar axis by a second diagonal. The light beam, thus, always emerges from the end of the polar axis, whatsoever part of the sky the telescope may be inspecting. Designs with similar properties can be devised for most other types of mountings although additional flat mirrors may be needed in some cases.

With alt-az mountings, the light beam may be directed along the altitude axis to one of the two Nasmyth foci on the side of the mounting. These foci still rotate horizontally as the telescope changes its azimuth, but this poses far fewer problems than the changing altitude and attitude of a conventional Cassegrain focus. On large modern telescopes, platforms of considerable size are often constructed at the Nasmyth foci allowing large ancillary instruments to be used. The fixed or semi-fixed foci of the Coudé and Nasmyth systems are a great advantage when bulky items of equipment, such as high dispersion spectrographs, are to be used because these can be permanently mounted in a nearby separate laboratory and the light brought to them, rather than have to have the equipment mounted on the telescope. The design also has several disadvantages; the field of view rotates as the telescope tracks an object across the sky, is tiny because of the large effective focal ratio (*f*25 to *f*40) which is generally required to bring the focus through the axes and finally, the additional reflections will cause loss of light.

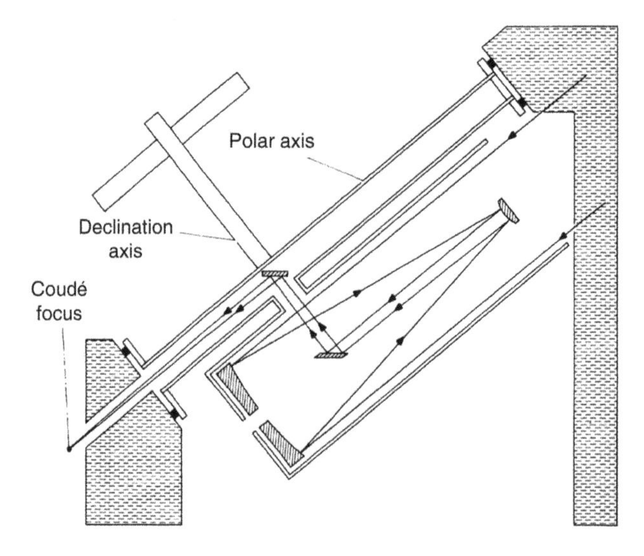

FIGURE 1.52 Coudé system for a modified English mounting.

The simplest of all designs for a telescope is a mirror used at its prime focus. That is, the primary mirror is used directly to produce the images and the detector is placed at the top end of the telescope. The largest of the older telescopes have a platform or cage which replaces the secondary mirror and which is large enough for the observer to ride in while he or she operates and guides the telescope from the prime focus position. The more modern large telescopes, when used at prime focus, simply have remotely controlled detectors and guiding. With smaller instruments too much light would be blocked, so that they must be guided using a separate guide telescope or use a separate detector monitoring another star (the guide star) within the main telescope's field of view. The image quality at the prime focus is usually poor even a few tens of arcseconds away from the optical axis because the primary mirror's focal ratio may be as short as *f*3 or less to keep the length of the instrument to a minimum. Thus, correcting lenses are essential to give acceptable images and reasonably large fields of view. These are similar to those used for correcting Cassegrain telescopes and are placed immediately before the prime focus.

A system that is almost identical to the use of the telescope at prime focus and which was the first design to be made into a working reflecting telescope is that due to Newton and hence, is called the 'Newtonian telescope'. A secondary mirror is used which is a flat diagonal placed just before the prime focus. This reflects the light beam to the side of the telescope from where access to it is relatively easy (Figure 1.53). The simplicity and cheapness of the design make it popular as a small telescope for the amateur market, but it is rarely encountered in telescopes larger than about 1 m. There are several reasons for its lack of popularity for large telescope designs; the main ones are that it has no advantage over the prime focus position for large telescopes because the equipment used to detect the image blocks out no more light than the secondary mirror, and the secondary mirror introduces additional light losses and the position of the equipment high on the side of the telescope tube causes difficulties of access and counterbalancing. The images in a Newtonian system and at prime focus are similar and are of poor quality away from the optical axis as shown in Figure 1.54.

As has already been noted, if n aberrations are to be corrected, then the optical design will generally need to involve n optical surfaces. Thus, three-mirror telescope designs are increasingly being considered for future telescope projects. The (many) individual design variants permitted by three optical surfaces are usually merged together as Korsch[68] telescopes or as Three-Mirror-Anastigmats (TMAs). These designs have close to zero spherical aberration, coma and astigmatism. The JWST is already in production and will use a Korsch-type design for its telescope with a 6.5-m segmented elliptical primary mirror, an hyperbolic secondary and an elliptical tertiary mirror. Other future applications of the designs will include the VRO, the ELT (see above) and ESA's Euclid space-based survey mission (Section 1.7.2.4).

A great variety of other catoptric (reflecting objective) telescope designs abound, but most have few or no advantages over the groups of designs which have been discussed. Some find specialist applications, for example the Gregorian design is similar to the Cassegrain except that the secondary

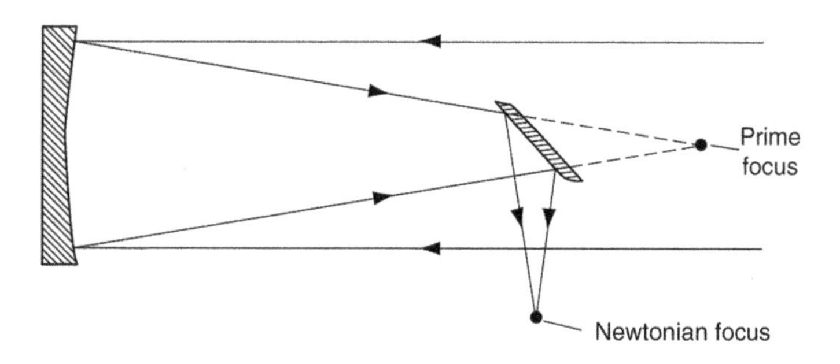

FIGURE 1.53 Newtonian telescope optical system.

[68] After Dietrich Korsch who developed a general set of solutions for three-mirror telescopes in 1972.

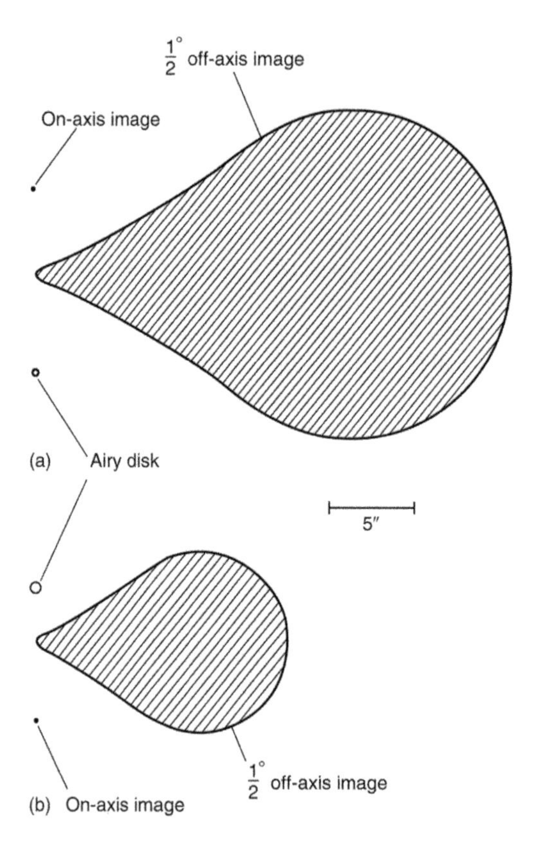

FIGURE 1.54 Images in Newtonian telescopes. (a) 1-m *f*4 telescope; (b) 0.5-m *f*8 telescope.

mirror is a concave ellipsoid and is placed after the prime focus. The two 8.4-m instruments of the LBT on Mount Graham, thus include Gregorian secondary mirrors for IR work.

The Pfund-type telescope uses a siderostat to feed a fixed paraboloidal main mirror that reflects light through a hole in the flat mirror of the siderostat to a focus behind it (i.e., it is essentially a prime-focus telescope). The Infrared Spatial Interferometer (ISI) interferometer (Section 2.5) used Pfund-type telescopes and also the wheelchair-accessible Wren-Marcario Accessible Telescope (WMAT) at the McDonald observatory uses a Pfund design to enable all its visitors to view the skies (Figure 1.55), but other examples are uncommon.

A few other specialised designs may also be built as small telescopes for amateur use because of minor advantages in their production processes, because of the challenge to the telescope maker's skills or because of space-saving reasons as in the folded Schiefspieglers and the Loveday design that utilises a double reflection from the primary mirror, but most such designs will be encountered rarely.

In the radio and microwave regions (Section 1.2) off-axis Gregorian designs are used (e.g., the Green Bank Radio telescope – Figure 1.74) because the secondary mirror is kept clear of the incoming beam of e-m radiation. For the same reason, variations on the off-axis Gregorian design, known as Dragonian[69] telescopes and using a concave paraboloidal primary mirror and a concave hyperboloidal secondary mirror are also to be found, such as that for the CCAT-prime instrument (Section 1.1.15.4).

Of the dioptric (refracting objective) telescopes, only the basic refractor using an achromatic doublet, or occasionally a triplet, as its objective is in any general use (Figure 1.56). Apart from the large refractors that were built towards the end of the nineteenth century, most refractors are now found as small instruments for the amateur market or as the guide telescopes of larger

[69] After the design's originators; Corrado Dragone, Yoshihiko Mizuguchi and Masataka Akagawa.

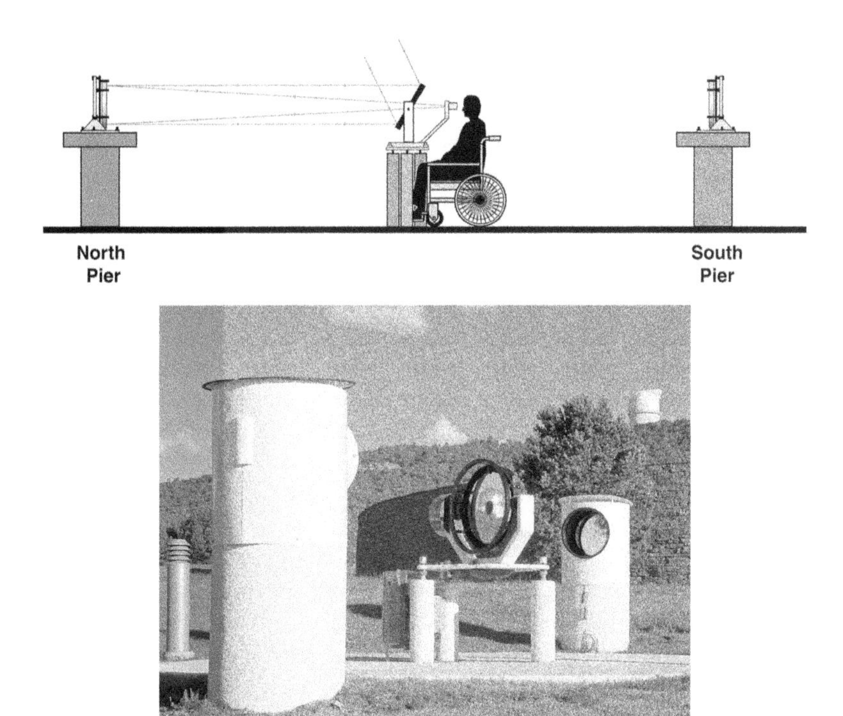

FIGURE 1.55 The Wren-Marcario accessible telescope (WMAT). Top, The optical layout of the instrument. The flat siderostat mirror is in front of the wheelchair user and reflects the light beam from the sky to one of the two fixed 0.46-m parabolic mirrors. The returned and focussed beam passes through the central hole in the siderostat mirror to the eyepiece and the observer. The siderostat mirror can be switched to feed either parabolic mirror, so that the whole sky may be observed. (Courtesy of Tim Jones/Mike Jones/McDonald Observatory). Bottom, The WMAT instrument (foreground): the flat siderostat mirror is in the centre and the two parabolic mirrors are housed in the two cylindrical towers (and the far mirror may be seen through the tower's aperture). (Courtesy of Frank Cianciolo/McDonald Observatory.)

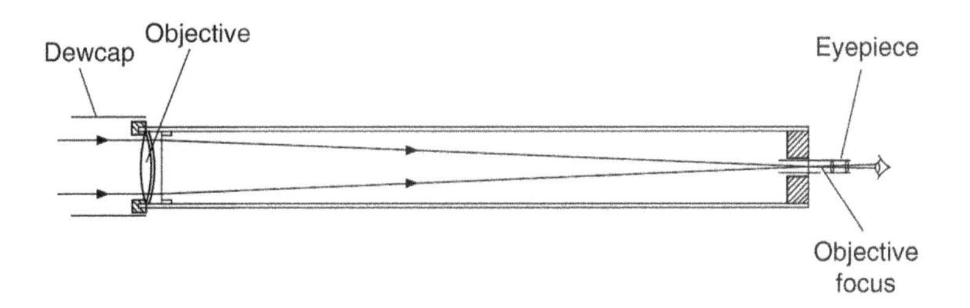

FIGURE 1.56 The astronomical refractor.

instruments. The enclosed tube and relatively firmly mounted optics of a refractor means that they need little adjustment once aligned with the main telescope.

Of course, there is no real difference between telescopes and cameras, though the former term implies, to most peoples' minds, a longish instrument, whilst the latter is expected to be much more compact and probably to have many more optical components than the two or three usually involved in a 'telescope'. Numerous spacecraft therefore carry imagers which are called 'cameras', but could equally well be called 'telescopes', and these designs are often lens-based. Even ground-based observing may use 'cameras' rather than 'telescopes'. In fact, the ground-based Wide Angle Search for Planets (WASP) exoplanet hunter actually uses off-the-shelf Canon™ 200-mm f1.8 telephoto camera lenses for its instrumentation. While the Transiting Exoplanet Survey Satellite (TESS, launched 2018) exoplanet-hunting spacecraft uses dioptric cameras, each with seven lenses making up 150-mm f1.4 optical systems which have fields of view of 24° × 24°.

The one remaining class of optical telescopes is the catadioptric group, of which the Schmidt camera is probably the best known. A catadioptric system uses both lenses and mirrors in its primary light gathering section. Very high degrees of correction of the aberrations can be achieved because of the wide range of variable parameters that become available to the designer in such systems. The Schmidt camera uses a spherical primary mirror so that coma is eliminated by not introducing it into the system in the first place! The resulting spherical aberration is eliminated by a thin correcting lens at the mirror's radius of curvature (Figure 1.57). The only major remaining aberration is field curvature, and the effect of this is eliminated through the use of a detector shaped to match the focal surface or through the use of additional correcting lenses (field flatteners). The correcting lens can introduce small amounts of coma and chromatic aberration but is usually so thin that these aberrations are negligible. Diffraction-limited performance over fields of view of several degrees with focal ratios as fast as $f1.5$ or $f2$ is possible.

The need to use a lens in the Schmidt design limits the sizes possible for them. Larger sizes can be achieved using a correcting mirror. The largest such instrument is the 4-m Large sky Area Multi-Object fibre Spectroscopic Telescope (LAMOST) instrument at the Xinglong Station of the Chinese National Astronomical Observatory. It operates using a horizontally fixed primary mirror fed by a coelostat (see below). Both these mirrors are mosaics of hexagonal segments. The coelostat mirror is actively controlled and also acts as the corrector. With up to 4,000 optical fibres, the instrument is one of the most rapid spectroscopic survey tools in the world, and it typically obtains some 1,500,000 spectra per year although its site is poor (only about one night in three is clear).

Other large Schmidt cameras still operating at the time of writing include the 1.34-m Alfred-Jensch instrument at Tautenburg and the Oschin Schmidt at Mount Palomar and the UK Schmidt camera at the Anglo-Australian Observatory (AAO) both of which have entrance apertures 1.2 m across. The Kepler space telescope (operational from 2009 to 2018) used a 0.95-m Schmidt in its search for exoplanets.

The Schmidt design suffers from having a tube length at least twice its focal length. Also, the Schmidt camera cannot be used visually because its focus is inaccessible. There are several modifications of its design, however, that produce an external focus while retaining most of the beneficial properties of the Schmidt. One of the best of these is the Maksutov design (Figure 1.58) which

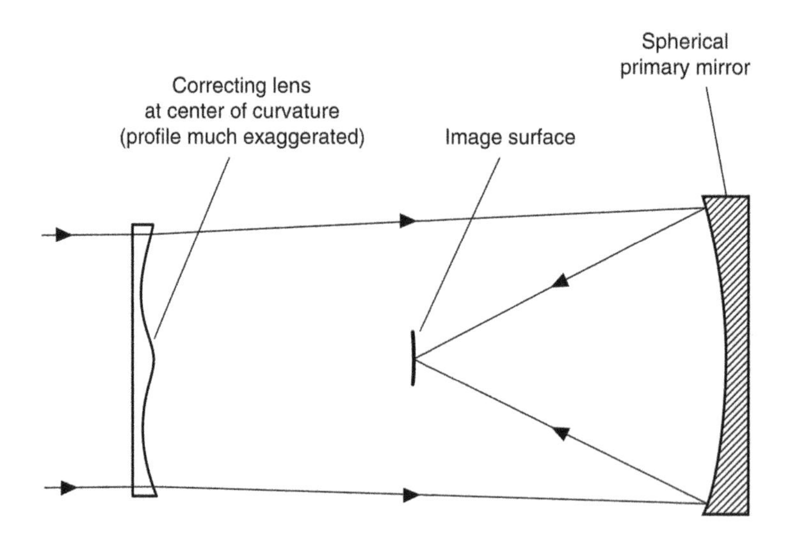

FIGURE 1.57 The Schmidt camera optical system.

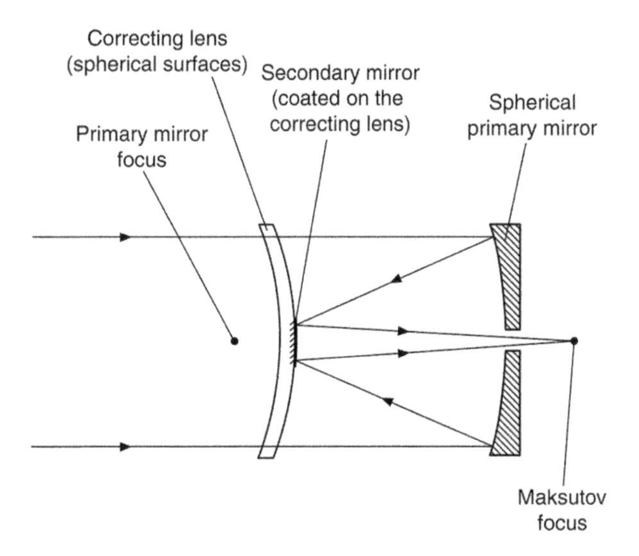

FIGURE 1.58 The Maksutov optical system.

originally also had an inaccessible focus, but which is now a kind of Schmidt-Cassegrain hybrid. All the optical surfaces are spherical and spherical aberration, astigmatism and coma are almost eliminated, while the chromatic aberration is almost negligible. A similar system is the Schmidt-Cassegrain telescope itself. This uses a thin correcting lens like the Schmidt camera and has a separate secondary mirror. The focus is accessible at the rear of the primary mirror as in the Maksutov system. Schmidt-Cassegrain telescopes are now available commercially in sizes up to 0.4-m diameter. They are produced in large numbers by several firms, at reasonable costs, for the amateur and education markets. They are also finding increasing use in specific applications, such as site testing and for professional astronomy.

Although it is not a telescope, there is one further system that deserves mention here and that is the coelostat. This comprises one or two flat mirrors that are driven so that a beam of light from any part of the sky is directed towards a fixed direction. They are particularly used in conjunction with solar telescopes whose extremely long focal lengths make them impossible to move. One mirror of the coelostat is mounted on a polar axis and driven at half sidereal rate. The second mirror is mounted and driven to reflect the light into the fixed telescope. Various arrangements and adjustments of the mirrors enable differing declinations to be observed. A related system is the siderostat (Figure 1.55) which uses just a single flat mirror to feed a fixed telescope. The flat mirror has to be driven at variable speeds in altitude and azimuth to track an object across the sky but even a small computer can now undertake the calculations needed for this purpose and so control the driving motors.

No major further direct developments in the optical design of large telescopes seem likely to lead to improved resolution because such instruments are already limited by the atmosphere. Better resolution, therefore, requires that the telescope be lifted above the Earth's atmosphere or that the distortions that the atmosphere introduces be overcome by a more subtle approach. Telescopes mounted on rockets, satellites and balloons are discussed later in this section, while the four ways of improving the resolution of Earth-based telescopes: interferometry, speckle interferometry, occultations and real-time compensation are discussed in Sections 2.5–2.7 and later in this section, respectively. A fairly common technique for improving resolution, deconvolution, is discussed in Section 2.1. This requires that the precise nature of the degradations of the image be known so that they can be removed. Since the atmosphere's effects are changing in a fairly random manner on a timescale of milliseconds, it is not an appropriate technique for application here, although it should be noted that one approach to speckle interferometry is in fact related to this method and that deconvolution may be needed on the compensated image.

A technique termed 'apodisation' (literally 'removal of the feet', see also Sections 2.5 and 4.1) gives an apparent increase in resolution. It is actually however a method of enabling the telescope to achieve its normal resolution and sometimes even reduces that resolution slightly. For optical telescopes, apodisation is usually accomplished through the use of masks over the entrance aperture of the telescope. These have the effect of changing the diffraction pattern (see previous discussion about the Airy disk and also Section 2.1). The nominal resolution given by the size of the central disk may then worsen, but the outer fringes disappear or become of a more convenient shape for the purpose in mind. A couple of examples will illustrate the process. When searching for a faint object next to a bright one, such as Sirius B next to Sirius A, or for an exoplanet next to a star, a square aperture with fuzzy edges, or a four-armed star aperture will produce fringes in the form of a cross and suppress the normal circular fringes. If the aperture is rotated so that the faint object lies in one of the regions between the arms of the cross, then it will have a much higher signal-to-noise ratio than if superimposed upon a bright fringe from a circular aperture and so be more easily detected.

Secondly, for observing low-contrast features on planets, especially with small to medium-sized telescopes, a mask that is a neutral density filter varying in opacity in a Gaussian fashion from clear at the centre to nearly opaque at the edge of the objective may be used. This has the effect of doubling the size of the Airy disk but also of eliminating the outer fringes completely. The resulting image will be significantly clearer because the low-contrast features on extended objects are normally swamped by the outer fringes of the point spread function (PSF; see Section 2.1). A professionally made neutral density filter of this type would be extremely expensive, but a good substitute may be made using a number of masks formed out of a mesh. Each mask is an annulus with an outer diameter equal to that of the objective and an inner diameter that varies appropriately in size. When they are all superimposed on the objective and provided that the meshes are *not* aligned, then they will obstruct an increasingly large proportion of the incoming light towards the edge of the objective. The overall effect of the mesh masks will, thus, be similar to the Gaussian neutral density filter.

Now although resolution may not be improved in any straightforward manner (but see discussion on real-time atmospheric compensation that follows), the position is less hopeless for the betterment of the other main function of a telescope, which is light gathering. The upper limit to the size of an individual mirror is probably being approached with the 8-m telescopes presently in existence. A diameter of 10–15 m for a monolithic metal-on-glass mirror would almost certainly be about the ultimate possible with present-day techniques and their foreseeable developments.[70] Greater light-gathering power may therefore be acquired only by the use of several mirrors, by the use of a single mirror formed from separate segments (see discussions elsewhere in this section), by aperture synthesis (Section 2.5) or perhaps in the future by using liquid mirrors (see above).

The accuracy and stability of surfaces, alignments and so on required for both the multi-mirror and the segmented mirror systems are a function of the operating wavelength. For IR and short microwave work, far less technically demanding limits are, therefore, possible than for telescopes operating in the visual region and such telescopes are correspondingly easier to construct. Specialist IR telescopes do, however, have other requirements that add to their complexity. First, it is often necessary to chop between the source and the background to subtract the latter. Some purpose-built IR telescopes such as the 3.8-m United Kingdom Infrared Telescope (UKIRT) oscillate the secondary mirror through a small angle to achieve this (a mass equal to that of the chopping mirror may need to be moved in the opposite direction to avoid vibrating the telescope). The UKIRT secondary mirror can also be moved to correct for flexure of the telescope tube and mounting and for buffeting by the wind. The increasing size of IR detector arrays however (see previous discussion) has now much reduced the necessity for chopping in this manner.

[70] An additional practical constraint on mirror size is that imposed by the need to transport the mirror from the manufacturer to the telescope's observing site. Even with 8-m mirrors, roads may need widening, obstructions levelling, bends have to have their curvature reduced and so on and this can add significantly to the cost of the project.

FIGURE 1.59 The 2.5-m Stratospheric Observatory for Infrared Astronomy (SOFIA) telescope seen through the open observing hatch in the side of its Boeing 747 aircraft in flight. (Courtesy of National Aeronautics and Space Administration [NASA]/Carla Thomas.)

Most telescopes designed for shorter wavelengths can also be used in the IR insofar as their observing conditions permit.

Ideally, though, for NIR observations, the telescope must be sited above as much of the atmospheric water vapour as possible, limiting the best sites to places like Mauna Kea on Hawaii, the Chilean altiplano or the Antarctic plateau. For MIR and some FIR work it may be necessary also to cool the telescope, or parts of it, to reduce its thermal emission and such telescopes have to be lifted above most of the atmosphere by balloon or aircraft or flown on a spacecraft. In fact, since UKIRT (built 1979), all major specialist IR telescopes have been or will be aircraft or space-based to eliminate *any* atmospheric absorptions.

Thus, SOFIA (Figure 1.59; see also Section 1.1.15.4) recently started observing at IR wavelengths and is a 2.5-m telescope flown on board a Boeing 747 aircraft which flies at altitudes up to 13,700 m. Its instruments can cover wavelengths from 300 nm to 200 μm. Similarly, Spitzer, the 3.5-m Herschel Space Observatory and the 2.4-m Wide Field Infrared Survey Explorer (WISE) instrument have operated or continue to operate from space. The JWST will be the next major IR telescope and the 2.4-m Wide Field Infrared Survey Telescope (WFIRST) spacecraft, currently planned to observe from 500 nm to 2 μm with a field of view of 0.28 deg^2, may be launched in the mid-2020s.

The surface accuracy requirements may also be relaxed if the system is required as a 'light bucket' rather than for imaging. Limits of ten times the operating wavelength may still give acceptable results when a photometer or spectroscope is used on the telescope, providing that close double stars of similar magnitudes and extended sources are avoided. The Observatoire de Haute Provence's Carlina telescope concept, for example, envisages utilising a natural bowl-shaped valley on the Plateau de Calern[71] and lining it with up to a hundred smallish spherical mirrors to synthesise an unfilled spherical mirror up to 100 m in diameter. The detectors and correcting optics would be suspended from a helium balloon at the focal point of the mirror. If operated with adaptive optics (see below) and used as an interferometer such an instrument might achieve milliarcsecond resolutions. To date, though, only a demonstration prototype with three 0.25-m mirrors separated by about 10 m has actually been built, though this has shown that the principle works.

[71] Often called a Large Diluted Telescope (LDT). Cf. The Arecibo radio telescope (Section 1.2), the Giant Magellan Telescope (Section 1.1.18.2) and multi-element interferometers (Section 2.5).

The atmosphere is completely opaque to radiation shorter than about 320 nm. Telescopes designed for UV observations, therefore, have to be launched on rockets or carried onboard spacecraft (see also Section 1.1.16). Except in the extreme ultraviolet ([EUV] ~6 to ~90 nm), where glancing optics are needed (see Section 1.3), conventional telescope designs are used. Aluminium suffices as the mirror coating down to 100 nm. At shorter wavelengths, silicon carbide may need to be used.

UV Transient Astronomy Satellite (ULTRASAT) is planned for a 2023 launch into a geostationary orbit and intended to be able to acquire transient UV sources (such as those from γ-ray bursters and gravitational wave events) very rapidly. It will use a variant of the Schmidt camera with a 15° field of view, a 100-megapixel mosaic of 2D-doped (see delta doping – Section 1.1.8.1), 2k × 4k UV-enhanced CCD detector arrays and cover the UV spectrum from 220 to 280 nm.

1.1.20 Telescopes in Space

The most direct way to improve the resolution of a telescope that is limited by atmospheric degradation is to lift it above the atmosphere, or at least above the lower, denser parts of the atmosphere. The three main methods of accomplishing this are to place the telescope onto an aircraft, a balloon-borne platform or an artificial satellite or sounding rocket. The Kuiper Airborne Observatory (KAO), a 0.9-m telescope on board a C141 aircraft that operated until 1995, and SOFIA (see Section 1.1.15.4) are examples of the first, but aircraft are limited to maximum altitudes of about 15 km.

Balloon-borne telescopes, of which the 0.9-m Stratoscope II is probably the best known but far from the only example, have the advantages of relative cheapness and that their images may be recorded by fairly conventional means. Their disadvantages include low maximum altitude (40 km) and short flight duration (about 100 days maximum). There have also been problems with crash landings of the instruments upon their return to the Earth's surface, with ensuing extensive damage or even their complete destruction. For example, Balloon-borne Large Aperture Submillimeter Telescope (BLAST), a 2-m diameter submillimetre telescope was so badly damaged after its third flight that it had to be completely re-built and High Energy, Replicated Optics (HERO), an X-ray telescope, was a total write-off after a crash in 2010. BLAST was recently launched on its fifth and final mission and Sunrise, a 1-m Gregorian solar telescope successfully observed the Sun in 2009 and 2013. Balloon-borne telescopes are often launched from near to the North or South poles where the wind patterns tend to cause them to move in circular paths around the pole. This both facilitates recovery and enables long duration observations of single objects to be undertaken. Thus, the E and B Experiment (EBEX) telescope was launched from Antarctica in 2012 and achieved a 26-day flight. It comprised a 1.4-m Dragone telescope with 1,320 TES detectors observing the CMB between 150 and 450 GHz (2 mm and 670 μm).

The recent development of super-pressure balloons (see also Section 1.5.2.1.2) may increase the usefulness of balloons as observing platforms. Most high-altitude balloons to date have been of the variable volume design; their internal pressure equals that of the surrounding atmosphere. Their volume, therefore, changes as their temperature and/or the external pressure changes. Their buoyancy thus also changes and so they move up and down within the Earth's atmosphere. Variable volume balloons have the advantage of being able to be made from very thin material because the balloon's skin is not under tension but can only keep to a (reasonably) constant altitude by venting gas or by adding it from gas bottles, and this limits their lifetime. Super-pressure balloons use a stronger skin and are at a slightly higher pressure than their surrounding atmosphere. They, thus, have a constant volume and so also a much more constant buoyancy. Their altitude is therefore much easier to maintain close to the desired level. In 2016, NASA achieved a duration of nearly 47 days for a super-pressure balloon flight launched from New Zealand. The currently under development, Super-pressure Balloon-borne Imaging Telescope (SuperBIT) mission, planned for a 100-day or longer mission in 2020, is expected to carry a 0.5-m Ritchey-Chrétien telescope imaging over the 300- to 900-nm region with a resolution of 250 milliarcsecond, a field of view of 0.5° and a hoped-for stability of 20 milliarcsecond.

Many small UV, visual, and IR telescopes have already been launched or are due for launch soon on satellites and mention has already been made of several of them: HST, Herschel, JWST, Spitzer and Gaia. Other examples include Kepler (2009–2018) with a 0.95-m telescope to search for exoplanets, High Precision Parallax Collecting Satellite (Hipparcos,[72] 1989–1993), an astrometric spacecraft carrying a 0.29-m telescope and WISE (2009–2011) with a 0.4-m telescope. Of course, in addition to these examples, most planetary exploration spacecraft such as Cassini-Huygens (1997–2017), Messenger (2004–2015), New Horizons (2006–to date) and the forthcoming Mars 2020 and ExoMars 2020 carry or will carry small optical telescopes.

In the future it is possible that ambitious plans for the direct imaging of exoplanets using numerous spacecraft to form interferometers may be realised, but at the time of writing all such proposals remain simply concept studies.

1.1.21 Mountings

The functions of a telescope (or other optical instrument) mounting are simple – to hold the optical components in their correct mutual alignment and to direct the optical axis towards the object to be observed. The problems in successfully accomplishing this to within the accuracy and stability required by astronomers are so great, however, that the cost of the mounting can be the major item in funding a telescope. We may consider the functions of a mounting under three separate aspects; first, supporting the optical components, secondly preserving their correct spatial relationships and thirdly acquiring and holding the object of interest in the field of view.

Mounting the optical components is largely a question of supporting them in a manner that ensures their surfaces maintain their correct shapes. Lenses may only be supported at their edges, and it is the impossibility of doing this adequately for large lenses that limits their practicable sizes to a little more than a metre. There is little difficulty with smaller lenses and mirrors because most types of mount may grip them firmly enough to hold them in place without at the same time straining them. However, large mirrors require careful mounting. They are usually held on a number of mounting points to distribute their weight and these mounting points, especially those around the edges, may need to be active so that their support changes to compensate for the different directions of the mirror's weight as the telescope moves. The active support can be arranged by systems of pivoted weights, especially on older telescopes or, more normally nowadays, by computer control of active supports. As discussed previously, some recently built telescopes and many of those planned for the future have active supports for the primary mirror that deliberately stress the mirror so that it retains its correct shape whatever the orientation of the telescope or the temperature of the mirror. Segmented mirror telescopes additionally need the supports to maintain the segments in their correct mutual alignments. On the 10-m Keck telescopes, for example, the active supports adjust the mirror segment's positions twice a second to maintain their places correctly to within ±4 nm.

The optical components are held in their relative positions by the telescope tube. With most large telescopes, the 'tube' is in fact an open structure, but the name is often still retained. For small instruments the tube may be made sufficiently rigid that its flexure is negligible. But this becomes impossible as the size increases. The solution then is to design the flexure so that it is identical for both the primary and secondary mirrors. The optical components then remain aligned on the optical axis but are no longer symmetrically arranged within the mounting. A structure used frequently some decades ago, before computer-aided design (CAD) became available, which allows this equal degree of flexure and also maintains the parallelism of the optical components is the Serrurier truss (Figure 1.60). However, this design is not without its problems. The lower trusses may need to be excessively thin or the upper ones excessively thick to provide the required amount of flexure. Even so, many of the large reflecting telescopes built during the twentieth century have used Serrurier truss designs for their tubes.

[72] N.B. although named in tribute to Hipparchus of Nicea (c. 190 BCE to c. 120 BCE) the spelling of the spacecraft name differs from that of the person.

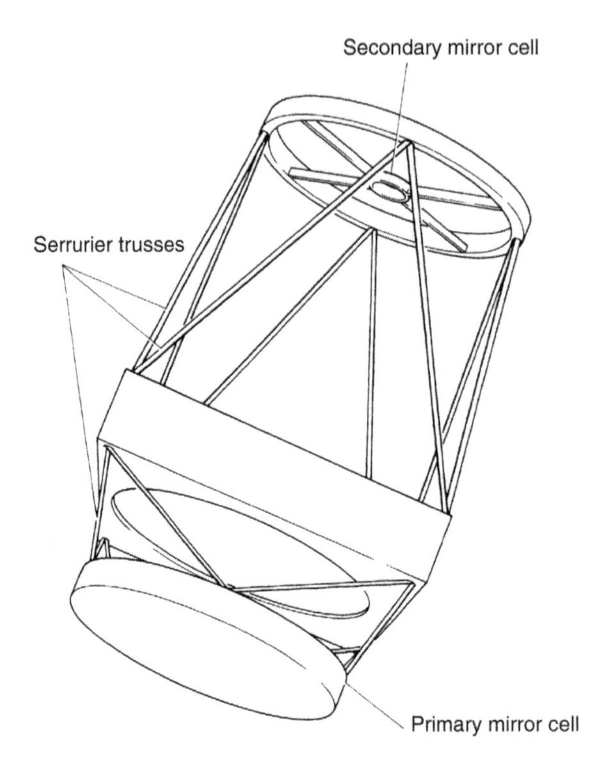

FIGURE 1.60 Telescope tube construction based on Serrurier trusses.

More recently CAD has enabled other truss systems to be developed. This, however, is mostly for economic reasons; the design objectives of such supports remain unchanged.

A widely used means of mounting the telescope tube so that it may be pointed at an object and then moved to follow the object's motion across the sky is the equatorial mounting. This is a two-axis mounting, with one axis, the polar axis, aligned parallel with the Earth's rotational axis and the other, the declination axis, perpendicular to the polar axis. The design has the enormous advantage that only a single constant velocity motor is required to rotate the mounting around the polar axis to track an object. It is also convenient in that angular read-outs on the two axes give the hour angle or right ascension and declination directly.

A large variety of different arrangements for the basic equatorial mounting exist, but these will not be reviewed here. Books on telescopes and most general astronomy books list their details by the legion and should be checked if the reader requires further information.

The alt-az mounting, which has motions in *alt*itude and *az*imuth, is the other main two-axis mounting system. Structurally it is a much simpler form than the equatorial and has, therefore, long been adopted for most large radio telescope dishes. Its drawbacks are that the field of view rotates with the telescope motion[73] and that it needs driving continuously in both axes and with variable speeds to track an object. For the last two decades, most new large optical telescopes have also used alt-az mountings and the reduction in price and increase in capacity of small computers means that smaller commercial telescopes even down to a few centimetres in size frequently have such a mounting as an option.

[73] Large alt-az mounted telescopes use field de-rotators, often in the form of counter-rotating Dove prisms, to overcome this problem. Some manufacturers of small telescopes also offer field de-rotators as accessories, but they add considerably to the instrument's cost. A cheaper solution to obtaining sharp time-exposure images using small alt-az telescopes is to note that there are periods when the field rotation ceases for any object whose declination is nearer to the equator than the observer's latitude. Details of how to find such opportunities for an observer at a particular latitude and for an object at a particular declination are given, for example, in the author's *Telescopes and Techniques*, third edition (2013).

Telescopes up to 1 m in diameter constructed for or by amateur astronomers are often mounted on a Dobsonian alt-azimuth mounting. This may be simply and cheaply constructed from sheets of plywood or similar material. The telescope is supported by two circles, which rest in two semi-circular cut-outs in a three-sided cradle shaped like an inverted Greek letter 'pi' (⊔). These allow the telescope to move in altitude. The cradle then pivots at its centre to give motion in azimuth. Telescopes on such mountings are usually undriven and are moved by hand to find and follow objects in the sky. Small telescopes on any alt-azimuth design of mounting can be given a tracking motion for a short interval of time by placing them on a platform that is itself equatorially driven. There are several designs for such equatorial platforms using sloping planes or inclined bearings and details may be found in modern books aimed at the amateur observer.

Large telescopes, whether on equatorial or alt-azimuth mountings, almost universally use hydrostatic bearings for the main moving parts. These bearings use a thin layer of pressurised oil between the moving and static parts and the oil is continuously circulated by a small pump. The two solid surfaces are, thus, never in direct contact and movement round the bearing is smooth and of low friction.

Several telescopes have fixed or partially fixed positions. These include liquid-mirror telescopes that always point near the zenith (see above) and the 11-m Hobby-Eberly and SALT telescopes (see Section 1.1.8.1) which can only move in azimuth. Tracking is then accomplished by moving the detector and any correcting optics to follow the motion of the image, or for short exposures, by moving the charges along the pixels of a CCD detector (TDI, see Section 1.1.8.1). Zenith-pointing telescopes are generally limited to a few degrees either side of the zenith, but the Hobby-Eberly telescope can point to objects with declinations ranging from $-11°$ to $+71°$. It accomplishes this because it points to a fixed zenith angle of 35° but can be rotated in azimuth between observations. There are also a few instruments mounted on altitude-altitude mountings. These use a universal joint (such as a ball and socket) and tip the instrument in two orthogonal directions to enable it to point anywhere in the sky.

With any type of mounting, acquisition and tracking accuracies of an arcsecond or better are required, but these may not be achieved especially for the smaller Earth-based telescopes. Thus, observers may have to search for their object of interest over an area of the sky tens to hundreds of arcseconds across after the initial acquisition and then guide, either themselves, or with an automatic system, to ensure that the telescope tracks the object sufficiently closely for their purposes. With balloon-borne and space telescopes such direct intervention is difficult or impossible. However, the reduction of external disturbances and possibly the loss of weight mean that the initial pointing accuracy is higher. For space telescopes tracking is easy because the telescope will simply remain pointing in the required direction once it is fixed, apart from minor perturbations such as light pressure, the solar wind, gravitational anomalies and internal disturbances from the spacecraft. Automatic control systems can, therefore, usually be relied upon to operate the telescope. Previous spacecraft such as the International Ultra-violet Explorer (IUE; 1978–1996), however, had facilities for transmitting a picture of the field of view, after the initial acquisition, to the observer on the ground, followed by corrections to the position to set onto the desired object.

With terrestrial telescopes the guiding may be undertaken via a second telescope which is attached to the main telescope and aligned with it, or a small fraction of the light from the main telescope may be diverted for the purpose. The latter method may be accomplished by beam splitters, dichroic mirrors, or by intercepting a fraction of the main beam with a small mirror, depending upon the technique that is being used to study the main image. The whole telescope can then be moved using its slow-motion controls to keep the image fixed, or additional optical components can be incorporated into the light path whose movement counteracts the image drift, or the detector can be moved around the image plane to follow the image motion or the charges can be moved within a CCD chip (TDI). Since far less mass needs to be moved with the latter methods, their response times can be much faster than that of the first method. The natural development of the second method leads on to the active surface control, which is discussed later in this section and it can go some way towards eliminating the effects of scintillation.

If guiding is attempted by the observer then there is little further to add, except to advise a plentiful supply of black coffee to avoid going to sleep through the boredom of the operation. Automatic guiding has two major problem areas. The first is sensing the image movement and generating the error signal, while the second is the design of the control system to compensate for this movement. Detecting the image movement has generally been attempted in two main ways: CCDs or other array detectors and quadrant detectors.

With a CCD guide, the guide image is read out at frequent intervals and any movement of the object being tracked is detected via software that also generates the correction signals. The guide CCD is adjacent to the main detector (which may or may not be a CCD itself), and thus, it guides on objects other than those being imaged. This is termed 'off-set guiding' and often has advantages over guiding directly on the object of interest. The off-set guide star can be brighter than the object of interest, and if the latter is an extended object, then guiding on it directly will, in any case, be difficult. However, for comets, asteroids and other moving objects, the object itself must be followed.

In some small CCD systems aimed at the amateur market, it is the main image that is read-out at frequent intervals and the resulting multiple images are then shifted into mutual alignment before being added together (often known as 'shift and add' – see also 'lucky imaging' in Section 1.1.22). This enables poor tracking to be corrected provided that the image shift is small during any single exposure. The shift and add technique also lies behind the proposed GravityCam instrument for the 3.6-m New Technology Telescope (NTT) although here it would be atmospheric movement of the images (Section 1.1.22), rather than poor tracking, that would be corrected. The instrument would obtain images at a rate of about 25 per second and these would be put into their best mutual alignments to produce the final image. It is estimated that the resolution would be improved by a factor of 2.5–3 in this manner.

A quadrant detector is one whose detecting area is divided into four independent sectors (i.e., a 2×2-pixel array), the signal from each of which can be separately accessed. If the image is centred on the intersection of the sectors, then their signals will all be equal. If the image moves, then at least two of the signals will become unbalanced and the required error signal can then be generated. A variant of this system uses a pyramidal 'prism'. The image is normally placed on the vertex and so is divided into four segments each of which can then be separately detected. When guiding on stars by any of these automatic methods, it may be advantageous to increase the image size slightly by operating slightly out of focus, scintillation 'jitter' then becomes less important.

The problem of the design of the control system for the telescope to compensate for the image movement should be handed to a competent control engineer. If it is to work adequately, then proper damping, feedback, rates of motion and so on, must be calculated so that the correct response to an error signal occurs without 'hunting' or excessive delays.

1.1.22 Real-Time Atmospheric Compensation

The Earth's atmosphere causes astronomers many problems: it absorbs light from the objects that you want to look at, it scatters light into the telescope from objects that you do not want to look at and its clouds, rain and snow are distinctly inconvenient at times.

Here though we are concerned with two other effects that arise from turbulence and inhomogeneity within the atmosphere – scintillation (or twinkling) and seeing. Scintillation is the rapid change in the brightness of a stellar image[74] as more or less light from the incoming beam is scattered out of the beam by irregularities within the atmosphere (Figure 1.61). Seeing is the slight movement of

[74] Small fragments of the images of extended objects, like planets, also undergo scintillation. However, because the overall image is made up from many such fragments, some of which will be brightening whilst others will be getting fainter, the scintillation is much less apparent. The widely believed maxim that you can distinguish a star from a planet with the unaided eye because the latter does not twinkle, arises from this phenomenon, but even planets will twinkle when the atmospheric turbulence is high enough.

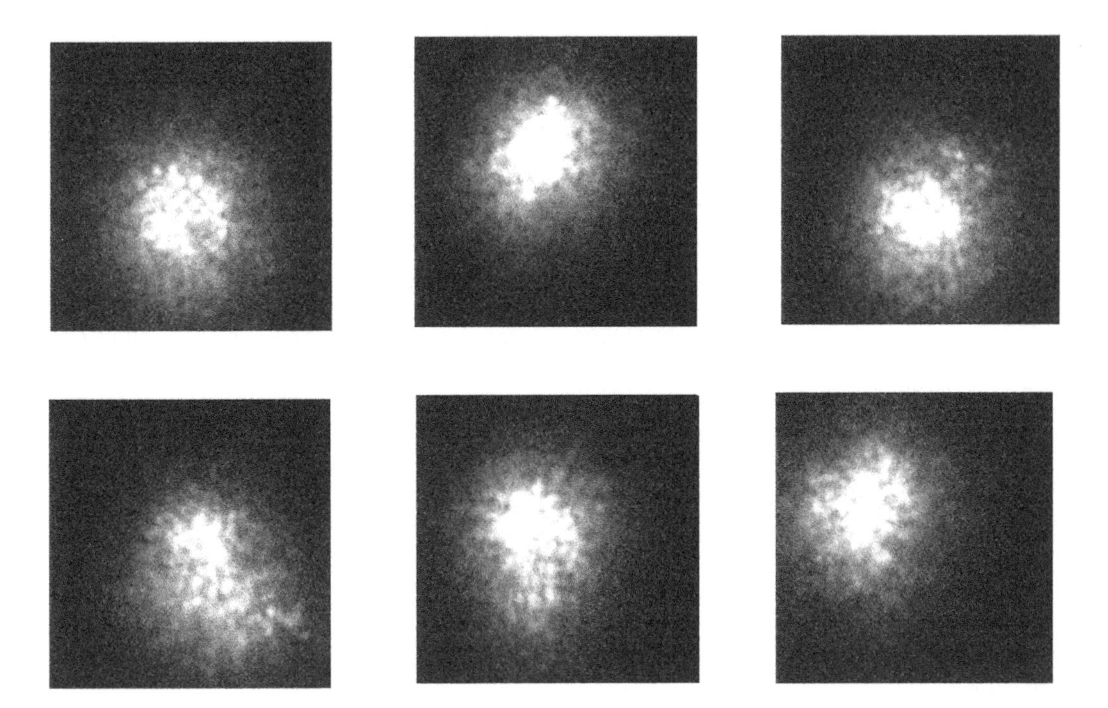

FIGURE 1.61 Scintillation and seeing. Images of a star obtained in July 2000 on the William Herschel Telescope (WHT). The changing brightness (scintillation) and the image movement (seeing) are both apparent. The smallest of the individual speckles forming the images (see Section 2.6) are diffraction-limited images of the star from the 4.2-m instrument. (Courtesy of Nick Wooder. The full animation can be seen at http://apod.nasa.gov/apod/ap000725.html.)

the stellar image away from its true position arising from refraction at boundaries between different layers within the atmosphere.[75] Scintillation and seeing both change on timescales ranging from a hundredth to several seconds. Images obtained with exposures longer than around 0.01 s will therefore be blurred into the 'seeing disk' whose diameter can be from 0.5 to 5″ or more, irrespective of the actual resolution of the telescope's optics. For about the last four decades, therefore, increasingly successful attempts have been made to reduce or eliminate the effects of scintillation and seeing so that large telescopes can perform to their theoretical capabilities. One procedure for doing this, known as Real-Time Atmospheric Compensation (or sometimes as Adaptive Optics), is discussed in this section.[76] Alternative approaches such as interferometry, speckle interferometry, and occultations are covered in Sections 2.5, 2.6, and 2.7, respectively.

The resolution of ground-based telescopes of more than a fraction of a metre in diameter is limited by the turbulence in the atmosphere. The maximum diameter of a telescope before it becomes seriously affected by atmospheric turbulence is given by Fried's coherence length, r_0,

$$r_o \approx 0.114\left(\frac{\lambda \cos z}{550}\right)^{0.6} m \tag{1.60}$$

[75] In the widely used Kolmogorov model of the atmosphere, the refractive index variations arise within thin layers of the atmosphere each of which has a different intensity for the turbulence.

[76] Related procedures applicable in the microwave and radio regions are discussed in Section 1.2.

where λ is the operating wavelength in nm and z is the zenith angle. Fried's coherence length, r_o, is the distance over which the phase difference is one radian. Thus, for visual work, telescopes of more than about 11.5 cm (4.5 in) diameter will always have their images degraded by atmospheric turbulence.

In an effort to reduce the effects of this turbulence, many large telescopes are sited at high altitudes or placed on board high-flying aircraft or balloons. The ultimate, though expensive solution, of course, is to orbit the telescope beyond the Earth's atmosphere out in space. Recently significant reductions in atmospheric blurring have been achieved through 'lucky imaging'. In this process, numerous short exposure images are obtained of the object (cf. Speckle Interferometry, Section 2.6 and Shift and Add, Section 1.1.21). The sharpest images are selected from their Strehl ratios (see below) and then combined to produce the final image. In this way the final image can approach the diffraction limit for 2- or 3-m telescopes at visible wavelengths. However, many of the images may need to be discarded, making the process expensive in telescope time. A combination of Adaptive Optics (see below) and Lucky Imaging was successfully tried on the Hale telescope in 2007 giving 35 milliarcsecond NIR images. A second-generation Adaptive Optics and Lucky Imaging (AOLI) instrument started operations in 2016 on WHT (and which is planned to be transferred to the 10.4-m Gran Telescopio Canarias [GTC]) and a much lighter-weight version of the instrument, Adaptive and Lucky Imaging Optics Lightweight Instrument (ALIOLI), is under development. It has recently been suggested that a combination of tip-tilt correctors and lucky imaging might enable natural guide stars as faint as 17.5^m to be used in more complete adaptive optics systems (see below).

An alternative approach to obtaining diffraction-limited performance for large telescopes, especially in the NIR and MIR, which is relatively inexpensive (when compared with the cost of the telescope[77]) and widely applicable, is to correct the distortions in the incoming light beam produced by the atmosphere. This atmospheric compensation is achieved through the use of adaptive optics. In such systems, one or more of the optical components can be changed rapidly and in such a manner that the undesired distortions in the light beam are reduced or eliminated. Adaptive optics is actually an ancient technique familiar to us all. That is because the eye operates via an adaptive optic system to keep objects in focus, with the lens being stretched or compressed by the ciliary muscle.

Adaptive optic systems, in some form, have been added to almost all major existing optical and NIR telescopes,[78] and they are automatically a part of the package whenever, now, new ground-based telescopes are being considered.

The efficiency of an adaptive optics system is measured by the Strehl ratio. This quantity is the ratio of the intensity at the centre of the corrected image to that at the centre of a perfect diffraction-limited image of the same source. The normalised Strehl ratio is also used. This is the Strehl ratio of the corrected image divided by that for the uncorrected image. A Strehl ratio of close to 100% (perfection) has been realised 10 μm wavelengths by the 6.5-m MMT using a deformable secondary mirror, while recently the LBT has achieved 95% at 5 μm and 60%–90% at 1.8 μm. For comparison the HST's mirror at 400 nm has a Strehl ratio of 60%.

[77] Additional, perhaps unexpected, costs may arise because better images may then require upgrades of the telescope mount and drive system, its optics and optical alignments, temperature control systems, etc.

[78] The most basic form of adaptive optics image correction is via a tip-tilt system (Section 1.1.22.1). A simple 20 Hz tip-tilt adaptive optics system was used between 1992 and 1995 on the then 85-year-old 1.5-m telescope at the Mount Wilson Observatory. Amateur astronomers may have noticed some commercial astronomical equipment suppliers advertising tip-tilt image correction systems. These, however, are for correcting tracking errors in the telescope's drive system (guiding, see Section 1.1.19.2), *not* for correcting image distortions arising from the atmosphere. Telescopes of about 1-m diameter seem, at the time of writing, to be about the smallest ones to be working with 'proper' adaptive optics systems. The Pomona College's 1-m teaching telescope, for example, has the KAPAO (unofficially derived from 'Kid-Assembled Pomona Adaptive Optics') adaptive optics system which has been largely developed internally and partially via students' project and thesis work. The current NPOI's up-grade will also involve 1-m adaptive optic telescopes (Section 2.5.4).

Adaptive optics is not a complete substitute for spacecraft-based telescopes, however, because in the visual IR and NIR, the correction only extends over a small area (the isoplanatic patch, see below – although multi-conjugate adaptive optics, Section 1.1.22.3, is improving this problem somewhat). Additionally, of course, ground-based telescopes are still limited in their wavelength coverage by atmospheric absorption.

There is often some confusion in the literature between adaptive optics and active optics. However, the most widespread usage of the two terms is that an adaptive optics system is a fast closed-loop system and an active optics system a more slowly operating open- or closed-loop system. The division is made at a response time of a few seconds. Thus, the tracking of a star by the telescope drive system can be considered as an active optics system that is open loop if no guiding is used and closed-loop if guiding is used. Large thin mirror optical telescopes and radio telescopes may suffer distortion because of buffeting by the wind at a frequency of a tenth of a hertz or so, they may also distort under gravitational loadings or thermal stresses and have residual errors in their surfaces from the manufacturing process. Correction of all of these effects would also be classified under active optics. There is additionally the term 'active support' that refers to the mountings used for the optical components in either an adaptive or an active optics system.

An atmospheric compensation system contains three main components, a sampling system, a wavefront sensor and a correcting system. We will look at each of these in turn.

1.1.22.1 Sampling System

The sampling system provides the sensor with the distorted wavefront or an accurate simulacrum thereof. For astronomical adaptive optics systems, a beam splitter is commonly used. This is just a partially reflecting mirror that typically diverts about 10% of the radiation to the sensor, while allowing the other 90% to continue on to form the image. A dichroic mirror can also be used which allows all the light at the desired wavelength to pass into the image while diverting light of a different wavelength to the sensor. However, atmospheric effects change with wavelength and so this latter approach may not give an accurate reproduction of the distortions unless the operating and sampling wavelengths are close together.

Because astronomers go to great lengths and expense to gather photons as efficiently as possible, the loss of even 10% to the sensor is to be regretted. Most adaptive optics systems now, therefore, use a guide star rather than the object of interest to determine the wavefront distortions. This becomes essential when the object of interest is a large extended object because most sensors need to operate on point or near-point images. The guide star must be close in the sky to the object of interest, or its wavefront will have undergone different atmospheric distortion (Figure 1.62). For solar work small sunspots or pores or just nearby areas of the solar surface or entire prominences can be used as the guide object. The Dunn solar telescope (Section 5.3.2), for example, has tested an adaptive optics system for imaging prominences off the edge of the solar limb using a Shack-Hartmann sensor and a different prominence as its natural guide 'star'.

The region of the sky over which images have been similarly affected by the atmosphere is called the 'isoplanatic area or patch'. It is defined by the distance over which the Strehl ratio improvement due to the adaptive optics halves. In the visible it is about 15″ across. The size of the isoplanatic patch scales as $\lambda^{1.2}$, so it is larger in the IR, reaching 80″ at 2.2 μm (K band – see Section 3.1). With several guide stars (Multi-Conjugate Adaptive Optics [MCAO] and Multi Object Adaptive Optics [MOAO] – see below) corrected areas of 1 to 2 square arc-minutes are currently possible and plans for the Multi Conjugate Adaptive Optics Relay (MAORY) instrument on the ELT envisage corrected areas in the NIR up to 40 square arc-minutes in area (Figure 1.63).

The small size of the isoplanatic area means that few objects have suitable guide stars. Less than 1% of the sky can be covered using real stars as guides, even in the IR. Recently,[79] therefore,

[79] The idea dates back to about 1980 but was originally classified because of its application to the Strategic Defense Initiative (Star Wars). It only started to be used astronomically from about the year 2000.

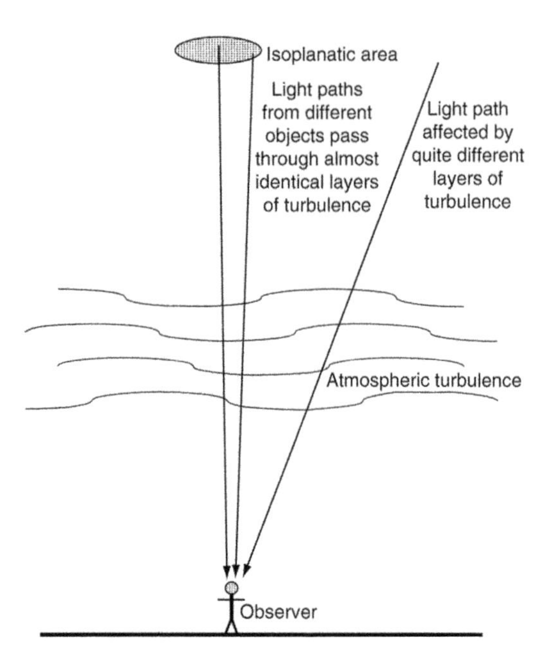

FIGURE 1.62 The isoplanatic area.

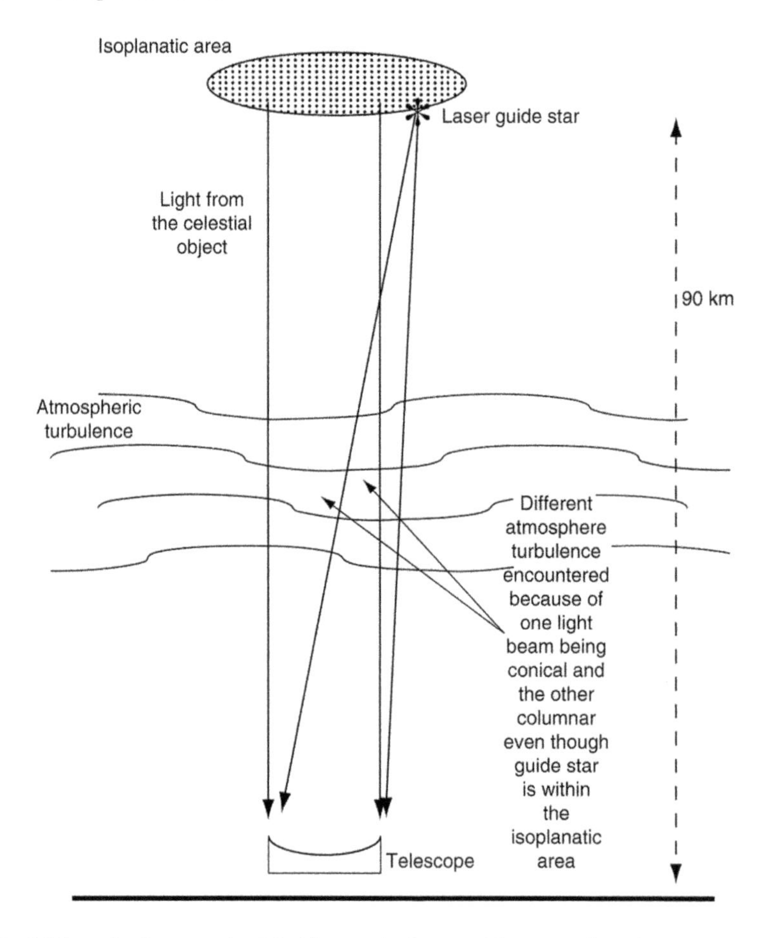

FIGURE 1.63 Light paths from a celestial object and a laser guide star to the telescope.

FIGURE 1.64 A single laser beam being sent from the Very Large Telescope's (VLT's) Yepun telescope to produce an artificial guide star. *See also* Figure 1.68. (Courtesy of European Southern Observatory/Y. Beletsky.)

artificial guide stars have been produced. This is accomplished using a powerful laser[80] pointed skywards. The laser is tuned to one of the sodium D line frequencies and excites the free sodium atoms in the atmosphere at a height of about 90 km (Figure 1.64). The glowing atoms appear as a star-like patch that can be placed as near in the sky to the object of interest as required.

Guide stars at lower heights and at other wavelengths can be produced through back scattering by air molecules of a laser beam. Because these latter guide stars are produced via Rayleigh scattering, they are sometimes called 'Rayleigh stars'. They can be of any wavelength, but a laser operating in the green or near UV is often chosen. The laser light can be sent out through the main telescope or more usually using an auxiliary telescope mounted on the main telescope.

Laser-produced guide stars have two problems that limit their usefulness. First, for larger telescopes, the relatively low height of the guide star means that the light path from it to the telescope differs significantly from that for the object being observed (the cone problem, Figure 1.63). At 1 μm, the gain in Strehl ratio is halved through this effect and in the visible it results in almost no improvement at all to the images. Secondly, the outgoing laser beam is affected by atmospheric turbulence, and therefore, the guide star moves with respect to the object. This limits the correction of the overall inclination (usually known as the 'tip-tilt') of the wavefront resulting in a blurred

[80] Continuous wave lasers with powers of up to 50 W may be used. These radiate sufficient energy to cause skin burns and retinal damage. Care, therefore, has to be taken to ensure that the beam does not intercept an aircraft. The Very Large Telescope (VLT), for example, uses a pair of cameras and an automatic detection system to close a shutter over the laser beam should an aircraft approach. The lasers are also powerful enough to damage the optics on board some spacecraft. Similar precautions, thus, need to be taken in this respect, although these are somewhat easier to accomplish because the positions of potentially vulnerable satellites in the sky can be predicted well ahead of time. On sites with many telescopes – such as Cerro Tololo, Mauna Kea, Paranal and Teide – the laser beam(s) from one telescope must also be kept from interfering with the other telescopes. At Paranal, for example, there is a computer-based 'traffic control system' which alerts operators to potential clashes.

image on longer exposures. A real star can, however, be used to determine the tip-tilt of the wavefront separately.[81] The wavefront sensor simply needs to detect the motion of the star's image and because the whole telescope aperture can be utilised for this purpose, faint stars can be observed. So, most objects have a suitable star sufficiently nearby to act as a tip-tilt reference.

An adaptive optics system using two or more (up to a dozen) guide stars, known as MCAO, eliminates the cone effect and produces an isoplanatic patch up to 120″ across. The guide stars can be artificial and/or real stars and are separated by small angles and detected by separate wavefront sensors. This enables the atmospheric turbulence to be modelled as a function of altitude. Two or three subsidiary deformable mirrors then correct the wavefront distortion. A recent variation on MCAO is MOAO with several deformable correcting mirrors. This latter system also uses several guide stars and may be able to correct over areas up to 7 or more arc-minutes across. However, the correction is only undertaken for a few selected smaller areas within the overall field of view, the selected smaller areas, of course, being chosen to coincide with the positions of the objects of interest to the observer.

It has recently been suggested that lights carried by drones might be used to provide artificial guide stars. Clearly though, these would be at much lower altitudes than the 90 km heights of the sodium-emission-type guide stars and also stringent control of their (changing) positions with respect to the telescope would be needed.

1.1.22.2 Wavefront Sensing

The wavefront sensor detects the residual and changing distortions in the wavefront provided by the sampler after reflection from the correcting mirror. The Shack-Hartmann (also known as the Hartmann-Shack or the Hartmann) sensor is in widespread use in astronomical adaptive optics systems. This uses a 2D array of small lenses (Figure 1.65). Each lens produces an image that is

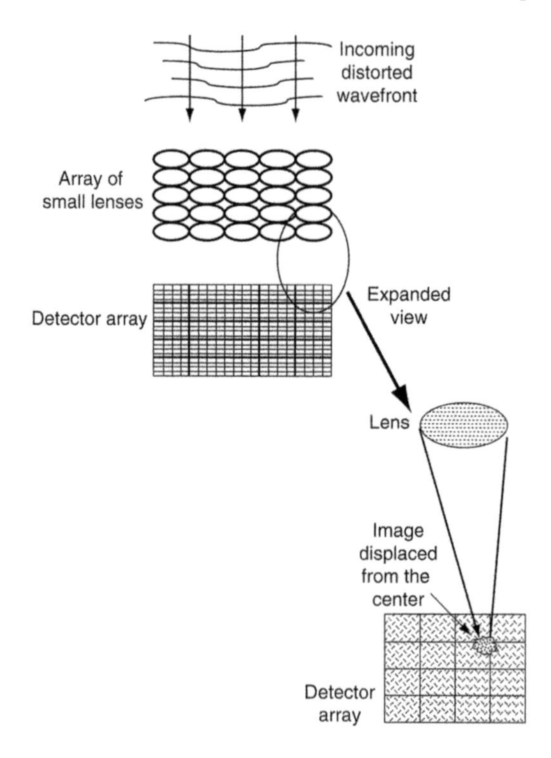

FIGURE 1.65 The Shack-Hartmann sensor.

[81] Adaptive optics with several guide stars (natural and/or artificial) is sometimes termed 'laser tomography'; this term is also widely applied in medical contexts though.

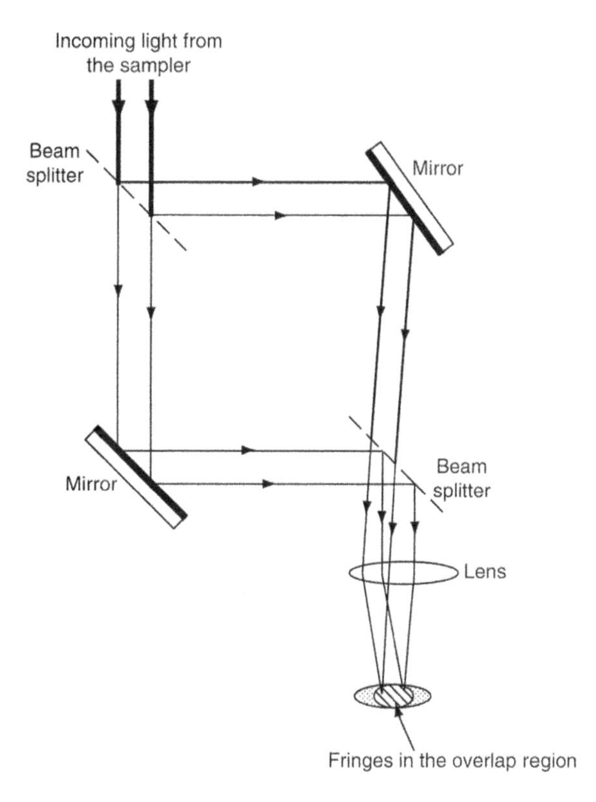

FIGURE 1.66 The shearing interferometer.

sensed by an array detector. In the absence of wavefront distortion, each image will be centred on each detector. Distortion will displace the images from the centres of the detectors and the degree of displacement and its direction is used to generate the error signal. An alternative sensor is based upon the shearing interferometer (Figure 1.66). This is a standard interferometer but with the mirrors marginally turned so that the two beams are slightly displaced with respect to each other when they are recombined. The deformations of the fringes in the overlap region then provide the slopes of the distortions in the incoming wavefront. The shearing interferometer was widely used initially for adaptive optics systems but has now largely been replaced by the Shack-Hartman sensor.

A new sensor has been developed recently, which is known as the 'curvature sensor'; it detects the wavefront distortions by comparing the illumination variations across slightly defocused images just inside and outside the focal point. A vibrating mirror is used to change the focus at kilohertz frequencies or two lenses are used; one just inside and one just outside the focal surface. Amongst other instruments curvature wavefront sensors are used on the VISTA and Mayall telescopes and four are planned to be used on the VRO.

Currently on solar telescopes like the 1.6-m Goode telescope at Big Bear Solar observatory, the 1.5-m GREGOR[82] telescope on Tenerife and the planned 4.2-m Daniel K. Inouye telescope on Hawaii, Shack-Hartman-type wavefront sensors are used with small parts of the solar surface itself forming the guide 'stars' for their single or multiple-corrector adaptive optic systems. In the future for solar work and perhaps for other extended sources, a Mach-Zehnder interferometer[83] may be usable to estimate the wavefront distortions from the interference patterns between the central and outer parts of the wavefront.

[82] A name.

[83] Essentially this is a double Michelson type interferometer (Section 1.6) that is used to determine the phase shift introduced into one of two light beams originating from a coherent source and arising from a sample placed within that light beam.

1.1.22.3 Wavefront Correction

In most astronomical adaptive optics systems, the correction of the wavefront is achieved by distorting a subsidiary mirror. Since the atmosphere changes on a timescale of 10 ms or so, the sampling, sensing and correction have to occur in a millisecond or less. In the simplest systems only the overall tip and tilt of the wavefront introduced by the atmosphere is corrected. That is accomplished by suitably inclining a plane or segmented mirror placed in the light beam from the telescope in the opposite direction (Figure 1.67). An equivalent procedure because the overall tilt of the wavefront causes the image to move, is 'shift and add' (Section 1.1.21). Multiple short exposure images are shifted until their brightest points are aligned and then added together. Even this simple correction, however, can result in a considerable improvement of the images.

More sophisticated approaches provide better corrections; either just of the relative displacements within the distorted wavefront or of both displacement and fine scale tilt. In some systems the overall tilt of the wavefront is corrected by a separate system using a flat mirror whose angle can be changed. Displacement correction would typically use a thin mirror capable of being distorted by up to several hundred piezoelectric or other actuators placed underneath it. The error signal from the sensor is used to distort the mirror in the opposite manner to the distortions in the incoming wavefront. The reflected wavefront is, therefore, almost flat. The non-correction of the fine scale tilt, however, does leave some small imperfections in the reflected wavefront. Nonetheless, currently operating systems using this approach can achieve diffraction-limited performance in the NIR for telescopes of 3- or 4-m diameter (i.e., about 0.2″ at 2 μm wavelength).

Most large telescopes now have an adaptive optics image correction system(s) available and many have MCAO and/or MOAO systems available or under construction at the time of writing.

The Subaru telescope's adaptive optics system (Adaptive Optics 188 [AO188]), for example, uses curvature wavefront sensors and a deformable mirror with 188 actuators in combination with a natural or a laser guide star and covers an isoplanatic area about 1 arc minute across. It can achieve angular resolutions down to about 50 milliarcseconds and Strehl ratios up to about 65%. It was used initially with the second-generation stellar coronagraph, High Contrast Instrument for the Subaru

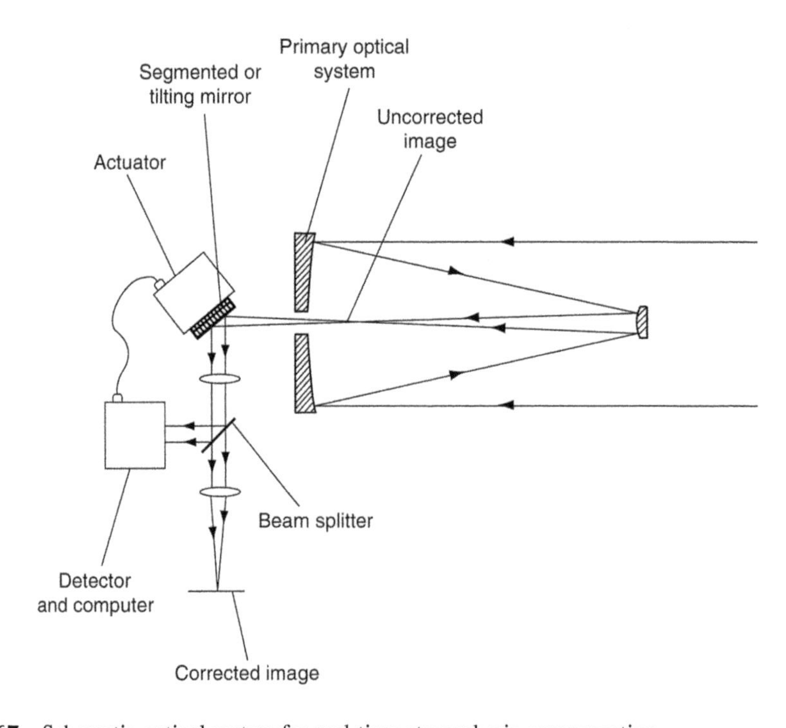

FIGURE 1.67 Schematic optical system for real-time atmospheric compensation.

next generation Adaptive Optics (HiCIAO; see Section 5.3) in an attempt to image extra-solar planets directly. The HiCIAO is now decommissioned. The MOAO system demonstrator, Raven, operated on the Subaru telescope in 2014 and 2015 using three natural stars and a single artificial star over a 3.5 arc-minute field of view and provided 150 milliarcsecond resolutions. Now, AO188 is working with the SCExAO instrument.

Gemini multi-conjugate adaptive optics system (GeMS) has been operating on the Gemini-South telescope since 2011 and provides a diffraction-limited field of view >60″ across in the NIR using five laser guide stars. Gemini North uses Altitude conjugate Adaptive Optics for the Infrared (ALTAIR) that can utilise either a single natural or a single artificial star, although the laser system for the latter is currently out of action (but is due to be reinstated at the time of writing). ALTAIR is based upon a 177-actuator deformable mirror and operates between 1 and 2.4 μm, sensing the wavefront deformations at visible wavelengths with a Shack-Hartmann detector. The 6.5-m Clay Magellan telescope is equipped with Magellan Adaptive Optics (MagAO) enabling it to obtain 20 milliarcsecond images in the visible. It uses an adaptive secondary mirror with 585 actuators, a pyramid wavefront sensor operating at 1 kHz and employs natural guide stars.

The GMT's adaptive optics system is planned to be based upon deformable secondary mirrors, each with more than 7,000 actuators operating at a kHz or more. Corrections will be able to be applied for ground-layer effects covering a field of view of over a sixth of a degree and for effects from higher in the atmosphere using a single natural star or six artificial stars plus a natural star and correcting a field of view up to 30″ across. The ESO VLT Yepun telescope's Adaptive Optics Facility (AOF) uses a 1.96-mm thick deformable secondary mirror with 1,170 actuators, four artificial guide stars (Figure 1.68) and operates at 1 kHz (see also Ground Layer Adaptive Optics Assisted by Laser [GRAAL]; Section 1.1.24).

FIGURE 1.68 The four 22-W artificial guide star laser beams for the Very Large Telescope (VLT) Yepun's Adaptive Optics Facility (AOF) system. *See also* Figure 1.64. (Courtesy of European Southern Observatory/G. Hüdepohl.)

The ELT will use MAORY (Figure 1.68) to correct for atmospheric distortions, and this instrument will use up to six artificial and up to three natural guide stars. Three deformable mirrors will be needed; one of which will be the 2.4-m fourth mirror of the telescope's main optical system. More than 8,000 actuators will be used to control the shape of the (nominally flat) fourth mirror. The instrument will operate over the 800 nm to 2.4 μm wavelength region and should provide a corrected area of the sky over 2 arc-minutes in diameter. For the TMT, it is planned to commission the MCAO Narrow Field Infrared Adaptive Optics System (NFIRAOS) instrument to provide diffraction limited images over a 34″ diameter field of view. It is expected to be based upon a pyramid-based Shack-Hartmann wavefront sensor and to have six artificial guide stars or to use a natural star or a small extended object. Two deformable mirrors will be required; it will be cooled to about −30°C and operate at about 800 Hz.

The artificial guide stars need to be as angularly small and stable as possible. To this end, the lasers producing GeMS' artificial stars themselves have their beams shaped by active optics. Two deformable mirrors are used with 120 actuators and the artificial stars' sizes are thereby reduced by about 15% and their brightness increased by about 40%.

A recent proposal to improve the correcting mirrors is for a liquid mirror based upon reflective particles floating on a thin layer of oil or ethylene glycol. The liquid contains nanometre-sized magnetic grains (a ferro-fluid), and its surface can be shaped by the use of small electromagnets, but this has yet to be applied to astronomical image correction. Potentially though such mirrors could offer strokes (the distance moved up or down) measured in millimetres instead of the few tens of μm offered by current flexible mirrors. At visual wavelengths, reductions of the uncorrected image size by about a factor of ten are currently being reached in the laboratory. Further improvements can sometimes be achieved by applying blind or myopic deconvolution (Section 2.1) to the corrected images after they have been obtained.

Correction of fine scale tilt within the distorted wavefront as well as displacement is now being investigated in the laboratory. It requires an array of small mirrors rather than a single 'bendy' mirror. Each mirror in the array is mounted on four actuators so that it can be tilted in any direction as well as being moved linearly. At the time of writing, it is not clear whether such systems will be applied to many large astronomical telescopes because at visual wavelengths few such telescopes have optical surfaces good enough to take advantage of the improvements that such a system would bring. Plans for future 50-m and 100-m telescopes however include adaptive secondary or tertiary mirrors up to 8 m in diameter, requiring up to 500,000 actuators.

Even when all the atmospheric corrections have been applied up to the current state-of-the-art levels, there remain some problems with the final images. In particular, the image may contain spurious features, generally termed 'Speckles' – though 'blobs' would give a clearer idea of their appearance (see also Section 2.7.4). The speckles can arise from several sources. The light path to the wavefront sensor must obviously differ that to the imaging system, even if only over a short distance. The error messages generated by the wavefront sensor may, therefore, not correct for all the wavefront distortions that are at the image, or it may correct for wavefront distortions that appear at the sensor but not at the image. Speckles may also arise from atmospheric distortions that are not corrected by the adaptive optics system or from imperfections in the optical system. In the future, the solution may be to use the data in the final image to generate the correcting signals.

All this section has, so far been concerned only with ground-based instruments, but future space-based instruments may also need wavefront corrections via adaptive optics. This may particularly apply to high contrast situations – such as imaging exoplanets close to their host stars (Section 2.7.4). It seems likely though, given the much harder problems that have needed to be solved for terrestrial adaptive optics, that suitable space-based instruments should be developable fairly easily in due course.

1.1.23 Future Developments

There are several telescopes significantly larger than 10 m either now in operation or under construction. The VLT is currently starting to act as an aperture synthesis system (Section 2.5.4) with sensitivity equal to that of a 16-m telescope and an unfilled aperture diameter of 100 m. For some years the Keck telescopes operated as an aperture synthesis system with a sensitivity equal to that of a 14-m telescope and an unfilled aperture diameter of 85 m, but their interferometer mode (not the telescopes) was mothballed in 2012.

The GMT has started construction with a planned completion date of 2025. It will be sited on Cerro Las Campanas in Chile and will comprise seven 8.4-m monolithic mirrors in a close packed array of 25.4 m across. Also under construction at the time of writing, the ELT is to have a 39-m diameter primary mirror formed from seven hundred and ninety-eight 1.4-m hexagonal segments (Figure 1.69). The construction of the instrument started in 2014 with ground breaking at Cerro Armazones in Chile (about 20 km away from the existing observatory complex at Paranal). First light is planned for around 2025.

The United States and Canada are pursuing the TMT project that is expected to have a 30-m primary mirror formed from 492 hexagonal segments and perhaps will start operations around 2027 (Figure 1.69). However, the project has run into legal difficulties and protests in getting planning permission for its construction on Hawaii. At the time of writing, it is not at all clear that the project will be able to go ahead, and it may need to be moved to another site; perhaps to one of the Canary isles.

With MCAO correction of the images (which might require 100,000 active elements), a 40-m class telescope would have a resolution of 1 milliarcsecond in the visible region. The limiting magnitude for such a telescope might be 35^m in the visible, not just because of its increased light grasp, but because a 1 milliarcsecond stellar image only has to equal the brightness of a 1-milliarcsecond-squared area of the background sky to be detectable. This would enable the telescope to observe Jupiter-sized exoplanets directly out to a distance of about 100 pc.

As a general guide to further into the future we may look to the past. Figure 1.70 shows the way in which the collecting area of optical telescopes has increased with time.

FIGURE 1.69 Top, An artist's concept of the Thirty Metre Telescope (TMT) in its dome. (Courtesy of the TMT project.). Bottom, A mock-up showing the scale of the mirror for the Extremely Large Telescope (ELT). (Courtesy of the European Southern Observatory.)

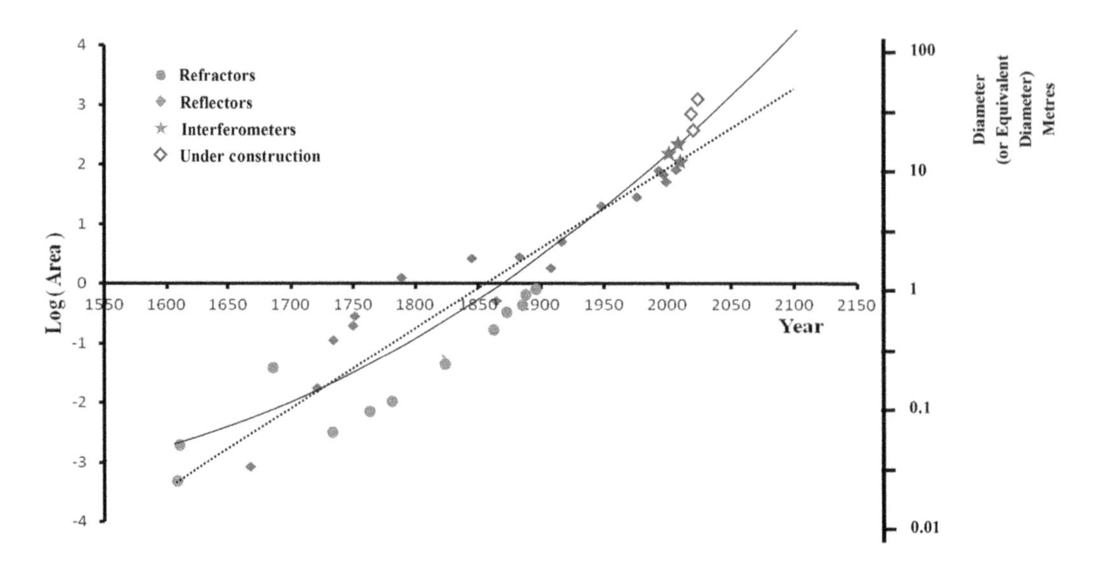

FIGURE 1.70 Optical telescope collecting area as a function of time. The logarithm to base ten of the area in square metres is plotted over the last four centuries for a selection of telescopes which were, in their time, the largest examples of their type. Linear (dotted line) and quadratic (solid line) curves have been fitted to the data. For multi-mirror telescopes (like the Giant Magellan Telescope [GMT]) and for interferometers (like the Very Large Telescope [VLT] interferometer [VLTI]), the diameter is that for the equivalent single circular aperture with the same total area.

Simple extrapolation of the trends shown there (dotted line) suggests that an optical telescope with a diameter of 100 m (or the equivalent total area) might be working around the year 2150. A quadratic extrapolation (solid line) suggests that target might be reached by 2080. However there appear to be no fundamental technical differences between constructing 10-m segmented mirrors like those for the Keck telescopes and 30-m or even 100-m segmented mirrors. It is just a case of doing the same thing more times and finding the increased funding required. As discussed previously, the fact that the largest fully steerable radio telescopes are about 100 m in diameter suggests that this sort of size is likely to be the upper limit for individual optical telescopes as well. Of course, interferometers and aperture synthesis systems (Section 2.5) conceivably could be made thousands of kilometres across like the radio very long baseline interferometers.

Traditionally the cost of a telescope is expected to rise as $D^{2.6}$ or thereabouts. Based upon the costs of the large telescopes built in the 1960s and 1970s, this would suggest a price tag in today's (2020) money for a 100-m optical telescope of $200–$350 billion. In 1992, though, the Keck I 10-m telescope was built for $100 million – a quarter of the cost suggested by the $D^{2.6}$ formula.

The estimated costs for the TMT and the ELT (see below) are currently around $1.5 billion and $1.3 billion, respectively. However, using the $D^{2.6}$ formula on the cost today ($180 million) of the Keck I telescope suggests possible estimates of $3 billion and $6.5 billion for these two instruments. The (closer-to-completion) GMT has a current cost estimate of $1 billion.

Taking today's (probably optimistic) costing though, puts the price tag on the 100-m telescope at around $50 billion to $75 billion. Since the JWST is now expected to come in at some $10–$12 billion, the price of the 100-m telescope is not necessarily prohibitive. In addition to the capital costs, however, the running costs of major telescopes need to be taken into account. These are usually put at around 5% per year of their capital costs – operating a 100-m telescope would thus be likely to cost some $2 billion per year.

For the scientific funding agencies, a developing consideration is that of cost effectiveness – is it better (from a scientific point of view) to spend the available money on bigger and/or better telescopes, or on better ancillary instruments, or on more sensitive, higher resolution, and/or lower noise detectors or on faster, bigger computers? Until the 1960s, bigger telescopes were the clear favourite. During the following few decades, better detectors won out. Now, many detectors are approaching the limits of their possible developments and larger telescopes (TMT, ELT, etc.) have returned to the top spot. However, once the 30-m to 40-m class optical telescopes are up and running, it seems likely that there will be no longer be any obvious front-runners for the best way to achieve further developments – and detailed cost analysis and trade-offs between the various options will have to become the standard approach.

The multi-mirror concept may be extended or made easier by the development of fibre-optics. Fused silica fibres can now be produced whose light losses and focal ratio degradation are acceptable for lengths of tens of metres. Thus, mirrors on independent mounts with a total area of hundreds of square metres which all feed a common focus will become technically feasible within a decade. The Optical Hawaiian Array for Nanoradian Astronomy ('OHANA)[84] project currently envisages linking the major telescopes on Hawaii (Keck 1, Keck 2, Subaru, Gemini North, 3.6-m Canada-France-Hawaii Telescope (CFHT)[85], UKIRT, and IRTF by fibre-optics to form a gigantic optical interferometer. At the time of writing the two Keck telescopes have been linked (some years ago) and recently, two 0.2-m moveable telescopes set up 300 m apart have further demonstrated the feasibility of the project.

A quite different future use of fibre-optics may be the suppression of atmospheric emission lines in the NIR as mentioned previously.

At the other end of the scale, there is still a use for small telescopes – some of them very small indeed. The WASP for example uses eight 0.11-m wide-angle lenses coupled to CCD cameras to search for exoplanets via transits and the Dragonfly telephoto array examines galaxies using forty-eight 0.14-m diameter f/2.8 lenses. Likewise, the decommissioned Trans-Atlantic Exoplanet Survey (TrES) comprised three 0.1-m instruments and monitored thousands of stars at a time. Another exoplanet finder though takes the record though for being the baby amongst all research instruments. The Kilo-degree Extremely Little Telescope (KELT) uses a 0.05-m camera lens – the same size as the telescope that Newton showed to the Royal Society in 1761 (Figure 1.29).

A technique that is in occasional use today but which may well gain in popularity is daytime observation of stars. Suitable filters and large focal ratios (which diminish the sky background brightness, but not that of the star, see the previous discussion) enable useful observations to be made of stars as faint as the seventh magnitude. The technique is particularly applicable in the IR where the scattered solar radiation is lower. If diffraction-limited 100-m telescopes are ever built, then with 10-nm bandwidths they would be able to observe stars (or other sub-milliarcsecond sources but not extended objects) down to a visible magnitude of 23^m during the day. Adaptive optics might be able to be used to assist such observations because the artificial guide stars might still be visible when viewed through very narrow-band filters centred on the sodium emission lines.

Lenses seem unlikely to make a comeback as primary light-gathering components (though see Section 1.1.19.2 and the VRO's corrector lenses). However, they are still used extensively in eyepieces and in ancillary equipment. Three major developments seem likely to affect these applications, principally by making lens systems simpler and/or cheaper. The first is the use of plastic to form the lens. High-quality lenses then can be cheaply produced in quantity by moulding. The second advance is related to the first, and it is the use of aspherical surfaces. Such surfaces have been used when they were essential in the past, but their production was expensive and time-consuming. The investment in an aspherical mould, however, would be small if that mould were then to produce thousands of lenses. Hence, cheap aspherical lenses could become available. The final possibility is for lenses whose refractive index varies across their diameter or with depth. Several techniques exist for introducing such variability to the refractive index of a material such as the diffusion of silver into glass and the growing of crystals from a melt whose composition varies with time. Lenses made from liquid crystal are now being made in small sizes. The refractive index in this material

[84] One nanoradian = 12 mas. 'OHANA means 'Thank You' in Hawaiian.

[85] Possibly to be replaced in the next decade by an 11-m telescope dedicated to spectroscopy and to be called the Maunakea Spectroscopic Explorer (MSE).

depends upon an externally applied electric field and can, thus, be changed rapidly. Such lenses may find application to wavefront correction in adaptive optics systems in due course.

At THz frequencies (tens to hundreds of microns wavelengths) lenses have recently been constructed from metal meshes embedded within a dielectric such as polypropylene. By using several meshes of variable grid sizes the refractive index can be varied radially resulting in a flat lens (sometimes called a 'Wood's lens'[86]). The Lüneburg lens (Section 1.2) is a radio-frequency version of a Wood's lens in the shape of a sphere and with the refractive index varying radially from its centre to its surface. Yet another design for a flat lens relies upon variable phase shifts to the radiation across its surface. It is potentially usable from the NIR to the submillimetre region and could be largely aberration-free. At present only available in the laboratory, it comprises a layer of gold a few nanometres thick deposited onto a silicon substrate. The gold layer is etched into closely spaced V-shaped ridges which act as antennas, receiving and then re-emitting the signal and so introducing a brief delay to it. The delays are tuned across the device's surface so that the radiation is brought to a focus, just as with a 'normal' lens. Highly corrected and relatively cheap lens systems may thus become available within a few years through the use of some or all of these techniques and possibly for use in many regions of the spectrum.

Space telescopes seem likely to follow the developments of terrestrially based telescopes towards increasing sizes, though if terrestrial diffraction-limited telescopes with diameters of several tens of metres or more become available, the space telescopes will only have advantages for spectral regions outside the atmospheric windows. Two concept proposals for the 2030s, prepared for the Astro2020 Decadal Survey and to follow the HST, are LUVOIR and HabEx (see also Origins – Section 1.1.15.3 and Lynx – Section 1.3.8).

The LUVOIR, which was originally called Advanced Technology Large Aperture Space Telescope (ATLAST), has a basically similar design to the JWST with up to a 15-m segmented mirror and would operate from the IR to the UV. It would orbit (like the JWST) at the Sun-Earth Lagrange L2 point with the aims of directly imaging exoplanets and to be able to observe objects some ten times fainter than the HST's limit. With a planned lifetime of two decades, the instrument would be designed from the beginning for robotic servicing and replacement of its instruments, etc. The L2 point will have the advantage for the UV observations of being outside the Earth's Geocorona and so having much lower background noise.

HabEx would also have a primary aim of directly imaging exoplanets. It would involve two co-orbiting spacecraft separated by about 76,000 km at the L2 position. These would be a 4-m UV, optical and IR telescope (possibly with the mirror being monolithic, not segmented) and a separate starshade some 50-m across. The two would act together to form a stellar coronagraph (Section 2.7.4) for exoplanet imaging, but the telescope would also be able to work independently of the starshade for more 'normal' observations.

Other developments may be along the lines of Laser Interferometer Space Antenna (LISA, see Section 1.6) with two or more separately orbiting telescopes forming very high-resolution interferometers.

A quite different approach that may yield great dividends in the future is the application of photonics to astronomical instrumentation (see, for example, PRAXIS in Section 1.1.17 and Figure 1.25). Photonics is the science of using e-m radiation to perform some of the functions traditionally associated with electronics. Most astronomical applications which can be classed under this heading, so far, are related to fibre-optics, but beam combiners for IR interferometers (Section 2.5) have recently been fabricated by laser writing onto chalcogenide glass (IR-transmitting glass containing sulphur, selenium, tellurium, etc.).

1.1.24 Observing Domes, Enclosures and Sites

Any permanently mounted optical telescope requires protection from the elements. Traditionally this has taken the form of a hemispherical dome with an aperture through which the telescope

[86] Robert Wood produced cylindrical lenses with variable refractive indices in 1905 via a gelatine dipping technique.

can observe. The dome can be rotated to enable the telescope to observe any part of the sky and the aperture may be closed to protect the telescope during the day and during inclement weather. Many recently built large telescopes have used cylindrical or other shapes for the moving parts of the enclosure for economic reasons; however, such structures are still clearly related to the conventional hemisphere. The dome however is an expensive item, and it can amount to a third of the cost of the entire observatory, including the telescope. Domes and other enclosures also cause problems through heating-up during the day so inducing convection currents at night and through the generation of eddies as the wind blows across the aperture. Future very large telescopes may, therefore, operate without any enclosure at all, just having a movable shelter to protect them when not in use. This will expose the telescopes to wind buffeting, but active optics can now compensate for that at any wind speed for which it is safe to operate the telescope.

The selection of a site for a major telescope is at least as important for the future usefulness of that telescope as the quality of its optics. The main aim of site selection is to minimise the effects of the Earth's atmosphere, of which the most important are usually scattering, absorption and scintillation.

Scattering by dust and molecules causes the sky background to have a certain intrinsic brightness, and it is this that imposes a limit upon the faintest detectable object through the telescope. The main source of the scattered light is artificial light and most especially street lighting. Thus, a first requirement of a site is that it be as far as possible from built-up areas. If an existing site is deteriorating due to encroaching suburbs, then some improvement may be possible for some types of observation by the use of a light pollution rejection (LPR) filter which absorbs in the regions of the most intense sodium and mercury emission lines. Scattering can be worsened by the presence of industrial areas upwind of the site or by proximity to some deserts, both of which inject dust into the atmosphere.

Absorption is due mostly to the molecular absorption bands of the gases forming the atmosphere. The two well-known windows in the spectrum, wherein radiation passes through the atmosphere relatively unabsorbed, extend from about 360 nm to 100 μm and from 10 mm to 100 m (Figure 1.20). But even in these regions there is some absorption, so that visible light is decreased in its intensity by 10%–20% for vertical incidence – and by much more as the zenith angle increases.

The IR region is badly affected by water vapour, OH and other molecules to the extent that portions of it are completely obscured. Thus, the second requirement of a site is that it be as high an altitude as possible to reduce the air paths to a minimum and that the water content be as low as possible. We have already seen how balloons and spacecraft are used as a logical extension of this requirement and a few high-flying aircraft are also used for this purpose. The possibility of using a fibre-optic Bragg grating to reduce the sky background in the NIR by a factor of up to 50 has been mentioned previously (Section 1.1.17).

Scintillation has been discussed in relation to real-time atmospheric compensation. It is the primary cause of the low resolution of large telescopes because the image of a point source is rarely less than half an arcsecond across as a result of this blurring (Figure 1.61). Thus, the third requirement for a good site is a steady atmosphere. The ground-layer effects (tens to hundreds of metres above ground) of the structures and landscape in the telescope's vicinity may worsen scintillation. A rough texture to the ground around the dome, such as low-growing bushes, seems to reduce scintillation when compared with that found for smooth (e.g., paved) and rough (e.g., tree-covered) surfaces. Adaptive optics systems may be designed specifically to correct ground-layer effects; the LBT's Pathfinder system, for example, used natural guide stars and 12 movable pyramid wavefront sensors. The current system, Advanced Rayleigh Guided Ground Layer Adaptive Optics (ARGOS), reduces the images' sizes by about a factor of two to three over a 4 arc-minute field of view using three Rayleigh artificial guide stars, and the whole adaptive optics system can reach the diffraction limit over a 30″ field of view. The VLT's Yepun telescope currently uses GRAAL (see also AOF Section 1.1.22.3) to correct for ground-layer turbulence employing one natural and four artificial guide stars located outside the main field of view and it operates at up to 1 kHz.

Great care also needs to be taken to match the dome; telescope and mirror temperatures to the ambient temperature or the resulting convection currents can worsen the scintillation by an order of magnitude or more. Cooling the telescope during the day to the predicted night-time temperature is helpful, provided that the weather forecast is correct. Forced air circulation by fans and thermal insulation; and siting any ancillary equipment that generates heat as far as possible from the telescope or away from the dome completely can also reduce convection currents.

These requirements for an observing site restrict the choice considerably and lead to a clustering of telescopes on the comparatively few optimum choices. Most are now found at high altitudes and on oceanic islands or with the prevailing wind from the ocean. Built-up areas tend to be small in such places, and the water vapour is usually trapped near sea level by an inversion layer. The long passage over the ocean by the winds tends to minimise dust and industrial pollution. The Antarctic plateau with its cold, dry and stable atmospheric conditions is also a surprisingly good observing site. The Arctic is also useful for radio studies of the aurora and so on with, for example, the European Incoherent Scatter (EISCAT) fixed and movable dishes operating at wavelengths of a few hundreds of millimetres. The EISCAT 3D is due to become operational in 2021 with five phased arrays, enabling three-dimensional (3D) studies of the ionosphere and upper atmosphere to be undertaken.

The recent trend in the reduction of the real cost of data transmission lines, whether these are via optic-fibres, cables or satellites, is likely to lead to the greatly increased use of remote control of telescopes. In a decade or two, most major observatories are likely to have only a few permanent staff physically present at the telescope wherever that might be in the world and astronomers would only need to travel to a relatively nearby control centre for his their observing shifts. Indeed, it seems quite likely that in some cases astronomers will be able to sit in their normal office and operate multi-metre telescopes thousands of kilometres away via the internet.

There are also already a few observatories with completely robotic telescopes, in which all the operations are computer-controlled, with no staff on site at all during their use. Many of these robotic instruments are currently used for long-term photometric monitoring programs or for teaching purposes. However, their extension to more exacting observations such as spectroscopy is now starting to occur; thus, the 2.4-m Automated Planet Finder (APF) is a spectroscopic instrument at Lick Observatory searching robotically for exoplanets a few times more massive than the Earth. The Robotic Adaptive Optics (Robo-AO) instrument, as its name suggests, is a robotic system that obtains high-resolution adaptive optics images automatically. It was initially deployed on the 1.5-m telescope at Mount Palomar, obtaining up to 250 diffraction-limited observations in a night. It uses an artificial UV guide star produced by Rayleigh scattering from a 355-nm laser. In 2015, Robo-AO was moved to the 2.1-m telescope at Kitt Peak, and it is, at the time of writing, being commissioned and upgraded on the 2.2-m University of Hawaii's 88-in (UH88) telescope. It is planned to use the instrument, or clones, or derivatives, on other medium-sized robotic telescopes in the future.

The Liverpool telescope on La Palma and the two Faulkes telescopes in Hawaii and Australia are 2-m robotic telescopes that can be linked to respond to Gamma-Ray Burst (GRB) alerts in less than 5 minutes. Five per cent of the observing time on the Liverpool telescope and most of the time on the Faulkes telescopes is devoted to educational purposes with students able to control the instruments via the internet. Other examples of robotic telescopes include the Rapid Eye Mount (REM) 0.6-m instrument designed for GRB observing and sited at La Silla and Panchromatic Robotic Optical Monitoring and Polarimetry Telescopes (PROMPT) that uses six 0.4-m telescopes covering the violet to NIR regions and is designed for GRB and supernova observations as well as for education. It started operations in 2006 and is now a part of the SKYNET robotic telescope network of 11 telescopes, which includes radio as well as optical instruments.

1.2 RADIO AND MICROWAVE DETECTION

1.2.1 Introduction

Radio astronomy is the oldest of the 'new' astronomies because it is now approaching a century since Karl Jansky first observed radio waves from the Milky Way galaxy. It has passed beyond the developmental stage wherein some of the other 'new' astronomies – neutrino astronomy, gravity wave astronomy, and so on – still remain. It, thus, has quite well-established instruments and techniques that are not likely to change overmuch.

The reason why radio astronomy developed relatively early is that radio radiation penetrates to ground level (Figure 1.20). For wavelengths from about 10 mm to 10 m, the atmosphere is almost completely transparent. The absorption becomes almost total at about 0.5-mm wavelength and between 0.5 and 10 mm there are a number of absorption bands that are mainly the result of oxygen and water vapour, with more or less transparent windows between the bands. The scale height for water vapour in the atmosphere is about 2 km, so that observing from high altitudes reduces the short-wave absorption considerably. Radiation with wavelengths longer than about 50 m again fails to penetrate to ground level, but this time the cause is reflection by the ionosphere.

This section is thus concerned with the detection of radiation with wavelengths longer than about 0.1 mm – that is; frequencies less than 3×10^{12} Hz or photon energies less than 2×10^{-21} J (0.01 eV). The detection of radiation in the 0.1 to a few mm wavelength region using bolometers is covered in IR detectors in Section 1.1. Also, this section is primarily concerned with individual radio telescopes, whilst two or more individual telescopes acting together as interferometers and aperture synthesis systems are covered in Section 2.5. However, there is inevitably some overlap and aspects of telescope arrays such as ALMA and Square Kilometer Array (SKA) are included in this section where it seems appropriate. The detection of the radio and microwave emission from cosmic ray showers in the Earth's atmosphere is covered in Section 1.4. Whilst the detection of neutrinos, via their induced radio emission, is covered in Section 1.5.

The unit of intensity that is commonly used at radio wavelengths as a measure of the intensity of point sources is the jansky (Jy)

$$1\,\mathrm{Jy} = 10^{-26}\,\mathrm{Wm^{-2}Hz^{-1}} \tag{1.61}$$

and detectable radio sources vary from about 10^{-6} to 10^{6} Jy. Most radio sources of interest to astronomers generate their radio flux as thermal radiation, when the Rayleigh-Jeans law gives their spectrum

$$F_\nu = \frac{2\pi k}{c^2} T \nu^2 \tag{1.62}$$

or as synchrotron radiation from energetic electrons spiralling around magnetic fields, when the spectrum is of the form

$$F_\nu \propto \nu^{-\alpha} \tag{1.63}$$

where F_ν is the flux per unit frequency interval at frequency, ν; α is called the spectral index of the source and is related to the energy distribution of the electrons. For many sources $0.2 \le \alpha \le 1.2$.

For extended sources, the unit of Jy sr^{-1}, is sometimes used, or more frequently, the brightness temperature. The latter is defined as the temperature of a black body that would emit the same intensity of radiation as the observed object at the selected frequency.

1.2.2 Detectors and Receivers

1.2.2.1 Detectors

The detection of radio signals is a three-stage process in which the radio signals are first collected and concentrated onto the detector, antenna or sensor[87] which converts the electromagnetic wave into an electrical signal and that then has to be processed until it is in a directly

[87] These are often quite tiny objects – millimetres to a metre or so in size – and are rarely noticed by casual observers of radio telescopes, who see only the huge structures that gather and concentrate the radio signals onto the detectors.

usable form. Coherent detectors, which preserve the phase information of the signal, are available for use over the whole of the radio spectrum, in contrast to the optical and IR detectors discussed in Section 1.1 which mostly respond only to the total power of the signal. The detection of submillimetre and millimetre radiation (FIR) via the use of bolometers is covered in Section 1.1.

The natures of the detectors vary with the frequencies being observed. Thus, at the lower frequencies (kHz to MHz), the sensor is normally a simple length of a conducting material. Many readers will have noticed the piece of wire or the extendable rigid metal rod coming out of the back of their transistor radios and which are the detectors for such devices. These are examples of monopole antennas, and they will typically best receive signals whose wavelengths are related to the length of the antenna. Thus, antennae whose lengths are ¼ or ½ of their operating wavelength are in frequent use, the latter having almost twice the sensitivity of the former.

Dipole antennae are more frequently encountered in low-frequency astronomical applications and also for television reception (Figure 1.79), with the half-wave dipole being the best-known example (Figure 1.71). The two halves of such a dipole are each a quarter of a wavelength long and it is placed at the focus of the parabolic dish of the radio telescope, with connection to the remainder of the system by coaxial cable. The basic half-wave dipole can only be optimised for one wavelength, and its sensitivity drops off rapidly away from that wavelength – to 50% at about 2.5% of the operating wavelength (i.e., the usable bandwidth of a basic half-wave dipole operating at 100 MHz is from about 97.5 to 102.5 MH. The highest sensitivity is perpendicular to the line of the dipole, and it drops to zero along the line of the dipole (Figure 1.75).

Many basic antennae have their conducting elements in more complex shapes than a linear rod. Spiral, helical, cone and sinuous antennae may be encountered, where the shape improves the bandwidth, reduces side lobes (see below), flattens the response or has other desirable advantages for the application involved.

In the GHz and higher frequency regions, a horn antenna is normally used to collect the radiation, usually with waveguides for the connection to the rest of the system, though plastic, quartz, metal grid and other lenses may be used at high frequencies. Recent developments in the design of horn antennas have enabled them to have much wider bandwidths than previous designs. These developments include using a corrugated internal surface for the horn and reducing its diameter in a series of steps or a smooth surface with step changes in the cone angle, giving a bandwidth that covers a factor of two in terms of frequency. Bandwidths with factors of nearly eight in frequency are possible with dielectric loaded horns. These are smooth-walled horns filled with an appropriate dielectric except for

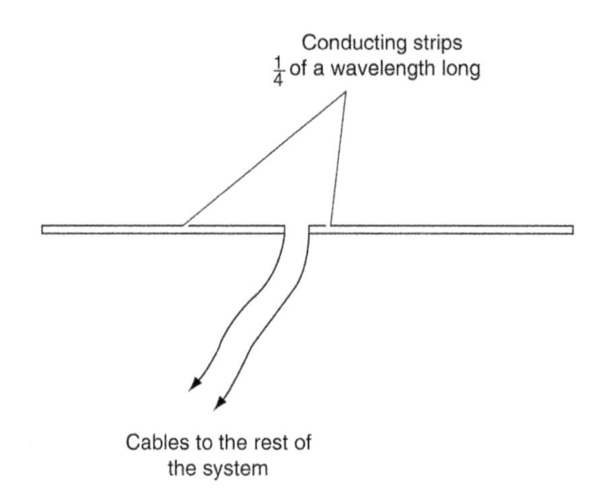

FIGURE 1.71 A half-wave dipole.

a small gap between the filling and the walls of the horn through which the wave propagates. At high frequencies, the feed horns may need cooling. The James Clerk Maxwell Telescope's (JCMT's) decommissioned SCUBA receiver, for example, operated at 350 and 680 GHz using small arrays of bolometers (Section 1.1.15.2) with conical feed horns (its replacement, SCUBA-2 does not use horns, but SCUBA-2's replacement, planned for 2022, is expected to use more 3,600 smooth-walled feed horns and MKID detectors – see below). The Brazilian Baryon Acoustic Oscillations in Neutral Gas Observations (BINGO) radio telescope has a crossed-Dragone design with two 40-m fixed dishes. It is currently under construction and will use fifty 1.5-m diameter feed horns for its 960 MHz to 1.26 GHz detectors.

Detectors and receivers at the higher frequencies are now normally based upon superconducting materials and so they need cooling to low temperatures – often to well below 1 K (Section 1.1.7). In an Superconductor-Insulator-Superconductor (SIS) device (aka STJs, Section 1.1), an electron in one superconducting film absorbs a photon, giving the electron enough energy to tunnel through the insulating barrier into the other superconducting film. This process, known as 'photon-assisted-tunnelling', produces one electron for every absorbed photon. Modern devices are based upon two niobium layers separated by an insulating region of aluminium oxide around 1-nm thick and the whole cooled to 4 K or less. Such SIS devices operate best over about the 100-GHz to 1-THz region (3 mm–300 μm), though SIS devices can be used up to 1.2 THz (250 μm) using niobium titanium nitride, and this may be extended in the future to 2 THz (150 μm).

Both the Nobeyama 45-m dish and ALMA use SIS devices as heterodyne receivers (see below) for operating around 100 GHz. The incoming signal is mixed with the local oscillator signal at the detector. The Four Beam Receiver System (FOREST) instrument on the Nobeyama telescope uses four 15″ beams and observes over the 80-to 116-GHz band with 4- to 12-GHz intermediate frequencies and ALMA's 73 Band 5 (163–211 GHz) receiver cartridges operate with niobium SIS devices and a 171–203 GHz local oscillator. Similarly, the JCMT's recently installed Nāmakanui instrument observes at 230 and 345 GHz with niobium/aluminium SIS detectors.

Even higher frequencies require the use of Schottky diodes (Section 1.1.15.3), TES devices (Section 1.1.15.3), HEBs, CEBs (Section 1.1.15.2) or MKIDS (Section 1.1.13). Often these devices can be made small and then thermally connected to much larger absorbers for the radiation of interest, thus reducing cosmic ray noise because efficient radiation absorbers may be fabricated which interact only poorly with the cosmic ray particles (see spider web absorbers, Section 1.1.15.3).

The OLIMPO[88] balloon-borne instrument, for example, was recently launched on a five-day mission from Svalbard with a 2.6-m mirror and lumped element MKID detectors covering the 190- to 480-GHz region for CMB observations; its original design, though, called for CEB detectors and these may be used on future flights. Balloon-borne Large Aperture Submillimeter Telescope-The Next Generation (BLAST-TNG), also a balloon-borne submillimetre telescope, will have more than 3,000 MKID detectors, cooled to 275 mK, operating at up to 1.2 THz and fed by a 2.5-m telescope. At the time of writing, it is due for its first launch shortly. SOFIA's German Receiver for Astronomy at THz frequencies (GREAT) receiver uses waveguide-fed niobium titanium nitride HEB mixer detectors for its observations over the 490 GHz to 4.74 THz waveband.

A recent development, which acts similarly to feed horns in that it concentrates radiation onto individual pixels within arrays, is the use of tiny antennas (nano-antennas) with thermal IR detectors (i.e., a frequency of a few tens of THz). These antennas are metal (currently gold is in use) square or cross shapes that are positioned above each pixel and which are smaller than the operating wavelength.

[88] Italian for Olympus.

1.2.2.2 Receivers

The signal from the sensor is carried to the receiver whose purpose is to convert the high-frequency electrical currents into a convenient form. The behaviour of the receiver is governed by five parameters: sensitivity, amplification, bandwidth, and receiver noise level and integration time.

The sensitivity and the other parameters are closely linked, for the minimum detectable brightness, $B_{\min}$, is given by

$$B_{\min} = \frac{2k\nu^2 K T_s}{c^2\sqrt{t\Delta\nu}} \tag{1.64}$$

where T_s is the noise temperature of the system, t is the integration time, $\Delta\nu$ is the frequency bandwidth, and K is a constant close to unity that is a function of the type of receiver. The bandwidth is usually measured between output frequencies whose signal strength is half the maximum when the input signal power is constant with frequency. The amplification and integration time are self-explanatory, so that only the receiver noise level remains to be explained. This noise originates as thermal noise within the electrical components of the receiver and may also be called 'Johnson' or 'Nyquist noise' (see also Section 1.1). The noise is random in nature and is related to the temperature of the component. For a resistor, the root-mean-square (RMS) voltage of the noise per unit frequency interval, $\bar{V}$, is given by

$$\bar{V} = 2\sqrt{kTR} \tag{1.65}$$

where R is the resistance and T is the temperature. The noise of the system is then characterised by the temperature T_s that would produce the same noise level for the impedance of the system. It is given by

$$T_s = T_1 + \frac{T_2}{G_1} + \frac{T_3}{G_1 G_2} + \ldots \frac{T_n}{G_1 G_2 \ldots . G_{n-1}} \tag{1.66}$$

where T_n is the noise temperature of the nth component of the system and G_n is the gain (or amplification) of the nth component of the system. It is usually necessary to cool the initial stages of the receiver with liquid helium to reduce T_s to an acceptable level. Other noise sources that may be significant include shot noise resulting from random electron emission, g-r noise due to a similar effect in semiconductors (Section 1.1), noise from the parabolic reflector or other collector that may be used to concentrate the signal onto the antenna, radiation from the atmosphere and last but by no means least, spill over from radio taxis, microwave ovens, and other artificial sources.

Many types of receiver exist; the simplest is a development of the heterodyne system employed in the ubiquitous transistor radio. The basic layout of a heterodyne receiver is shown in block form in Figure 1.72. The pre-amplifier operates at the signal frequency and will typically have a gain of 10–1000. It is often mounted close to the feed and cooled to near absolute zero to minimise its contribution to the noise of the system (see above). The most widely used amplifiers today are based upon cooled gallium arsenide and indium phosphide high electron mobility transistors (HEMTs; also known as heterostructure field effect transistors [HFETs]) in which the current-carrying electrons are physically separated from the donor atoms. The current-carrying electrons are restricted to a thin (10 nm) layer of undoped material producing a fast, low-noise device. Above 40 GHz (7.5 mm), the mixer must precede the pre-amplifier to decrease the frequency before it can be amplified; a second lower-frequency local oscillator is then employed to reduce the frequency even further.

The local oscillator (LO) produces a signal that is close to but different from the main signal in its frequency. Thus, when the mixer combines the main signal and the local oscillator signal, the beat frequency between them (the Intermediate Frequency [IF]), is at a much lower frequency than that of the original signal. The relationship is given by

$$\nu_{\mathrm{SIGNAL}} = \nu_{\mathrm{LO}} \pm \nu_{\mathrm{IF}} \tag{1.67}$$

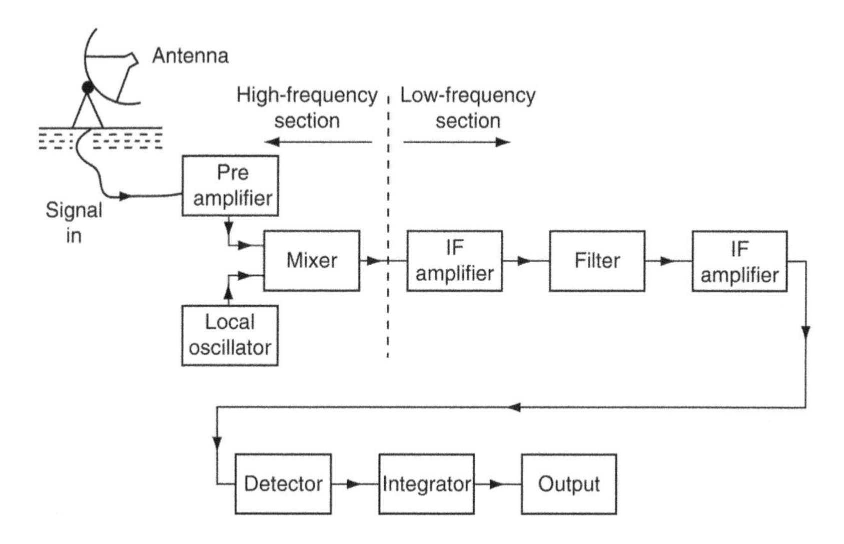

FIGURE 1.72 Block diagram of a basic heterodyne receiver.

where ν_{SIGNAL} is the frequency of the original signal (i.e., the operating frequency of the radio telescope), ν_{LO} is the local oscillator frequency and ν_{IF} is the intermediate frequency. Normally, at lower frequencies, only one of the two possible signal frequencies given by Equation 1.67 will be picked up by the feed antenna or passed by the pre-amplifier. At high frequencies, both components may contribute to the output.

Heterodyne detectors have now been developed to operate at submillimetre wavelengths and one such was the Heterodyne Instrument for the Far Infrared (HIFI) that was carried on board the Herschel spacecraft (Figure 1.73). HIFI was a spectrometer that observed over seven wavebands in the region from 480 GHz to 1.91 THz (625–160 μm). Its IF was generated by a simple high(ish) frequency oscillator and that emission then stepped up in frequency by using a chain of frequency multipliers,[89] until the desired frequency was reached.

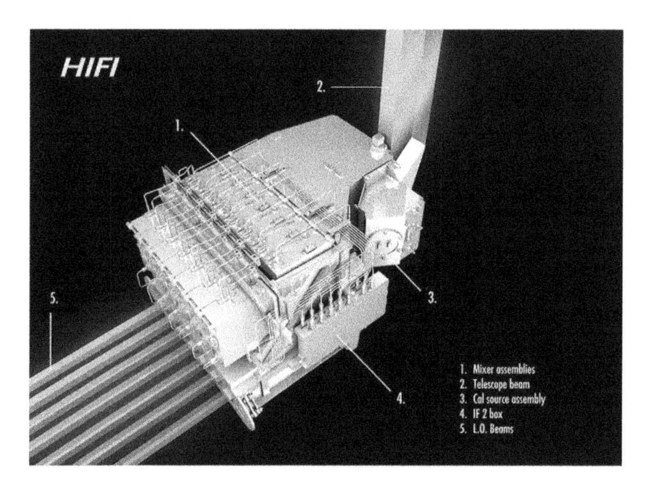

FIGURE 1.73 An artist's impression of Heterodyne Instrument for the Far Infrared (HIFI). (Courtesy of European Space Agency.)

[89] A frequency multiplier deliberately distorts the input signal, so producing higher harmonics of the basic frequency and these may then be filtered to leave just the higher frequency.

Heterodyne receivers for multi-THz frequencies have only recently started to be used. This advance has arisen through the development of quantum cascade lasers (QCLs). These can emit their radiation in the THz region and operate via stacks of two types of semiconductors which form a series of quantum wells,[90] the laser emission arising as electrons move from one quantum well to another. QCLs can be easily tuned by manufacturing different energy quantum wells. If an appropriate frequency QCL is used as the LO for a THz signal, then the IF can be in the GHz region and so detected by more 'normal' instruments (which themselves are likely to be of a heterodyne design). The ground-based planetary atmosphere spectroscope, MILAHI (see also Section 1.1.15.4), for example, uses uncooled QCLs covering the range 29–39 THz as its LOs.

The power of the intermediate frequency emerging from the mixer is directly proportional to the power of the original signal. The IF amplifiers and filter determine the pre-detector bandwidth of the signal and further amplify it by a factor of 10^6–10^9. The detector is normally a square-law device; that is to say, the output voltage from the detector is proportional to the square of the input voltage. Thus, the output *voltage* from the detector is proportional to the input *power*. In the final stages of the receiver, the signal from the detector is integrated, usually for a few seconds, to reduce the noise level. Then it is fed to an output device, usually an analogue-to-digital input to a computer for further processing. Further advances in speed and noise reduction come from combining the sensor and much of the electronics onto monolithic microwave integrated circuits (MMICs). With MMICs and other coherent devices, there is a noise source, termed the 'quantum limit' (see also shot noise, Section 1.1) that arises from fluctuations in the number of photons (quanta) being collected. The equivalent temperature is given by

$$T_{\text{Quantum limit}} = \frac{h\nu}{k} \approx 5 \times 10^{-11}\ \nu\ K \tag{1.68}$$

Thus, at GHz frequencies the quantum limit temperature is generally too low to be significant, but at THz frequencies it rises from 50 K to 50,000 K and becomes the dominant noise source. MMICs currently perform at about five to ten times the quantum limit for frequencies less than 150 GHz (2 mm), and it is hoped to reduce this to two or three times the quantum limit in the near future. MMICs are cheap and are used for example in SETI to construct low-noise amplifiers, filters and local oscillators.

The basic heterodyne receiver has a high system temperature and its gain is unstable. The temperature may be lowered by applying an equal and opposite voltage in the later stages of the receiver and the stability of the gain may be greatly improved by switching rapidly from the antenna to a calibration noise source and back again, with a phase-sensitive detector (Section 3.1) to correlate the changes. Such a system is then sometimes called a 'Dicke radiometer'. The radiometer works optimally if the calibration noise source level is the same as that of the signal, and so it may be further improved by continuously adjusting the noise source to maintain the balance and it is then termed a 'null-balancing Dicke radiometer'. Since the signal is only being detected half the time, the system is less efficient than the basic receiver, but using two alternately switched receivers will restore its efficiency. The value of T_s for receivers varies from 10 K at metre wavelengths to 10,000 K at millimetre wavelengths. The noise sources must therefore have a comparable range and at long wavelengths are usually diodes, whilst at the shorter wavelengths a gas discharge tube inside the waveguide and inclined to it by an angle of about 10° is used.

Receivers are generally sky-background-limited, just like terrestrial optical telescopes. The Earth's atmosphere radiates at 100 K and higher temperatures below a wavelength of about 3 mm. Only between 30 and 100 mm does its temperature fall as low as 2 K. Then, at longer wavelengths, the galactic emission becomes important, rising to temperatures of 10^5 K at wavelengths of 30 m.

Spectrographs at radio frequencies can be obtained in several different ways. In the past, the local oscillator has been tuned, producing a frequency-sweeping receiver, or the receiver had been

[90] Quantum wells use a sandwich of three layers of semiconductors. The energy of the conduction electrons in the two outer layers (e.g., aluminium arsenide) is higher than that for the middle layer (e.g., gallium arsenide).

a multichannel device so that it registered several different discrete frequencies simultaneously. For pulsar observations such filter banks may have several hundred channels over a bandwidth of a few MHz. Today, most radio spectroscopy is carried out by auto-correlation, even at the highest frequencies. Successive delays are fed into the signal that is then cross-correlated with the original signal in a computer. The spectrum is obtained from the Fourier transform (Section 2.1.1) of the result. The polarisation of the original signal is determined by separately detecting the orthogonal components and cross-correlating the electrical signals later within the receiver.

Alternatively, the radio signal may be converted into a different type of wave and the variations of this secondary wave studied instead. This is the basis of the Acousto-Optical radio Spectrometer (AOS). The radio signal is converted into an ultrasonic wave whose intensity varies with that of the radio signal and whose frequency is also a function of that of the radio signal. In the first such instruments water was used for the medium in which the ultrasound propagated and the wave was generated by a piezoelectric crystal driven either directly from the radio signal or by a frequency-reduced version of the signal. More recently materials such as fused silica, lithium niobate and lead molybdate have replaced water as the acoustic medium to improve the available spectral range. A laser illuminates the cell containing the acoustic medium and a part of the light beam is diffracted by the sound wave. The angle of diffraction depends upon the sound wave's frequency, whilst the intensity of the diffracted light depends on the sound wave's intensity. Thus, the output from the device is a fan beam of light, the position within which ultimately depends upon the observed radio frequency and whose intensity at that position ultimately depends upon the radio intensity. The fan beam may then simply be detected by a linear array of optical detectors or by scanning and the spectrum inferred from the result. The AOS initially found application to the observation of solar radio bursts (Section 5.3) at metre wavelengths but is now employed more in the submillimetre and IR regions because auto-correlation techniques have replaced it at the longer wavelengths and are even starting to do so in the millimetre region.

A major problem at all frequencies in radio astronomy is interference from artificial noise sources. In theory, certain regions of the spectrum are reserved partially or exclusively for use by radio astronomers. An up-to-date listing of the reserved frequencies may be found on the Committee on Radio Astronomy Frequencies (CRAF) website at www.craf.eu/. But leakage from devices such as microwave ovens, incorrectly tuned receivers and illegal transmissions often overlap into these bands. The use of highly directional aerials (see the next subsection) reduces the problem to some extent. But it is likely that radio astronomers will have to follow their optical colleagues to remote parts of the globe or place their aerials in space if their work is to continue in the future (Section 1.2.3.2). Even the latter means of escape may be threatened by solar power satellites with their potentially enormous microwave transmission intensities.

1.2.3 Radio Telescopes

The detector and receiver, whilst they are the main active portions of a radio detecting system, are far less physically impressive than the large structures that serve to gather and concentrate the radiation and to shield the antenna from unwanted sources (Figure 1.74). Before going on, however, to the consideration of these large structures that form most people's ideas of what comprises a radio telescope, we must look in a little more detail at the physical background of antennae.

The theoretical optics of light and radio radiation are identical, but different traditions within the two disciplines have led to differences in the mathematical and physical formulations of their behaviours. Thus, the image in an optical telescope is discussed in terms of its diffraction structure (Figure 1.33 for example), whilst that of a radio telescope is discussed in terms of its polar diagram. However, these are just two different approaches to the presentation of the same information. The polar diagram is a plot in polar coordinates of the sensitivity or the voltage output of the telescope, with the angle of the source from the optical axis. (Note, the polar diagrams discussed herein are all far-field patterns, that is, the response for a distant source, near-field patterns for the aerials may differ

FIGURE 1.74 The world's largest fully steerable radio telescope – the 100 m × 110 m Robert C. Byrd Green Bank Telescope. (Courtesy of National Radio Astronomy Observatory/Associated Universities Inc. [NRAO/AUI].)

from those shown here.) The polar diagram may be physically realised by sweeping the telescope past a point source or by using the telescope as a transmitter and measuring the signal strength around it.

The simplest antenna, the half-wave dipole (Figure 1.71), accepts radiation from most directions and its polar diagram is shown in Figure 1.75. This is only a cross section through the beam pattern; the full 3D polar diagram may be obtained by rotating the pattern shown in Figure 1.75 about the dipole's long axis and, thus, has the appearance of a toroid (ring doughnut) that is filled in to the centre. The polar diagram and hence, the performance of the antenna, may be described by four parameters: the beam width at half-power points (BWHP), beam width at first nulls (BWFN), the gain and the effective area. The first nulls are the positions either side of the optical axis where the sensitivity of the antenna first decreases to zero and the BWFN is just the angle between them. Thus, the value of the BWFN for the half-wave dipole is 180°. The first nulls are the direct equivalent of the first fringe minima in the diffraction pattern of an optical image, and for a dish aerial type of radio telescope, their position is given by Equation 1.33; thus,

$$\text{BWFN} = 2 \times \frac{1.22\lambda}{D} \tag{1.69}$$

The Rayleigh criterion of *optical* resolution may thus be similarly applied to *radio* telescopes; two point sources are resolvable when one is on the optical axis and the other is in the direction of

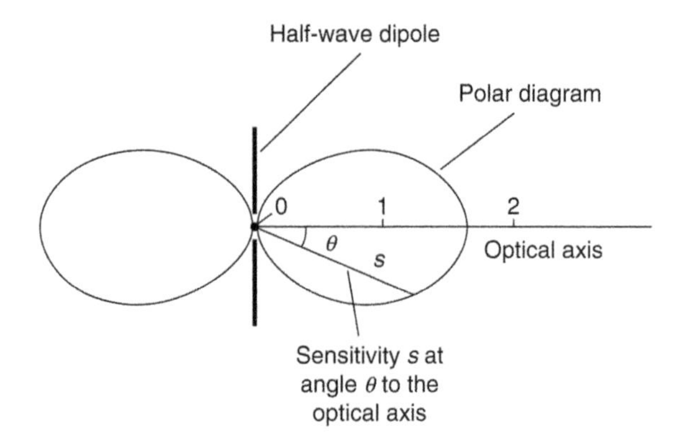

FIGURE 1.75 Polar diagram of a half-wave dipole.

a first null. The half-power points may be best understood by regarding the radio telescope as a transmitter; they are then the directions in which the broadcast power has fallen to one-half of its peak value. The BWHP is just the angular separation of these points. For a receiver, they are the points at which the output voltage has fallen by a factor of the $\sqrt{2}$, and hence, the output power has fallen by half. The maximum gain or directivity of the antenna is also best understood in terms of a transmitter. It is the ratio of the peak value of the output power to the average power. In a receiver it is a measure of the output from the system compared with that from a comparable (and hypothetical) isotropic receiver. The effective area of an antenna is the ratio of its output power to the strength of the incoming flux of the radiation that is correctly polarised to be detected by the antenna, that is,

$$A_e = \frac{P_\nu}{F_\nu} \tag{1.70}$$

where A_e is the effective area, P_ν is the power output by the antenna at frequency ν, and F_ν is the correctly polarised flux from the source at the antenna at frequency, ν. The effective area and the maximum gain, g, are related by

$$g = \frac{4\pi}{c^2}\nu^2 A_e \tag{1.71}$$

For the half-wave dipole, the maximum gain is about 1.6, and so there is little advantage over an isotropic receiver.

The performance of a simple dipole may be improved by combining the outputs from several dipoles that are arranged in an array. In a collinear array, the dipoles are lined up along their axes and spaced at intervals of half a wavelength (Figure 1.76). The arrangement is equivalent to a diffraction grating and so the sensitivity at an angle θ to the long axis of the array, $s(\theta)$, is given by

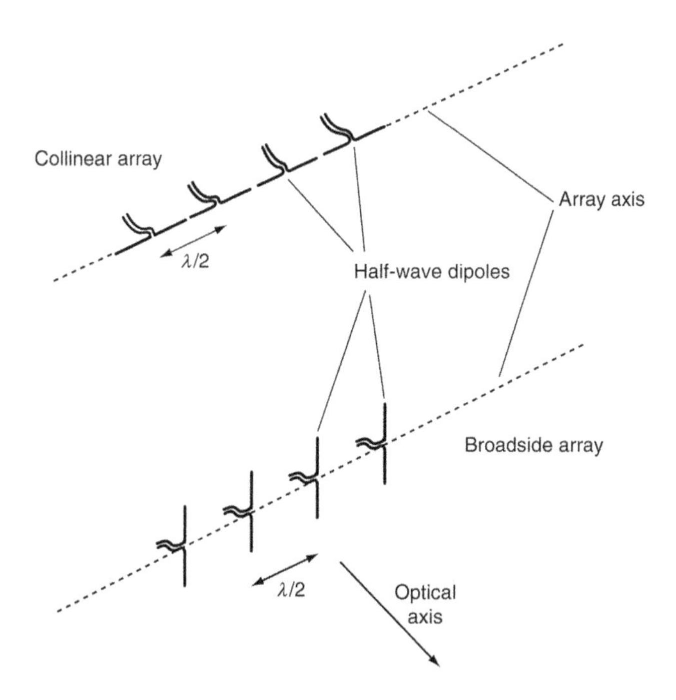

FIGURE 1.76 Dipole arrays.

$$s(\theta) = s_o \left(\frac{\sin(n\pi \sin\theta)}{\sin(\pi \sin\theta)} \right) \tag{1.72}$$

where n is the number of half-wave dipoles and s_o is the maximum sensitivity (cf. Equation 4.8). Figure 1.77 shows the polar diagrams for one, two, and four dipole arrays; their 3D structure can be obtained by rotating these diagrams around a vertical axis so that they become lenticular toroids. The resolution along the axis of the array, measured to the first null, is given by

$$\alpha = \sin^{-1}\left(\frac{1}{n}\right) \tag{1.73}$$

The structure of the polar diagrams in Figure 1.77 shows a new development. Apart from the main lobe whose gain and resolution increase with n as might be expected, a number of smaller side lobes have appeared. Thus, the array has sensitivity to sources that are at high angles of inclination to the optical axis. These side lobes correspond precisely to the fringes surrounding the Airy disc of an optical image (Figure 1.35, etc.). Although the resolution of an array is improved over that of a simple dipole along its optical axis, it will still accept radiation from any point perpendicular to the array axis. The use of a broadside array in which the dipoles are perpendicular to the array axis and spaced at half-wavelength intervals (Figure 1.76) can limit this 360° acceptance angle somewhat. For a four-dipole broadside array, the polar diagram in the plane containing the optical and array axes is given by polar diagram number II in Figure 1.77, while in the plane containing the optical axis and the long axis of an individual dipole, the shape of the polar diagram is that of a single dipole (number I in Figure 1.77) but with a maximum gain to match that in the other plane. The 3D shape of the polar diagram of a broadside array thus resembles a pair of squashed balloons placed end to end. The resolution of a broadside array is given by

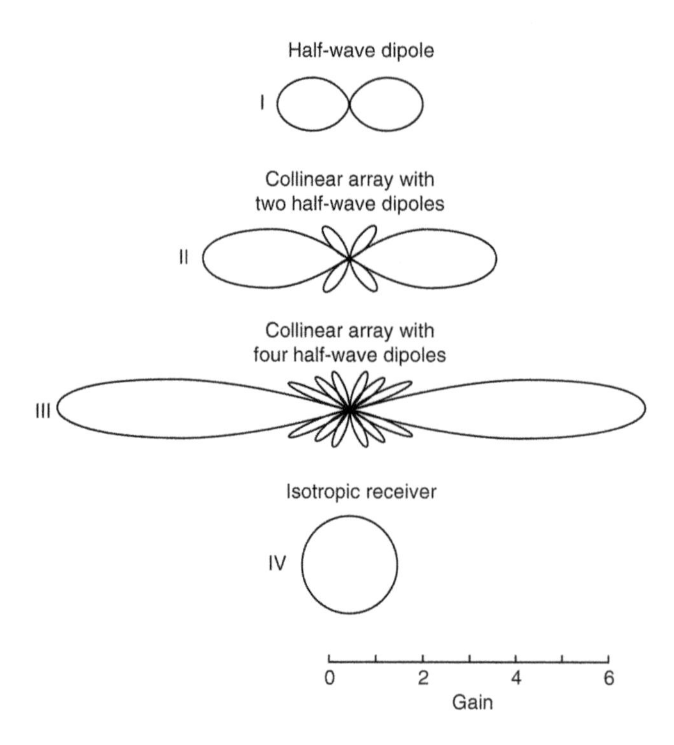

FIGURE 1.77 Polar diagrams for collinear arrays.

$$\alpha = \sin^{-1}\left(\frac{2}{n}\right) \tag{1.74}$$

along the array axis and is that of a single dipole (i.e., 90°), perpendicular to this. Combinations of broadside and collinear arrays can be used to limit the beam width further if necessary.

With the arrays as shown, there is still a two-fold ambiguity in the direction of a source that has been detected; however narrow the main lobe may have been made, because of the forward and backward components of the main lobe. The backward component may easily be eliminated, however, by placing a reflector behind the dipole. This is simply a conducting rod about 5% longer than the dipole and unconnected electrically with it. It is placed parallel to the dipole and about one-eighth of a wavelength behind it. For an array, the reflector may be a similarly placed electrically conducting screen. The polar diagram of a four-element collinear array with a reflector is shown in Figure 1.78 to the same scale as the diagrams in Figure 1.77. It will be apparent that not only has the reflector screened out the backward lobe, but it has also doubled the gain of the main lobe. Such a reflector is termed a 'parasitic element' because it is not a part of the electrical circuit of the antenna. Similar parasitic elements may be added in front of the dipole to act as directors. These are about 5% shorter than the dipole. The precise lengths and spacings for the parasitic elements can only be found empirically because the theory of the whole system is not completely understood. With a reflector and several directors, we obtain the parasitic or Yagi antenna, familiar from its appearance on so many rooftops as a television aerial. A gain of up to 12 is possible with such an arrangement and the polar diagram is shown in Figure 1.79. The Rayleigh resolution is 45°, and it has a bandwidth

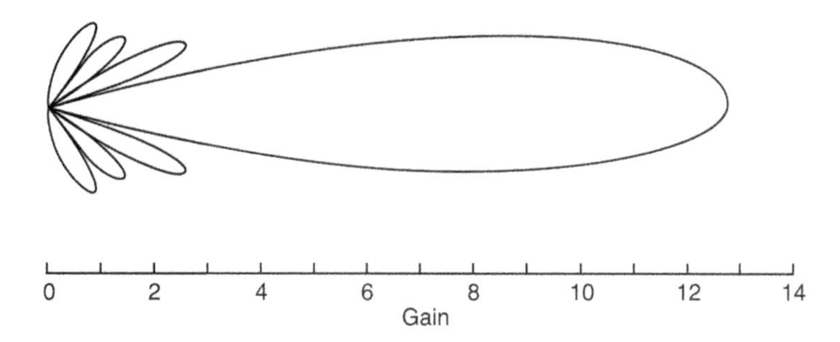

FIGURE 1.78 Polar diagram for a four-element collinear array with a mesh reflector.

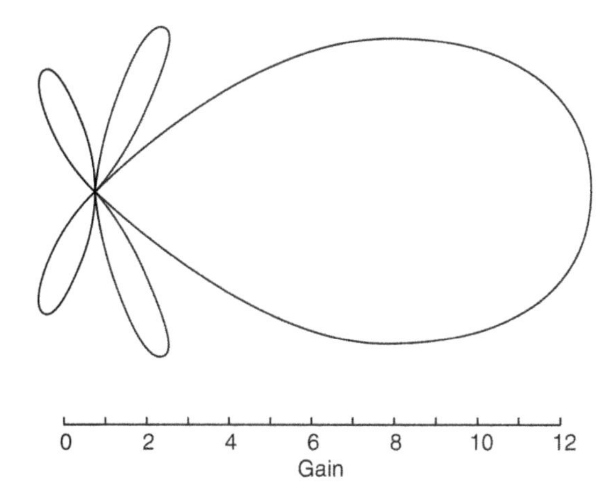

FIGURE 1.79 Polar diagram of a parasitic aerial.

of about 3% of its operating frequency. The main use of parasitic antennae in radio astronomy is as the receiving element (sometimes called the 'feed') of a larger reflector such as a parabolic dish, although arrays of parasitic aerials are used for longer wavelength observation.

The use of a single dipole or even several dipoles in an array is the radio astronomy equivalent of a naked-eye observation. The Netherlands-centred Low Frequency Array (LOFAR) though, which observes over the 10-to 240-MHz (30–1.2 m) band, uses simple dipole antennas. At the time of writing, LOFAR has 51 stations and about 20,000 antennas, the majority being located in the Netherlands with the rest distributed across Europe from Ireland to Poland (a 2,000-km baseline). Two more stations should be added (in Latvia and Italy) by 2022. Its effective collecting area is equivalent to that dish more than 600 m in diameter. The antennae at individual stations operate as phased arrays (see below), and then their data is combined to form an aperture synthesis system (Section 2.5.4). The Long Wavelength Array (LWA; 10–85 MHz, 256 antennas in New Mexico) and the low-frequency part of the SKA (50–350 MHz – see below) will also use simple dipole receivers.

However, just as in optical astronomy, the observations may be greatly facilitated by the use of some means of concentrating the signal from over a wide area onto the antenna. The most familiar of these devices are the large parabolic dishes that are the popular conception of a radio telescope. These are directly equivalent to an optical-reflecting telescope. They are usually used at the prime focus or at the Cassegrain focus or recently as off-axis Gregorians (Section 1.1).

The gain may be found roughly by substituting the dish's area for the effective area in Equation 1.71. The Rayleigh resolution is given by Equation 1.15. The size of the dishes is so large because of the length of the wavelengths being observed; for example, to obtain a resolution of 1° at a wavelength of 0.1 m requires a dish 7-m across, which is larger than most optical reflectors for a resolution more than 10^4 times poorer. The requirement on surface accuracy is the same as that for an optical telescope; deviations from the paraboloid to be less than $\lambda/8$ if the Rayleigh resolution is not to be degraded. Now, however, the longer wavelength helps because it means that the surface of a telescope working at 0.1 m, say, can deviate from perfection by more than 10 mm without seriously affecting the performance.[91] In practice a limit of $\lambda/20$ is often used, for as we have seen, the Rayleigh limit does not represent the ultimate limit of resolution.

These less stringent physical constraints on the surface of a radio telescope ease the construction problems greatly; more importantly however, it also means that the surface need not be solid because a wire mesh with spacings less than $\lambda/20$ will function equally well as a reflector. The weight and wind resistance of the reflector are thus reduced by large factors. The Giant Metrewave Radio Telescope (GMRT) in India thus comprises thirty 45-m parabolic dishes. The dishes' reflecting surfaces are a lightweight stainless-steel wire mesh with the 'holes' ranging from 10 to 20 mm across. Since the instrument operates between 50 MHz and 1.4 GHz (wavelengths 6-m to 0.2-m), the wire mesh is as effective as would be a solid reflector.

At the shorter radio wavelengths, though, a solid reflecting surface may be more convenient and at short wavelengths (<1 mm), active surface control to retain the accuracy is used along somewhat similar lines to the methods discussed in Section 1.1 for optical telescopes, the shape of the mirror being monitored holographically. The dishes are usually of small focal ratio, *f*0.5 is not uncommon and the reason for this is so that the dish acts as a screen against unwanted radiation as well as concentrating the desired radiation.

The feed antennae for radio telescope dishes may be made of such a size that they intercept only the centre lobe of the telescopes' responses (i.e., the Airy disc in optical terms). The effects of the side lobes are then reduced or eliminated. This technique is known as 'tapering the antenna' and is the same as the optical technique of apodisation (Section 4.1). The tapering function may take several forms and may be used to reduce background noise as well as eliminating the side lobes.

[91] Interferometers working in the THz region, such as ALMA, are affected by atmospheric scintillation and just like their optical counterparts need real-time atmospheric compensation to reach their diffraction limited resolutions (see Section 2.5).

Tapering reduces the efficiency and resolution of the dish, but this is usually more than compensated for by the improvement in the shape of the response function.

Fully steerable dishes up to 100-m across have been built such as the Green Bank (Figure 1.74) and Effelsberg telescopes, while the Arecibo telescope in Puerto Rico[92] is a fixed dish 300-m across. This latter instrument acts as a transit telescope and has some limited ability to track sources and look at a range of declinations by moving the feed antenna around the image plane. Such fixed telescopes may have spherical surfaces to extend this facility and use a secondary reflector to correct the resulting spherical aberration. China's Five Hundred Metre Aperture Spherical Telescope (FAST) instrument, like Arecibo, is a fixed dish, but made up from 4,450 eleven-metre individual segments[93] that can individually be angled to enable zenith distances of up to 60° to be observed. The feed is housed in a cabin that can be moved around the focal surface by adjusting its cable suspensions. Although the instrument's dish is 500 m in diameter, only a circular region 300-m across (or less) is used during any single observation, and this area is then adjusted to be parabolic in shape. FAST achieved first light in 2016 and currently observes over the 70-MHz to 3-GHz waveband. A proposal some years ago, called Kilometer-square Area Radio Synthesis Telescope (KARST), to build up to 30 more FAST-type instruments working in combination has effectively been superseded by the SKA.

In the microwave region, the largest individual dishes are currently the 50-m Large Millimeter Telescope (LMT) which started operations in 2006 and is located near Mexico City observing from 75 to 350 GHz, the 45-m telescope at Nobeyama in Japan (1–150 GHz) and the 30-m IRAM instrument on Pico Veleta in Spain and operating from 100 to 330 GHz.

Arrays, such as ALMA, Northern Extended Millimeter Array (NOEMA) and the Submillimeter Array have higher resolutions and (sometimes) larger collecting areas than the single dishes. Thus, ALMA has fifty-four 12-m and twelve 7-m dishes – a collecting area equivalent to a single 90-m dish The ALMA dishes can be moved to provide a maximum baseline of 16 km. Its best angular resolution can thus reach 20 milliarcseconds at 230 GHz or 43 milliarcseconds at 110 GHz.

The NOEMA, equivalent to a 52-m single dish with its twelve 15-m dishes, is sited on the Plateau de Bure in France and obtained its first observations in 2015. Its dishes can be moved along a 2,000-m twin track and it operates between 70 and 400 GHz with a best angular resolution of 100 milliarcseconds. Similarly, the Submillimeter Array on Mauna Kea uses eight 6-m dishes – equivalent to a single 17-m dish in area – with a maximum baseline of 500 m and operates over 180–420 GHz waveband.

Deep synoptic array (DSA)-10 and DSA-110 are prototypes for the DSA-2000 Astro2020 concept proposal. The DSA-10 has ten commercial 4.5-m dishes and mounts, with manual adjustment of their positions and observes around 1.4 GHz. The DSA-110 is currently under construction and will have 110 motor-driven 4.75-m dishes. DSA-2000 would have two thousand 5-m dishes, operate from 700 MHz to 2 GHz and produce ready-to-use radio images as its direct output (as opposed to interferometer visibility functions). The DSA-10 and DSA-110 costs are less than $10 per antenna.

While not designed to be an array, the Event Horizon Telescope (EHT) deserves some mention here (see also Section 2.5.4). The EHT comprises (at the time of writing) ten instruments, some of which themselves are arrays made up of many individual telescopes. The contributors to the EHT include the already mentioned ALMA (Chile), APEX (Chile), IRAM (Spain), JCMT (Hawaii), LMT (Mexico) and the Submillimeter Array (Hawaii). In addition, there are Atacama Submillimeter Telescope Experiment (ASTE; 10 m, Chile), the Heinrich Hertz Submillimeter Telescope (12 m, Arizona), the South Pole Telescope (10 m, Antarctica) and the 12-m ALMA prototype Telescope (Arizona), all of which combine to form many Very Long Baseline Interferometers (VLBIs, see Section 2.5.4). The Greenland telescope (12 m, due to be re-erected soon, near Greenland's summit) and NOEMA are due to be

[92] At the time of writing, funding for both the Green Bank and Arecibo instruments is under threat. Green Bank is now largely funded by its users on a 'pay-as-you-go' basis, and although Arecibo recently received a $6 million upgrade, hurricane Maria caused some $14 million worth of damage to it in 2017.

[93] Cf. Figures 1.30 and 1.47 and discussions of segmented-mirror optical telescopes in Section 1.1.

added to the system in the next few years. The distribution of the component instruments forming the EHT are such that maximum baselines may approach the diameter of the Earth (12,700 km), so operating at 230 GHz, the EHT's angular resolution can be around 25 microarcseconds,[94] resulting in April 2019, in the widely publicised 'image of the black hole[95] at the centre of M87'. The EHT operates by recording the signals at each observing instrument together with the time from an atomic clock and then bringing the recordings physically together for processing (Section 2.5.4).

With a single feed, the radio telescope is a point-source detector only. Images have to be built up by scanning (Section 2.4) or by interferometry (Section 2.5). Scanning used to be accomplished by pointing the telescope at successive points along the raster pattern and observing each point for a set interval. This practice, however, suffers from fluctuations in the signal due to the atmosphere and the electronics. Current practice is, therefore, to scan continuously or 'on-the-fly'. The atmospheric or electronic variations are then on a longer timescale than the changes resulting from moving over a radio source and can be separated out.

True imaging can be achieved through the use of cluster or array feeds (radio cameras). These are simply multiple individual feeds arranged in a suitable array at the telescope's focus. Each feed is then the equivalent of a pixel in a CCD or other type of detector. The number of elements in such cluster feeds is rising, though most are still smaller than their optical counterparts, but for example, the various SCUBA instruments on the JCMT have risen from 91 detectors at ~750 GHz or 37 at ~400 GHz for the original SCUBA (decommissioned in 2004) to SCUBA-2 (currently being phased out) which uses eight 1,280 arrays, while its replacement (expected for 2022) will have 3,600 detector arrays. The LMT will similarly exchange its current 270 GHz receiver (Aztronomical Thermal Emission Camera [AzTEC] – commissioned 2005) which has a 144-element silicon nitride micromesh bolometer array, for TolTEC[96] in the near future. The latter using 6,300 lumped element MKID detectors across three wavebands and will have a 4 minute-of-arc field of view. FAST has just had a 19-beam feed horn receiver installed covering the 1.05- to 1.45-GHz region. For the APEX telescope, the Submillimeter APEX Bolometer Camera (SABOCA) instrument (Figure 1.80),

FIGURE 1.80 The 39-beam Submillimeter APEX Bolometer Camera (SABOCA) instrument which preceded Atacama Pathfinder Experiment's (APEX's) current 295-beam Large APEX Bolometer Camera (LABOCA; *see* Figure 1.22). (Courtesy of European Southern Observatory.)

[94] The angular size of an astronaut's thumb when he or she is walking on the surface of the Moon, as seen by an observer on the Earth.

[95] Actually, of its shadow, which, at 10^{11} km, is about 2½ times the size of the black hole's event horizon.

[96] The derivation of this name is obscure.

which operated at 850 GHz with 39 bolometer detectors until 2015 was replaced by the current LABOCA (Figure 1.22) with 295 bolometer detectors (Figure 1.22) and operating at 345 GHz with an 11.4 arc-minute field of view.

Many other systems have been designed to fulfil the same function as a steerable paraboloid but which are easier to construct. The best known of these are the multiple arrays of mixed collinear and broadside type or similar constructions based on other aerial types. They are mounted onto a flat plane which is oriented East-West and which is tiltable in altitude to form a transit telescope. Another system such as the 600-m RATAN telescope uses an off-axis paraboloid that is fixed and which is illuminated by a tiltable flat reflector. Alternatively, the paraboloid may be cylindrical and tiltable itself around its long axis. For all such reflectors, some form of feed antenna is required. A parasitic antenna is a common choice at longer wavelengths, while a horn antenna, which is essentially a flared end to a waveguide, may be used at higher frequencies.

A quite different approach is used in the Mills Cross type of telescope. These uses two collinear arrays oriented North-South and East-West. The first provides a narrow fan beam along the North-South meridian, while the second provides a similar beam in an East-West direction. Their intersection is a narrow vertical pencil beam, typically 1° across. The pencil beam may be isolated from the contributions of the remainders of the fan beams by comparing the outputs when the beams are added in phase with when they are added out of phase. The in-phase addition is simply accomplished by connecting the outputs of the two arrays directly together. The out-of-phase addition delays one of the outputs by half a wavelength before the addition, and this is most simply done by switching in an extra length of cable to one of the arrays. In the first case, radiation from objects within the pencil beam will interfere constructively, while in the second case there will be destructive interference. The signals from objects not within the pencil beam will be mutually incoherent and so will simply add together in both cases. Thus, looking vertically down onto the Mills Cross, the beam pattern will alternate between the two cases shown in Figure 1.81. Subtraction of the one from the other will then just leave the pencil beam.

The pencil beam may be displaced by an angle θ from the vertical by introducing a phase shift between each dipole. Again, the simplest method is to switch extra cable into the connections between each dipole, the lengths of the extra portions, L, being given by

$$L = d \sin \theta \qquad (1.75)$$

where d is the dipole separation. The pencil beam may thus be directed around the sky as wished. In practice, the beam is only moved along the North-South plane, and the telescope is used as a transit telescope because the alteration of the cable lengths between the dipoles is a lengthy procedure.

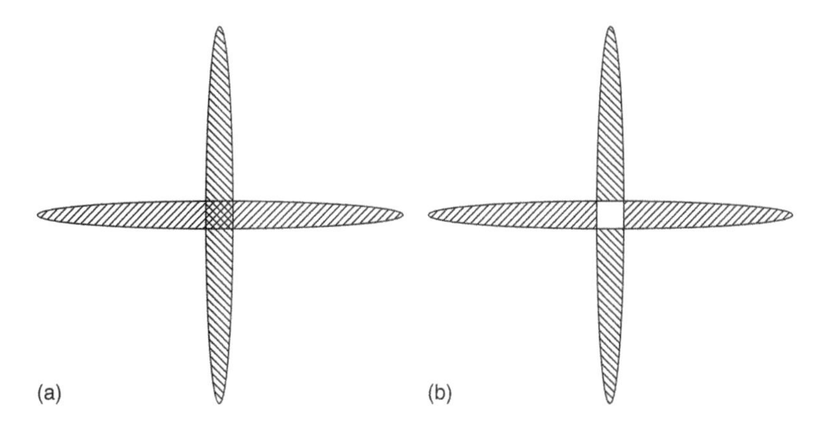

FIGURE 1.81 Beam patterns of a Mills Cross radio telescope. (a) with the beams added together (i.e., in phase); (b) with the beams subtracted from each other (i.e., 180° out of phase).

The resolution of a Mills Cross is the same as that of a parabolic dish whose diameter is equal to the array lengths. The sensitivity, however, is obviously much reduced from that of a dish because only a tiny fraction of the aperture is filled by the dipoles. As well as being cumbersome to operate, the Mills Cross suffers from the disadvantage that it can only operate at a single wavelength unless its dipoles are all changed. Furthermore, confusion of sources may arise if a strong source is in one of the pencil beams formed by the side lobes because it will have the same appearance as a weak source in the main beam. An alternative and related system uses a single fan beam and observes the object at many position angles. The structure of the source is then retrievable from the observations in a relatively unambiguous manner.

The Mills Cross design was mostly used early in the history of Radio Astronomy, but it serves as an introduction to the way in which its more modern equivalent, the phased array, operates.

The phasing of two dipoles, as we have seen, can be used to alter the angle of the beam to their axis. By continuously varying the delay, the lobe can be swept across the sky or accurately positioned onto an object and then moved to track that object. This is an important technique for use with interferometers (Section 2.5) and for solar work. It also forms the basis of a phased array. This latter instrument is basically a Mills Cross in which the number of dipoles has been increased until the aperture is filled. It provides great sensitivity because a large area may be covered at a relatively low cost. This type of phased array achieved its moment of glory in 1967 with Jocelyn Bell-Burnell's discovery of the first pulsar, since then, though, it has gradually given way to more sophisticated instruments. The huge modern arrays of simple antennae, such as LOFAR, LWA and SKA for example (previously in this section), however, are clearly its descendants.

Recently, the phased array has started to make a comeback as the detector part of radio cameras. A large individual radio dish can be converted to a multi-beam system by using a small phased array at its focus (a phased array feed – cf. multi-detector array feeds). Most of the fourteen 25-m dishes of the Westerbork Synthesis Radio Telescope (WSRT; see also Section 2.5.4) have had their previous feeds replaced with aperture tile in focus (Apertif) radio cameras, for example. The WSRT now has an array of 37 receiving beams and observes an area of 8 square degrees at an angular resolution of 15″ in a single observation, covering the 1.0 to 1.75 GHz frequencies with a bandwidth of 300 MHz. Similarly, the thirty-six 12-m antennas of the Australian SKA Pathfinder (ASKAP) instrument now each have 36 receiving beams thanks to their new phased array feeds. These have expanded the field of view to 30 square degrees using 188 individual receivers.

A radically different approach to concentrating the radiation is based upon refraction. The Lüneburg lens has yet to find much application to radio astronomy, although at one time it was being considered for the SKA. The Lüneburg lens is a solid sphere within which the refractive index increases linearly inwards from unity at the surface (other rates of change of the refractive index, such as a power law, may also be used). Using a central refractive index of 2, the focus is on the surface of the lens. Since there is no axis of symmetry, the lens can be used to observe in many directions simultaneously, simply by having numerous feeds distributed around it. Many materials potentially can be used for the lens' construction, but to date, high-density polystyrene is the one that has been used in practice. Lüneburg lenses have also been investigated for use at optical wavelengths to produce fish-eye sky cameras and, while the idea seems to be feasible, it has yet to be implemented.

More conventional lenses for use at long wavelengths have, however, been used in a few instruments. The material for such lenses could be (for example): acrylic glass, aluminium oxide (alumina), high density polyethylene (HDPE) or polytetrafluoroethylene (PTFE) which have refractive indices at 60 GHz of 1.6, 3.0, 1.5 and 1.5, respectively. Thus Background Imaging of Cosmic Extragalactic Polarisation (BICEP)-3, observing the CMB from Antarctica at 95 GHz, uses 0.58-m alumina lenses with epoxy resin anti-reflection coatings and 2,560 detectors. Its successor (expected for 2020) is the BICEP Array that will use four refracting telescopes with 0.65-m alumina lenses, more than 30,000 detectors and operate at four wavebands from 30 to 270 GHz. The earlier Keck array, also sited at the South Pole, observed the CMB at 95 and 150 GHz with five telescopes using 0.25-m HDPE lenses.

A recent laboratory development is of the water-drop antenna lens. This is a waveguide which has an inner surface shaped somewhat like the ripples produced on the surface of water, when something (e.g., a drop of water) is dropped into it. Its ease of fabrication (it can be made from metal) make this an interesting possibility for future use. The present example operates at 30 GHz.

The Simons observatory's main instrument, which is currently under construction,[97] will be a 6-m Dragone-type telescope with some 30,000 TES detectors operating over the 27- to 275-GHz frequency range. There will also, however, be three smaller refracting telescopes working in support of it, with 0.42-m apertures and using a total of another 30,000 detectors for the same frequencies. The small telescopes will each use three silicon lenses with metamaterial[98] anti-reflection coatings.

Very low frequency radiation (<30 MHz) is largely reflected by the Earth's ionosphere[99] and so does not penetrate down to ground-based radio telescopes. Observing astronomical sources at these frequencies, therefore, requires the receivers to be lifted into space. Thus, China's Chang'e-4[100] lunar farside lander (2019) required a relay spacecraft in the Earth-Moon Lagrange L2 position to return its data to Earth. The relay, named Queqiao,[101] also carries the Netherlands-China Low-Frequency Explorer (NCLE) instrument to explore the 80-kHz to 80-MHz frequency band for astronomical sources using three 5-m monopole antennae. Recently, difficulty was experienced in extending these antennae from their rolled-up storage state and so they are currently operating at reduced sensitivity.

The Mercury Magnetospheric Observatory (one-half of the BepiColombo spacecraft), currently journeying towards Mercury and planned to arrive in 2025, is carrying as a part of its plasma wave instrument, a radio receiver designed to operate up to a frequency of ~3 MHz. While in the near future, three 6.5-m long monopole wire antennae will be carried by the Solar Orbiter spacecraft (2020 launch) to pick up frequencies from (almost) DC up to several tens of MHz. The proposed DARE[102] mission (possible launch date in the early 2020s) is expected to use the Moon as a shield from the radio emissions from the Earth and the Sun whilst observing the highly red-shifted hydrogen 21-cm (1.4 GHz) line over the 40- to 120-MHz region. It will orbit the Moon at an altitude of 125 km, use four conical dipole antennas and be shielded from both the Sun and the Earth for between 0% and 50% of each four-hour orbit – allowing, over its expected two-year lifetime, some 1,000 hours of totally shielded observing.

Few low frequency (but above 30 MHz) spacecraft-borne telescopes have been used because suitable ground-based instruments are generally available. However, great improvements in the resolution and sensitivity of radio telescopes may be obtained through the use of interferometers and aperture synthesis systems (Section 2.5). Spacecraft carrying radio telescopes have, therefore, been launched to form part of such instruments in combination with Earth-based telescopes – so achieving baselines many times the size of the Earth and hence milli- or microarcsecond angular resolutions. We therefore here, for completeness (but see also Section 2.5), mention the Japanese Halca spacecraft (1997–2003) which carried an 8-m dish operating at low GHz frequencies as part of an aperture synthesis system and the Russian-led RadioAstron program with a 10-m diameter space radio dish, Spektr-R,[103] operating from 325 MHz to 22 GHz. When combining with ground-based instruments the baselines for these instruments could be up to 21,000 and 390,000 km, respectively.

---

[97] At Cerro Toco, 340 km NE of Paranal. First light expected in 2021.

[98] A material whose physical structure changes its intrinsic (solid) properties. The structures involved often mimic a bee's honeycomb.

[99] The same frequencies when used by terrestrial broadcasters (who call 30 MHz High Frequency or even Very High Frequency) reflect off the inside layers of the ionosphere and so can return to the surface (skip), to be received well beyond the horizon distance of the broadcasting station.

[100] Moon Goddess in Chinese.

[101] Magpie Bridge – from a Chinese folk story – the bridge is the Milky Way.

[102] Dark Ages Radio Explorer.

[103] Spectrum – Radio – Note that it is not the same as the Spektr-RG spacecraft (Section 1.1.8.1). The spacecraft has recently stopped responding to control commands, but its scientific payload is still working.

For the future, there are the proposed Chinese Space Millimetre Wavelength VLBI Array (SMVA) and the cosmic microscope (CM) missions with telescopes in space of 10-m and 30-m diameters, respectively, and baselines of up to 90,000 km, operating perhaps from 30 MHz to 40 GHz and so giving angular resolutions down to 20 milliarcseconds.

A number of spacecraft carrying microwave detectors have been launched, most of which (e.g., JWST, Spitzer, WFIRST and WISE) have already been mentioned because they also observe in the FIR (Section 1.1.15). In addition, there are the Cosmic Background Explorer (COBE) Satellite (1989–1993) which carried Dicke radiometers operating at 31.5, 53, and 90 GHz, the Planck spacecraft (2009–2012) designed to observe the cosmic microwave background radiation and used HEMTs and MMICs in its low frequency (30–70 GHz) instrument and Wilkinson Microwave Anisotropy Probe (WMAP) (2001–2010) mission which observed from 22 to 90 GHz using HEMTs.

1.2.3.1 Construction

The large dishes that we have been discussing pose major problems in their construction. Both the gravitational and wind loads on the structure can be large and shadowing of parts of the structure can lead to inhomogeneous heating and hence, expansion and contraction-induced stresses. The current Green Bank radio telescope (Figure 1.74) replaces an previous 90-m dish that collapsed catastrophically in 1988 whilst in use, although fortunately without anyone being injured.

The worst problem is due to wind because its effect is highly variable and the forces can be large – 1.5×10^6 N (150 tonnes) for a 50-m dish facing directly into a gale-force wind for example. A rough rule of thumb to allow scaling the degree of wind distortion between dishes of different sizes, is that the deflection, Δe, is given by

$$\Delta e \propto \frac{D^3}{A} \tag{1.76}$$

where D is the diameter of the dish and A is the cross-sectional area of its supporting struts. Thus, doubling the diameter of the design of a dish would require the supporting members' sizes to be increased by a factor of eight if it were to still work at the same wavelength. There are only two solutions to the wind problem: to enclose the dish or to cease using it when the wind load becomes too great. Some smaller dishes, especially those working at short wavelengths where the physical constraints on the surface accuracy are most stringent, are enclosed in radomes or space-enclosing structures built from non-conducting materials. But this is not usually practicable for the larger dishes. These must generally cease operating and be parked in their least wind-resistant mode once the wind speed rises above 10–15 m/s (35–55 km/h).

The effects of gravity are easier to counteract. The problem arises from the varying directions and magnitudes of the loads placed upon the reflecting surface as the orientation of the telescope changes. Three approaches have been tried successfully for combating the effects of gravity. The first is the brute force approach, whereby the dish is made so rigid that its deformations remain within the surface accuracy limits. Except for small dishes, this is an impossibly expensive option. The second approach is to compensate for the changing gravity loads as the telescope moves. The surface may be kept in adjustment by systems of weights, levers, guy ropes, springs or more recently by computer-controlled hydraulic jacks. The final approach is much more subtle and is termed the 'homological transformation system'. The dish is allowed to deform, but its supports are designed so that its new shape is still a paraboloid, although in general an altered one, when the telescope is in its new position. The only active intervention that may be required is to move the feed antenna to keep it at the foci of the changing paraboloids.

Alternatively, the dish may be fixed in position, like the Arecibo and FAST instruments (see the beginning of this section) and then the gravitational stresses are constant and so easily overcome. The disadvantage of a fixed instrument of only pointing to one part of the sky (usually the zenith) may be partially overcome by moving the instrument's feed assembly around the focal surface, so that FAST can observe up to zenith distances of 60°.

There is little that can be done about inhomogeneous heating, other than painting all the surfaces white so that the absorption of the heat is minimised. Fortunately, it is not usually a serious problem in comparison with the first two, though the Green Bank telescope has recently been enabled to undertake daytime observing, following the installation of a laser-based system to monitor its surface shape and then correct the thermal distortions.

The supporting framework of the dish is generally a complex, cross-braced skeletal structure, whose optimum design requires a computer for its calculation. This framework is then usually placed onto an alt-az mounting (Section 1.1) because this is generally the cheapest and simplest structure to build and also because it restricts the gravitational load variations to a single plane, so making their compensation much easier. A few, usually quite small, radio telescopes do have equatorial mountings, though.

1.2.3.2 Future

The Large Scale Polarization Explorer (LSPE) project will involve both ground-based and balloon-borne instruments to study the polarisation of the CMB on large angular scales. The ground-based component (Survey Tenerife Polarimeter [STRIP]), to be sited on Tenerife, will involve a 1.5-m Dragone-type telescope with 49 polarimeters operating at 43 and 95 GHz. The balloon-borne component (Short Wavelength Instrument for the Polarization Explorer [SWIPE]) will be launched to circle the North Pole, will carry 330 spider-web TES bolometers cooled to 300 mK and observe at 140, 220, and 240 GHz. It will use a 0.5-m refracting telescope and measure polarisation using a half-wave plate and a rotating wire-grid polariser (Section 5.2). The initial operations of LSPE are expected to start at about the time of writing.

CCAT-p (Section 1.1.15.4) is planned to begin operating in 2021. It will be sited at Cerro Chajnantor and be able to observe from 100 GHz to 1.5 THz. The three mirrors for its crossed-Dragone design will be fabricated from aluminium segments with adjustable (but not active) positions to attain a surface accuracy of about 10 μm. A previous proposal that CCAT-p should be a prototype for a 25-m successor (CCAT) has now been abandoned.

The SKA is currently under development with a number of preliminary studies and projects already completed. The main sites for the $2 billion instrument have been chosen as the Murchison Radio Astronomy Observatory in Western Australia and the Karoo desert in South Africa's Northern Cape Province. Additionally, there will also be outlier stations in a number of other Southern Hemisphere sites. The SKA is scheduled for completion[104] around 2030 and will be a phased array of many small receivers whose total collecting area will be 1 km^2 (10^6 m^2). It will cover the spectral region from 70 MHz to 30 GHz and have a maximum baseline of 3,000 km.

It is currently planned that three types of instruments will be used to cover this spectral region: dipoles from 70 to 200 MHz (4.3–1.5 m), phased array tiles for 200 to 500 MHz (1.5 m to 600 mm) and some three thousand 15-m dishes for the 500-MHz to 30-GHz (600–10 mm) region. Its field of view will range from about 200 deg^2 at the lower frequencies to 1 deg^2 at the high frequencies where its angular resolution will reach 2 milliarcsecond. Phased array feeds (see above) will be used for the individual instruments at the higher frequencies.

The thirty-six 12-m antennas of the ASKAP instrument were completed in 2012, acting both as a major radio telescope in their own right and as a test bed for possible SKA developments. At lower frequencies the Murchison Wide Field Array with 128 phased array tiles is also operational and was upgraded in 2018 to 256 antennae at ten times the original sensitivity and with improved angular resolution. In South Africa, the seven 13.5-m dishes of the Karoo Array Telescope-7 (KAT-7)

[104] It seems likely that the SKA will never be 'finished'; it will always be having updates, new sites added and/or old sites decommissioned.

prototype were completed in 2010 and the 64 dishes of the follow-on Meer Karoo Array Telescope (MeerKAT)[105] array began operations in 2018.

The Hydrogen Intensity and Real Time Analysis Experiment (HIRAX) array is currently under construction for a 2022 start. Its aim is to investigate, amongst other things, dark energy looking for baryon acoustic oscillations (Section 1.7.2.4). It will comprise 1,024 dishes which can be manually realigned (saving costs on motors, drives, etc.), when needed, to observe a new part of the sky. The dishes will be 6 m in diameter and of short focal ratio (f0.25) so that each dish's feed is shielded from all the other dishes, and most other sources of interference by the dish itself. It is to be installed at the SKA's Karoo site and observe over the –400–800 MHz region. Also nearing completion is the 12-m Large Latin American Millimeter Array (LLAMA) telescope, which is located about 180 km SE of ALMA and is of similar design to the ALMA antennas. It will cover the 35 GHz to 1 THz frequency region and be able to act as a VLBI with ALMA and other similar instruments.

A 110-m steerable radio telescope, to be located in the Tian Shan Mountains in Northwest China, is currently planned for a possible 2023 start. The Qitai radio Telescope (QTT) is expected to operate from 150 MHz to 115 GHz and would then become (by a few square metres) the world's largest steerable radio dish. Atacama Large Aperture Submillimeter Telescope (AtLAST) is a concept proposal (for Astro 2020) for a 50-m dish to observe the 35- to 950-GHz region with a 1° field of view perhaps for construction in Chile in the late 2020s.

Other developments are likely to be in the area of increasingly sophisticated integrated circuit technology that leads to smaller and less power-hungry detectors, amplifiers, correlators and so on which in turn allows the building of larger array detectors.

In the more distant future, increasing levels of radio noise from artificial terrestrial sources may force radio telescopes to the far side of the Moon (see DARE, above). However, it seems unlikely that any such project will be possible before there is a substantial and permanently occupied colony (or colonies) on the Moon – by which time the radio noise from artificial lunar sources may be just as bad as that on the Earth. It is possible, though, that before the Moon becomes colonised, remote/robotic instruments and techniques, whose sophistication is improving rapidly, will be able to operate on the Moon without human intervention. Thus, a NASA Probe-class mission concept has been included in the Astro2020 decadal survey which envisages a low-frequency interferometer array being erected robotically on the Lunar farside. Called FARSIDE,[106] it would operate over the 150-kHz to 40-MHz range with 128 antennas with an effective collecting area of up to 18.5 km^2. The antennae would be deployed by a rover and linked to a central base station. It would be able to scan 1,400 channels and cover the whole visible hemisphere every minute. The data would be sent back to Earth via a relay spacecraft[107] in lunar orbit. Setting up the antennae might take about four months from arrival on the Moon and the cost is put at about $1.3 billion.

One development which is currently underway though, which may reduce radio noise, is the increasing use of fibre-optic cables for data transmission. The popular demand for faster and faster broadband access seems likely to lead to the installation of fibre-optic lines throughout most of the world within the next decade or two. Once such lines are in place there will be little need for many, most and/or all of the present-day radio and TV broadcasting stations, and so the radio spectrum *may* become quieter.

[105] 'Meer' is Afrikaans for 'more' and MeerKat is an extension of KAT-7. The meerkat is also a species of mongoose living in the Karoo desert.

[106] Farside Array for Radio Science Investigations of the Dark ages and Exoplanets – and No, the scheme was not suggested by Gary Larson.

[107] A separate proposal is for the Lunar Gateway – a multi-purpose permanent spacecraft in lunar orbit that could act as a general communications hub, holding area for other spacecraft and instruments and even as a habitat for humans.

1.3 X-RAY AND GAMMA-RAY DETECTION

1.3.1 Introduction

The electromagnetic spectrum comprises radiation of an infinite range of wavelengths, but the intrinsic nature of the radiation is unvarying. There is, however, a tendency to regard the differing wavelength regions from somewhat parochial points of view, and this tends to obscure the underlying unity of the processes that may be involved. The reasons for these attitudes are many; some are historical hangovers from the ways in which the detectors for the various regions were developed; others are more fundamental in that different physical mechanisms predominate in the radiative interactions at different wavelengths. Thus, high-energy gamma rays may interact directly with nuclei, at longer wavelengths we have resonant interactions with atoms and molecules producing electronic, vibrational and rotational transitions, while in the radio region currents are induced directly into conductors. However, primarily, the reason may be traced to the academic backgrounds of the workers involved in each of the regions. Because of the previous reasons, workers involved in investigations in one spectral region will tend to have different biases in their backgrounds compared with workers in a different spectral region. Thus, there will be different traditions, approaches, systems of notation and so on with a consequent tendency to isolationism and unnecessary failures in communication. The discussion of the point source image in terms of the Airy disc and its fringes by optical astronomers and in terms of the polar diagram by radio astronomers, as already mentioned, is one good example of this process and many more exist.

It is impossible to break out from the straitjacket of tradition completely, but an attempt to move towards a more unified approach has been made in this work by dividing the spectrum much more broadly than is the normal case. We only consider three separate spectral regions, within each of which the detection techniques bear at least a familial resemblance to each other. The overlap regions are fairly diffuse with some of the techniques from each of the major regions being applicable. We have already discussed two of the major regions, radio and microwaves and optical and IR. The third region, X-rays and gamma rays, is the high-energy end of the spectrum and is the most recent area to be explored. This region also overlaps to a considerable extent, in the nature of its detection techniques, with the cosmic rays which are discussed in the next section. None of the radiation discussed in this section penetrates down to ground level, so its study had to await the availability of observing platforms in space or near the top of the Earth's atmosphere. Thus, significant work on these photons has only been possible since the 1960s and many of the detectors and observing techniques are still under development.

The high-energy electromagnetic spectrum is fairly arbitrarily divided into:

- The extreme UV (EUV or XUV region): 100–10 nm wavelengths (12–120 eV photon energies)
- Soft X-rays: 10–1 nm (120 eV to 1.2 keV)
- X-rays: 1–0.1 nm (1.2–12 keV)[108]
- Hard X-rays: 0.1–0.01 nm (12–120 keV)
- Soft γ rays: 0.01–0.001 nm (120 keV to 1.2 MeV)
- γ rays: less than 0.001 nm (greater than 1.2 MeV).

We shall be primarily concerned with the last four regions in this section (i.e., with wavelengths less than 10 nm and photon energies greater than 120 eV). Frequency is rarely used as a working unit in this part of the spectrum, but for completeness, we are concerned with frequencies higher than 30 PHz.

---

[108] It is fairly common, over this part of the spectrum, to find object's x-ray luminosities expressed in 'Crab' units: 1 Crab is the intensity of the Crab nebula (M1) at the same x-ray wavelength. At a few keV, 1 Crab $\approx 1.5 \times 10^8$ eV m^{-2} s^{-1}, but this conversion factor varies with wavelength and so the Crab unit is generally best avoided.

The main production mechanisms for high-energy radiation include electron synchrotron radiation, the inverse Compton effect, free-free radiation and pion decay, while the sources include the Sun, supernova remnants, pulsars, bursters, binary systems, cosmic rays, the intergalactic medium, galaxies, Seyfert galaxies and quasars. Absorption of the radiation can be by ionisation with a fluorescence photon or an Auger electron produced in addition to the ion and electron, by Compton scattering, or in the presence of matter, by pair production. This latter process is the production of a particle and its anti-particle and not the pair production process discussed in Section 1.1.4.1 which was simply the excitation of an electron from the valence band. The interstellar absorption in this spectral region varies roughly with the cube of the wavelength, so that the higher-energy radiation can easily pass through the whole galaxy with little chance of being intercepted. At energies under about 2 keV, direct absorption by the heavier atoms and ions can be an important process. The flux of the radiation varies enormously with wavelength. The solar emission alone at the lower energies is sufficient to produce the ionosphere and thermosphere on the Earth. At 1-nm wavelength (1.2 keV), for example, the solar flux is 5×10^9 photons $m^{-2}\ s^{-1}$, while the total flux from all sources for energies above 10^9 eV is only a few photons per square metre per day.

1.3.2 Detectors

1.3.2.1 Geiger Counters

The earliest detection of high-energy radiation from a source other than the Sun took place in 1962 when soft X-rays from a source that later became known as Sco X-1 were detected by large area Geiger counters flown on a sounding rocket. Geiger counters are no longer used as primary detectors, but variants of them have been developed into much more sophisticated detectors of both high energy e-m radiation and high energy subatomic particles (Sections 1.4 and 1.5) and are in widespread use today. We start, therefore, with a quick review of the operating principles of the Geiger counter.

Two electrodes inside an enclosure are held at such a potential difference that a discharge in the medium filling the enclosure is on the point of occurring. The entry of ionising radiation[109] triggers this discharge (cf. APDs), resulting in a pulse of current between the electrodes that may then be amplified and detected. The electrodes are usually arranged to be the outer wall of the enclosure containing the gas and a central coaxial wire (Figure 1.82). The medium inside the tube is typically argon at a low pressure with a small amount of an organic gas, such as alcohol vapour, added.

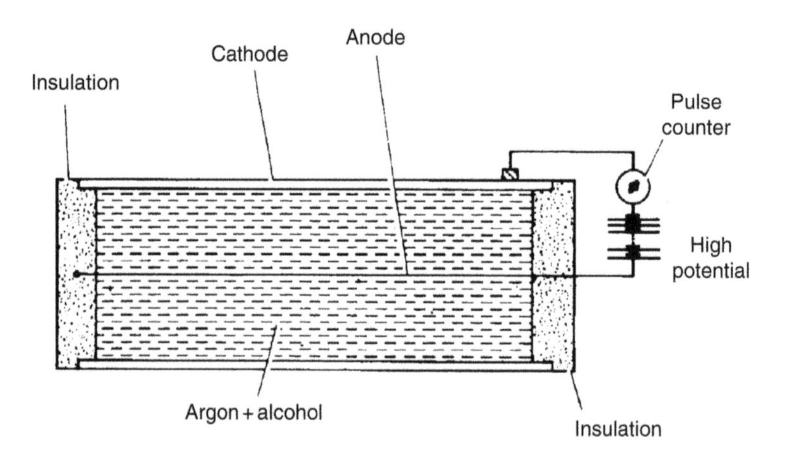

FIGURE 1.82 A typical arrangement for a Geiger counter.

[109] The production of electrons by ionising radiation is sometimes called the 'photoelectric effect' (cf. Section 1.1.4.1).

The electrons produced in the initial ionisation are accelerated towards the central electrode by the applied potential; as these electrons gain energy, they cause further ionisation, producing more electrons, which in turn are accelerated towards the central electrode and so on. The amplification factor can be as high as 10^8 electrons arriving at the central electrode for each one in the initial ionisation trail.

A problem with Geiger counters is that the avalanche of electrons rapidly saturates, so that the detected pulse is independent of the original energy of the photon. Another disadvantage, which also applies to many of the other detectors discussed in this and the following section, is that a response to one event leaves the detector inoperative for a short interval known as the 'dead time'. In the Geiger counter, the cause of the dead time is that a discharge lowers the potential between the electrodes; so that it is momentarily insufficient to cause a second avalanche of electrons should another X-ray enter the device. The length of the dead time is typically 200 μs.

1.3.2.2 Proportional Counters

These devices are closely related to Geiger counters and are also known as known as 'gas-filled ionisation detectors' (see also Section 1.3.2.5). They are in effect Geiger counters operated at less than the trigger voltage. By using a lower voltage, saturation of the pulse is avoided and its strength is then *proportional* to the energy of the original interaction. The gain of the system operated in this way is reduced to about 10^4 or 10^5, but it is still sufficient for further detection by conventional amplifiers, etc. Provided that all the energy of the ionising radiation is absorbed within the detector, its original total energy may be obtained from the strength of the pulse and we have a proportional counter.

At low-photon energies, a window must be provided for the radiation. These are typically made from thin sheets of plastic, mica, or beryllium, and absorption in the windows limits the detectors to energies above a few hundred electron volts. When a window is used, the gas in the detector has to be continuously replenished because of losses by diffusion through the window.

At high-photon energies the detector is limited by the requirement that all the energy of the radiation be absorbed within the detector's enclosure. To this end, proportional counters for high-energy detection may have to be made quite large. About 30 eV on average are required to produce one ion-electron pair, so that a 1 keV photon produces about 36 electrons and a 10 keV photon about 360 electrons. The spectral energy resolution to 2 1/2 standard deviations from the resulting statistical fluctuations of the electron numbers is thus about 40% at 1 keV and 12% at 10 keV. The quantum efficiencies of proportional counters approach 100% for energies up to 50 keV.

The position of the interaction of the X-ray along the axis of the counter may be obtained through the use of a resistive anode. The pulse is abstracted from both ends of the anode and a comparison of its strength and shape from the two ends then leads to a position of the discharge along the anode. The anode wires are thin, typically 20 μm across, so that the electric field is most intense near to the wire. The avalanche of electrons thus develops close to the wire, limiting its spread and giving a precise position. The concept may easily be extended to a 2D grid of anodes to allow genuine imaging. Spatial resolutions of about a tenth of a millimetre are possible. In this form, the detector is called a 'position-sensitive proportional counter'. They and their closely related variants are also known as 'multi-wire chambers', time-protection chambers (TPCs, see Section 1.5) or gas pixel detectors (GPDs),[110] especially in the context of particle physics (see also drift chambers, Section 1.4).

Proportional counters and their derivatives are also intrinsically sensitive to the polarization of the detected radiation. The direction of the initial track of the electrons is (roughly) in the same direction as the electric field of the X-ray photon. Determining the directions of the tracks within the detector therefore reveals the state of polarisation of the incoming X-rays; if the tracks are randomly distributed then the radiation is unpolarised, but a non-random distribution shows that the radiation as a whole is polarised.

---

[110] TPCs and GPDs differ only in their readout mechanisms.

Many gases can be used to fill the detector; argon, carbon dioxide, dimethyl ether (DME), methane, xenon and mixtures thereof at pressures near that of the atmosphere, are among the commonest ones (liquid argon as the filler has also been recently suggested for γ-ray detection). The inert gases are generally to be preferred because with their single atoms, there is then no possibility of the loss of energy into the rotation or vibration of multi-atom molecules.

The Indian Astrosat spacecraft carries two instruments based upon proportional counters; Large Area X-ray Proportional Counter (LAXPC) and Scanning Sky Monitor (SSM). The LAXPC uses three aligned proportional counters to cover the 3- to 80-keV region with a field of view 1°. Each of the counters has an area of ~0.27 m^2 and uses xenon at twice atmospheric pressure as its medium. The response time is just 10 μs. The SSM also uses three proportional counters but each with a linear coded mask collimator (Section 1.3.4.1) made up from six arrays of slits. Each counter covers a 10° × 90° area with a best angular resolution of ~10 arc-minutes and observes the 2- to 10-keV region. The instrument uses 25-μm carbon-coated quartz fibres as its resistive anodes, with a position on the wire resolution of 700 μm, and it has a response time of 1 ms.

TPC detectors were envisaged for the Polarimeter for Astrophysical Relativistic X-ray Sources (PRAXyS) spacecraft (a NASA phase A study) to operate within the 2- to 10-keV region. A recent development of the TPC, known as a micro-mesh gaseous structure (Micromegas) detector (Section 1.5), was to have been flown on board the Spektr-RG spacecraft as the detector for the Lobster-eye Wide Field X-ray Telescope (LWFT; see below for lobster-eye collimators); however, that instrument had to be dropped for budgetary reasons. Micromegas are, though, currently in use at Conseil Européeen pour la Recherche Nucléarire (CERN) as the detectors for the dark matter instrument CERN Axion Solar Telescope (CAST; see Section 1.7.2.1.7) and also will be used in the next-generation detector, International Axion Observatory (IAXO).

A DME-filled GPD with beryllium windows was used on board the 2018 technology demonstrator CubeSat, PolarLight, to observe the polarisation of 2- to 6-keV X-rays, and they are expected to be used on board the Chinese Enhanced X-ray Timing and Polarimetry (eXTP) mission, which has a possible launch date of 2025 and on NASA's Imaging X-ray Polarimetry Explorer (IXPE)with a 2021 possible launch date.

1.3.2.3 Scintillation Detectors

The ionising photons do not necessarily knock out only the outermost electrons from the atom or molecule with which they interact. Electrons in lower-energy levels may also be removed. When this happens, a 'hole' is left behind into which one of the higher electrons may drop, with a consequent emission of radiation. Should the medium be transparent to this radiation, the photons may be observed and the medium may be used as an X-ray detector. Each interaction produces a scintilla of light (i.e., a flash), from which the name of the device is obtained. There are many materials that are suitable for this application. Commonly used ones include sodium iodide doped with an impurity such as thallium and caesium iodide doped with sodium or thallium. For these materials the light flashes are detected by a photomultiplier (Figure 1.83). There is no dead time for a scintillation detector and the strength of the flash depends somewhat upon the original photon energy, so that some spectral resolution is possible. The noise level, however, is quite high because only about 3% of the X-ray's energy is converted into detectable radiation, with a consequent increase in the statistical fluctuations in their numbers. The spectral resolution is thus about 6% at 1 MeV.

Sodium iodide or caesium iodide are useful scintillators for X-ray energies up to several hundred keV and the recently developed, europium-doped strontium iodide can reach at least 6 MeV. High-purity germanium scintillators have energy resolutions better that 1% at 100 keV but need cryogenic cooling. Organic scintillators such as stilbene ($C_{14}H_{14}N_2$) can be used up to 10 MeV and bismuth germanate ($Bi_4Ge_3O_{12}$ or BGO) for energies up to 30 or more MeV. Organically doped plastics are also used and are to be found in the surface stations of the Telescope Array cosmic ray observatory in Utah (Section 1.4).

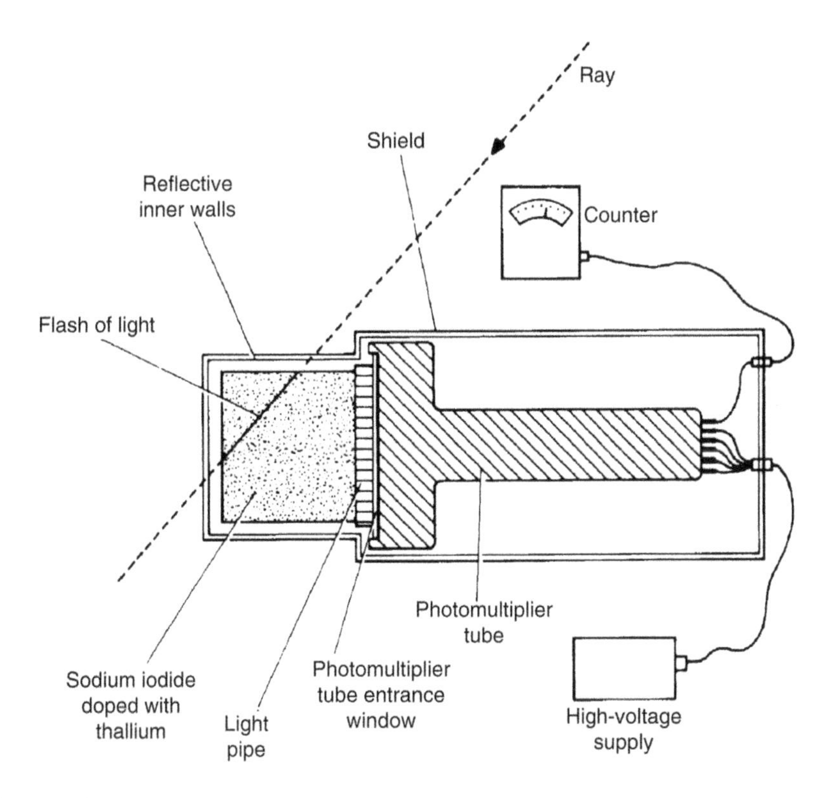

FIGURE 1.83 Schematic experimental arrangement of a scintillation counter.

Scintillator-based detectors have largely been superseded by other devices in recent years for spacecraft-borne detectors but are operating on the still functioning Fermi and International Gamma-Ray Astrophysics Laboratory (INTEGRAL) spacecraft and were in use, until 2015, on Suzaku. The Gamma-ray Burst Monitor (GBM) on board the Fermi gamma-ray space telescope employs ten NaI and two BGO scintillators to detect γ-ray bursts over the range from 150 keV to 30 MeV. While INTEGRAL uses BGO and plastic scintillators as active shields (see below) for two of its instruments. Gadolinium silicate and bismuth germinate scintillators were used on the Suzaku spacecraft for observing over the 30- to 600-keV region.

Discrimination of the X-ray's arrival direction can be obtained by using sodium iodide and caesium iodide in two superimposed layers. The decay time of the pulses differs between the two compounds so that they may be separately identified and the direction of travel of the photon inferred. This arrangement for a scintillation detector is frequently called a 'phoswich' detector. Several gases such as argon, xenon, nitrogen and their mixtures can also be used as scintillators and combined with an optical system to produce another imaging device. The X-rays and cosmic rays (Section 1.4) may be distinguished by the differences in their resulting pulses shapes. The X-rays are rapidly absorbed and their pulses are sharp and brief. Cosmic ray particles will generally have a much longer path and so their pulses will be comparatively broader and smoother than the X-ray pulse.

Scintillation detectors are still used at the higher energy end of this region. Thus, the 2018 balloon-borne Polarized Gamma ray Observer (PoGO+) instrument carried a 61-pixel array of plastic scintillators surrounded by a BGO anti-coincidence shield and operated over the 20- to 180-keV hard X-ray to soft γ-ray region. Sweden's Innovation Satellite (InnoSat) spacecraft platform is proposed for a possible future hard X-ray polarimeter, Satellite Polarimeter for High Energy X-rays (SPHiNX), which would also use plastic scintillator detectors and have a Pb/Sn/Cu shield, observing up to 600 keV.

A recent development in medical physics, which has yet to find astronomical applications, is of an X-ray scintillator which can be sprayed into position using an aerosol can. The scintillator is in the form of small ceramic particles embedded within a conductive plastic. A photodiode is used to detect the scintillations. It seems likely to be much cheaper than present devices and to enable large detectors with complex shapes to be fabricated easily.

1.3.2.4 Pair Production Detectors

An electron's mass, in energy terms, is about 500 keV. The γ rays with energies above about 1 MeV thus have sufficient energy to convert themselves into an electron and its anti-particle, a positron. This process is called 'pair production' (Section 1.3). The γ rays may thus be detected indirectly by detecting the electron and positron resulting from their pair production. Generally, the electron/positron pair will be moving fast enough, after their creation, to be detected by charged particle detectors (Section 1.4).

The detection of γ rays via their pair production is mentioned in respect of Fermi's LAT instrument in Section 1.3.2.7. NASA's Advanced Energy Pair Telescope (AdEPT) proposal, with a launch date of perhaps 2023, will be based around a 4-m^3 TPC filled with an argon and carbon disulphide mix at 150 kPa. It will observe the 2- to 500-MeV region with an angular resolution of about 30 arc-minutes.

1.3.2.5 Gas Scintillation Proportional Counters

A combination of the previous two types of detector leads to a significant improvement in the low-energy spectral resolution. Resolutions that are as good as 8% at 6 keV have been achieved in practice with these devices. The X-radiation produces ion-electron pairs in an argon- or xenon-filled chamber. The electrons are then gently accelerated until they cause scintillations of their own in the gas. These scintillations can then be observed by a conventional scintillation counter system. These have been favoured detectors for launch on several of the more recent X-ray satellites because of their good spectral and positional discrimination, though in some cases their operational lifetimes in orbit have been rather short.

1.3.2.6 Compton Interaction Detectors

Very-high-energy photons produce electrons in a scintillator through the Compton effect and these electrons can have sufficiently high energies to produce scintillations of their own. Two such detectors separated by a metre or so can provide directional discrimination when used in conjunction with pulse analysers and time-of-flight measurements to eliminate other unwanted interactions. The Compton Telescope (COMPTEL) instrument on board the Compton Gamma-Ray Observatory (CGRO, 1991–2000), used an organic liquid scintillator and a NaI scintillator separated by 1.5 m and both viewed by photomultipliers. The γ-rays in the 800-keV to 30-MeV range were detectable over a one steradian area of the sky with the photons being scattered in the liquid scintillator before being completely absorbed in the NaI. A plastic scintillator was also used as an active shield (see below). The angular resolution of COMPTEL ranged from 1.7° to 4.4° and its energy resolution from 5% to 8%. A balloon-borne Compton γ ray detector based upon a stilbene scintillator, together with a cerium-doped lanthanum bromide calorimeter, was flown successfully, albeit briefly, in 2014. The Tiangong-2 space station (2016–2019) carried the Polar[111] γ ray polarimeter whose detection mechanism was Compton scattering in a plastic scintillator to study GRBs over the 10- to 500-keV region.

Recently, Compton effect telescopes have been developed using stacks of Silicon Strip Detector (SSD) and CdTe strip detectors (see below). The CdTe detector stack lies underneath the SSD stack and CdTe detectors also surround the sides of the stack. The SSDs detect the Compton electrons produced by scattering within them and the CdTe detectors absorb the γ-rays.

[111] A name.

Japan Aerospace Exploration Agency's (JAXA's) Hitomi spacecraft's soft γ-ray detector used six such Compton effect telescopes for observing the 10- to 600-keV region over an area of the sky ranging from 0.5 deg^2 to 10 deg^2. Its energy resolution was 4,000 and an active well shield of BGO surrounded each stack of detectors. Unfortunately, the spacecraft broke-up two days after a successful launch in 2016 and before the instruments could be used.

1.3.2.7 Solid-State Detectors

Solid-state detectors have several advantages that suit them particularly for use in satellite-borne instrumentation: wide range of photon energies detected (from 1 keV to over 1 MeV), simplicity, reliability, low power consumption, high stopping power for the radiation, room temperature operation (some varieties), no entrance window needed, high counting rates possible, etc. They also have an intrinsic spectral sensitivity because, provided that the photon is absorbed completely, the number of electron-hole pairs produced is proportional to the photon's energy. About 5 eV are need to produce an electron-hole pair in CZT[112] (see below), so that the spectral sensitivity to 2 1/2 standard deviations is potentially 18% at 1 keV and 0.5% at 1 MeV. The main disadvantages of these detectors are that their size is small compared with many other detectors, so that their collecting area is also small and that unless the photon is stopped within the detector's volume, the total energy cannot be determined. This latter disadvantage, however, also applies to most other detectors.

There are a number of different varieties of these detectors (surveyed below) so that suitable ones may be found for use throughout most of the X- and gamma-ray regions.

The details of the operating principles for CCDs for optical detection are to be found in Section 1.1. They are also, however, becoming increasingly widely used as primary detectors at EUV and X-ray wavelengths. The CCDs become insensitive to radiation in the blue and UV parts of the spectrum because of absorption in the electrode structure on their surfaces. They regain sensitivity at shorter wavelengths as the radiation is again able to penetrate that structure ($\lambda < 10$ nm or so, energy > 120 eV). With optical CCDs, the efficiency of the devices may be improved by using a thin electrode structure, an electrode structure that only partially covers the surface (virtual phase CCDs) or by illuminating the device from the back. Typically, 3.65 eV of energy is needed to produce a single electron-hole pair in silicon. So, unlike optical photons which can only produce a single such pair, the X-ray photons can each produce many pairs. The CCDs are thus used in a photon-counting mode for detecting X-rays and the resulting pulses are proportional to the X-ray's energy, giving CCDs an intrinsic spectral resolution (Section 4.1) of around 10 to 50. The Spektr-RG spacecraft carries the Extended Roentgen Survey with an Imaging Telescope Array (eROSITA) instrument, an imager for X-rays from 300 eV to 10 keV which uses seven 384 × 384-pixel deep depletion frame-store CCDs.

Swept charge devices (SCDs) are a non-imaging variant of the CCD with a high sensitivity and good spectral resolution. The geometry of the readout system on an SCD is such that the individual charges across the device are added together before they are registered. The two Chandrayaan spacecraft both employed SCD arrays within their X-ray fluorescence spectrometers. The Chandrayaan-2 Large Area Soft x-ray Spectrometer (CLASS) instrument on Chandrayaan-2 uses 16 SCDs each 20 mm^2 which have a spectral resolution of 30 at 6 keV.

Detectors also closely related to CCDs and known as SSDs, consist of a thick layer (up to 5 mm) of a semiconductor, such as silicon. More recently, CZT detectors and cadmium plus tellurium (CdTe strip detectors) have become popular because these latter materials have higher stopping powers for the radiation. The semiconductor in each case has a high resistivity by virtue of being back-biased to a hundred volts or so. The silicon surface is divided into 20- to 25-μm wide strips by thin layers doped with boron. Aluminium electrodes are then deposited on the surface between these layers. When an X-ray, gamma-ray or charged particle passes into the material, electron-hole pairs are produced via the photoelectric effect, Compton scattering or collisional ionisation. The pulse of current is detected as the electron-hole pairs are collected at the electrodes on the top and bottom of the semiconductor slice.

[112] Cadmium (about 90%) and Zinc (about 10%) doped with Tellurium.

Processing of the pulses can then proceed by any of the usual methods and a positional accuracy for the interaction of ±10 μm can be achieved. The Focusing Optics Solar X-ray Imager (FOXSI) instrument has been used in recent sounding rocket studies of the Sun and uses CdTe strip detectors to image the Sun over a field of view of about 13 arc-minutes at between 4 and 15 keV. The CdTe strip detectors were also used in the ill-fated Hitomi spacecraft's hard X-ray imager.

Silicon Drift Detectors (SDDs), developed for the Large Hadron Collider, are similar to SSDs and also to deep depletion CCDs. They comprise a disk of high purity *n*-doped silicon up to 10-mm across and 0.5-mm thick. The X-rays penetrating into the disk produce electron-hole pairs. The holes are attracted to negatively charged *p*-type regions (cathodes) on the flat surfaces of the disk and eliminated. The electrons are attracted to a single central anode and encouraged to move in that direction by a decreasing negative charge on the cathodes from the edge to the centre of the disk. The voltages involved are low (100 V or less), so there is no intrinsic amplification of the cloud of electrons through collisions, but there are sufficient numbers in the original cloud for the pulse to be detected when the electrons reach the anode.

The SSDs are used for the Fermi spacecraft's Large Area Telescope (LAT). These provide positional and directional information on the gamma rays. The silicon strips are arranged orthogonally in two layers. A gamma ray, or the electron-positron-pair produced within a layer of lead by the gamma ray, will be detected within one of the strips in each layer. The position of the ray is then localised to the crossover point of the two strips. By using a number of such pairs of layers of silicon strips piled on top of each other, the direction of the gamma ray can also be determined. The LAT uses 16 such 'towers', each comprising 16 layers of silicon strip pairs and lead plates, enabling it to determine gamma-ray source positions to within 0.5–5 arc-minutes. The possible ESA Large Observatory for x-ray Timing (LOFT) mission for the mid-2020s is envisaged as using up to 2,000 SSDs covering some 10 m^2 for its Large Area Detector (LAD) instrument and observing a 1° field of view over the 2- to 30-keV X-ray region.

Germanium may be used as a sort of solid proportional counter. A cylinder of germanium cooled by liquid nitrogen is surrounded by a cylindrical cathode and has a central anode (Figure 1.84).

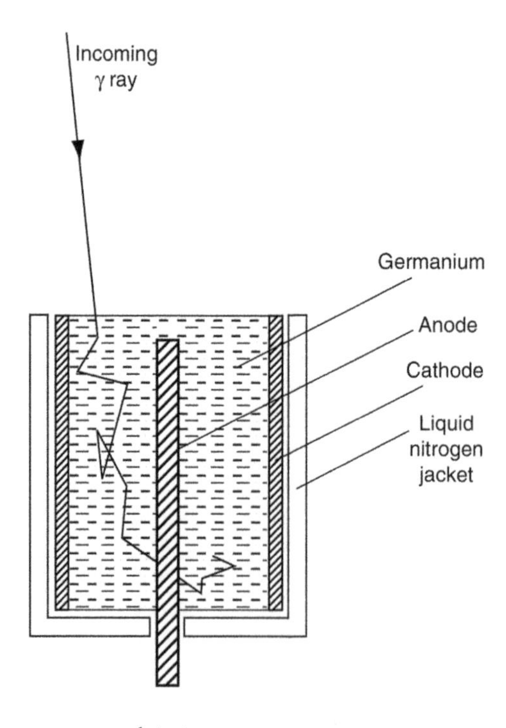

FIGURE 1.84 Germanium gamma-ray detector.

A gamma ray scatters off electrons in the atoms until its energy has been consumed in electron-hole pair production. The number of released electrons is proportional to the energy of the gamma ray, and these are attracted to the anode where they may be detected. The spectral resolution is very high – 0.2% at 1 MeV, so that detectors of this type are especially suitable for gamma-ray line spectroscopy. Other materials that may replace the germanium include germanium doped with lithium, cadmium telluride and mercury-iodine. At lower energies (0.4–4 keV), silicon-based solid-state detectors may be used similarly. Their energy resolution ranges from 4% to 30%. INTEGRAL's Spectrometer for INTEGRAL (SPI) uses 19 germanium crystals with plastic and BGO scintillators as active shields and a tungsten-coded mask, for spectroscopy with a spectral resolution of 500 at 1 MeV.

A rather different type of solid-state detector from the previous ones is the microcalorimeter in which the detection occurs when the absorption of a photon induces a temperature rise in the absorbing material (i.e., a thermal-type detector; see Table 1.1). Microcalorimeters have good intrinsic spectral resolution (*R* between 200 and 1,000; see Section 4.1), though so far, their principal use has been for detection only of low to medium-energy X-rays (up to 1 keV).

There are three components to a microcalorimeter: an absorber, a temperature sensor, and a thermal sink. The energy of the X-ray photon is converted into heat within the absorber, and this change is picked up by the sensor. There is then a weak thermal link to the heat sink so that the temperature of the absorber gradually returns to its original value. Typical response times are a few milliseconds with up to 80 ms recovery time, so microcalorimeters are not suitable for detecting high fluxes of X-rays. The microcalorimeter needs to be cooled so that its stored thermal energy is small compared with that released from the X-ray photon. A wide variety of materials can be used for all three components of the device with the main requirement for the absorber being a low thermal capacity and a high stopping rate for the X-rays.

Microcalorimeters can be formed into arrays – Suzaku, for example, carried a 32-pixel microcalorimeter comprising mercury telluride absorbers combined with silicon thermistors and cooled to 60 mK. They are also planned for use on the Advanced Telescope for High Energy Astrophysics (ATHENA) mission, expected to be launched in the early 2030s into the Sun-Earth Lagrange L2 position. An array of some 4,000 microcalorimeters is currently expected to be used for the X-ray Integral Field Unit (X-IFU) imaging spectroscope with gold/bismuth absorbers and molybdenum/gold TES temperature detectors. It will use micropore optics (see below) and cover the 200-eV to 12-keV spectral region.

1.3.2.8 Microchannel Plates

For EUV and low-energy X-ray amplification and imaging, there is an ingenious variant of the PMT (Section 1.1): the microchannel plate (MCP). The devices are also known as MAMAs and additionally are used as optical components for imaging X-rays when they are often called micro-pore optics (see lobster-eye collimator, below).

When acting as a detector, the MCP comprises a thin plate pierced by numerous tiny holes, each perhaps only about 10 mm across or less. Its top surface is an electrode with a negative potential of some few thousand volts with respect to its base. The top is also coated with a photoelectron emitter for the X-ray energies of interest. An impinging photon releases one or more electrons that are then accelerated down the tubes. There are inevitably collisions with the walls of the tube during which further electrons are released and these in turn are accelerated down the tube and so on (Figure 1.85). As many as 10^4 electrons can be produced for a single photon and this may be increased to 10^6 electrons in future devices when the problems caused by ion feedback are reduced. The quantum efficiency can be up to 20%. The electrons spray out of the bottom of each tube, where they may be detected by a variety of the more conventional imaging systems, or they may be fed into a second MCP for further amplification.

Early versions of these devices often used curved channels (Figure 1.85). Modern versions now usually employ two plates with straight channels, the second set of channels being at an

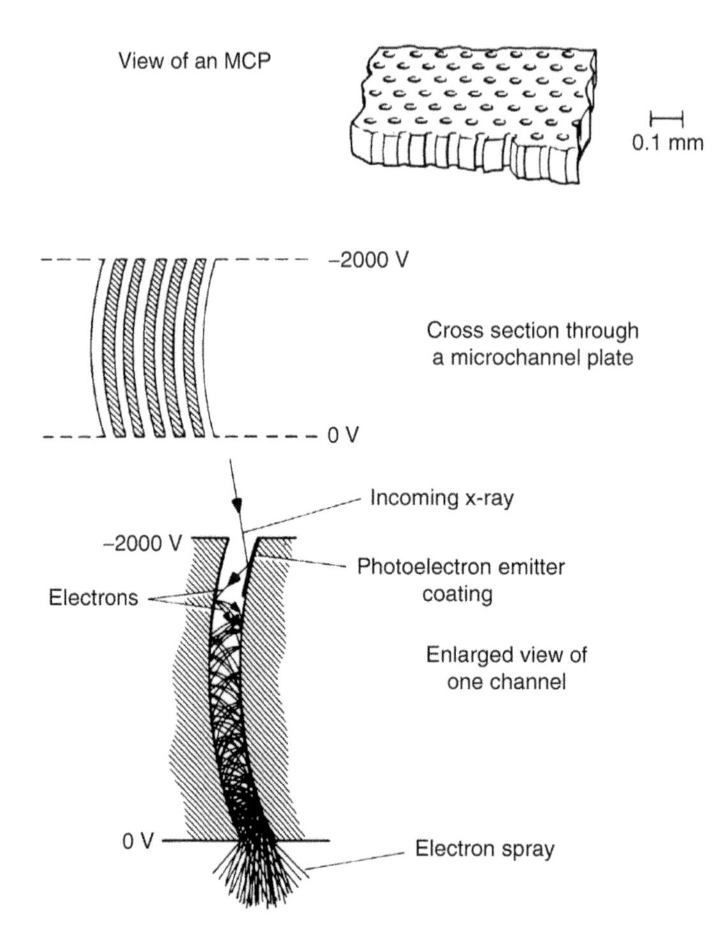

FIGURE 1.85 Schematic view of the operation of a microchannel plate (MCP).

angle to the direction of the first set (known as a 'chevron microchannel plate'). A third plate may even be added to produce a Z-stack microchannel plate. The plates are currently manufactured from a billet of glass that has an acid-resisting shell surrounding an acid-soluble core. The billet is heated and drawn out until it forms a glass 'wire' about 1-mm across. The wire is cut into short sections, and the sections stacked to form a new billet. That billet in turn is heated and drawn out and so on. The holes in the final plate are produced by etching away the acid-soluble portions of the glass. Holes down to 6 μm in diameter may be produced in this fashion and the fraction of the plate occupied by the holes can reach 65%. Plates can be up to 0.1-m^2. A single x-ray results in about 30 million electrons that are collected by a grid of wires at the exit holes from the second-stage plate.

As examples, the high-resolution X-ray camera on board the Chandra spacecraft (still operating after surviving in space four times longer than planned!) uses a 93-mm^2 chevron MCP detector, with 69 million 10-μm holes and can provide a resolution of 0.5″. The planetary space probe, BepiColombo, currently on its way to Mercury, carries MCPs as detectors for its UV spectrometer. The photoemitters are CsI for the extreme UV region (55–155 nm) and CsTe for the UV region (145–315 nm) and they are 25 × 40 mm in size.

The MCPs can also be used in optical and near UV; however, they are then used simply to amplify the photoelectrons produced from a photocathode deposited onto the entrance window of the detector and are thus similar to image intensifiers (Section 2.3).

1.3.2.9 Čerenkov Detectors

Čerenkov radiation and the resulting detectors are referred to in detail in Section 1.4. For X-ray and gamma radiation their interest lies in the detection of particles produced by the Compton interactions of the very-high-energy photons. If those particles have a velocity higher than the velocity of light in the local medium then Čerenkov radiation is produced. Čerenkov radiation produced in this manner high in the Earth's atmosphere is detectable. Thus, the Very Energetic Radiation Imaging Telescope Array System (VERITAS) instrument, located in Arizona, employs four 12-m reflectors, each made up from 350 mirror segments, while Major Atmospheric Gamma Imaging Čerenkov (MAGIC) telescope on La Palma uses two 17-m multi-segment reflectors and HESS in Namibia has four 12-m dishes plus a single 28-m dish. Major Atmospheric Čerenkov Experiment (MACE) is currently being installed at an altitude of 4,300 m at Hanle in India. It comprises a 21-m dish made up from 356 actively controlled segments and a 1,088 PMT camera. It will detect cosmic rays up to energies of 5 TeV and more.

The Čerenkov Telescope Array (CTA) is a project for a much larger installation that is currently just starting construction at its northern site on La Palma (Canary Islands) and which, in recent years, has had prototype telescopes for all its designs built and tested. It will use three sizes of dishes, all with segmented mirrors. The small (4-m) telescopes will be the most numerous, with 70 currently being expected to be needed. They will have fields of view of ~9° and be used at the CTA's southern site, close to ESO's Paranal observatory complex in Chile. The medium and large dishes (12-m and 23-m diameter, respectively) will be used at both CTA sites. There will be eight of the large telescopes; four at each site and 40 of the medium-sized instruments – 25 in the South and 15 in the North. The cameras will all be based upon PMTs, conventional ones for the larger instruments and SiPMTs for the small dishes and with all the cameras having around 2,000 pixels. The CTA's first light is expected in about 2022 and it is expected to observe cosmic ray γ rays with energies ranging upwards from 10 GeV.

Very-high-energy primary γ rays (10^{15} eV and more) may also produce cosmic ray showers and so be detectable by the methods discussed in Section 1.4, especially through the fluorescence of atmospheric nitrogen.

1.3.2.10 Future Possibilities

The CCD variant CMOS-APS (Section 1.1.16), which reads-out the pixels on an individual basis, is likely to replace the CCD in many future X-ray space missions because it can be hardened to radiation and can provide high count rates. However, it has the disadvantages of only being available in small arrays and there are low production yields. Monolithic CMOS detectors, sometimes known as Silicon-On-Insulator (SOI) devices, are therefore also a future possibility. These have a thick silicon layer as the detecting element insulated by an oxide layer and followed by a thin silicon layer comprising the CMOS read-outs. The layers are firmly bonded together and the detection layer is linked to the read-out layer by connections passing through the oxide layer. The read-out can be event-driven – that is, pixels are only read-out immediately after they have made a detection. The JAXA's Focusing on Relativistic Universe and Cosmic Evolution (FORCE) mission, a proposal for launch in the mid-2020s, is currently expected to use event-driven monolithic CMOS detectors for observing the 1- to 80-keV region at high angular resolution.

The details of the operating principles of STJs and MKIDs may be found in Section 1.1. They are however also sensitive to X-rays and, indeed, the first applications of STJs were for X-ray detection. Since an X-ray breaks about a thousand times as many Cooper pairs as an optical photon, their spectral discrimination is about 0.1%. The need to cool them to below 1 K and the small sizes of the arrays currently available means that they have yet to be flown on a spacecraft.

Another type of detector altogether is a solid analogue of the cloud or bubble chamber (Section 1.4). A superconducting solid would be held at a temperature fractionally higher than its critical temperature or in a non-superconducting state and at a temperature slightly below the critical value.

Passage of an ionising ray or particle would then precipitate the changeover to the opposing state, and this in turn could be detected by the change in a magnetic field (see TES detectors, Section 1.1).

The Russian spacecraft, the Gamma Astronomical Multifunctional Modular Apparatus (GAMMA)-400, which is currently under construction and (perhaps) due for launch in 2025, will be a γ-ray instrument that combines almost all the individual detectors that we have just discussed. It will be observed in the 100-MeV to 1-TeV range with angular resolutions ranging from 2° to 0.1°, with an energy resolution of about 1% and with a 2.5 sr field of view. To accomplish this, it will use SSDs interlaced with tungsten strips, time of flight (ToF) scintillators, an imaging microcalorimeter comprising SSDs and tungsten layers, two more scintillators, a BGO microcalorimeter and more scintillators for active shielding.

1.3.3 Shielding

Few of the detectors which we have just reviewed are used in isolation. Usually, several will be used together in modes that allow the rejection of information on unwanted interactions. This is known as active or anti-coincidence shielding of the detector. The range of possible configurations is wide and a few examples will be suffice to give an indication of those in current use.

1. The germanium solid-state detectors must intercept all the energy of an incoming photon if they are to provide a reliable estimate of its magnitude. If the photon escapes, then the measured energy will be too low. A thick layer of a scintillation crystal, therefore, surrounds the germanium. A detection in the germanium simultaneously with two in the scintillation crystal is rejected as an escapee.
2. The solid angle viewed by a germanium solid-state detector can be limited by placing the germanium at the bottom of a hole in a scintillation crystal (a 'Well' active shield). Only those detections it not occurring simultaneously in the germanium and the crystal are used and these are the photons that have entered down the hole. Any required degree of angular resolution can be obtained by sufficiently reducing the size of the entrance aperture.
3. Primary detectors are surrounded by scintillation counters and Čerenkov detectors to discriminate between gamma-ray events and those resulting from cosmic rays.
4. A sodium iodide scintillation counter may be surrounded, except for an entrance aperture, by one formed from caesium iodide to eliminate photons from the unwanted directions.
5. The Fermi spacecraft uses segmented plastic scintillator tiles for its anti-coincidence shielding. These are read out by PMTs.
6. Another detector or some other part of the instrument or spacecraft may be used as a passive (below) or active shield for the detector. Thus, the ground-based cosmic ray detector (Section 1.4) concept, Muon Array with RPCs for Tagging Air Showers (MARTA), envisages placing the muon detector below the calorimeter and using the latter as the active shield.

Recently, nuclear power sources on board other spacecraft have caused interference with γ–ray observations. These pulses have to be separated from the naturally occurring ones by their spectral signatures.

Passive shields are also used, and these are just layers of an absorbing material that screen out unwanted rays. They are especially necessary for the higher-energy photon detectors to reduce the much greater flux of lower-energy photons. Generally, the higher the atomic number (Z) of the material making up the shield, the better is its protection. The mass involved in an adequate passive shield, however, is often a major problem for spacecraft-based experiments with their tight mass budgets arising from their launcher capabilities. Some passive shields use combinations of several different materials to provide protection against differing types or differing energies of the radiation or particles. This is called 'Graded-Z shielding' and will be used, for example, on the planned

ATHENA spacecraft's wide field imager. This is expected to involve a 40-mm thick aluminium proton shield followed by high-Z material (e.g., tungsten or molybdenum) to exclude hard X-rays and then low-Z material (e.g., aluminium or beryllium) to shield against low-energy electrons.

1.3.4 Imaging

Imaging of high-energy photons is a difficult task because of their extremely penetrating nature. Normal designs of telescope are impossible because in a reflector the photon would just pass straight through the mirror, while in a refractor it would be scattered or unaffected rather than refracted by the lenses. At energies under a few keV, forms of reflecting telescope can be made which work adequately, but at higher energies the only options are occultation, collimation and coincidence detection.

1.3.4.1 Collimation

A collimator is simply any device that physically restricts the field of view of the detector without contributing any further to the formation of an image. The image is obtained by scanning the system across the object. Almost all detectors are collimated to some degree, if only by their shielding. The simplest arrangement is a series of baffles (Figure 1.86) that may be formed into a variety of configurations, but all of which follow the same basic principle. Their general appearance leads to their name, honeycomb collimator, even though the cells are usually square rather than hexagonal. They can restrict angles of view down to a few arc-minutes, but they are limited at low energies by reflection off their walls and at high energies by penetration of the radiation through the walls. At high energies, the baffles may be formed from a crystal scintillator and pulses from there used to reject detections of radiation from high inclinations (cf. active shielding).

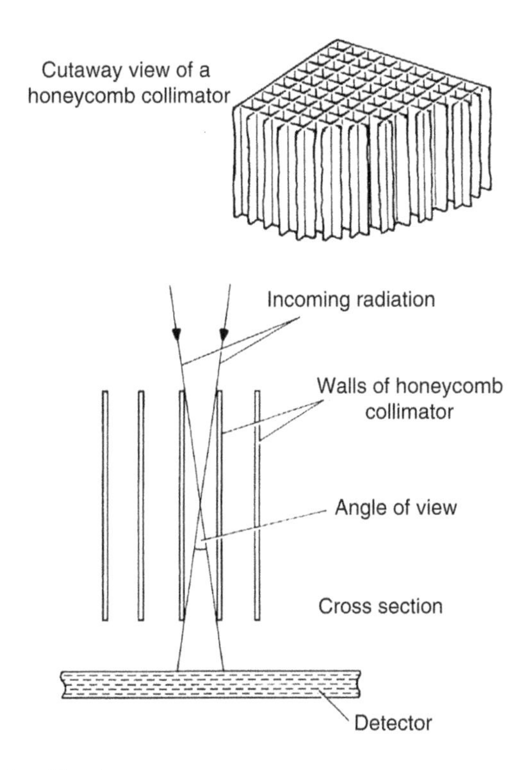

FIGURE 1.86 Honeycomb collimator.

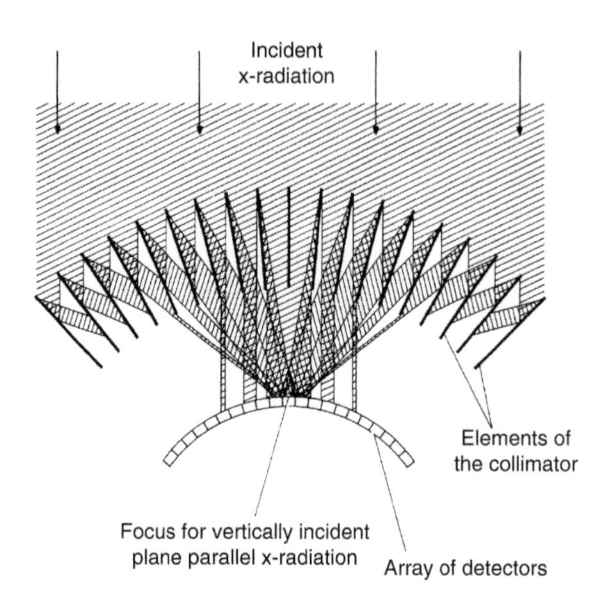

FIGURE 1.87 Cross section through a 'lobster eye' focusing wide-angle x-ray collimator.

At the low energies the glancing reflection of the radiation can be used to advantage and a more truly imaging collimator produced. This is called a 'lobster eye'[113]-focusing collimator (also called micropore optics) and is essentially a honeycomb collimator curved into a portion of a sphere (Figure 1.87), with a position-sensitive detector at its focal surface. The imaging is not of high quality, and there is a high background from the unreflected rays and the doubly reflected rays (the latter not shown in Figure 1.87). But it is a cheap system to construct compared with others that are discussed later in this section, and it has the potential for covering wide fields of view (tens of degrees) at high resolutions (arcseconds).

Early in the development of X-ray astronomy, honeycomb and lobster-eye collimators were physically large (holes of several millimetres in size) and fabricated from aluminium (or another material) plates. Today, the devices are constructed by fusing together thousands of tiny glass rods in a matrix of a different glass. The composite is then heated and shaped by slumping onto a spherical mould and the glass rods etched away to leave the required hollow tubes in the matrix glass. This is the same process as that used to produce MCPs (see above) and the only difference is that for a lobster-eye collimator, the channels are of square cross section, whereas they are circular in MCPs. Multi-layer coatings (see below) may be applied to enhance reflectivity and increase the waveband covered.

BepiColombo's Mercury Imaging X-ray Spectrometer (MIXS) instrument will look for X-ray fluorescence from Mercury's surface in the 500-eV to 7.5-keV region using two lobster-eye collimators with differing spatial resolutions when the spacecraft arrives at the planet in 2025. The joint European-Chinese spacecraft, Solar Wind Magnetosphere Ionosphere Link Explore (SMILE), scheduled for a 2023 launch, will carry the Soft X-ray Imager (SXI) instrument. This is a wide-field (15.5° × 26.5°) micropore lobster-eye imager using CCD detectors for the 200-eV to 2.5-keV energy range.

Another system that is known as a 'modulation collimator', 'rotation collimator' or 'Fourier transform telescope' uses two or more parallel gratings that are separated by a short distance (Figure 1.88). Since the bars of the gratings alternately obscure the radiation and allow it to pass

[113] The device mimics the way that lobsters' eyes work.

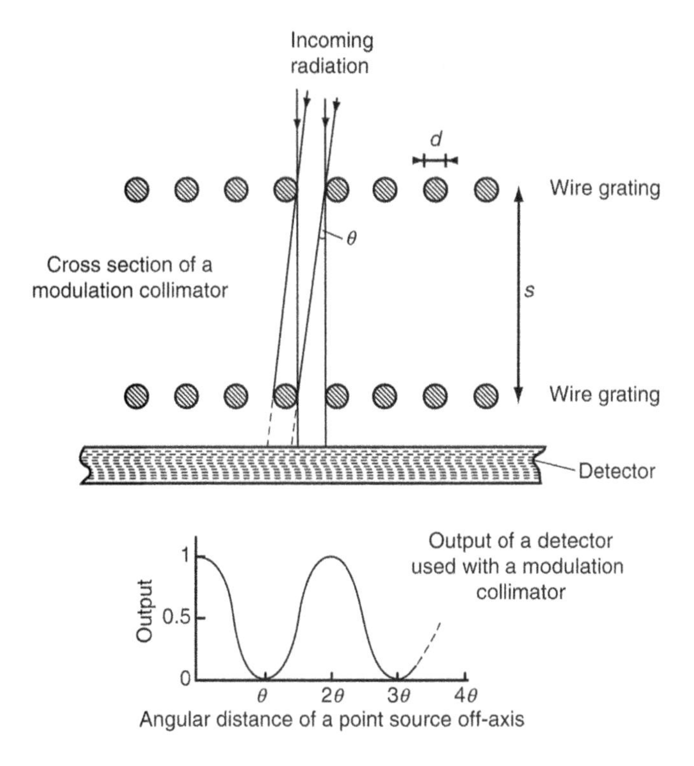

FIGURE 1.88 A modulation collimator.

through, the output as the system scans a point source is a sine wave (Figure 1.88). The resolution is given by the angle α

$$\alpha = \frac{d}{s} \tag{1.77}$$

To obtain unambiguous positions for the sources, or for the study of multiple or extended sources, several such gratings of different resolutions are combined. The image may then be retrieved from the Fourier components of the output (cf. aperture synthesis, Section 2.5). Two such grating systems at right angles can give a 2D image. With these systems, resolutions of a few tens of arcseconds can be realised even for the high-energy photons.

The Reuven Ramaty High Energy Solar Spectroscopic Imager (RHESSI) spacecraft (decommissioned in 2018) used modulation collimators to study the X-ray emission from solar flares. There were five pairs of coarse grids and four pairs of fine grids and the detectors were germanium crystals. Spectral resolutions of 1–5 keV were obtained over the range 100 keV to 17 MeV with angular resolutions ranging from 2″ to 36″.

A third type of collimating imaging system is a simple pinhole camera. A position-sensitive detector such as a resistive anode proportional counter or a MCP is placed behind a small aperture. The quality of the image is then just a function of the size of the hole and its distance in front of the detector. Unfortunately, the small size of the hole also gives a low flux of radiation so that such systems are limited to the brightest objects. A better system replaces the pinhole with a mask formed from clear and opaque regions. The pattern of the mask is known, so that when sources cast shadows of it onto the detector, their position and structure can be reconstituted in a similar manner to that used for the modulation collimator. The technique is known as 'coded mask imaging', and resolutions of 10 arc-minutes or better can be achieved. Because only about half the aperture is obscured, the technique obviously uses the incoming radiation much more efficiently than the pinhole camera.

A somewhat related technique, known as 'Hadamard mask imaging', is discussed in Section 2.4. The image from a coded aperture telescope is extracted from the data by a cross correlation between the detected pattern and the mask pattern. The disadvantage of coded mask imaging is that the detector must be nearly the same size as the mask.

Examples of coded mask X-ray and γ ray instruments include INTEGRAL's SPI (Section 1.3.2.7) which has a 127 hexagonal element coded mask. 64 of the mask's elements are clear and 63 are opaque and made from 30-mm thick tungsten. SPI operates from 20 keV to 8 MeV. While INTEGRAL's twin x-ray instruments, Joint European X-ray Monitor (JEM-X) 1 and 2, which cover the region from 3 to 35 keV, use masks with 22,500 hexagonal elements. The opaque elements are 0.5-mm thick tungsten plates. The still operational Neil Gehrels Swift Observatory spacecraft uses a coded mask with an area of 3.2 m^2, allied with nearly 33,000 CZT detectors (see solid-state detectors, above) covering an area of 0.52 m^2 in its Burst Alert Telescope (BAT). The BAT detects gamma-ray bursts and it has a field of view of two steradians; thus, it is able to pinpoint the position of a gamma-ray burst to within 4′ within 15 seconds of its occurrence. While the possible future Brazilian spacecraft, Lattes-1, may carry a coded mask X-ray instrument, Monitor e Image Ador de Raios X (MIRAX). MIRAX is expected to have four wide-field–coded mask telescopes with 64-pixel CZT detector arrays and to detect transient sources over the 5–200 keV spectral region.

A combination of an X-ray imager with a coded mask might be able to improve the imager's basic angular resolution by two or three times. The suggestion is to insert the coded mask into the focussed X-ray beam but slightly away from the best focus position. The slightly fuzzy image is then processed by the coded mask cross correlation as usual. Such a system has yet to be tried in practice.

1.3.4.2 Coincidence Detectors

A telescope, in the sense of a device that has directional sensitivity, may be constructed for use at any energy and with any resolution, by using two or more detectors in a line and by rejecting all detections except those which occur in both detectors and separated by the correct flight time. Two separated arrays of detectors can similarly provide a 2D imaging system.

1.3.4.3 Occultation

Even though when it is not a technique that can be used at will on any source, the occultation of a source by the Moon or another object can be used to give precise positional and structural information. The use of occultations in this manner is discussed in detail in Section 2.7. The technique is less important now than in the past because other imaging systems are available, but it was used in 1964, for example, to provide the first indication that the X-ray object associated with the Crab nebula was an extended source.

1.3.4.4 Reflecting Telescopes

At energies below about 100 keV, photons may be reflected with up to 50% efficiency off metal surfaces, when their angle of incidence approaches 90°. Mirrors in which the radiation just grazes the surface (grazing or glancing incidence optics) may therefore be built and configured to form reflecting telescopes. The angle between the mirror surface and the radiation has to be small though: less than 2° for 1-keV X-rays, less than 0.6° for 10-keV X-rays and less than 0.1° for 100-keV X-rays. Several systems have been devised, but the one which has achieved most practical use is formed from a combination of annular sections of deep paraboloidal and hyperboloidal surfaces (Figure 1.89) and known as a 'Wolter type I telescope' after its inventor. Other designs that are also due to Wolter are shown in Figure 1.90 and have also been used in practice, although less often than the first design. Small adjustments to the parabolic and hyperbolic mirror shapes results in the Wolter-Schwarzschild design which has reduced optical aberrations compared with the basic Wolter designs. A simple paraboloid can also be used, though the resulting telescope then tends to be excessively long.

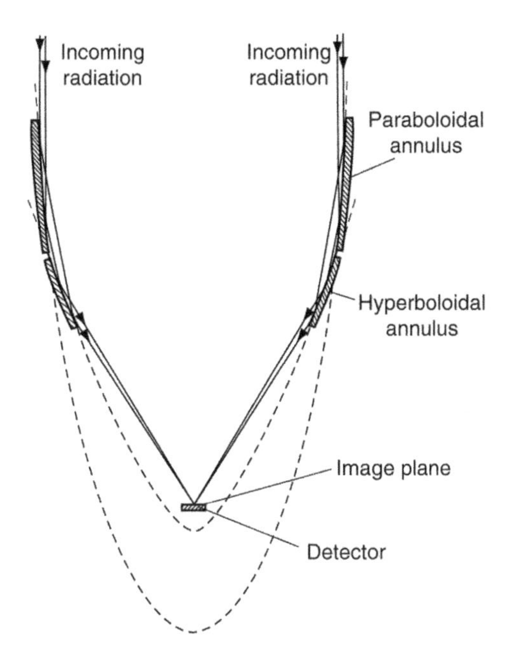

FIGURE 1.89 Cross section through a grazing incidence x-ray telescope.

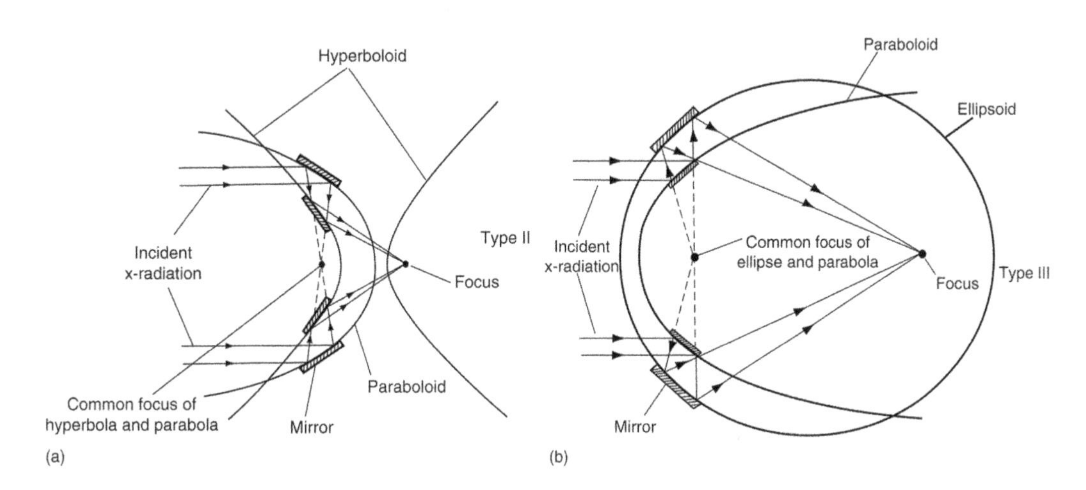

FIGURE 1.90 Cross sections through alternative designs for grazing incidence X-ray telescopes: (a) Wolter Type II design and (b) Wolter Type III design.

The aperture of such telescopes is a thin annulus because only the radiation incident onto the paraboloidal section is brought to the focus. To increase the effective aperture and hence, the sensitivity of the system, several such confocal systems of differing diameters may be nested inside each other (Figure 1.91).

The limit of resolution is due to surface irregularities in the mirrors, rather than to the diffraction limit of the system. The irregularities are about 0.3 nm in size for the best of the current production techniques (i.e., comparable with the wavelength of photons of about 1 keV). The mirrors are produced by electro-deposition of nickel onto a mandrel of the inverse shape to that required for the mirror, the mirror shell then being separated from the mandrel by cooling. A coating of a dense metal such as gold, iridium or platinum may be applied to improve the X-ray reflectivity. Recently, large blocks of single crystal silicon have become available, and these can be made into segments

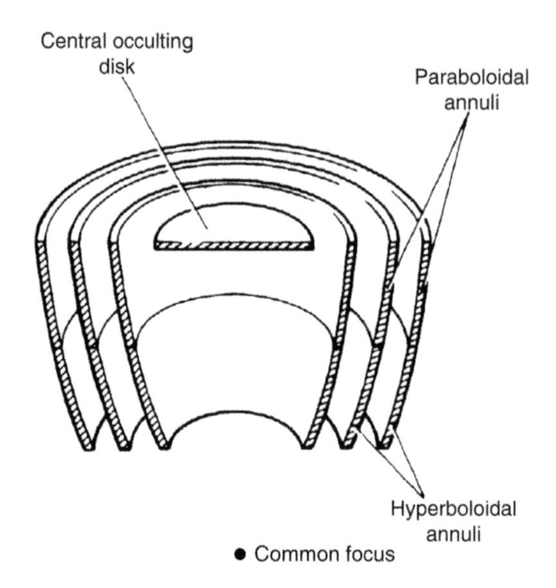

FIGURE 1.91 Section through a nested grazing incidence x-ray telescope.

for very high-quality, light weight, X-ray grazing incidence mirrors by direct machining and polishing. The segments are then assembled into the complete mirrors and these can be as little as 0.5-mm thick. Many such thin mirrors can then be nested to form the final telescope.

At energies of a few keV, glancing incidence telescopes with low but usable resolutions can be made using foil mirrors. The incident angle is less than 1°, so a hundred or more mirrors are needed. They may be formed from thin aluminium foil with a lacquer coating to provide the reflecting surface. The low angular resolution of around 1′ means that the mirror shapes can be simple cones rather than the paraboloids and hyperboloids of a true Wolter telescope, and so the fabrication costs are much reduced.

Three venerable X-ray spacecraft with Wolter-mirror type telescopes, Chandra, Neils Gehrels Swift Observatory and X-ray Multi-mirror Mission (XMM)-Newton (launched 1999, 2004, and 1999, respectively) are still operating at the time of writing. Chandra's X-ray telescopes have four nested Wolter type-1 shells with iridium coatings and a maximum diameter of 1.2 m, giving angular resolutions down to 0.5″ for X-rays up to 10 keV. Swift has 12 nested shells. A position-sensitive detector at the focal plane then produces images with resolutions as high as an arcsecond, and it operates over the 200-eV to 10-keV energy range. The XMM-Newton spacecraft has a total of 58 nested telescope shells with gold coatings giving a total collecting area of about half a square metre. Its angular resolution is about 20″, and it also operates up to 10 keV.

There are seven Wolter-I telescopes for the eROSITA instrument on the recently launched Spektr-RG spacecraft. Each has 54 nested shells with an outer diameter of 358 mm and a focal length of 1.6 m. Operating in the 500-eV to 10-keV region the shells are fabricated from nickel and have a gold reflective coating. The Neutron star Interior Composition Explorer (NICER)[114] instrument, attached to the ISS, uses 56 X-ray telescopes each fabricated from 24 nested parabolic grazing-incidence aluminium foil mirrors with 1.05-m focal lengths. It uses silicon drift detectors to observe from 200 eV to 12 keV.

A concept proposal for the Astro2020 decadal survey is called Advanced X-ray Imaging Satellite (AXIS). It would be a probe-class mission for a late 2020s launch with rapid slewing to enable it

[114] It recently was able to produce the first surface map of a pulsar – PSR J0030+0451 – the spatial resolution of the map is about 1 km and with the pulsar's distance about 340 pc, this corresponds to an angular resolution of about 20 pico-arcseconds or the thickness of a human hair seen at the distance of Jupiter.

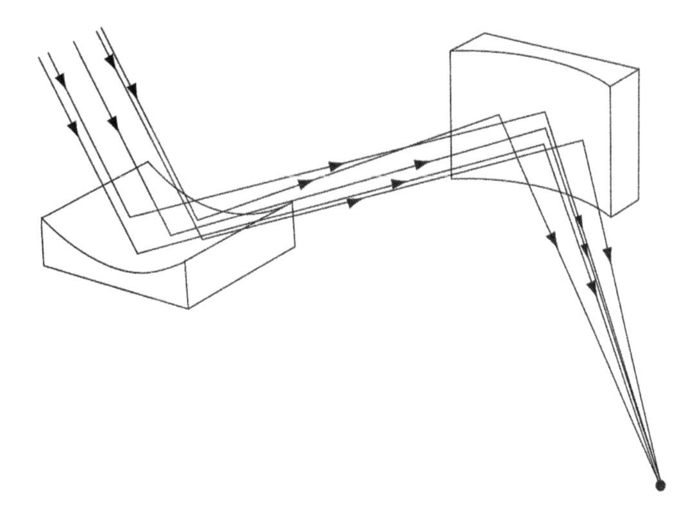

FIGURE 1.92 X-ray imaging by a pair of orthogonal cylindrical mirrors.

to catch transient sources and with some 50 times Chandra's sensitivity. It would use 200 nested Wolter-Schwarzschild mirrors fabricated from single-crystal silicon segments and might operate in conjunction with ATHENA (whose micropore optics will have Wolter type 1 cross-sectional shapes; see Section 1.3.2.7).

An alternative glancing incidence system, known as the 'Kirkpatrick-Baez design', that has fewer production difficulties is based upon cylindrical mirrors. Its collecting efficiency is higher but its angular resolution is poorer than the 3D systems as discussed previously. Two mirrors are used which are orthogonal to each other (Figure 1.92). The surfaces can have a variety of shapes, but two paraboloids are probably the commonest arrangement. The mirrors can again be stacked in multiples of the basic configuration to increase the collecting area. The basic Kirkpatrick-Baez design suffers from significant aberrations, but designs in which these are greatly reduced have recently been devised and which can also be fabricated by stress-deforming flat sheets of glass, so are much cheaper than any of the Wolter designs to manufacture. The design may thus see astronomical applications in the near future.

At lower energies, in the EUV and soft X-ray regions, near normal incidence reflection with efficiencies up to 20% is possible using multilayer coatings. These are formed from tens, hundreds or even thousands of alternate layers of (for example) tungsten and carbon, aluminium and gold or magnesium and gold, each about 1-nm thick. The reflection is essentially monochromatic, with the wavelength depending upon the orientation of the crystalline structure of the layers and upon their thickness. Reflection of several wavelengths may be accomplished by changing the thickness of the layers through the stack. The thickest layers are at the top and reflect the longest, least penetrating wavelengths, while further down, narrower layers reflect shorter, more penetrating wavelengths.

Alternatively, several small telescopes may be used each with a different wavelength response, such the SDO's AIA instrument which uses four 200-mm Cassegrain telescopes to cover seven EUV regions, from 9.4 to 170 nm, using multi-layer reflection coatings and filters. Yet another variation, as in the Transition Region And Corona Explorer (TRACE) spacecraft (1998–2010), different multilayer coatings may be applied in each of four quadrants of the optics. Telescopes of relatively conventional design are used with these mirrors and direct images of the Sun at wavelengths down to 4 nm can be obtained.

Multi-layer coatings and grazing-incidence optics can be combined to improve reflectivity and extend the usage to higher energies – perhaps to 300 keV in the future. The coatings are alternate layers of high and low atomic number materials such as molybdenum and silicon or tungsten and boron carbide. The layers vary in thickness so that a wide range of X-ray wavelengths are reflected. The balloon-borne High Energy Focusing Telescope (HEFT), for example, used 72 conical-approximation Wolter-nested glass mirror shells in each of its three telescopes each of which also used multi-layer reflection coatings. While Nuclear Spectroscopic Telescope Array (NuSTAR) uses two grazing incidence telescopes to observe up to 79 keV. The telescopes each have 133 concentric shells and have graded density multi-layer reflective coatings. The hard X-ray telescopes on Hitomi would have used 213-nested Wolter-I shells with potassium-carbon depth graded multi-layer reflective coatings.

At energies of tens or hundreds of keV, Bragg reflection (as in the multi-layer coatings) and Laue diffraction can be used to concentrate the radiation and to provide some limited imaging. For example, the Laue diffraction pattern of a crystal comprises a number of spots into which the incoming radiation has been concentrated (Figure 1.93). With some crystals, such as germanium and copper, only a few spots are produced. A number of such crystals can be mutually aligned so that one of the spots from each crystal is directed towards the same point (Figure 1.94), resulting in a crude type of lens. With careful design, efficiencies of 25%–30% can be achieved at some wavelengths. As with the glancing incidence telescopes, several such ring lenses can be nested to increase the effective area of the telescope. A position-sensitive detector such as an array of germanium crystals (Figure 1.84) can then provide direct imaging at wavelengths down to 0.001 nm (1 MeV). A study for a possible future space mission using a Laue lens suggests that the focal length of the lens would have to be some 500 m. Thus, two space craft – one to carry the lens and one to carry the detectors – flying in tandem would be needed. The γ-rays with energies as high as 1 MeV might be imaged by such a system. A recent laboratory development, though, has produced a Laue lens fabricated from bent GaAs and Ge crystals which operates up to 600 keV and whose focal length could be as short as 20 m.

The X- and γ-rays, like other forms of electromagnetic radiation, have their paths bent by gravitational fields. Far-away galaxies can thus act as lenses to brighten even more distant x-ray sources. Chandra has used this effect to detect a dwarf galaxy that is 2.9 Gpc away from us and which is lensed by the Phoenix cluster of galaxies whose distance is 1.7 Gpc.

At the time of writing, a radical new approach to the refractive imaging of X-rays has just been announced. The device has been developed for medical purposes and is distantly related to the

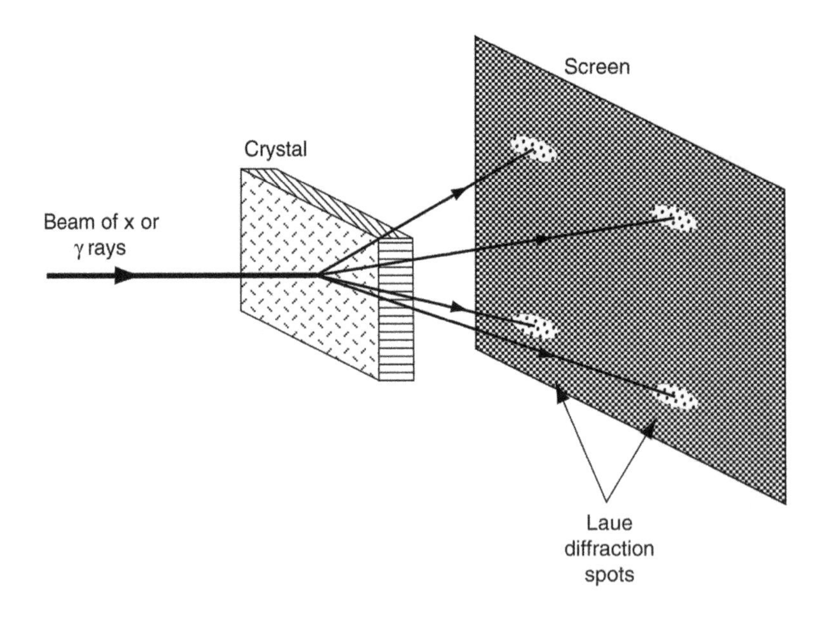

FIGURE 1.93 Laue diffraction of X-rays and γ rays.

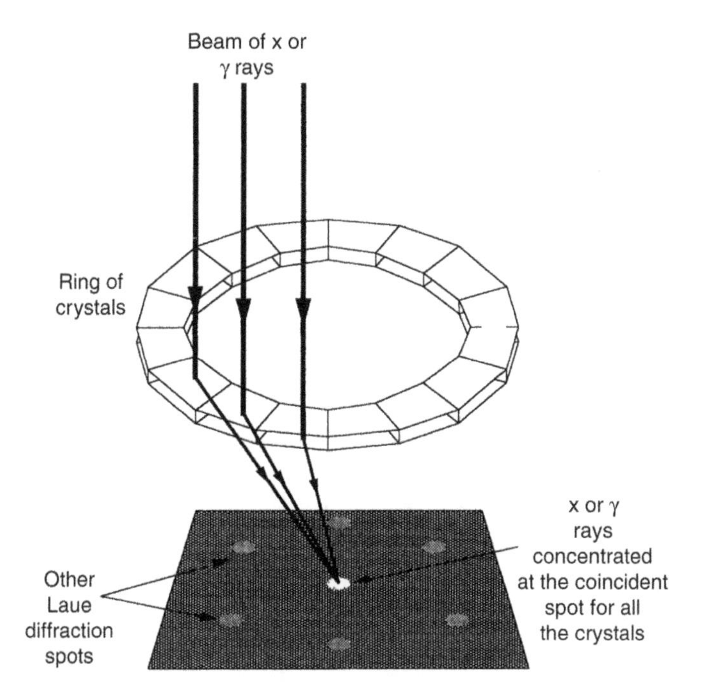

FIGURE 1.94 A Laue diffraction lens for X-rays or γ rays (only three crystals shown illuminated for clarity).

(probably more familiar) Fresnel lens[115] used at optical wavelengths. The device is called a 'stacked prism lens' and that name describes it exactly.

Its basic unit is a right-angled plastic prism which refracts the X-rays passing through it by a small angle (Figure 1.95). The prism forms a part of an annulus with a triangular cross section and several different-sized such annuli are mounted concentrically in a horizontal plane to form a set. Several such sets, with decreasing numbers of annuli, are then stacked vertically.

X-rays near the centre of the beam thus pass through only one prism (Figure 1.95) and so are deflected through only a small angle towards the optical axis. X-rays near the edge of the beam pass through several prisms and are deflected by a larger angle. With suitable prism shapes and sizes, and numbers and distributions of the annuli, the device can be made to bring the whole of the original beam to a focus. By using large-scale integrated circuit manufacturing techniques to create the prism annuli sets, the annuli can be made numerous and narrow, so that the focal area is small and the spatial resolution of the X-ray image high. Focal lengths of a metre or less may be possible. Stacked prism lenses, if produced in large quantities commercially for medical applications, could become relatively cheap and so also form the basis for lightweight, large scale astronomical X-ray telescopes.

1.3.5 Resolution and Image Identification

At the low-energy end of the X-ray spectrum, resolutions and positional accuracies comparable with those of Earth-based optical systems are possible, as we have seen. There is therefore usually little difficulty in identifying the optical counterpart of an X-ray source, should it be bright enough to be visible. The resolution rapidly worsens, however, as higher energies are approached. The position

[115] Phase Fresnel lenses and zone plates have been examined in the past as possible x-ray lenses but produce focal lengths of a million km or so – which even today, using twin spacecraft, would be difficult to implement.

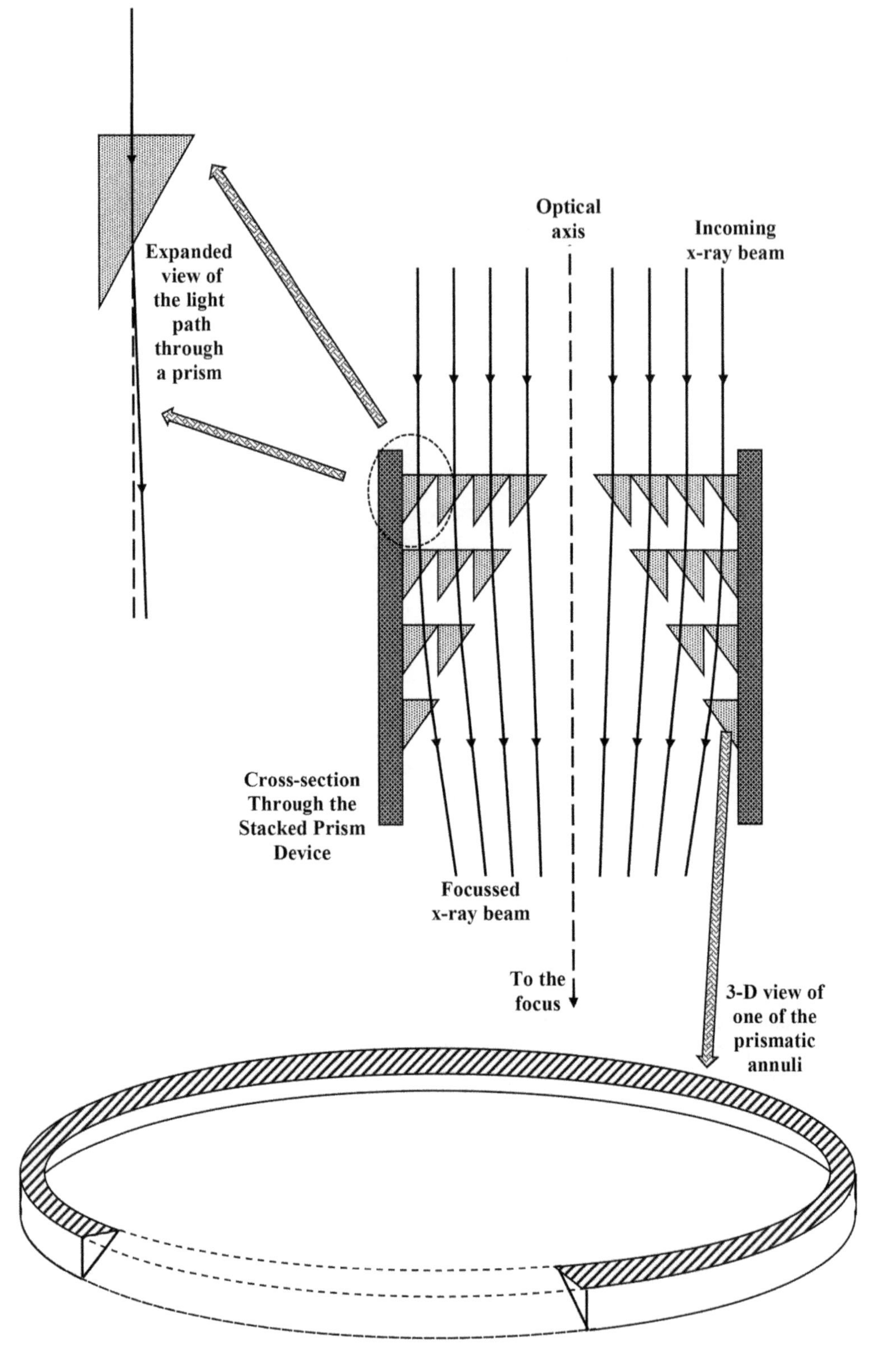

FIGURE 1.95 A schematic cross section through a stacked prism X-ray lens.

of a source that has been detected can therefore only be given to within quite broad limits. It is customary to present this uncertainty in the position of a source by specifying an Error Box. This is an area of the sky, usually rectangular in shape,[116] within which there is a certain specified probability of finding the source. For there to be a reasonable certainty of being able to find the optical counterpart of a source unambiguously, the error box must be smaller than about 1 min^2 of arc. Since high-energy X-ray and γ-ray imaging systems have resolutions measured in tens of arc-minutes or m more, such identification is not usually possible (though the hard X-ray imager on Hitomi might have achieved 1.7′ resolution at 60 keV). The exception to this occurs when an 'unusual' optical object lies within the error box. This object may then, rather riskily, be presumed also to be the X-ray source. Examples of this are the Crab nebula and Vela supernova remnant which lie in the same direction as two strong sources at 100 MeV and are therefore assumed to be those sources, even though the positional uncertainties are measured in many degrees.

1.3.6 Spectroscopy

Many of the detectors that were discussed previously are intrinsically capable of separating photons of differing energies. The STJ detector (Section 1.1 and others) has an intrinsic spectral resolution in the X-ray region that potentially could reach 10,000 (i.e., ±0.1 eV at 1 keV). However, the need to operate those devices at lower than 1 K seems likely to prevent their use on spacecraft for some time to come. Microcalorimeters though, which also need to operate at low temperatures, have been flown on the Suzaku mission and are planned to be used on other spacecraft (see above). They have intrinsic spectral resolutions of up to 1,000. At photon energies above about 10 keV, it is only this inherent spectral resolution which can provide information on the energy spectrum, although some additional resolution may be gained by using several similar detectors with varying degrees of passive shielding; the lower energy photons will not penetrate through to the most highly shielded detectors. The hard X-ray imager for the Hitomi spacecraft would have employed the detectors themselves in this manner. It had four stacked SSDs placed above a single CdTe strip detector. The SSDs absorb (and detect) X-rays of less than 30 keV so that only the higher energy photons penetrate to the CdTe detector. The instrument was housed at the bottom of a BGO well active shield.

At low and medium energies, the gaps between the absorption edges of the materials used for the windows of the detectors can form wideband filters. Devices akin to the more conventional idea of a spectroscope, however, can only be used at the low end of the energy spectrum and these are discussed elsewhere.

1.3.6.1 Grating Spectrometers

Gratings may be either transmission or grazing incidence reflection. The former may be ruled gratings in a thin metal film which has been deposited onto a substrate transparent to the X-rays, or they may be formed on a pre-ruled substrate by vacuum deposition of a metal from a low angle. The shadow regions where no metal is deposited then form the transmission slits of the grating. The theoretical background for X-ray gratings is identical with that for optical gratings and is discussed in detail in Section 4.1. Typical transmission gratings have around 1,000 lines per millimetre. The theoretical spectral resolution (Section 4.1) is between 10^3 and 10^4 but is generally limited in practice from 50 to 100 by other aberrations.

Reflection gratings are also similar in design to their optical counterparts. Their dispersion differs, however, because of the grazing incidence of the radiation. If the separation of the rulings is d then, from Figure 1.96 (top), we may easily see that the path difference, ΔP, of two rays that are incident onto adjacent rulings is

$$\Delta P = d\left[\cos\theta - \cos(\theta+\varphi)\right] \tag{1.78}$$

[116] See Section 1.6 though for lenticular error 'boxes' for gravitational wave sources.

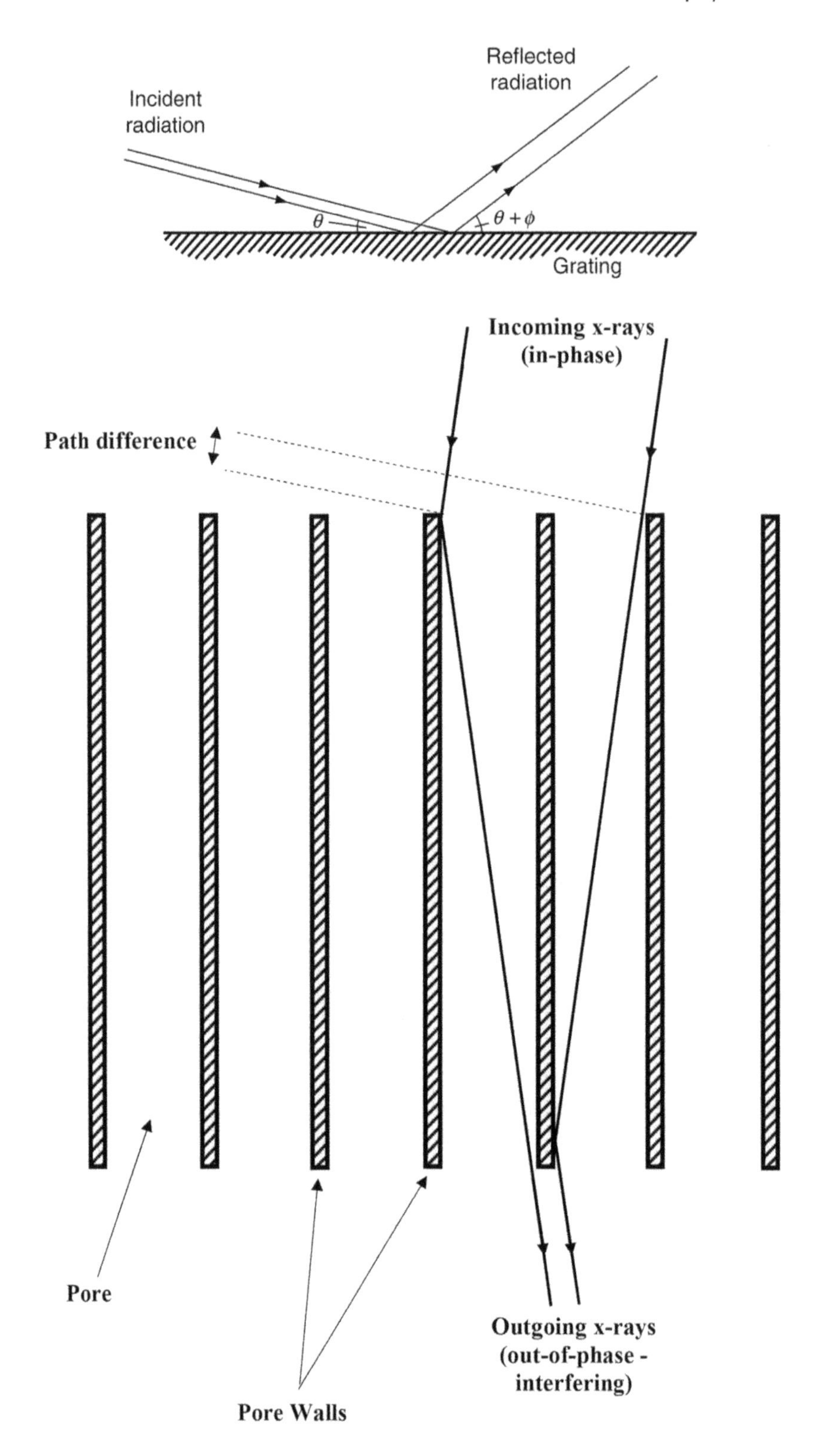

FIGURE 1.96 Top, Optical paths in a grazing incidence reflection grating. Bottom, Optical paths in a section through a CAT grating.

We may expand this via the Taylor series and neglect powers of θ and φ higher than two since they are small angles, to obtain

$$\Delta P = \frac{1}{2} d\left(\varphi^2 - 2\theta\varphi\right) \tag{1.79}$$

In the *m*th order spectrum, constructive interference occurs for radiation of wavelength λ if

$$m\lambda = \Delta P \tag{1.80}$$

so that

$$\varphi = \left(\frac{2m\lambda}{d} + \theta^2\right)^{1/2} - \theta \tag{1.81}$$

and

$$\frac{d\varphi}{d\lambda} = \left(\frac{m}{2d\lambda}\right)^{1/2} \tag{1.82}$$

where we have neglected θ^2, since θ is small. The dispersion for a glancing incidence reflection grating is, therefore, inversely proportional to the square root of the wavelength, unlike the case for near-normal incidence, when the dispersion is independent of wavelength. The gratings may be plane or curved and may be incorporated into many different designs of spectrometer (Section 4.2). Resolutions of up to 10^3 are possible for soft X-rays, but again this tends to be reduced by the effects of other aberrations.

Recently, Critical Angle Transmission (CAT) glancing incidence gratings have started to be produced in laboratories. These have the conventional wavelength/dispersion relationship and are micropore optical components identical in their basic construction to a honeycomb collimator (Figure 1.86). The sides of the pores are smoothed to better than the nm level and the X-rays illuminate them at an angle smaller than the critical angle (i.e., the X-rays are totally externally reflected). The depth of the device is arranged so that X-ray being reflected from the bottom of one pore coincides with the X-ray being reflected from top of the next pore along (Figure 1.96, bottom). Thus, when the two components of the X-ray beam emerge from the CAT, they have a path difference and interference occurs as it would in a normal diffraction grating. When the wavelength of the X-rays is such that the path difference is a whole number of wavelengths, then constructive interference ensures that particular wavelength emerges from the CAT and may then be detected conventionally. Spectral resolutions of 5,000 to 10,000 may be possible around 1 keV.

Detection of the spectrum, once produced, may be by scanning the spectrum over a detector (or vice versa) or by placing a position-sensitive detector in the image plane so that the whole spectrum is detected in one go. The XMM-Newton spacecraft, for example, uses two reflection grating arrays. Each array has 182 individual gratings and they are incorporated into a Rowland-circle (Figure 4.6) spectroscope. CCDs are used as the detectors and spectral resolutions up to 800 are achieved over the 330-eV to 2.5-keV region.

A recent development is the production by photo-lithographic techniques of off-plane grazing incidence gratings. These are placed in the converging beam of X-rays from (say) a Wolter-I telescope (N.B. *not* at the focus). The rulings on the grating are parallel to the X-ray beam and converge at the same rate. The result is a spectrum at the focal plane in place of the normal image. Currently gratings with up to 6,000 grooves per mm can be fabricated and potentially lead to soft X-ray spectrometers with spectral resolutions up to 3,000 and areas of 0.1 m^2 – a factor of ten improvement for both specifications compared with the spectrometers on XMM-Newton and Chandra.

1.3.6.2 Bragg Spectrometers

Distances ranging from 0.1 to 10 nm separate the planes of atoms in a crystal. This is comparable with the wavelengths of X-rays,[117] and so a beam of X-rays interacts with a crystal in a complex manner. The details of the interaction were first explained by the Braggs (father and son). Typical radiation paths are shown in Figure 1.97. The path differences for rays such as *a*, *b* and *c*, or *d* and *e*, are given by multiples of the path difference, ΔP, for two adjacent layers

$$\Delta P = 2d \sin\theta \tag{1.83}$$

There will be constructive interference for path differences that are whole numbers of wavelengths. So that the reflected beam will consist of just those wavelengths, λ, for which this is true;

$$M\lambda = 2d \sin\theta \tag{1.84}$$

The Bragg spectrometer uses a crystal to produce monochromatic radiation of known wavelength. If a crystal is illuminated by a beam of X-rays of mixed wavelengths, at an approach angle, θ, then only those X-rays whose wavelength is given by Equation 1.84 will be reflected. The first-order reflection ($M = 1$) is by far the strongest and so the reflected beam will be effectively monochromatic with a wavelength of

$$\lambda_\theta = 2d \sin\theta \tag{1.85}$$

The intensity of the radiation at that wavelength may then be detected with, say, a proportional counter and the whole spectrum scanned by tilting the crystal to alter θ (Figure 1.98). An improved version of the instrument uses a bent crystal and a collimated beam of X-rays so that the approach angle varies over the crystal. The reflected beam then consists of a spectrum of all the wavelengths (Figure 1.99), and this may then be detected in a single observation using a position-sensitive detector.

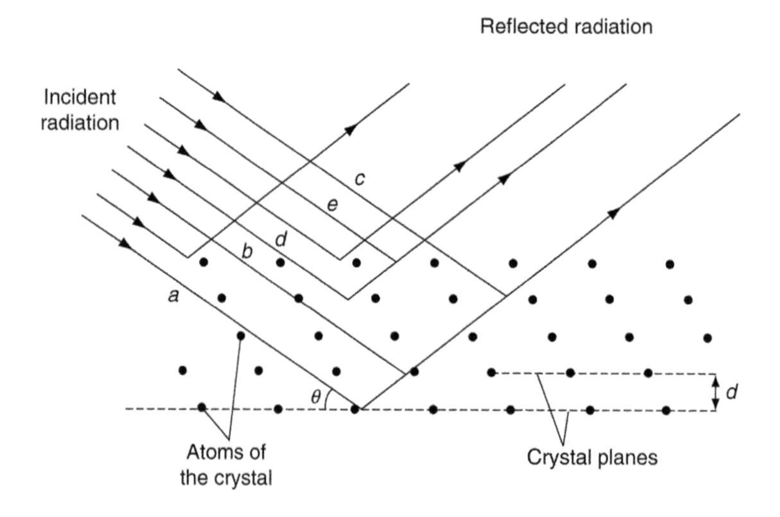

FIGURE 1.97 Bragg reflection.

[117] A 1-nm wavelength photon has an energy of 1.2 keV.

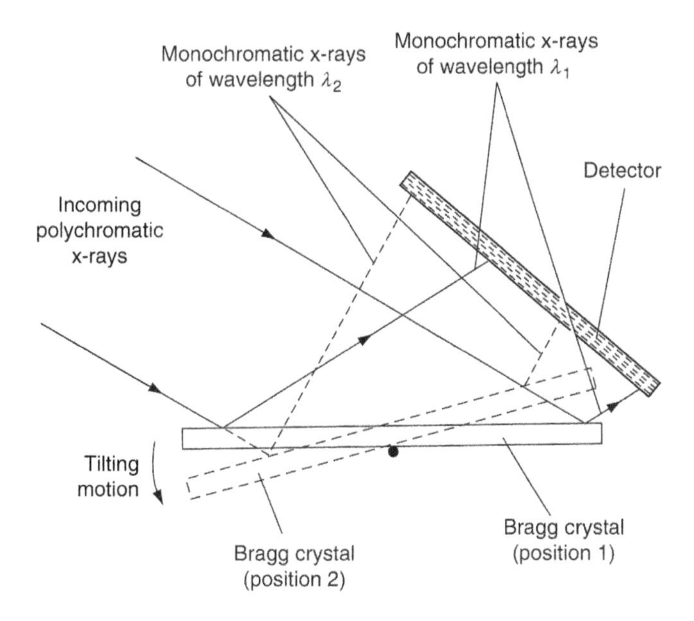

FIGURE 1.98 Scanning Bragg crystal X-ray spectrometer.

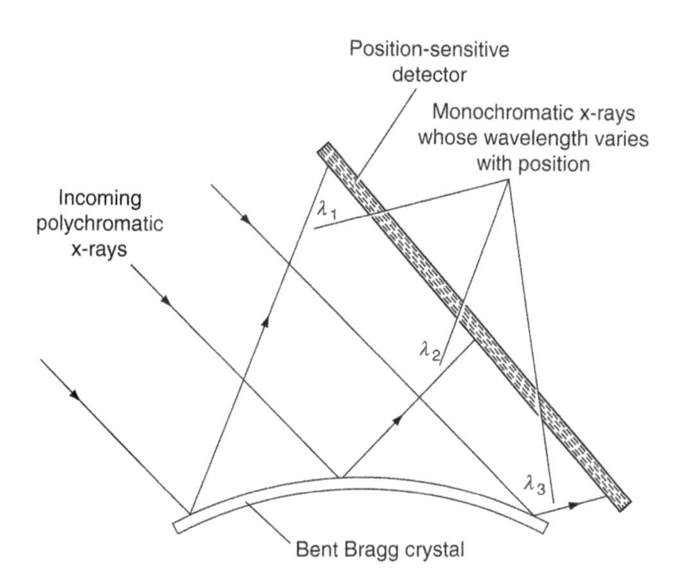

FIGURE 1.99 Bent Bragg crystal X-ray spectrometer.

The latter system has the major advantage for satellite-borne instrumentation of having no moving parts and so has a significantly higher reliability. It also has good time resolution. High spectral resolutions are possible up to 10^3 at 1 keV – but large crystal areas are necessary for good sensitivity, and this may present practical difficulties on a satellite. Many crystals may be used. Among the commonest are lithium fluoride, lithium hydride, tungsten disulphide, graphite, and potassium acid phthalate (KAP).

Many variants upon the basic spectrometer can be devised. It may be used as a monochromator and combined with a scanning telescope to produce spectroheliograms, and for faint sources, it may be adapted for use at the focus of a telescope and so on.

The Russian instrument package, KORTES,[118] is expected to be mounted on the ISS around 2021 and it includes the Solar Polarisation in X-rays (SolpeX) soft X-ray spectrometer/polarimeter. SolpeX includes eight flat Bragg crystals which rotate rapidly to produce a spectrum over the 52–920 eV range with a best time resolution of 100 ms.

1.3.7 Polarimetry

Bragg reflection of x-rays is polarisation dependent. For an angle of incidence of 45°, photons that are polarized perpendicularly to the plane containing the incident and reflected rays will be reflected, while those polarised in this plane will not. Thus, a crystal and detector at 45° to the incoming radiation and which may be rotated around the optical axis will function as a polarimeter. The efficiency of such a system would be low, however, because of the narrow energy bandwidth of Bragg reflection. This is overcome by using many randomly orientated small crystals. The crystal size is too small to absorb radiation significantly. If it should be aligned at the Bragg angle for the particular wavelength concerned, however, that radiation will be reflected. The random orientation of the crystals ensures that overall, many wavelengths are reflected and we have a polarimeter with a broad bandwidth.

A second type of polarimeter looks at the scattered radiation due to Thomson scattering in blocks of lithium or beryllium. If the beam is polarised, then the scattered radiation is asymmetrical, and this may be measured by surrounding the block with several pairs of detectors.

Recently, X-ray polarimeters have been based upon the intrinsic polarization sensitivity of proportional counters and their derivatives such as TPCs (see above). The now cancelled Gravity Extreme Magnetism Small explorer (GEMS) spacecraft was due to carry a TPC-based polarimeter for the 2–10 keV region as its main instrument. The GEMS was also due to carry a Bragg reflection polarimeter for 500-eV X-ray observations. The future ISS instrument, SolpeX, will include a Bragg polarimeter operating at 3 keV that will use two rotating, bent silicon crystals with 2048 × 2048-pixel CMOS detectors.

1.3.8 Observing Platforms

The Earth's atmosphere completely absorbs X-ray and γ-ray radiation, so that all the detectors and other apparatus discussed previously, with the exception of the Čerenkov detectors, have to be lifted above at least 99% of the atmosphere. The three systems for so lifting equipment are balloons, rockets and spacecraft.

Spacecraft give the best results in terms of their height, stability and the duration of the mission. Their cost is high, however, and there are weight and space restrictions to be complied with; nonetheless there have been quite a number of satellites, many of which have been mentioned previously, launched for exclusive observation of the X-ray and γ-ray region and appropriate detectors have been included on a great many other missions as secondary instrumentation.

Balloons can carry heavy equipment far more cheaply than satellites, but only to a height of about 40 km, and this is too low for many of the observational needs of this spectral region. Their mission duration is also comparatively short being a few days to a month for even the most sophisticated of the self-balancing versions. Complex arrangements have to be made for communication and for retrieval of the payload because of the unpredictable drift of the balloon during its mission. The platform upon which the instruments are mounted has to be actively stabilised to provide sufficient pointing accuracy for the telescopes, etc. Thus, there are many drawbacks to set against the lower cost of a balloon. The HEFT and PoGO+ missions have already been mentioned as examples of balloon-carried X-ray instruments.

[118] Derivation unknown.

Sounding rockets are even cheaper, and they can reach heights of several hundred kilometres without difficulty. But their flight duration is measured only in minutes and the weight and size restrictions are even tighter than for spacecraft. Rockets still, however, find uses as rapid-response systems for events such as solar flares. A balloon- or spacecraft-borne detector may not be available when the event occurs or may take some time to bear onto the target, whereas a rocket may be held on standby on Earth, during which time the cost commitment is minimal and then launched when required at only a few minutes notice. The High resolution Coronal imager (Hi-C), for example, was launched in 2012 on a Black Brant rocket to observe the Sun at 64 eV and obtained 165 images with 0.2″ resolution using a CCD camera during its five minutes of observing time.

1.4 COSMIC RAY DETECTORS

1.4.1 Background

Cosmic rays comprise two quite separate populations of particles, primary cosmic rays and secondary cosmic rays. The former are the true cosmic rays[119] and consist mainly of atomic nuclei in the proportions 84% hydrogen, 14% helium, 1% other nuclei and 1% electrons and positrons plus γ rays. The nature of a given primary cosmic ray particle can be found in several ways, including the height at which Extensive Air Shower (EAS, see below) particle production peaks. For the same original energy, the particle production will peak lowest in the atmosphere for γ rays, then progressively higher as the atomic mass of the particle increases (i.e., the EAS will peak higher in the atmosphere for an iron nucleus than for a proton and both will be higher than that for the γ ray).

Primary cosmic rays are moving at velocities from a few per cent of the speed of light to 0.999, 999, 999, 999, 999, 999, 999, 9c. That is, the energies per particle range from 10^6 to a few times 10^{20} eV (10^{-13} to 50 J) with the bulk of the particles having energies near 10^9 eV, or a velocity of 0.9c. These are obviously relativistic velocities for most of the particles, for which the relationship between kinetic energy and velocity is

$$\mathrm{v} = \frac{(E^2 + 2Emc^2)^{1/2}c}{E + mc^2} \tag{1.86}$$

where E is the kinetic energy of the particle and m is the rest mass of the particle. The flux in space of the primary cosmic rays of all energies is about 10^4 particles m^{-2} s^{-1}. A limit, the GZK cut-off,[120] in the cosmic-ray energy spectrum is expected above about 4×10^{19} eV (6 J). At this energy, the microwave background photons are blue-shifted to gamma rays of sufficient energy to interact with the cosmic ray particle to produce pions. This will slow the particle to energies lower than GZK limit in around 100 million years. However, surprisingly, higher-energy cosmic rays *are* observed, implying that there must be one or more sources of ultra-high-energy cosmic rays within 100 million light-years (30 million pc) of the Earth. The Pierre Auger cosmic ray observatory (see below) confirmed in 2007 that the most likely sources of cosmic ray particles with energies above the GZK limit were relatively nearby active galactic nuclei, although the mechanism for accelerating the particles to such enormous energies remains a mystery.

Secondary cosmic rays are produced by the interaction of the primary cosmic rays with nuclei in the Earth's atmosphere. A primary cosmic ray has to travel through about 800 kg m^{-2} of matter

[119] The term 'ray' in this context is thus a misnomer arising from before the time that their true nature was understood. However, the use is unlikely to be corrected now.

[120] Named for Ken Greisen, Timofeevich Zatsepin and Vadim Kuzmin.

on average before it collides with a nucleus. A column of the Earth's atmosphere 1 m^2 in cross section contains about 10^4 kg, so the primary cosmic ray usually collides with an atmospheric nucleus at a height of 30–60 km. The interaction results in numerous fragments, nucleons, pions, muons, electrons, positrons, neutrinos, gamma rays, etc., and these in turn may still have sufficient energy to cause further interactions, producing more fragments and so on. For high-energy primaries (>10^{11} eV), some of the secondary particles will survive down to sea level[121] and be observed as secondary cosmic rays. At even higher energies (>10^{13} eV), large numbers of the secondary particles (>10^9) survive to sea level and a cosmic ray shower or EAS is produced. At altitudes higher than sea level, the secondaries from the lower-energy primary particles may also be found. Interactions at all energies produce electron and muon neutrinos as a component of the shower. Almost all of these survive to the surface, even those passing through the whole of the Earth, though some convert to tau neutrinos during that passage. The detection of these latter particles is considered in Section 1.5.

We also consider in this section the detection of very high-energy γ rays (Section 1.3) via the Čerenkov radiation that they induce in the Earth's atmosphere because they are detected by the same instruments as those for observing very-high-energy cosmic rays.

1.4.2 Detectors

Both primary and secondary cosmic rays may be detected by similar instruments because both are composed of very energetic subatomic particles (or γ rays). The difference lies in where the detections occur – in space or high in the Earth's atmosphere for the primary rays and by ground-based or low altitude instruments for the secondary rays.

As usual, therefore, for spacecraft- or balloon-borne instruments, the emphasis in their design lies in minimising their mass and their power consumption and in using sophisticated remote control and robotic systems. The Alpha Magnetic Spectrometer (AMS-02) detector (see below) on board the ISS, though, masses more than 6.7 tonnes and consumes up to 2.5 kW. Furthermore, it recently required several astronauts to conduct lengthy repairs in space over several long space walks. The repairs included sawing through and rejoining several stainless-steel pipes making up part of the cooling system and replacing pumps. By contrast, the Pierre Auger (ground-based) cosmic ray observatory uses 12 tonnes for *each* of its main detector elements (and there are 1,660 of them spread over 3,000 km^2).

The methods of detecting cosmic rays may be divided into:

1. Real-time methods. These observe the particles instantaneously and produce information on their direction as well as their energy and composition. They are basically the standard detectors of nuclear physicists, and there is considerable overlap between these detectors and those discussed in Section 1.3. They are generally used in conjunction with passive and/or active shielding so that directional and spectral information can also be obtained.
2. Residual track methods. The path of a particle through a material may be found some time (hours to millions of years) after its passage.
3. Indirect methods. Information on the flux of cosmic rays at large distances from the Earth or at considerable times in the past may be obtained by studying the consequent effects of their presence.

[121] At sea level the radiation exposure coming from secondary cosmic rays is around 0.4 mSv per year. The normal total background exposure is 3 mSv per year. The cosmic ray component will increase though with altitude – reaching perhaps 5 mSv per year for aircraft crews who work full-time on high-flying aircraft. To put these rates in context, the recommended maximum occupational exposure limit is 20 mSv per year and exposure to 1 Sv in a single short exposure is likely to induce temporary radiation sickness. A 5-Sv dose in a single short exposure is likely to cause 50% fatalities amongst those exposed.

1.4.2.1 Real-Time Methods

1.4.2.1.1 Scintillation Detectors

See Section 1.3 for further discussion of these devices. There is little change required for them to detect cosmic ray particles. Plastic sheet scintillator detectors, with areas of 3 m^2 and with PMTs as detectors, are used for the 507 surface stations of the Telescope Array Project in Utah. They were also used as triggers and active shields for the Payload for Anti-Matter Exploration and Light nuclei Astrophysics (PAMELA) experiment (2006–2016) onboard the Resurs-DK No.1 spacecraft and are currently a part of the AMS-02's ToF trigger mechanism.

Scintillator-based muon detectors using a small piece of plastic scintillator and an SiPMT can now be made as small as a mobile phone and such instruments may find astronomical applications in the near future.

1.4.2.1.2 Čerenkov Detectors

1.4.2.1.2.1 Background When a charged particle is moving through a medium with a speed greater than the local speed of light in that medium, it causes the atoms of the medium to radiate. This radiation is known as 'Čerenkov radiation', and it arises from the abrupt change in the electric field near the atom as the particle passes by. At sub-photic speeds, the change in the field is smoother and little or no radiation results. The radiation is concentrated into a cone spreading outward from the direction of motion of the particle (Figure 1.100), whose half angle, θ, is

$$\theta = \tan^{-1}\left[\left(\mu_\nu^2 \frac{\mathrm{v}^2}{c^2} - 1\right)^{1/2}\right] \tag{1.87}$$

where μ_ν is the refractive index of the material at frequency, ν, and v is the particle's velocity ($\mathrm{v} > c/\mu_\nu$). Its spectrum is given by

$$I_\nu = \frac{e^2 \nu}{2\varepsilon_o c^2}\left(1 - \frac{c^2}{\mu_\nu^2 \mathrm{v}^2}\right) \tag{1.88}$$

where I_ν is the energy radiated at frequency ν per unit frequency interval, per unit distance travelled by the particle through the medium. The peak emission depends upon the form of the variation of refractive index with frequency. For a proton in air, peak emission occurs in the visible when its energy is about 2×10^{14} eV.

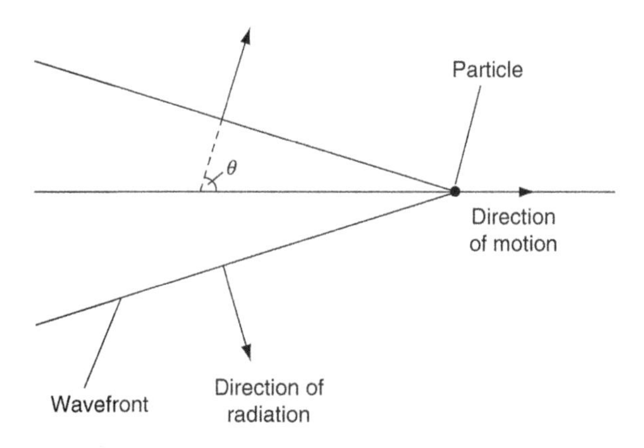

FIGURE 1.100 Čerenkov radiation.

1.4.2.1.2.2 Detectors Čerenkov detectors are similar to scintillation detectors and are sometimes called that, although the flashes of visible radiation are produced by different physical mechanisms in the two detectors. A commonly used system employs a tank of pure water surrounded by PMTs for the detection of the heavier cosmic ray particles, while high-pressure carbon dioxide is used for the electrons and positrons. With adequate observation and the use of two or more different media, the direction, energy and possibly the type of the particle may be deduced.

The Pierre Auger cosmic ray array in Argentina uses both water-based Čerenkov detectors and atmospheric fluorescence detectors (see below). There are 1,660 Čerenkov detectors each containing 12 tonnes of water and the flashes are detected by PMTs. The detectors, which are solar powered, are spread out in a grid over a 3,000-km^2 area of grazing land. The proposed Pierre Auger North, to be sited (perhaps) in Colorado, was originally expected to be up to three times the size of the southern array, but at the time of writing it has only a few Čerenkov detectors in Utah, sited with some of the Telescope Array Project's instruments and working in collaboration with them.

A major upgrade of the southern Pierre Auger observatory is currently underway, which will see the installation of 4-m^2 plastic scintillators on top of all the Čerenkov detectors, the replacement of the PMT detectors with SiPMTs and the addition of underground muon detectors to the system. The data gathered by the Pierre Auger observatory is partially publicly available. The 1% of the data is placed into the Public Event Explorer which is at http://labdpr.cab.cnea.gov.ar/ED-en/index.php and is freely available over the internet.

A related Čerenkov-based detector is called the Ring Imaging Čerenkov (RICH) detector because the Čerenkov radiation spreads out in a cone and thus appears as a ring when intercepted by a flat detector. The Čerenkov radiation is produced within a thin plate of (for example) sodium fluoride which is called a 'radiator' and is detected by an array of PMTs or other photon sensors. Similar devices that only detect the burst of Čerenkov radiation with a single PMTs and so do not image the ring are also in use. The balloon-borne Super Trans-Iron Galactic Element Recorder (SuperTIGER) instrument, last flown in 2020, for example, uses the latter type of Čerenkov detector with acrylic plastic and aerogel as its radiators, alongside scintillation detectors, while AMS-02's RICH detector uses a sodium fluoride radiator surrounded by aerogel with 680 multi-anode PMT photon sensors and can identify the nuclei in primary cosmic rays up to iron.

As mentioned previously (Čerenkov detectors in Section 1.3), the flashes produced in the atmosphere by the primary cosmic rays and high energy γ-rays can be detected. A large light 'bucket' and a PMT are needed, preferably with two or more similar systems observing the same part of the sky so that non-Čerenkov events can be eliminated by anti-coincidence discrimination. At least two images are also needed so that the direction in space of the incoming particle or photon can be determined. For the high energy γ rays, the tracks may be expected to point directly back to their points of origin (interstellar magnetic fields mean that the track directions for charged particles are unrelated to their points of origin). In this way, the 10-m Whipple telescope (decommissioned in 2013) found TeV γ-rays coming from the Crab nebula in 1989.

The HESS array in Namibia has four 12-m and a single 32.6-m × 24.3-m dish. Each of the smaller dishes is made up from three hundred and eighty-two 0.6-m circular mirrors forming overall spherical reflectors. The larger dish is parabolic and built up from eight hundred and seventy-five 0.9-m hexagonal segments. The detectors for the smaller dishes are arrays of 960 eight-stage PMTs.

An intriguing aside to Čerenkov detectors arises from the occasional flashes seen by astronauts when in space. These are thought to be Čerenkov radiation from primary cosmic rays passing through the material of the eyeball. Thus, cosmic ray physicists could, potentially, observe their subjects directly.

It is also just possible that they could listen to them as well! A large EAS hitting a water surface will produce a 'click' sound that is probably detectable by the best of the current hydrophones. Unfortunately, a great many other events produce similar sounds so that the cosmic ray clicks are likely to be well buried in the noise. Nonetheless the highest-energy cosmic ray showers might even be audible directly to a skin diver. Recently, several attempts have been made to detect these acoustic pulses and those from ultra-high-energy neutrinos, but so far without success. Hydrophones have been included though, in several water/ice neutrino detectors including the ANTARES Modules for the Acoustic Detection Under the Sea (AMADEUS) addition to Astronomy with a Neutrino Telescope and Abyssal Environmental Research (ANTARES; (2008–2016, 36 hydrophones) and 28 hydrophones for South Pole Acoustic Test Setup (SPATS) on IceCube (Section 1.5).

1.4.2.1.3 Solid-State Detectors

These have been discussed in Section 1.3. Little change to their operation is needed for cosmic ray detection. Their main disadvantages are that their size is small compared with many other detectors, so that their collecting area is also small and that unless the particle is stopped within the detector's volume, only the rate of energy loss may be found and not the total energy of the particle. This latter disadvantage, however, also applies to most other detectors. The PAMELA experiment launched in 2006 uses a number of different types of detectors as detectors trackers, active shields and detectors to find primary cosmic ray anti-protons and positrons with energies up to 270 GeV. Amongst these are silicon detectors, plastic sheet scintillator detectors and tungsten-silicon micro-calorimeters (Section 1.3).

AMS-02 was taken to the ISS by the space shuttle in 2011. Its silicon tracker instrument is based upon nine layers of SSDs (Section 1.3.2.7), its calorimeter (Section 1.3.2.7) comprises nine layers of lead and scintillators and its Transition Radiation Detector (TRD; see Section 1.4.2.1.4) uses 5,248 Kapton straw tubes with a fleece radiator. AMS-02, though, additionally contains a strong permanent magnet. The magnet bends the paths of lower-energy-charged particles enabling them to be identified. It is searching in particular for anti-helium nuclei. Cosmic ray electrons and positrons up to 20 TeV are the main targets of the Calorimetric Electron Telescope (CALET) instrument recently installed on the ISS. The CALET uses calorimeters as its detectors, a thin calorimeter made up from alternating tungsten plates and scintillating fibres for imaging and a thicker lead tungstate calorimeter for measuring the cosmic ray's energy.

1.4.2.1.4 Proportional Counters

These and their various developments (TPCs, micromegas, etc.) were also discussed in Section 1.3. They are commonly called 'drift chambers' when used as cosmic ray detectors (not to be confused with SDDs, although these can also be used for cosmic ray detection). A TRD is a drift chamber which is attached to a slab of material called a 'radiator'. The radiator is a composite of, for example, polypropylene sheets and polymethacrylamide foam. High-energy particles passing through the radiator induce x-radiation within it which is then detected within the drift chamber alongside the electrons produced directly within the chamber by the particle. TRDs are used on AMS-02 (Section 1.4.2.1.3).

A variety of proportional counter called a 'Streamer Chamber' is also used for cosmic ray detection. It operates via a short high voltage pulse that produces a line of ionised particles (the streamer) along the track of the cosmic ray. The track is then imaged directly from the light emitted as the ions and electrons recombine. The device has much in common with the Wilson cloud chamber which shows the paths of charged particles via the drops of liquid that form along the ionised track of the particle.

1.4.2.1.5 Radar

The Telescope Array project added a bistatic radar system (Section 2.8) to its fluorescence telescopes in 2012 because this could operate at any time and not just on dark nights. Telescope Array Radar (TARA) radiates 25 kW at 54 MHz using Yagi antennas and the ionised trail through the atmosphere from an EAS reflects the beam down to an array of dipole antennae, 40 km away, for reception. The signal should take the form of a chirp (see also Section 1.6) with the frequency rising from the start to the finish of the event. At the time of writing, there do not seem to be any published details of actual EAS detections by TARA.

1.4.2.2 Residual Track Detectors

1.4.2.2.1 Photographic Emulsions

In a sense the use of photographic emulsion for the detection of ionising radiation is the second-oldest technique available to the cosmic ray physicist after the electroscope, since Henri Becquerel discovered the existence of ionising particles in 1896 by their effect upon a photographic plate. However, the early emulsions only hinted at the existence of the individual tracks of the particles and were far too crude to be of any real use. For cosmic ray detection, blocks of photographic emulsion with high silver bromide content were exposed to cosmic rays at the tops of mountains, flown on balloons or sent into space on some early spacecraft. The cosmic rays could be picked up and their nature deduced to some extent by the tracks left in the emulsion after it had been developed. This type of detector is little used now.

1.4.2.2.2 Ionisation Damage Detectors

These provide a selective detector for nuclei with masses above about 150 amu. Their principle of operation is allied to that of the nuclear emulsions in that it is based upon the ionisation produced by the particle along its track through the material. The material is usually a plastic with relatively complex molecules. As an ionising particle passes through it, the large complex molecules are disrupted, leaving behind short, chemically reactive segments, radicals, etc. Etching the plastic in sodium hydroxide (for example) reveals the higher chemical reactivity along the track of a particle and a conical pit develops along the line of the track. By stacking many thin layers of plastic, the track may be followed to its conclusion. The degree of damage to the molecules and, hence, the characteristics of the pit which is produced, is a function of the particle's mass, charge and velocity. A particular plastic may be calibrated in the laboratory so that these quantities may be inferred from the pattern of the sizes and shapes of the pits along the particle's track. Cellulose nitrate and polycarbonate plastics are the currently favoured materials. The low weight of the plastic and the ease with which large-area detectors can be formed make this a particularly suitable method for use in space when there is an opportunity for returning the plastic to Earth for processing, as for example, when the flight is a manned one or a high-altitude balloon is used.

Similar tracks may be etched into polished crystals of minerals such as feldspar and rendered visible by infilling with silver. Meteorites and lunar samples can thus be studied and provide data on cosmic rays which extend back into the past for many millions of years. The majority of such tracks appear to be attributable to iron group nuclei, but the calibration is uncertain. Because of the uncertainties involved, the evidence has so far been of more use to the meteoriticist in dating the meteorite than to the cosmic ray astronomer.

1.4.2.3 Indirect Detectors

1.4.2.3.1 100-MeV γ-rays and keV X-Rays

Primary cosmic rays occasionally collide with nuclei in the interstellar medium. The chance of this happening is only about 1 in 1,000 if the particle were to cross the galaxy in a straight line; nonetheless, it happens sufficiently often to produce detectable results. In such collisions π° mesons will frequently be produced and these will decay rapidly into two gamma rays, each with an energy of about 100 MeV.

The π° mesons may also be produced by the interaction of the cosmic ray particles and the 3 K microwave background radiation. This radiation when 'seen' by a 10^{20} eV proton is Doppler-shifted to a gamma ray of 100-MeV energy and neutral pions result from the reaction

$$p^+ + \gamma \rightarrow p^+ + \pi^\circ \tag{1.89}$$

Inverse Compton scattering of starlight or the microwave background by cosmic ray particles can produce an underlying continuum around the 100-MeV line emission produced by the pion decay.

Gamma rays with energies as high as these are little affected by the interstellar medium so that they may be observed by spacecraft (Section 1.3) wherever they may have originated within the galaxy. The 100-MeV gamma ray flux thus gives an indication of the cosmic ray flux throughout the galaxy (and beyond). The detection of gamma rays is covered in Section 1.3.

Lower-energy cosmic rays have been detected indirectly from the x-ray emission which results from their interactions with the interstellar medium. Data mining the XMM-Newton's archive has revealed the 6.0- and 7.4-keV X-ray emissions from iron in the vicinity of the Arches star cluster close to the centre of the Milky Way. The star cluster is moving relative to its local interstellar medium at some 200 km/s and the cosmic rays are thought to originate in interactions between the cluster and the interstellar medium.

1.4.2.3.2 Radio Emission

Cosmic ray electrons are only a small proportion of the total flux and the reason for this is that they lose significant amounts of energy by synchrotron emission as they interact with the galactic magnetic field. This emission lies principally between 1 MHz and 1 GHz (300 m–300 mm) and is observable as diffuse radio emission from the galaxy. However, the interpretation of the observations into electron energy spectra and so on is not straightforward and is further complicated by the lack of a proper understanding of the galactic magnetic field.

Primary cosmic rays have also recently been detected from their low-frequency emissions as they collide with the Earth's atmosphere (see Askaryan emission, Section 1.5.2.1.2). The LOFAR array (Sections 1.2 and 2.5) has detected bright flashes from cosmic rays that occur about once a day and last for a few tens of nanoseconds as has Antarctic Impulsive Transient Antenna (ANITA; see Section 1.5.2.1.2). The SKA should be able to detect the radio emissions from high-energy cosmic ray interactions within the lunar regolith.

The charged particles in a cosmic ray shower emit MHz synchrotron radiation (sometimes called 'geosynchrotron radiation'; see also Section 1.2) as their paths are affected by the Earth's magnetic field. LOFAR Prototype Station (LOPES), for example, has been detecting geosynchrotron pulses since 2006. In its current version it uses three orthogonal monopole antennae (called a 'tripole'; see also Section 1.2.2) to pick up all the electric field components of the radio wave. There are also currently proposals for the construction of a radio detector for cosmic rays in the radio-quiet Antarctic continent. The recently completed Auger Engineering Radio Array (AERA), sited alongside the Pierre Auger observatory, has 150 antennae covering 17 km^2 and observes over the 30- to 80-MHz waveband. It is thus able to detect, from their γ-ray emissions, cosmic ray particles with energies above 10^{17} eV.

EASs also emit GHz radiation from the plasma formed along the particle's track through the atmosphere. Microwave Detection of Air Showers (MIDAS), now also located at the Pierre Auger observatory site, comprises a 5-m dish with a 53-pixel detector which gives it a 10° × 20° field of view. It can detect cosmic ray particles with energies in excess of 10^{18} eV. MIDAS' operating frequency lies within the extended C-band (3.4–4.2 GHz) used by satellite TVs, so that almost all the equipment needed is available in inexpensive commercially mass-produced form.

The potential for the construction of large arrays relatively cheaply in the future is therefore excellent. Furthermore, these receivers can operate almost all the time – unlike the fluorescence detectors (below) which can only operate during moonless periods.

1.4.2.3.3 Fluorescence

The highest-energy EASs are detectable via weak fluorescent light from atmospheric nitrogen. This is produced through the excitation of the nitrogen by the electron-photon component of the cascade of secondary cosmic rays. The equipment required is a light bucket and a detector (cf. Čerenkov detectors above and in Section 1.3) and detection rates of a few tens of events per year for particles in the 10^{19}–10^{20} eV[122] range are achieved by several automatic arrays. Fluorescence detectors require dark sites and moonless and cloudless nights and so are not in continuous operation.

The fluorescent emissions (and also Čerenkov air emissions) enable the EAS to be imaged thus greatly aiding the determination of the primary particle's incoming direction and the height of the particle production peak (so, therefore, also helping to identify the primary's nature – see above). At the higher energies (>10 TeV), current array detectors can pinpoint the incoming primary's direction to a few arc-minutes and the CTA should improve on this by about a factor of two or three.

The Fly's Eye detector in Utah used sixty-seven 1.5-m mirrors feeding PMTs on two sites 4 km apart to monitor the whole sky. It operated from 1982 to 1992 and in 1991 it recorded the highest-energy cosmic ray yet found: 3.2×10^{20} eV. This is well above the GZK limit (see above). Fly's Eye was upgraded to High Resolution (HiRes) which had a baseline of 12.5 km, with 64 quadruple mirror units feeding 256 PMTs. HiRes in turn is now has now been superseded by the Telescope Array which has thirty-eight 5- to 7-m^2 telescopes each with 256 PMTs as their detectors and located in three stations sited at the tops of hills some 35 km apart. The Pierre Auger observatory has four fluorescence detectors separated by about 50 km, each using six 4-m mirrors and feeding cameras containing 440 PMTs.

There have been several proposals and attempts to detect cosmic rays from space using the Earth's entire atmosphere as the detector. The instruments would look down towards the Earth and catch the fluorescent emissions over a wide area. Thus, the Lomonosov spacecraft (2016–2018) carried the Tracking Ultra-Violet Set-Up (TUS) instrument for this purpose. The TUS could observe an area of about 6,400 km^2 in the near UV with an optical telescope comprising a single Fresnel segmented mirror and a 256-pixel PMT detector. It made a number of detections of high-energy cosmic rays before ceasing to operate in 2017.

1.4.2.3.4 Solar Cosmic Rays

High fluxes of low-energy cosmic rays can follow the eruption of a large solar flare. The fluxes can be intense enough to lower the Earth's ionosphere and to increase its electron density. This in turn can be detected by direct radar observations, through long-wave radio communication fade-outs or through decreased cosmic radio background intensity as the absorption of the ionosphere increases or through intense auroral emissions.

1.4.2.3.5 Carbon-14

The radioactive isotope ${}^{14}_{6}\mathrm{C}$ is produced from atmospheric ${}^{14}_{7}\mathrm{N}$ by neutrons from cosmic ray showers via the reaction:

$$ {}^{14}_{7}\mathrm{N} + n \rightarrow {}^{14}_{6}\mathrm{C} + p^{+} \tag{1.90}$$

The isotope has a half-life of 5,730 years and has been studied intensively as a means of dating archaeological remains. Its existence in ancient organic remains shows that cosmic rays have been present in the Earth's vicinity for at least 20,000 years. The flux seems, however, to have varied markedly at times from its present-day value, particularly between about 4,000 and 1,000 years BCE. But this is probably attributable to increased shielding of the Earth from the low-energy cosmic rays at times of high solar activity, rather than to a true variation in the number of primary cosmic rays.

---

[122] Cosmic ray particles with energies of 10^{20} eV and above, occur at a rate of about 10^{-17} to 10^{-16} m^{-2} s^{-1} – i.e., about once a year over an area of 100 km^2.

1.4.3 Arrays

Primary cosmic rays may be studied by single examples of the detectors that we have considered previously. The majority of the work on cosmic rays, however, is on the secondary cosmic rays, and for these, a single detector is not informative. The reason is that the secondary particles from a single high-energy primary particle have spread over an area of 10 km^2 or more by the time they have reached ground level from their point of production some 50 km up in the atmosphere. To deduce anything about the primary particle that is meaningful, the secondary shower must hence be sampled over a significant fraction of its area. Thus, arrays of detectors are usually used rather than single ones (see also Section 1.5 for the detection of neutrinos produced by cosmic rays). Plastic or liquid scintillators (Section 1.3), water Čerenkov and fluorescence detectors are frequently chosen as the detectors but any of the real-time instruments can be used, and these are typically spread out over an area of hundreds to thousands of square kilometres. The resulting several hundred to thousands of individual detectors are all then linked to a central computer for data analysis.

A second reason for using arrays of detectors is that the flux of cosmic rays at the highest energies is low. Primary cosmic ray particles with energies around at 10^{12} eV have a flux of about one particle per square metre per second, and at 10^{16} eV the flux is a few particles per square metre per year, while at 10^{20} eV the flux falls to less than one particle per square kilometre per century. Thus, to have any chance of catching the highest-energy (and perhaps the most interesting) cosmic ray particles, arrays have to be spread over hundreds or thousands of square kilometres.

The Pierre Auger observatory, for example, is a 3,000-km^2 array in Argentina and so may catch a few of the highest energy particles per year. It has 1,660 water-based Čerenkov detectors distributed over its area and twenty-four 12-m^2 optical telescopes to detect nitrogen fluorescence at distances of up to 30 km. Latin American Giant Observatory (LAGO) is a linked network of water-based Čerenkov detectors at a number of different sites in nine countries across South America and Antarctica. The Telescope Array in Utah uses three fluorescence detectors and five hundred and seven 2 × 3-m plastic scintillators spread over a 700-km^2 area.[123]

The long-established Tibet Air Shower γ ray (ASγ) array, sited at the YangBaJing International Cosmic Ray observatory, some 4,300 m up in the Himalayas, currently has 789 scintillation detectors based upon 30-mm thick, and 0.5-m^2 plastic sheets plus PMTs as photon detectors on the surface of the ground. It is spread over an area of around 400 hectares. Its total sensitive area is some 400 m^2. Another 120 surface scintillation detectors may soon be added, almost doubling the total area covered to nearly 800 hectares. It also uses 24 Čerenkov water-based underground muon detectors with a combined collecting area of 4,500 m^2. It has detected many $\geq$100 TeV particles, including the current record holder for a cosmic ray particle from an identified astronomical source: a 450-TeV particle from the Crab nebula. IceCube (Section 1.5.2), although built for neutrino detection, can also detect the EASs from cosmic rays when used with its Čerenkov surface detector array (called 'IceTop').

The Siberian complex, Tunka Advanced Instrument for Cosmic Ray Physics and γ Astronomy (TAIGA), sited about 50 km from Lake Baikal houses five arrays. It uses (variously) scintillation, radio, Čerenkov and air-Čerenkov detectors in arrays covering up to 5 km^2 (possibly with expansion to 100 km^2 in the future) for the detection of all types of cosmic rays.

The High Altitude Water Čerenkov (HAWC)[124] Observatory achieved its first detections in 2017 and is designed to detect TeV γ-rays from their secondary particle products. It has 300 tanks each

[123] There are possible plans to expand this array which may be firmed up in the near future.

[124] Not to be confused with SOFIA's HAWC+.

containing 330 tonnes of water in a dense array covering two hectares. The Čerenkov emissions are detected by four PMTs in each tank. The first water-Čerenkov γ-ray detectors have recently been installed at Large High Altitude Air Shower Observatory (LHAASO) in Sichuan. The instrument occupies 2.2 hectares and currently uses 900 Čerenkov units together with two Čerenkov atmospheric telescopes and 80 muon detectors.

The analysis of the data from such arrays is difficult. Only a small sample, typically less than 0.01%, of the total number of particles in the shower is normally caught. The properties of the original particle have then to be inferred from this small sample. However, the nature of the shower varies with the height of the original interaction, and there are numerous corrections to be applied as discussed below. Thus, nor the observations are fitted to a grid of computer-simulated showers.[125] The original particle's energy can usually be obtained within fairly broad limits by this process, but its further development is limited by the availability of computer time and by our lack of understanding of the precise nature of these extraordinarily high-energy interactions.

Arrays, or at least several telescopes, are also needed for the atmospheric Čerenkov detectors,[126] so that pairs of stereoscopic images may be obtained and used to plot the track's direction in space accurately. The five telescopes of the HESS systems have already been mentioned. Other currently active instruments include MAGIC (Figure 1.101) sited on La Palma in the Canary Islands which, since 2009, has had two 17-m reflectors each feeding 596 PMTs and VERITAS in Arizona with four 12-m reflectors each feeding 499 PMTs.

For the future, the planned CTA (Section 1.3.2.9) will have more than 100 telescopes and sites in both the Northern and Southern Hemispheres. Prototypes for its three designs of antennae are already being constructed and tested. In space, the JEM-EUSO instrument is currently planned to go onto the ISS in the early 2020s. It will look downwards to pick up Čerenkov events over an area up to 10^6 km^2 using a 2.5-m mirror and some 6,000 PMT tubes. A possible future concept design (although a few prototype antennae have been built and operated for the last two years) is Fluorescence Detector Array of Single Pixel Telescopes (FAST).[127] This might have some 60 observing stations and be spread over a 17,000-km^2 area. The antennae might be 1.6-m segmented mirror dishes with PMTs as the photon detectors.

FIGURE 1.101 The two MAGIC telescopes on the Roque de los Muchachos, La Palma. (Courtesy of Robert Wagner, Max-Planck-Institut für Physik.)

[125] At the time of writing, Cosmic Ray Simulations for 'K'ascade (CORSIKA) and Geometry and Tracking (GEANT4) are widely used computer codes for this purpose.

[126] The possible use of air Čerenkov arrays as intensity interferometers is discussed in Section 2.5.

[127] Not to be confused with China's Five Hundred Metre Aperture Spherical Telescope (FAST).

1.4.4 Correction Factors

1.4.4.1 Atmospheric Effects

Secondary cosmic rays are produced within the Earth's atmosphere and so changes may affect the observations. The two most important variations are caused by air mass and temperature.

The air mass depends upon two factors – the zenith angle of the axis of the shower and the barometric pressure. The various components of the shower are affected in different ways by changes in the air mass. The muons are more penetrating than the nucleons and so the effect upon them is comparatively small. The electrons and positrons come largely from the decay of muons and so their variation tends to follow that of the muons. The corrections to an observed intensity are given by

$$I(P_o) = I(P)e^{K(P-P_o)/P_o} \tag{1.91}$$

$$I(0) = I(\theta)e^{K(\sec\theta-1)} \tag{1.92}$$

where P_o is the standard pressure, P is the instantaneous barometric pressure at the time of the shower, $I(P_o)$ and $I(P)$ are the shower intensities at pressures, P_o and P, respectively, $I(0)$ and $I(\theta)$ are the shower intensities at zenith angles of zero and θ, respectively and K is the correction constant for each shower component. K has a value of 2.7 for the muons, electrons and positrons and 7.6 for the nucleons. In practice a given detector will also have differing sensitivities for different components of the shower and so more precise correction factors must be determined empirically.

The atmospheric temperature changes primarily affect the muon and electron components. The scale height of the atmosphere, which is the height over which the pressure changes by a factor of e^{-1}, is given by

$$H = \frac{kR^2T}{GMm} \tag{1.93}$$

where R is the distance from the centre of the Earth, T the atmospheric temperature, M the mass of the Earth and m the mean particle mass for the atmosphere. Thus, the scale height increases with temperature, and so a given pressure will be reached at a greater altitude if the atmospheric temperature increases. The muons, however, are unstable particles and will have a longer time in which to decay if they are produced at greater heights. Thus, the muon and hence also the electron and positron intensity decreases as the atmospheric temperature increases. The relationship is given by

$$I(T_o) = I(T)e^{0.8(T-T_o)/T} \tag{1.94}$$

where T is the atmospheric temperature, T_o the standard temperature and $I(T_o)$ and $I(T)$ are the muon (or electron) intensities at temperatures, T_o and T, respectively. Since the temperature of the atmosphere varies with height, Equation 1.94 must be integrated up to the height of the muon formation to provide a reliable correction. The temperature profile will, however, be only poorly known, so it is not an easy task to produce accurate results.

1.4.4.2 Solar Effects

The Sun affects cosmic rays in two main ways. Firstly, it is itself a source of low-energy cosmic rays whose intensity varies widely. Secondly, the extended solar magnetic field tends to shield the Earth from the lower-energy primaries. Both these effects vary with the sunspot cycle and also on other timescales and are not easily predictable.

1.4.4.3 Terrestrial Magnetic Field

The Earth's magnetic field is essentially dipolar in nature. Lower-energy-charged particles may be reflected by it and so never reach the Earth at all, or particles that are incident near the equator may be channelled towards the poles. There is, thus, a dependence of cosmic ray intensity on latitude (Figure 1.102). Furthermore, the primary cosmic rays are almost all positively charged, and they are deflected so that there is a slightly greater intensity from the West. The vertical concentration (Equation 1.92 and Figure 1.103) of cosmic rays near sea level makes this latter effect only of importance for high-altitude balloon or satellite observations.

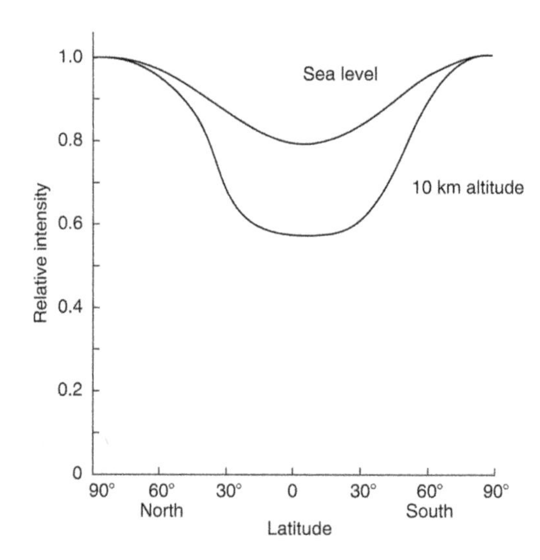

FIGURE 1.102 Latitude effect on cosmic rays.

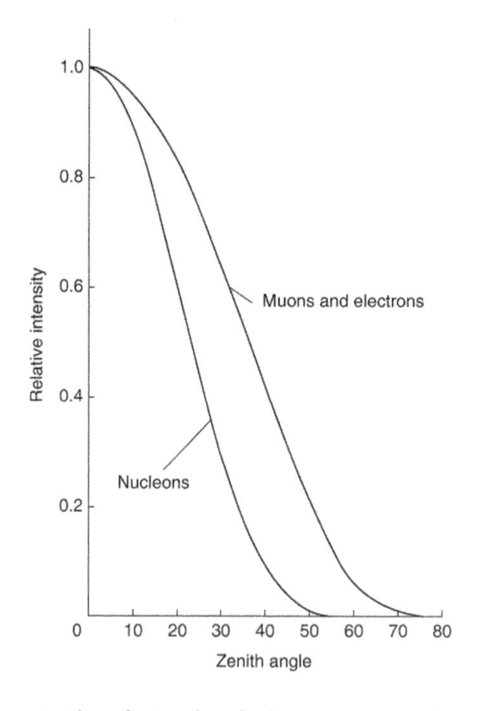

FIGURE 1.103 Zenithal concentration of extensive air shower components as given by Equation 1.92.

1.5 NEUTRINO DETECTORS

1.5.1 Background

Neutrino astronomy is somewhat in the doldrums at the moment. Only '2½' identified astronomical sources of neutrinos have ever been observed: the Sun and supernova 1987A in the LMC. The '½' being a single 300-TeV neutrino detected by IceCube (see below) which is likely to have come from the blazar Texas Survey of Radio Sources (TXS)[128] 0506 + 056 (see below). Furthermore, the solar neutrino problem (see below) was solved two and a half decades ago, by comparison, there are only minor details left to be learnt about the Sun from neutrino observations.

Since 2012, IceCube has been detecting a few individual neutrinos each year with extremely high energies; nearly 300 TeV in one case. We now know that neutrinos have mass, but that its rest value is likely to be less, perhaps very much less, than ~0.3 eV (~5 × 10^{-37} kg)[129]. Experimental limits on the mass are from 0.02 to 1 eV, whilst theoretical calculations set a maximum of 0.086 eV. Since neutrinos have mass, their velocities must be less than the speed of light (or gravity, see Section 1.6). However, even if the mass is as 'high' as 0.3 eV, a 1-PeV neutrino originating from near the edge of the visible universe (say about 4 Gpc away) would lag behind a simultaneously emitted photon by about only 30 fs (3 × 10^{-14} s), while a 40-MeV neutrino from a supernova at that same distance would lag by about 10 s. Thus, if an event emits both e-m radiation and neutrinos at the same instant, then we should expect to receive the two bursts close to simultaneously. However, even the intense event resulting from the coalescence of two neutron stars in 2017 that was 'just' 40-Mpc distant from us (Section 1.6) did not seem to be associated with any of these extreme energy neutrinos.[130]

As mentioned, though, a single neutrino seems likely to have originated from the blazar TXS 0506 + 056 (~1.75-Gpc away from us, in Orion). This was detected with an energy of ~290 TeV by IceCube on 22 September 2017. Simultaneously, outbursts from the blazar at γ-ray, X-ray, optical, and radio wavelengths were found. A surplus of a dozen or so neutrinos from the blazar's direction was later uncovered in previous records. The probability of these observations being coincidental is around 0.02%.

Possible origins of such ultra-high-energy neutrinos are still being sought (their individual energies are up to 10^8 times those from Supernova 1987A, so supernovae seem unlikely to be their sources), but their uniform distribution shows that the sources are not terrestrial nor within the solar system, perhaps not even within the Milky Way galaxy. One possibility, though, is that they could be produced during interactions between CMB microwave photons and cosmic rays (Section 1.4), whose highest observed energies can exceed 100 EeV (10^{20} eV). For TXS 0506 + 056, a suggested possible origin for the neutrino is via a collision between material in the blazar's relativistic jets and surrounding gas clouds.

Despite this disheartening background, neutrino detectors are still thriving. This is partly to study neutrinos in their own right, partly to study cosmic rays via the neutrinos that they produce during their interactions with the Earth's atmosphere and partly to continue the lookout for neutrinos from astrophysical sources. As detector sensitivities improve, more types of astronomical objects may become detectable. Thus, IceCube is expected soon to be able to pick up neutrinos from Active Galactic Nuclei (AGN), our own galaxy and may be able to detect GRBs. There should also be relic neutrinos[131] left over from the Big Bang (expected to decouple from other particles at a temperature of ~10^{10} K, about 1 s after the start of the Big Bang and now to have a temperature of ~1.9 K

---

[128] It has recently been suggested that this is not a Blazar but a 'normal' radio quasar.

[129] Particle masses are given hereinafter in this section, as is the conventional practice of subatomic physicists, in energy terms through the relationship $e = m\,c^2$. For comparison, the rest mass of the electron (Appendix A) is 511,000 eV (9.11 × 10^{-31} kg).

[130] GW170817/GRB 170817A – See Section 1.6. As noted there, the coalescences of the even more massive black holes are expected to result mainly in gravitational wave emissions.

[131] Sometimes called the 'Cosmic Neutrino Background' (CND).

with an electron-neutrino [see below] density of 10^6 to 10^7 m^{-3}). Plus, of course, another nearby supernova could happen at any moment, and perhaps neutrino detectors could give an early warning of its occurrence.[132] IceCube has also been searching for neutrinos resulting from the decay of dark matter (Section 1.7), so far without success, though this also remains as a possibility for the future. Thus, there is an ongoing astronomical interest in current and forthcoming neutrino detectors.

Wolfgang Pauli postulated the existence of the neutrino[133] in 1930 to retain the principle of conservation of mass and energy in nuclear reactions. It was necessary to provide a mechanism for the removal of residual energy in some beta-decay reactions. From other conservation laws, the neutrino's properties could be defined quite well; zero charge, zero or very small rest mass, zero electric moment, zero magnetic moment and a spin of one-half. More than a quarter of a century passed before the existence of this hypothetical particle was confirmed experimentally (in 1956). The reason for the long delay in its confirmation lies in the low probability of the interaction of neutrinos with other matter (neutrinos interact via the weak nuclear force only). Thus, out of 10 billion neutrinos originating at the centre of the Sun, only one will interact with any other particle during the whole of their 700,000-km journey to the surface of the Sun. The interaction probability for particles is measured by their cross-sectional area for absorption, σ, given by

$$\sigma = \frac{1}{\lambda N} \tag{1.95}$$

where N is the number density of target nuclei and λ is the mean free path of the particle. Even for high-energy neutrinos, the cross section for the reaction

$$\nu + {}^{37}_{17}\mathrm{Cl} \rightarrow {}^{37}_{18}\mathrm{Ar} + e^{-} \tag{1.96}$$

that was originally used for their detection (Section 1.5.2.3) is only 10^{-46} m^2 (Figure 1.104), so that such a neutrino would have a mean free path of over 1 parsec even in pure liquid ${}^{37}_{17}\mathrm{Cl}$.

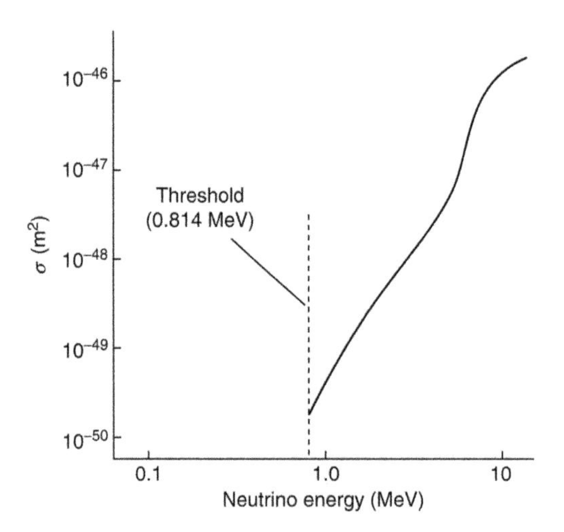

FIGURE 1.104 Neutrino absorption cross sections for the ${}^{37}_{17}\mathrm{Cl} \rightarrow {}^{37}_{18}\mathrm{Ar}$ reaction.

[132] Calculations suggest that current and planned neutrino detectors could pick up an increased neutrino flux from a pre-supernova up to two hours before the explosion and out to a distance of 1 kpc. To put this 'hope' in perspective there has been no known supernova closer than 2 kpc (the Crab) for at least the last 2,000 years. Several neutrino detectors are used to operate the Supernova Early Warning System (SNEWS). This however, whilst providing an alert to a supernova some hours ahead of when its light and other em emissions become observable, would be triggered by the neutrino emissions from the actual supernova.

[133] Not to be confused with the neutralino (Section 1.7) or another group of hypothetical particles called the 'sterile neutrinos' (Section 1.7).

Three varieties (or flavours) of neutrino are known, plus (possibly)[134] their anti-particles. The electron neutrino, ν_e, is the type originally postulated to save the β reactions and which is, therefore, involved in the archetypal decay: that of a neutron

$$n \rightarrow p^+ + e^- + \tilde{\nu}_e \tag{1.97}$$

where $\tilde{\nu}_e$ is an anti-electron neutrino. The electron neutrino is also the type commonly produced in nuclear fusion reactions and, hence, to be expected from the Sun and stars. For example, the first stage of the proton-proton cycle is

$$p^+ + p^+ \rightarrow {}^2_1\mathrm{H} + e^+ + \nu_e \tag{1.98}$$

and the second stage of the carbon cycle

$${}^{13}_{7}\mathrm{N} \rightarrow {}^{13}_{6}\mathrm{C} + e^+ + \nu_e \tag{1.99}$$

and so on.

The other two types of neutrino are the muon neutrino, ν_μ, and the tau neutrino, ν_τ. These are associated with reactions involving the heavy electrons called 'muons' and 'tauons'. For example, the decay of a muon

$$\mu^+ \rightarrow e^+ + \nu_e + \tilde{\nu}_\mu \tag{1.100}$$

involves an anti-muon neutrino amongst other particles. The muon neutrino was detected experimentally in 1962, but the tau neutrino was not found until 2000.

The first astronomical neutrino detector started operating in 1968 (Section 1.5.2.3), and it quickly determined that the observed flux of solar neutrinos was too low by a factor of 3.3 compared with the theoretical predictions. This deficit was confirmed later by other types of detectors such as Soviet-American Gallium Experiment (SAGE), Gallium Experiment (GALLEX) and Kamiokande and became known as the solar neutrino problem. However, these detectors were only sensitive to electron neutrinos. Then, in 1998, the Super-Kamiokande detector found a deficit in the number of muon neutrinos produced by cosmic rays within the Earth's atmosphere. Some of the muon neutrinos were converting to tau neutrinos during their flight time to the detector. The Super Kamiokande, Sudbury Neutrino Observatory (SNO) and Oscillation Project with Emulsion Tracking Apparatus (OPERA) detectors have since confirmed that all three neutrinos can exchange their identities (Figure 1.105)[135]. The solar neutrino problem has thus disappeared because two-thirds of the electron neutrinos

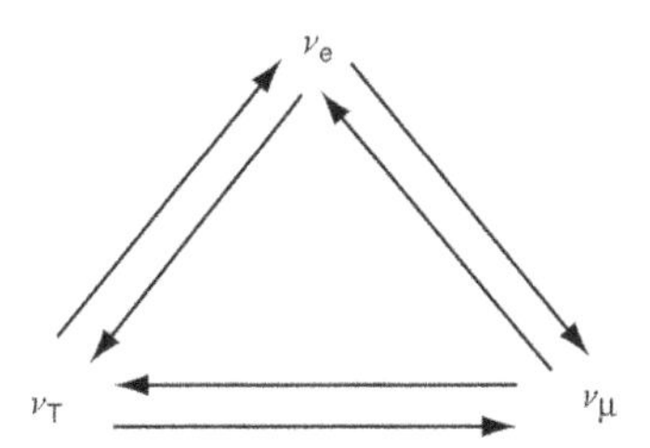

FIGURE 1.105 Neutrino oscillation.

[134] It is possible that a neutrino forms its own anti-particle. But only if its rest mass is zero. Such particles are known as Majorana particles.

[135] The inter-conversion of neutrino types (neutrino oscillation) is well confirmed experimentally, but its theoretical basis is less certain. The oscillations are usually taken to imply that neutrinos have mass; however, there is also the possibility that they arise from Lorentz invariance violations. Lorentz invariance requires the laws of physics to be the same for all observers moving at constant velocities with respect to each other and underlies Special Relativity. The possibility of Lorentz invariance being violated is the subject of much current research and speculation, but further details are beyond the scope of this book and are left to interested readers to research for themselves.

produced by the Sun have changed to muon and tau neutrinos by the time that they reach the Earth and so were not detected during the early experiments. The current IceCube neutrino detector and more probably its 2022–2023 upgrade (Section 1.5.2) are soon likely to detect high-energy astrophysical tau neutrinos arising from oscillations of the high-energy electron neutrinos already detected.

The flux of solar neutrinos (of all types) at the Earth is now measured to be about 6×10^{14} m^{-2} s^{-1}, in good agreement with the theoretical predictions. The experiments also show that neutrinos have mass (see above), although only upper limits have been determined so far. The tightest constraints on neutrino masses come from cosmology and suggest that the combined masses of all three varieties of neutrino to be no more than 0.3 eV (the mass of an electron for comparison is 511 keV).

1.5.2 Neutrino Detectors

Neutrinos from the Sun were first detected in 1968 by Davis' $^{37}_{17}$Cl-based radio chemical experiment[136] (Section 1.5.2.3). Today's main second-generation instruments and most of those being considered for the future are based upon observing the Čerenkov radiation in water or ice produced by the super-photic charged subatomic particles resulting from neutrino interactions. The reason for the use of water or ice instead of chlorine is that the former are cheap, or even, in the case of water in the oceans or ice in the Antarctic ice cap, free. The neutrino interactions with matter are so weak that vast amounts of material have to be used if they are to be observed with reasonable frequency. Davis used more 600 tonnes of the chlorine-containing compound tetrachloroethene in his detectors. IceCube (see below) monitors a cubic kilometre (10^9 tonnes) of ice but is likely to be regarded by researchers within a decade or so as a 'small' detector. A useful measure for comparing neutrino detector performances is the neutrino effective area. This is the area of a hypothetical detector with 100% efficiency of detection of neutrinos that would detect the same number of neutrinos from the far side of the Earth as does the instrument in question. For the current IceCube configuration (Section 1.5.2.1) , this varies from ~100 m^2 at 100 TeV to ~1 m^2 at 1 TeV (cf. the actual surface area monitored by IceCube is about 10^6 m^2).

1.5.2.1 Direct Čerenkov Detectors

1.5.2.1.1 Optical Detectors

After Davis' first neutrino detector, the next working neutrino telescopes did not appear for nearly two decades. Neither of these instruments, the Kamiokande[137] detector buried 1 km below Mount Ikenoyama in Japan nor the Irvine-Michigan- Brookhaven (IMB) detector, 600 m down a salt mine in Ohio, were built to be neutrino detectors.[138] Both were originally intended to look for proton decay and were only later converted for use with neutrinos. The design of both detectors was similar (Figure 1.106), they differed primarily only in size (Kamiokande, 3,000 tonnes and IMB, 8,000 tonnes).

The principle of their operation was the detection of Čerenkov radiation from the products of neutrino interactions. These may take direct two forms, electron scattering and inverse beta-decay. In the former process, a collision between a high-energy neutrino and an electron sends the latter off at a speed in excess of the speed of light in water (225,000 km s^{-1}) and in roughly the same direction as the neutrino. All three types of neutrinos scatter electrons, but the electron neutrinos are 6.5 times more efficient at the process. In inverse beta-decay, an energetic positron is produced via the reaction

$$\tilde{\nu}_e + p \rightarrow n + e^+ \tag{1.101}$$

[136] Raymond Davis, 1914–2006, awarded 2002 Nobel Prize for Physics.

[137] Kamioka Neutrino Detector.

[138] The large depths of these and other neutrino detectors are needed so that the overlying material acts as a shield against non-neutrino interactions.

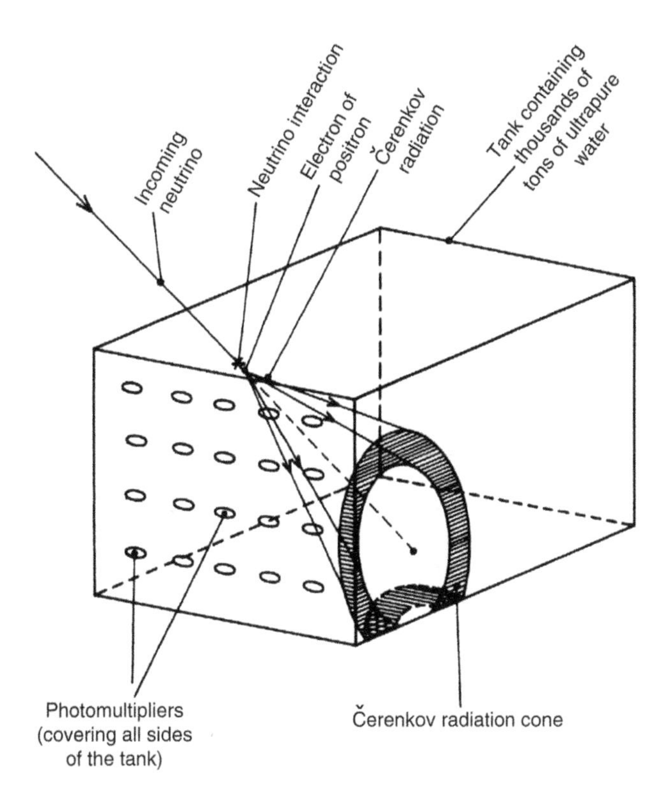

FIGURE 1.106 Principle of neutrino detection by water-based detectors.

Inverse beta-decays are about 100 times more probable than the scattering events, but the positron can be emitted in any direction and thus, gives no clue to the original direction of the incoming neutrino.

Both positron and electron, travelling at super-photic speeds in the water, emit Čerenkov radiation (Section 1.4) in a cone around their direction of motion. That radiation is then picked up by arrays of PMTs that surround the water tank on all sides. The pattern of the detected radiation can be used to infer the energy of the original particle and in the case of a scattering event, also to give some indication of the arrival direction. The minimum detectable energy is around 5 MeV due to background noise.

Both detectors were fortunately in operation in February 1987 and detected the burst of neutrinos from the supernova in the LMC (SN1987A). This was the first and so far, only[139] detection of neutrinos from a clearly identified astronomical source other than the Sun, and it went far towards confirming the theoretical models of supernova explosions. Both IMB and Kamiokande have now ceased operations,[140] but a number of detectors operating on the same principle are functioning or are under construction.

Super-Kamiokande contains 50,000 tonnes of pure water and uses 13,000 PMT tubes in a tank buried 1 km below Mount Ikenoyama in Japan. Super-Kamiokande is able to detect muon neutrinos as well as electron neutrinos. The muon neutrino interacts with a proton to produce a relativistic muon. The Čerenkov radiation from the muon produces a well-defined ring of light. The electron resulting from an electron neutrino scattering event by contrast generates a much fuzzier ring of light. This is because the primary electron produces gamma rays that in turn produce electron-positron pairs. The electron-positron pairs then create additional Čerenkov radiation cones, thus blurring the ring resulting from the primary electron.

[139] Except, perhaps, for the blazar TXS 0506+056 – Section 1.5.1 and below.

[140] The IMB PMTs were reused as an active shield for Super-Kamiokande.

This ability to distinguish between electron and muon neutrinos enabled Super-Kamiokande to provide the first evidence of neutrino oscillations (see background section above). Muon neutrinos produced in the Earth's atmosphere by high-energy cosmic rays should be twice as abundant as electron neutrinos.[141] But in 1998 Super-Kamiokande found that although this was the case for neutrinos coming from above (a distance of about 60 km), there were roughly equal numbers of muon neutrinos and electron neutrinos coming from below (i.e., having travelled 12,000 km across the Earth). Since there were no extra electron neutrinos, some of the muon neutrinos must be oscillating to tau neutrinos during their flight time across the Earth to the detector. This result was confirmed in 2000 by a 30% shortfall in the number of muon neutrinos observed from an artificial source at the KEK laboratory at Tsukuba some 250 km away from the detector. Further confirmation of the oscillation of all three types of neutrino has recently come from comparison of the Super-Kamiokande and SNO results and the Kamioka Liquid Scintillator Antineutrino Detector (KamLAND) and OPERA experiments. The designer of the Kamiokande and Super-Kamiokande detectors, Matatoshi Koshiba, shared the Nobel Physics Prize in 2002 for his work. Hyper Kamiokande is a planned replacement for Super-Kamiokande due to start operations around 2025 and with some 20 times Super-Kamiokande's volume of water (almost a million tonnes).

Man-made water containers significantly larger than Hyper Kamiokande are probably impractical. However, much greater quantities of water (or ice, which is equally good) may be monitored using parts of lakes, the sea or the Antarctic ice cap. These types of neutrino telescope are generally most sensitive to the high-energy (TeV) neutrinos because of the large spacing (tens of metres) between their PMTs. For such water-based Čerenkov telescopes, PMTs attached to long cables are anchored to the sea or lake bed and float in the water. Thus, Lake Baikal in Southeast Siberia is to host the Gigaton Volume Detector (GVD) which is scheduled for completion in the 2020s. This is expected to comprise 27 clusters of eight strings with 48 PMTs on each string. Currently, three detector strings have been deployed. The area covered will be about 2 km^2, and the detectors will range in depth from 600 to 1,300 m below the surface. The total volume to be monitored will thus be 1.5 km^3. The GVD will also primarily detect upwardly moving muons. GVD is an upgrade of the NT200+[142] instrument which commenced observations in 2005. NT200+ currently monitors a volume of about 0.01 km^3 using 228 PMTs on 11 strings.

Cubic Kilometre Neutrino Telescope (KM3NeT) is a planned development from existing neutrino detectors, such as ANTARES.[143] ANTARES was completed in 2008 and is some 40 km off the coast near Toulon. It is 150-m across by 300-m high and uses 900 PMTs to observe a volume of 0.01 km^3 at a depth of 2.5 km. Its pointing accuracy, as shown by its detection of the cosmic ray shadow of the Moon, is 0.73°. KM3NeT will also be sited in the Mediterranean at depths of up to 3.5 km. It will have three sites, offshore from France, Greece and Italy. Its detectors will comprise optical modules each containing 31 PMTs with 18 optical modules to a string and 115 strings to a detector block. Astroparticle Research with Cosmics in the Abyss (ARCA) – offshore from Italy will have two such detector blocks and be optimised for TeV to PeV neutrino detection with string separations of 90 m. Oscillation Research with Cosmics in the Abyss (ORCA), offshore from France, has been in operation since 2017. It has one detector block and is optimised for GeV atmospheric neutrino detection with string separations of 20 m. The ultimate instrument (KM3NeT 2.0) is planned to have six detector blocks and to monitor about 1 km^3 of water. IceCube, ANTARES, KM3NeT and the Lake Baikal instruments collaborate to form the Global Neutrino Network (GNN).

---

[141] The cosmic ray interaction first produces (amongst other particles) pions. A positively charged pion usually then decays in 26 ns to a positive muon and a muon neutrino and the muon decays in 2.2 μs to a positron, a muon anti-neutrino and an electron neutrino. A negative pion likewise decays to an electron, a muon neutrino, a muon anti-neutrino and an electron anti-neutrino. There should thus be two muon neutrinos to each electron neutrino in an EAS (Section 1.4).

[142] NT stands for 'Neutrino Telescope'.

[143] Other earlier water-based detectors include Deep Underwater Muon and Neutrino Detector project (DUMAND)–cancelled, Megaton Mass Physics (MEMPHYS), Neutrino Ettore Majorana Observatory and Neutrino Mediterranean Observatory (NEMO) and Neutrino Extended Submarine Telescope with Oceanographic Research (NESTOR).

A neutrino detector based upon 1,000 tonnes of heavy water (D_2O) operated at Sudbury in Ontario from 1999 to 2007. Known as SNO, the heavy water was contained in a 12-m diameter acrylic sphere immersed in an excavated cavity filled with 7,000 tonnes of highly purified normal water. The normal water shielded the heavy water from radioactivity in the surrounding rock. It was located 2 km down within the Creighton copper and nickel mine and used 9600 PMTs to detect Čerenkov emissions within the heavy water. Neutrinos may be detected through electron scattering as with other water-based detectors, but the deuterium in the heavy water provides two other mechanisms whereby neutrinos may be found. The first of these senses just electron neutrinos. An electron neutrino interacts with a deuterium nucleus to produce two protons and a relativistic electron

$$\nu_e + {}^2_1\mathrm{H} \rightarrow p^+ + p^+ + e^- \tag{1.102}$$

and the electron is observed via its Čerenkov radiation. This mechanism is called the 'charged current reaction' because the charged W boson mediates it. The second mechanism is mediated by the neutral Z boson and is, thus, called the 'neutral current reaction'. In it, a neutrino of any type simply splits the deuterium nucleus into its constituent proton and neutron. The neutron is thermalised in the heavy water and eventually combines with another nucleus with the emission of gamma rays. The gamma rays in turn produce relativistic electrons via Compton scattering, and these are finally detected from their Čerenkov radiation. Although the neutron can combine with a deuterium nucleus, the capture efficiency is low so that 75% of the neutrons will escape from the detector. The SNO detector could, therefore, add two tonnes of salt to the heavy water to enable the neutron to be captured more easily by a ${}^{35}_{17}\mathrm{Cl}$ nucleus (converting it to ${}^{36}_{17}\mathrm{Cl}$) and reducing to 17% the number of escaping neutrons. The thermalised neutrons can also be detected using ${}^3_2\mathrm{He}$ proportional counters (Section 1.3). The neutron combines with the ${}^3_2\mathrm{He}$ to produce a proton and tritium and these then ionise some of the remaining gas to give an output pulse.

Since SNO could detect the number of electron neutrinos separately from the total for all types of neutrino, it provided the definitive data for solving the solar neutrino problem (see above). The standard solar model predicts that SNO should detect about 30 charged current or neutral current reactions per day and about three electron scattering events. However, the reality of the oscillation of neutrinos between their different type and, hence, the solution to the solar neutrino problem was demonstrated by the first results from SNO. It was initially operated in the charged current mode (i.e., without salt being added to the heavy water) to detect just the solar electron neutrinos. This gave a flux that was lower than that measured by Super-Kamiokande because the latter also detects a proportion of the muon and tau neutrinos as well as the electron neutrinos. Some solar electron neutrinos must, therefore, have oscillated to the other types during their journey from the centre of the Sun. The SNO then made measurements of the total solar neutrino flux using the neutral current mode (i.e., with salt added to the heavy water) and in April 2002 finally demonstrated that the total neutrino flux is as predicted by the standard solar model. The SNO ceased operations in May 2007 and the heavy water was drained from its tank to be reused elsewhere.

IceCube, which started operations in 2011 in Antarctica and followed on from the Antarctic Muon and Neutrino Detection Array (AMANDA),[144] is currently the largest neutrino detector. It uses photomultipliers to detect the Čerenkov radiation produced within the ice. Each of its 5,160 PMTs is housed in a transparent pressure vessel[145] along with its electronics. Sixty of these sensors at a time are suspended from a cable that is lowered down a hole in the ice (Figure 1.107) to a depth between 1.45 and 2.45 km below the surface.[146] There are 86 such cables spread over an area of about 1 km^2 at 125-m spacings. IceCube therefore monitors about 0.9 km^3 of ice for Čerenkov radiation. IceCube

[144] AMANDA was decommissioned in 2009.

[145] Also known as Digital Optical Modules (DOMs).

[146] The detectors are placed at such depths because the ice there is clear and almost free from air bubbles. Also the layer of ice above the detectors shields them from most of the charged particles in extensive air showers (except for the muons).

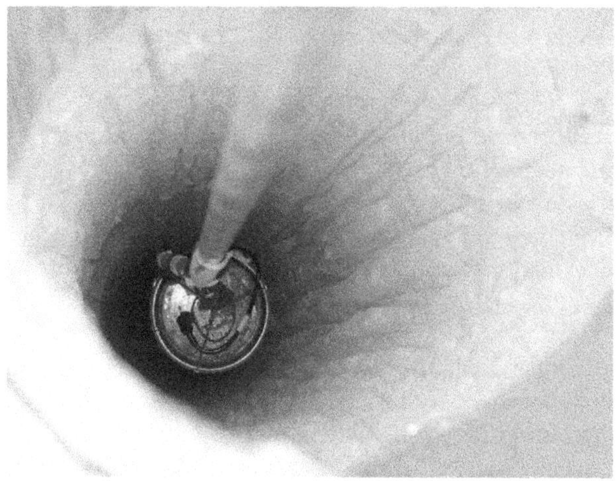

FIGURE 1.107 Top, One of the pressure spheres containing the photomultiplier and electronics for the IceCube neutrino telescope ready to be lowered into position. (Courtesy of IceCube Collaboration/National Science Foundation [NSF]). Bottom, The IceCube pressure sphere on its way down through the hole drilled into the Antarctic icecap. (Courtesy of IceCube Collaboration/NSF.)

also monitors a 0.002 km^3 volume with more closely spaced detectors. Called 'DeepCore', this latter instrument extends IceCube's sensitivity down to 10 GeV. Additionally, EASs are monitored by IceTop, an array of 160 tanks each containing 2.5 tonnes of ice on the surface of the ice cap, above the main IceCube instrument and equipped as Čerenkov detectors.

IceCube is most sensitive to muons and hence to muon neutrinos. Muons, however, are also a large component of EASs and there are around a million muons produced by cosmic rays for every muon produced by neutrinos. IceCube, therefore, ignores the muons moving downwards that have come from cosmic rays in the atmosphere above the instrument. Instead it selects muons that are moving upwards. These will have been produced by muon neutrinos that have come from the opposite side of the Earth and interacted to produce the muons only when close to or inside the monitored volume (thus, in effect using the whole Earth as a filter or shield). The neutrinos coming from the other side of the Earth, however, still mostly originate from cosmic ray interactions. Typically, IceCube detects around three upwardly moving neutrinos per hour. IceCube has so far detected the shielding effect of the Moon on cosmic ray protons and a slight anisotropy in the distribution of secondary cosmic ray muons. Neutrinos from more distant astronomical sources can only be identified if there is a noticeable change in the flux or the energy distribution or if a particular direction in space suddenly seems to be favoured, although, as mentioned previously, the occasional high-energy neutrinos (TeV to PeV) that are detected most probably come from distant deep space sources such as, perhaps, the blazar TXS 0506 + 056.

The proposed IceCube-Gen2 upgrade, due to take place in 2022–2023, will increase the volume of ice monitored to about 8 km^3 using an additional 120 cables at 240-m spacings and with 80 sensors per cable. A slightly irregular elliptical area some 5 km along its major axis will then be covered and will be surrounded by circular areas some 12- and 20-km across, containing cosmic ray detectors (Section 1.4) and radio neutrino detectors (see below), respectively. This upgrade should increase the neutrino detection rate by about a factor of ten and improve on IceCube's sensitivity by about a factor of five. Neutrinos with energies up to the EeV region should be detectable.

Precision IceCube Next Generation Upgrade (PINGU) is a proposed infill to the existing IceCube instrument that will be able to monitor neutrinos with energies of a few GeV and will aim to study neutrino properties as well as looking for solar dark matter weakly interacting massive particles (WIMPs; see Section 1.7) and perhaps investigating the Earth's core. It would be at the centre of DeepCore with 60 sensors at 5-m spacings. The existing IceCube instrument would be used as an active shield against cosmic ray muons. A prototype for a possible large-scale detector in the Pacific (Strings for Absorption Length in Water [STRAW]) using six detector modules on two mooring strings has been operating since 2018.

1.5.2.1.2 Radio Detectors

Although the peak Čerenkov emission produced by multi-TeV particles is in the optical region, radio waves are produced as well (Equation 1.88), mostly in the 100-MHz to 1-GHz (3-m to 300-mm) region as predicted by Gurgen Askaryan in 1962. At the temperature of the Antarctic ice cap (about 220 K), the ice is almost completely transparent to metre-wavelength radio waves and so these can be detected by receivers at the surface. Experiments in Antarctica and Greenland have shown attenuation lengths[147] in the ice of about 1 km at 75 MHz and 1.5 km at 300 MHz; for comparison, alternative materials suggested for these types of detections (see below) have attenuation lengths of about 60–70 m at 300 MHz (rock salt) and 200 m at 100 MHz (lunar regolith).

Askaryan Radio Array (ARA) is currently being deployed near the South Pole to detect 10-PeV to 10-EeV neutrinos. It is planned to install a total of 37 stations over a 100-km^2 area and so to monitor thousands of cubic kilometres of the ice cap. At the time of writing, three of the stations are operating with three more under construction. Each station has 16 antennas at a depth of 200 m and operates over the 200- to 850-MHz band. Antarctic Ross Ice-shelf Antenna Neutrino Array (ARIANNA) could be up to 1,000 km^2 in area and use a higher density of antennas than ARA. It is now in the planning stages for possible construction on the Ross ice-shelf. A prototype with eight sets of four dipole antennas has been operating since 2015 over the 50-MHz to 1-GHz region. The planned Greenland Neutrino Observatory (GNO), which has had a prototype detector installed, is also intended to detect high-energy neutrinos via their radio Askaryan emissions. The similar Radio Neutrino Observatory (RNO) is currently a proposal under consideration to operate in conjunction with IceCube.

---

[147] The distance over which absorption in the material reduces the radio wave's electrical field strength by about 63% ($= 1 - 1/e$).

A radio receiver borne by a high-altitude balloon would enable the entire Antarctic ice cap to be watched: a volume of around 1.5 million km^3. In 2006, 2008, and 2014 just such a receiver was flown. Called ANITA, it operated over the 200-MHz to 1.2-GHz band with up to 48 antennae. During the flights at a height of 35 km, it picked up millions of events. Most of these events though, were background non-neutrino pulses. In the future, the Extravolt Antenna (EVA) is hoped to improve on ANITA by a factor of 100 using a part of the surface of its super-pressure balloon as a reflector to increase its sensitivity.

Another radio-based neutrino detector will potentially monitor a volume of up to 20 million km^3. This instrument is to look for radio pulses produced by neutrino interactions within the Earth's Moon. The pulses originate via the Askaryan mechanism within the lunar rocks and may be expected to be detectable from any part of the Moon's visible surface. The fourteen 25-m dishes of the WSRT have recently conducted a, so far unsuccessful, search for these pulses. The Goldstone and Parkes radio telescopes have searched for lunar Askaryan emissions in the past, and it is planned to search for them in the near future with LOFAR and in due course with the SKA. A related proposal, called Passive Radio Ice Depth Experiment (PRIDE), would be to detect the neutrino Asakaryan radio emissions from the ice on outer planet moons, such as Enceladus, Europa or Ganymede, using radio receivers on board nearby fly-by or orbiting spacecraft. The primary objective of the concept, however, would be to measure the ice sheets' thicknesses rather than to study the neutrinos.

1.5.2.2 Indirect Čerenkov Detectors

High-energy neutrinos can interact with other subatomic particles to produce EASs (Section 1.4), either directly from the interaction or following the decay of particles produced within the interaction. These showers may then be detected from their Čerenkov radiation. The EASs resulting from cosmic rays are to be found high in the atmosphere and generally moving in a downwards direction. If an EAS were to be found at low altitude or moving horizontally or even in an upwards direction, then it almost certainly must have originated from a neutrino interaction. Such detections are currently being sought, but so far without success using existing cosmic ray observatories such as the Pierre Auger Observatory and the Telescope Array (Section 1.4). The partially completed Taiwan Astroparticle Radiowave Observatory for Geo-Synchrotron Emissions (TAROGE) array is intended to detect cosmic rays with energies above 3×10^{18} eV by detecting the radio emissions from EASs both directly and after reflection from the sea. It uses dipole array antennas and operates from 110 to 300 MHz.

The planned Giant Radio Array for Neutrino Detection (GRAND) in China will comprise some 200,000 antennas covering about 200,000 km^2 and probably be sited in the mountainous region between Lake Baikal and the Himalayas. Its aim will be to detect EASs resulting from the decay of tauons in the Earth's atmosphere from their radiation over the 50- to 200-MHz region. The tauons are in turn produced by high-energy tau neutrino interactions below the ground. A 300-antenna prototype is under construction at the time of writing.

1.5.2.3 Radiochemical Detectors

These detectors operate on a quite different principle from that of the Čerenkov detectors. The neutrino is absorbed by a suitable atomic nucleus, changing it to the nucleus of a radioactive isotope of a different element. The radioactive decay is then separately detected and acts to show the occurrence of the initial neutrino interaction. These detectors are small in volume compared with the ones just discussed because the material involved is expensive (sometimes *very* expensive) compared with water or ice. However, they are still in use and also have interest as the first type of detector to register extraterrestrial neutrinos.

The first neutrino telescope to be built was designed to detect electron neutrinos through the chlorine-to-argon reaction given in Equation 1.96. The threshold energy of the neutrino for this reaction is 0.814 MeV, so that 80% of the neutrinos from the Sun that are detectable by this experiment arise from the decay of boron to beryllium (Figure 1.104) in a low probability side chain of the proton-proton (pp) reaction

$$^{8}_{5}\mathrm{B} \rightarrow {}^{8}_{4}\mathrm{Be} + e^{+} + \nu_{e} \tag{1.103}$$

The full predicted neutrino spectrum of the Sun, based upon the conventional astrophysical models and Borexino data, is shown in Figure 1.108.

The chlorine-based detector operated from 1968 to 1998 under the auspices of Ray Davis Jr and his group and so is now largely of historical interest, although many of the operating principles and the precautions that it needed apply to other types of neutrino detector. It consisted of a tank containing more than 600 tonnes of tetrachloroethene (C_2Cl_4) located 1.5 km down in the Homestake gold mine in South Dakota. About one chlorine atom in four is the $^{37}_{17}\mathrm{Cl}$ isotope; so on average, each of the molecules contains one of the required atoms, giving a total of about $2 \times 10^{30}\, ^{37}_{17}\mathrm{Cl}$ atoms in the tank. Tetrachloroethene was chosen rather than liquid chlorine for its comparative ease of

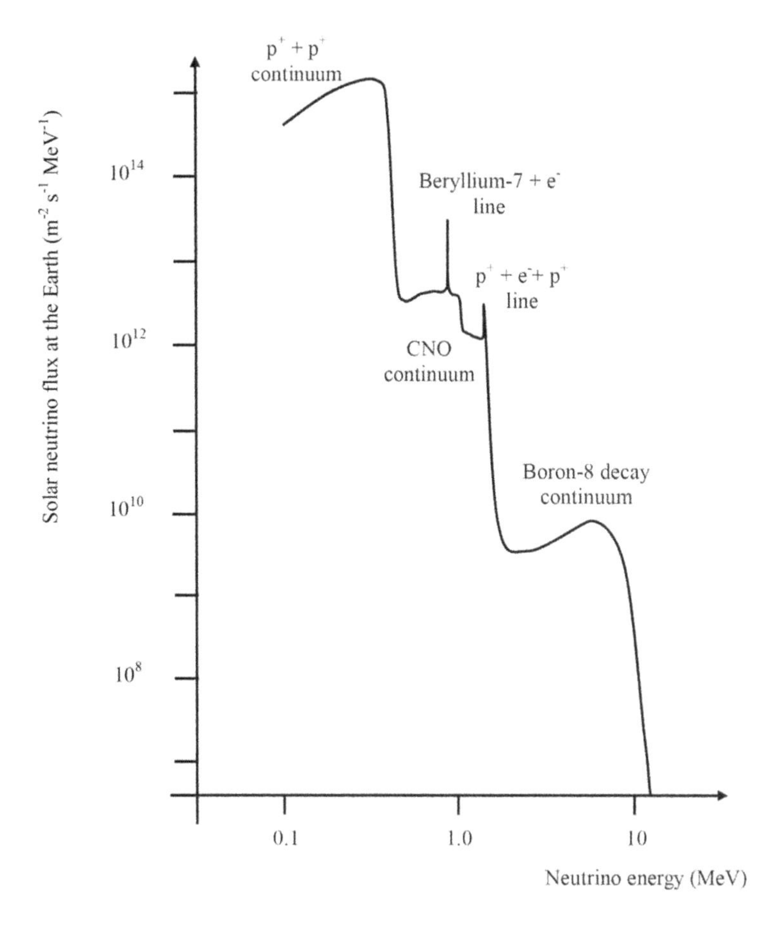

FIGURE 1.108 Solar neutrino spectrum. Only 1% of solar neutrinos have energies above 2 MeV. Note: A predicted low probability branch of the set of proton-proton (pp) reactions is the direct combination of a proton with a $^{3}_{2}He$ nucleus, which is known as the Hep reaction. It should produce neutrinos up to the pp reactions' energy limit of 26.7 MeV, but these have yet to be detected.

handling, and because of its cheapness, it is a common industrial solvent. The interaction of a neutrino (Equation 1.96) produces a radioactive isotope of argon. The half-life of the argon is 35 days, and it decays back to $^{37}_{17}Cl$ by capturing one of its own inner orbital electrons, at the same time ejecting a 2.8-keV electron. The experimental procedure was to allow the argon atoms to accumulate for some time, then to bubble helium through the tank. The argon was swept up by the helium and carried out of the tank. It was separated from the helium by passing the gas stream through a charcoal cold trap. The number of $^{37}_{18}Ar$ atoms was then counted by detecting the 2.8-keV electrons as they were emitted.

The experimental procedure had to be carried out with extraordinary care if the aim of finding a few tens of atoms in 10^{31} were to succeed (similar precautions have to be taken for all neutrino detectors). Preeminent amongst the precautions were:

1. Siting the tank 1.5-km below the ground to shield it against cosmic rays and natural and artificial radiation sources.
2. Surrounding the tank with a thick water jacket to shield against neutrons.
3. Monitoring the efficiency of the argon extraction by introducing a known quantity of $^{36}_{18}Ar$ into the tank before it is swept with helium and determining the percentage of this recovered.
4. Shielding the sample of $^{37}_{18}Ar$ during counting of its radioactive decays and using anti-coincidence techniques to eliminate extraneous pulses.
5. Use of long integration times to accumulate the argon in the tank.
6. Measurement of and correction for the remaining noise sources.

The Homestake detector typically caught one neutrino every other day. This corresponds to a detection rate of 2.23 ± 0.22 SNU.[148] The expected rate from the Sun is 7.3 ± 1.5 SNU, and the apparent disparity between the observed and predicted rates prior to the confirmation of the oscillation of neutrino types gave rise to the previously mentioned solar neutrino problem. In 2002, Ray Davis shared the Nobel Physics Prize for his work on neutrinos.

There have been two radiochemical neutrino detectors based upon gallium. The SAGE detector used 60 tonnes of liquid metallic gallium buried beneath the Caucasus Mountains at the Baksan laboratory and detected neutrinos via the reaction

$$^{71}_{31}Ga + \nu_e \rightarrow ^{71}_{32}Ge + e^- \tag{1.104}$$

The germanium product of the reaction is radioactive, with a half-life of 11.4 days. It is separated out by chemical processes on lines analogous to those of the chlorine-based detector. The detection threshold is only 0.236 MeV so that the p-p neutrinos (Figure 1.108) are detected directly. The other gallium-based detector, Gallium Neutrino Observatory (GNO), was based upon the same detection mechanism as SAGE, but used 30 tonnes of gallium in the form of gallium trichloride ($GaCl_3$). It replaced the GALLEX experiment and was located within the Gran Sasso Tunnel under the Apennines, 150 km East of Rome. Both detectors have measured a solar neutrino flux of around 50% of that predicted by the standard solar model and are now decommissioned. Other radiochemical bases for past neutrino detectors or possible future neutrino detectors include the conversion of $^{7}_{3}Li$ to $^{7}_{4}Be$, $^{37}_{19}K$ to $^{37}_{18}Ar$, $^{115}_{49}In$ to $^{115}_{50}Sn$, $^{127}_{53}I$ to $^{127}_{54}Xe$ and $^{176}_{70}Yb$ to $^{176}_{71}Lu$.

Another detector, perhaps best covered in this section, is based upon the interaction of neutrinos with tritium (see also Karlsruhe Neutrino Experiment [KATRIN] in Section 1.5.2.7). Called Princeton Tritium Observatory for Light, Early Universe Massive Neutrino Yield (PTOLEMY), it is hoped that it will detect the neutrinos left over from the Big Bang (Section 1.5.1) and also to be

[148] Solar Neutrino Unit; 1 SNU is 10^{-36} captures per second per target atom.

able to measure their masses. Tritium $\left({}_1^3\mathrm{H}\right)$ is radioactive and undergoes β-decay to helium-3 (${}_2^3\mathrm{He}$) with a half-life of 12.3 years. Normally, such a decay produces an electron and an anti-electron neutrino that share the released energy of 18.6 eV. At maximum, therefore, the electron's energy can be 18.6 eV. However, a related reaction is for the tritium nucleus to capture an incoming electron neutrino, converting to only two particles; an ${}_2^3\mathrm{He}$ nucleus and an electron. The electron then has all the 18.6 eV from the reaction *plus* the incoming neutrino's energy (kinetic energy plus mass converted to energy – about 0.1 eV). PTOLEMY aims to measure the electron's energy precisely (±0.15 eV) using cryogenic calorimeters (Sections 1.2–1.4) in the hope of finding some of these electrons with energies above the beta-decay limit. It is currently being tested with 10 μg of tritium, and it is estimated that detection rates of several per year may be achieved with 100 g or so of tritium.[149]

1.5.2.4 Scintillator-Based Detectors

High-energy photons and particles resulting from neutrino interactions may be observed using scintillation detectors (Section 1.3). The Soudan II detector (1989–2001) was located 700-m below the surface in an old iron mine at Soudan in Minnesota and used 1,000 tonnes of alternating sheets of steel and a plastic scintillator (sometimes called a 'Tracking Calorimeter'). It was originally built to try and detect the decay of protons but also detected the muons resulting from muon neutrino interactions within the steel plates. Additionally, in the Soudan mine, are the two tracking calorimeters forming the Main Injector Neutrino Oscillation Search (MINOS) far detector. Each comprises more than 3,000 tonnes of alternating 25-mm thick steel and scintillator sheets. Together with the 1,000 tonne MINOS near detector at Fermilab, it operates to refine the measurements of the parameters of neutrino oscillations by observing a beam of muon neutrinos produced at Fermilab (in Illinois, some 730 km away).

Boron Experiment (Borexino) is an experiment currently aimed at detecting the solar neutrino emission line at 863 keV arising from the decay of ${}_4^7\mathrm{Be}$ to ${}_3^7\mathrm{Li}$ (Equation 1.103). It detects all types of neutrinos via their scattering products within 300 tonnes of a benzene-type scintillator observed by 2,200 PMTs and has a detection threshold of 250 keV. The scintillator is manufactured from ancient oil deposits so that as many as possible of the originally present radioactive carbon-14 nuclei[150] have decayed away, so reducing the instrument's background noise levels. The active scintillator is surrounded by 1,000 tonnes of a liquid buffer and that in turn by 2,400 tonnes of water. Additionally, 400 PMTs pointing outwards into the water buffer provide an active shield. It is located in the Grand Sasso tunnel in Italy, started operating in 2007 and is still working at the time of writing. In 2014, Borexino detected neutrinos from the primary p-p reaction (Equation 1.98) in which two protons combine to form a deuteron. Since more than 99% of the Sun's energy comes from the conversion of hydrogen to helium starting with this reaction, observing the resulting up to 420-keV neutrinos gives a measure of the energy currently being generated by the Sun. The energy produced at the centre of the Sun typically takes 10,000 years to percolate to the solar surface where we may then observe it as (mainly) e-m radiation. Any long-term disparity between the energy currently being generated by the Sun and that currently being radiated out into space would indicate future changes in the Sun that would be likely to have significant consequences for ourselves and other terrestrial life forms. Rapid changes, though, would be smoothed out by the 10,000 years taken for the energy to reach the Sun's surface.

The KamLAND experiment in Japan uses two nested plastic spheres, 13 and 18 m in diameter surrounded by 2000 PMTs. The spheres are filled with isoparaffin mineral oil and the inner one is also doped with a liquid scintillator. KamLAND detects anti-electron neutrinos from nuclear

[149] Practical note: tritium's current cost price is about $30,000 per gram. So the full scale experiment would cost $250,000 per year just in radioactive decays alone.

[150] Carbon-14 is continually being produced in the Earth's atmosphere through the bombardment of nitrogen-14 nuclei by cosmic ray particles. Two nearly simultaneous carbon-14 decays would appear in the detector like a single solar p-p neutrino interaction.

reactors some 175 km away. It started scientific operations in 2002 and had a container 320 kg of liquid xenon added at its centre in 2011 to aid its search for double beta-decay[151] reactions (KamLAND-ZEN). It is proposed to add NaI(Tl) crystals in the future to enable it to search for dark matter particles (Section 1.7 – KamLAND-PICO). It is still in action at the time of writing.

SNO+ is currently under construction reusing much of the old SNO equipment. It will have 800 tonnes of liquid scintillator surrounded by a water bath and some 10,000 PMTs to detect solar and terrestrial neutrinos. The scintillator will be linear alkyl benzene which is used in the manufacture of toilet soap, and so is available at low cost and in large quantities. SNO+ is also intended, amongst other things, to search for neutrino-less double beta-decay.

Several past, current or planned neutrino detectors are based upon the scintillations produced in liquid argon TPCs (Section 1.3.2.2) by the collision products from neutrino interactions. Imaging Cosmic and Rare Underground Signals (ICARUS) T600, for example, used 760 tonnes of liquid argon and was situated in the Gran Sasso tunnel in Italy. It could detect neutrinos via high-energy electrons produced either from electron scattering events or from the conversion of argon to potassium. It operated from 2010 to 2014 before being extensively overhauled and moved to Fermilab where it now operates along with the 170-tonne argon-based neutrino detector, Micro Booster Neutrino Experiment (MicroBooNE). These two detectors will soon be joined by four more, each with 17,000 tonnes of liquid nitrogen and together will be known as Deep Underground Neutrino Experiment (DUNE) which will be sited 1.5 km below ground 1,300 km away from Fermilab at Sanford.

Neutrino detectors similar to ICARUS T600 but using scintillator liquids other than argon include Jiangmen Underground Neutrino Observatory (JUNO), Low Energy Neutrino Astronomy (LENA), and NuMI Off-axis ν_e Appearance (NOvA).[152]

JUNO is planned to use some 20,000 tonnes of linear alkyl benzene as its liquid scintillator (cf. SNO+), to use PMTs as its optical detectors, to be sited in Southern China about 200 km West of Hong Kong and 0.7 km below ground level. It is currently hoped that JUNO will commence operations around 2020.

The LENA detector is in the early stages of planning and could also use 50,000 to 70,000 tonnes of linear alkyl benzene contained in an underground tank 100-m high and 30 m in diameter. PMTs may be used to detect the scintillations. Possible sites include the Pyhäsalmi mine in Finland and the Laboratoire Souterrain de Modane in the Frejus road tunnel between France and Italy. In either case a chamber for the instrument would need to be specially excavated. LENA is or was one possible candidate for the concept instrument, Large Apparatus studying Grand Unification and Neutrino Astrophysics (LAGUNA).

NOνA has been operating since 2014 with detectors near Fermilab and in Minnesota and has the objective of observing muon neutrinos converting to electron neutrinos. Fibre-optic cables collect the scintillation flashes from the small cells making up the detectors (~20,000 cells in the near detector and ~350,000 in the far detector). It uses nearly 9,000 tonnes of mineral oil containing 4.1% pseudocumene (1,2,4-Trimethybenzene) as its scintillator. Two wavelength shifters (cf. CCD short wave sensitivity, see Section 1.1.8) are used to convert the pseudocumene's UV emissions to wavelengths of about 430 nm and these are then picked up by APDs (Section 1.1.9).

A new neutrino laboratory (Yemilab), located 1,000-m underground within the Handuk iron mine in Korea, is currently under construction. It is intended to include within it a several kilo-tonne neutrino detector, using a water-based liquid scintillator.

---

[151] A radioactive decay process wherein two electrons and/or positrons are emitted from the nucleus as two protons transform simultaneously into neutrons or two neutrons into protons. The reaction can also result in the emission of two anti-electron neutrinos or none (neutrinoless double beta-decay), though the latter possibility has yet to be observed.

[152] NuMI (Neutrinos at Main Injector) is a neutrino source at Fermilab.

Although not actually a scintillator-based detector, the planned Indian Neutrino Observatory (INO) in Tamil Nadu, is perhaps best mentioned here. It will be a 50,000-tonne tracking calorimeter using iron plates and with resistive plate chambers in place of the scintillator sheets. Resistive plate chambers are a variant on drift chambers (Section 1.4) using high resistivity plastic plates for the anode and cathode. The electron avalanche is to be picked up by a separate system of metallic strips. Environmental considerations are currently delaying this project.

1.5.2.5 Acoustic Detectors

The highest-energy cosmic rays (Section 1.4) have energies measured in several joules. The Pierre Auger cosmic ray observatory has recently shown that these particles probably originate from AGNs within 30–40 mega-parsecs or so of the Earth. It is possible that such AGNs may also produce neutrinos with energies comparable to those of the cosmic ray particles (10^{20} eV). If such a neutrino were to interact with an atom on the Earth, much of its energy would be converted into heating a small volume of the material around the interacting particles. The rapid heat rise would then generate a pulse f sound at 10–50 kHz frequencies and which would have a bipolar profile. An array of sensitive microphones could then detect that acoustic pulse, and it could be distinguished from the myriad of other pulses by its characteristic shape. Volumes of water, ice or salt of a 100+ km^3 could thus be monitored for high-energy neutrinos (and also cosmic rays; see Section 1.4).

No neutrinos or cosmic rays have yet been detected by this technique but hydrophones were added to NT200+ in 2006. In 2008, AMADEUS was added to ANTARES and operated until 2016 with 36 hydrophones. SPATS is currently operating at IceCube with four strings of seven detectors. It is planned to add hydrophones to both the GVD and KM3NeT detectors in the near future.

1.5.2.6 Indirect Detectors

For neutrinos, at the moment, only one indirect detector exists and that is the influence that neutrinos have upon the CMB. Results from the WMAP spacecraft suggest that when the CMB originated (when the Universe was just 380,000 years old), neutrinos formed around 10% of the mass of the Universe compared with less than 1% today.

1.5.2.7 Other Types of Detectors

KATRIN is closely related to PTOLEMY (Section 1.5.2.3) and is based upon the beta-decay of tritium. Like PTOLEMY, it detects the energy of the electron emitted in the decay; however, unlike PTOLEMY it is searching for electrons with energies close to, but *less* than, the 18.6-eV reaction energy limit. This is because it uses the normal beta-decay route of tritium whose products are the electron *and* an anti-electron neutrino. KATRIN is attempting to measure the neutrino's mass and this (in energy terms) should be given by the gap between the 18.6-eV reaction limit and the highest-energy electrons that are detected. To date, KATRIN has placed an upper limit of ~1 eV on the neutrino mass.

The Cryogenic Underground Laboratory for Rare Events (CUORE) experiment was designed to search for neutrino-less double beta-decay (cf. KamLAND and SNO+; see Section 1.5.2.4). It used TeO_2 crystals as both its source of electrons and as its bolometer detectors. The instrument is currently being upgraded. As an interesting fact, CUORE used lead from historical Roman sources to construct the shield for its detectors. This has a much lower level of radioactivity than recently produced lead, but the use has resulted in objections from historians and archaeologists.

Other direct interaction detectors which are based upon different principles have recently been proposed, for example, the detection of the change in the nuclear spin upon neutrino absorption by $^{115}_{49}In$ and its conversion to $^{115}_{50}Sn$. Up to 20 solar neutrinos per day might be found by a detector based upon 10 tonnes of superfluid helium held at a temperature below 0.1 K. Neutrinos would deposit their energy into the helium leading to the evaporation of helium atoms from the surface of the liquid. Those evaporated atoms would then be detected from the energy (heat) that they deposit into a thin layer of silicon placed above the helium. Alternatively, electrons might be stored within a superconducting ring and coherent scattering events detected by the change in the current. With this latter system detection rates for solar neutrinos of one per day may be achievable with volumes for the detector as small as a few millilitres. Also, for the future, there is the possibility of detecting neutrino emissions from gravitational wave events (Section 1.6), and additionally, some dark matter detectors, such as Dark matter WIMP search with liquid xenon (DARWIN; see Section 1.7.2.1.1), might be able to detect neutrinos through coherent elastic neutrino-nucleus scattering.

1.6 GRAVITATIONAL RADIATION

1.6.1 The Quite Remarkable GW150914[153]

When the first edition of this book was written nearly 40 years ago, the equivalent paragraph to this one began: 'This section differs from the previous ones because none of the techniques that are discussed have yet indisputably detected gravitational radiation'. That situation continued to be the case throughout the second to sixth editions, but finally, with this seventh edition, confirmed detections of gravitational waves have been made and we may start this section on a MUCH more positive note![154]

The remit of this book is to discuss the principles and practices of the hardware used by astrophysicists and for gravitational waves these are covered in the following sections. However, the opening up of a completely new branch of observational astronomy occurs only rarely[155] and so an outline of the events of 14 September 2015 and afterwards is included here for interest and to set those principles and practices in context.

The first definite detection of a gravity wave was made on the 14 September 2015. This was 99 years after Einstein published a prediction of their possible existence based upon his General theory of Relativity (GR), although suggestions that gravity waves might exist had been made in previous decades. Einstein, however, did have second thoughts about them, and so the theoretical basis for gravitational waves was not firmly established until the late 1950s. Only after the theory had been understood was serious funding to be made available for the development and construction of Gravitational Wave Detectors (GWDs) – so leading to the waves' discovery 60 years later.

Many events throughout the Universe are producing gravity waves, either continuously or in bursts (discussion of such waves' natures are covered in Section 1.6.2). The 'loudest' such events, however, have long been expected to be the collision and amalgamation of two black holes – the bigger the black holes, the better! Thus, while the actual detection of GW150914 was a remarkable event and a much-anticipated surprise, it was, however, *not* a surprise in that it resulted from the coalition of two black holes. The black holes had masses of some 36 and 29 times that of the Sun and an estimated distance of about 400 Mpc. The mass of the resulting single black hole is about 62 $M_\odot$ so that some 3 $M_\odot$[156] were converted into the gravity waves and other emissions during the event.

[153] The notation is GW for Gravitational Wave followed by the date of its detection expressed as YYMMDD – that is, 14 September 2015 in this case.

[154] A similar statement, however, remains in the book, but now it is for dark matter and dark energy detectors (Section 1.7).

[155] The last such occasion was for neutrinos from outside the solar system (23 February 1957) when neutrinos were detected from Supernova 1987A in the large Magellanic cloud (Section 1.5). Arguably the direct detection of extra-terrestrial subatomic particles, atoms and ions is slightly more recent (3 November 1957 – *Sputnik* 2 and/or 1 February 1958 – *Explorer* 1). But the *indirect* detection of these particles dates back to Størmer's work on the aurora 50 years previously. Indirect proof of the existence of gravitational waves was almost certainly demonstrated by Hulse and Taylor's 1975 discovery and later observations of PSR1913+16 several decades before GW150914; see Section 1.6.3.4. A claimed detection of the signature of gravitational waves from the Big Bang in the CMB (Section 1.2) in 2014 was later withdrawn.

[156] This is ~6×10^{30} kg – equivalent to ~5×10^{47} J and sufficient to keep the Sun going at its present rate for about 10,000 times its present age (and thus for about 5,000 times its actually expected total lifetime).

Given its probable distance, the black hole merger actually occurred around about the time that multi-cellular life was emerging on the Earth (i.e., about the middle of the Proterozoic geological age or ~1,300 million years ago).

The signal was picked up by both AdvLIGO instruments with a 6.9-ms delay between the two detections arising from their physical separation.[157] The signal lasted for about 200 ms and swept up from a frequency of about 35–250 Hz.[158] The possibility that this was a false alarm (i.e., not a gravity wave; see also Section 1.1.17.1) was about once in 200,000 years (or a significance level of 5.1σ; see Section 1.1.17.4).

With just two GWDs in operation,[159] the direction of GW150914 could only be placed to be somewhere in the sky within an annulus about 46° in radius, a few degrees wide and centred on about RA 7^h, Dec −50° (near the Puppis/Carina border and just to the West of Canopus).[160] This area is determined by the delay time of about 6.9 ms between the two receptions of the gravitational wave and the linear separation of 3,002 km between the two AdvLIGO sites. Assuming gravitational waves travel at the speed of light (300,000 km/s – something confirmed to very high levels of accuracy by observations of the sixth gravitational-wave event, GW170817 – see below), in 6.9 ms they will have moved 2,070 km – and 2,070/3,002 = 0.69 = $\cos^{-1}$ (46°).[161] The curvature of the Earth means that the planes containing the delay lines of the AdvLIGO interferometers (see below) are inclined to each other at an angle of about 27°. The phase changes in the gravitational waves induced by this inclination enable the location of GW150914 in the sky to be limited to the southern and eastern portions of the whole possible position annulus, with the likeliest area quite close to the LMC.[162] Nonetheless the area of the sky which has a 90% probability of including the event still covers some 600 deg^2 (similar in size to the whole of the constellation of Orion). Thus, the point of origin of GW150914 could not be linked to being within any particular galaxy.

In addition to being the first detection of gravitational waves, however, it is also remarkable that GW150914 was the first event wherein we have received signals of any sort generated *actually* by black holes themselves – previous black hole ‘discoveries’ are in fact based upon e-m radiation emitted by objects interacting with assumed black holes. Even more remarkably, GW150914 provided the first evidence for the existence of black holes with masses a few tens of times the mass of the Sun – previously only black holes with masses below about 10 $M_\odot$ or upwards of 10^5 $M_\odot$ had been known.

The two AdvLIGO GWDs continued to operate periodically by themselves for the next two years, detecting three more black hole binary mergers: GW151226, GW170104 and GW170608. These black holes’ original masses ranged between 7 $M_\odot$ and 31 $M_\odot$, their distances away from us, from ~440 to ~1,000 Mpc and their possible positions in the sky covered areas ranging from 400 to 1,000 deg^2.[163] Then on 1 August 2017, the European GWD, AdvVirgo, sited in Tuscany, restarted

[157] One detector (LIGO Hanford Observatory [LHO]) is near Hanford in Washington state and the second (LIGO Livingston Observatory [LLO]) is near Livingston in Louisiana. The straight-line distance between the two (i.e., *not* following the curved surface of the Earth but penetrating through the Earth’s outer layers) is 3,002 km. The 2017 Nobel Prize for Physics was awarded to Barry C. Barish, Kip S. Thorne and Rainer Weiss for this discovery and the two LIGO sites are now designated as historic physics sites.

[158] The signal, converted into audio, can be heard via several internet sites such as www.soundsofspacetime.org, at the time of writing.

[159] GEO600 was operating at the time but not in observational mode. In any case, GEO600 would not have been sensitive enough to detect the signal.

[160] This position is the instantaneous direction in the sky of the line joining the two LIGO detectors at the time that GW150914 was received.

[161] The uncertainties in the delay time is +5 ms to −4 ms: corresponding to delays ranging from 6.5 ms to 7.4 ms and to angular radii for the position annulus ranging from 49.5° to 42.3°.

[162] GW150914, however did *not* occur within the LMC, whose distance is 50 kpc – about 0.01% of the actual distance to the colliding black holes.

[163] Another event on the 12 October 2015 – Ligo/Virgo Transient (LVT) 151012 had a 13% chance of arising from noise and so was not scientifically acceptable as being genuine (though these would be good betting odds!).

scientific operations following its upgrade from Virgo. With three GWDs widely separated over the Earth, reasonably precise positions for the gravitational wave events could be obtained and so other, associated events, such as neutrino and e-m radiation emissions could also be sought.

The scientists' luck was in. Within less than three weeks of three GWDs being in operation, on 14 and 17 August 2017, two more gravitational wave events were detected by all three instruments. GW170814 originated as the merger of a 25 $M_\odot$ black hole with a 30 $M_\odot$ black hole. It was about 540 Mpc away from us, on the edge of the Fornax/Horologium border and to be found within a 60-deg^2 area centred on about RA 3^h, Dec −45° although it could not be pinpointed to within a specific galaxy.

The next event, occurring three days later, GW170817, was the closest gravitational wave event detected up to that date – just 40 Mpc away from us. Furthermore, it did not arise from the merger of two black holes, but from the merger of two neutron stars with masses of~1.3 $M_\odot$ and ~1.5 $M_\odot$, respectively[164] and its signal lasted for about 100 seconds. The three GWDs were able to place its sky position with 90% probability to lie within a 28-deg^2 area about 10° south of Spica.

However, just 1.7 seconds after GW170817 was detected, the Fermi and INTEGRAL γ-ray spacecraft picked-up a 2-sec long γ-ray burst (GRB170817A).[165] The detectors on these spacecraft cover large areas of the sky in a single observation, so they could not locate the GRB source precisely; GW170817 *was*, though, within their fields of view *and* short GRBs are thought to result from neutron star mergers. Thus, it seemed likely that the gravitational waves and γ rays both originated from the same neutron star merger.[166] Eleven hours later and optical and radio astronomers[167] joined the fray by observing a transient optical event in the lenticular galaxy New General Catalogue (NGC) 4993[168] (Figure 1.109). This galaxy is positioned at RA 13^h 09^m, Dec −23° 23′ – just a couple of degrees from the centre of the 90% probability region for GW170817. The e-m radiation transient event was observed from the radio to the X-ray regions and showed a cloud of neutron-rich material that was fast moving and cooling rapidly – exactly what would be expected following a neutron star merger. The remarkable coincidences between these three events would convince most people already that they had originated together. The clincher, though, was the distance of NGC 4993 – at 40 Mpc it was exactly the same as that calculated for the GW170817 event.

With the GW170817 gravitational waves and the GRB gamma rays assumed to originate at the same place and time, it is simple to work out their relative speeds through space. A difference in travel time of 1.7 seconds over a 40-Mpc distance gives the speed of gravitational waves to be the same as that of light (or gamma rays) to within about 1 part in 10^{15} (a few times 0.000 000 1 m/s in a value of 300,000,000 m/s).[169] Thus, the assumption that the speed of gravity and the speed of light are the same, mentioned a few paragraphs previously, is amply supported.

---

[164] Recent work has suggested that such systems might originate from a binary star whose two components originally each have masses in excess of ~10 $M_\odot$. The first component to explode as a supernova leaves behind a neutron star. That neutron star then strips most of the outer layers from the second component. When the second component becomes a supernova, therefore, little material remains to be expelled and the system can remain gravitationally stable.

[165] The Fermi spacecraft also detected a 1-s long x-ray burst about half a second after GW 150914. The chances that the two events were related was about 99.8% – well into the normally accepted significant area for a result (Section 1.1.17.4). However black hole mergers are not expected to lead to much e-m radiation emission. The reality of the association is now generally regarded as 'not confirmed'.

[166] The probability of them *not* originating from the same physical event was about 0.000 005%.

[167] For the first 55 gravitational wave events, no neutrinos *clearly* associated with a gravitational wave event have been detected (Section 1.5). For one event, GW191216ap, IceCube did detect a neutrino event candidate, but this occurred at least 8 s before the gravitational wave event. Nonetheless, the possibility of the two events being associated is put at between 79% and 99% (σ = 1.26 to 2.52).

[168] The discoverers *were* searching the probable location of GW170817 for just such a transient, but it was still a major task; 28-deg^2 is about 110 times the area of the full moon and is 100,000 to 1,000,000 times the typical field of view of an optical telescope. Neutron star mergers are expected to emit significant amounts of e-m radiation, whilst black hole mergers are not. Neutron star mergers are also expected to emit neutrinos (Section 1.5). These were not detected for GW170817, but current and future neutrino detectors could potentially detect some of these events out to a distance of 300 Mpc or so.

[169] Actually, the speed of gravity is found from the data to lie between (c – 900 nm/s) and (c + 210 nm/s).

FIGURE 1.109 NGC4993. The optical counterpart of GW170817 is the bright object above and slightly to the left of the galaxy's core. (Courtesy of European Southern Observatory [ESO]/J. D. Lyman, A. J. Levan, N. R. Tanvir.)

Recently, closer examination of the data has shown tentative evidence of a gravitational wave 'echo' from GW170817 arising from the 2.6 to 2.7 $M_\odot$ rotating black hole that resulted from the merger. The echo had a frequency of ~72 Hz and occurred about 1 s after the main pulse, it is currently estimated to have a significance of about 4.6σ. Additionally, re-examination of previous data from Chandra, Fermi and other spacecraft have shown a GRB in January 2015 that had similar characteristics to GRB 170817A, and so may indicate an earlier double neutron star coalescence that occurred before LIGO had restarted its operations and so any associated gravitational wave emissions could not then have been observed.

In April 2019, during the early days of LIGO/Virgo's O3 observing run, gravitational waves from a second binary neutron star merger (GW190425z) were detected together with two weak γ-ray bursts (delayed by 0.5 and 5.9 seconds and observed by INTEGRAL). Unfortunately, the gravitational waves were only registered by the LIGO Livingston detector, and the INTEGRAL observations only detect the γ-rays without indicating their direction, so the position of the gravitational wave event remains very uncertain (the 90% probability area for the event covers nearly 20% of the whole sky, although the *non-detection* of the GRB by Fermi does reduce this area somewhat). It is possible, nonetheless, to deduce that the binary neutron star merger occurred at a distance of approximately 150 Mpc (about four times further away than GW170817), that its FAR is around once in 70,000 years and that the chances of the GRB *not* being associated with the gravitational wave event is about 0.001%. To date, however, no gravitational wave events, other than these three possibilities, have been associated with *any* other form of signal. In total, to the end of January 2020, there have been 56 observed gravitational wave events, with the types of events probably leading to them being;

Binary black hole mergers – 40
Black hole-neutron star mergers – 6
Binary neutron star mergers – 5
Small mass black hole (masses of about 2 or 3 $M_\odot$ – known as the 'Mass Gap') binary mergers – 4
Unknown – 1

1.6.2 Introduction

The basic concept of a gravitational wave[170] is simple; if an object with mass changes its position then, in general, its gravitational effect upon another object will change. The information on that changing gravitational field will propagate outwards at the speed of light (Section 1.6.1) in the form of fluctuations in the space-time continuum and be experienced by other masses as jittering tidal forces. The propagation of the changing field obeys equations analogous to those for electromagnetic radiation, provided that a suitable adaptation is made for the lack of anything in gravitation that is equivalent to positive and negative electric charges. Hence, we speak of gravitational radiation and gravitational waves.

Like e-m waves, gravitational waves can potentially be of any frequency from 0 to ∞, with no intrinsic difference in their basic natures over that range. Like e-m waves, though, we divide them up into convenient bands based upon their varying interactions with matter and e-m radiation. At the moment there are no names for the different frequency bands of gravitational waves that are the equivalent of (say) X-rays, the visible region, microwaves and so on, and neither is there much agreement between different workers on how to divide up the gravitational wave spectrum. Following the practice adopted in this book for the e-m spectrum, we shall therefore use a broad division for the gravitational wave spectrum into

Extremely low frequency (ELF-GW) – 0 to ~1 pHz – arising mostly from the Big Bang and cosmological scale events;
Low frequency (LF-GW) – ~1 pHz to ~1 μHz – arising mostly from binary and rotating objects;
Medium frequency (MF-GW) – ~1 μHz to ~1 MHz – arising mostly from binary coalescences and exploding stars, and
High frequency (HF-GW) – ~1 MHz to ∞ – arising mostly from plasma instabilities and (perhaps) the basic structure of space-time and/or the Big Bang.

There is also a division into continuous and transient sources. The ELF and LF sources listed are likely to be continuous, whilst the higher frequency sources are likely to be transient. The continuous sources are also likely to be much fainter than the transient sources, but, because their waves are emitted over long periods of time, they may be observable via different data-processing techniques such as integration. There may also be a stochastic higher frequency background arising from fainter but numerous transient events; throughout the whole of the visible universe, it is estimated that neutron star mergers occur several times a minute and black hole mergers happen several times an hour. This background, though, will probably be much more difficult to detect.

The expected frequencies of gravitational waves for currently and potentially observable astronomical sources are anticipated to run from a few kHz for collapsing or exploding objects to a few nanohertz for pulsars and binary star systems (LF-GW to MF-GW). A pair of binary objects coalescing into a single object will emit a burst of waves whose frequency rises rapidly with time: a characteristic 'chirp' if we could hear it. Events during the early stages of the Big Bang are likely to leave gravitational waves with frequencies (now) ranging down to attohertz or less and to higher than 1 Hz. An improvement in GWD sensitivity[171] over that of AdvLIGO and AdvVirgo by a factor of about 1,000 to 10,000 might enable this 'Cosmic Gravitational Wave Background' to be detected at the higher frequencies. However, it seems unlikely that the attoHz signals (ELF-GW) will ever be directly detected by humankind because a single cycle takes as long as the current age of the Earth

---

170 The term 'gravity wave' is also used to describe oscillations in the Earth's atmosphere arising from quite different processes. There is not usually much risk of confusion.

171 An improvement in sensitivity by a factor x should lead to an improvement in the detection rate by x^3 because the latter depends upon the *volume* of the Universe that can be observed.

(4.5 Ga) to take place. Indirect detections may be possible though, via, for example, the gravitational waves' polarisation effects on the cosmic microwave background radiation.

As mentioned in Section 1.6.1, the concept of the gravitational wave dates back over a century. Although Einstein's GR theory is widely known to explain gravitational phenomena, there *are* other theories – such as scalar-tensor theory and C-field theory. However, GR has passed all its observational tests with flying colours to date, and so we will not further concern ourselves here with its alternatives (although they also predict the existence of gravitational waves but sometimes with different properties from those predicted by GR). The GR theory of gravitational waves then, forbids dipole radiation and so quadrupole radiation is the first allowed mode of gravitational radiation. There are also only two polarisation states[172] for GW radiation under GR.

The detection problem for gravity waves is best appreciated with the aid of a few order-of-magnitude calculations on their expected intensities.[173] The quadrupole gravitational radiation from a binary system of moderate eccentricity ($e \leq 0.5$),[174] is given in GR by

$$L_G \approx \frac{2\times 10^{-63} M_1^2 M_2^2 \left(1+30e^3\right)}{(M_1+M_2)^{2/3} P^{10/3}} \quad W \tag{1.105}$$

where M_1 and M_2 are the masses of the components of the binary system, e is the orbital eccentricity, and P is the orbital period. Thus, for a typical dwarf nova system with

$$M_1 = M_2 = 1.5 \times 10^{30}\text{kg} \tag{1.106}$$

$$e = 0 \tag{1.107}$$

$$P = 10^4\text{s} \tag{1.108}$$

we have

$$L_G = 2\times 10^{24} \text{ W} \tag{1.109}$$

But for an equally typical distance of 250 pc, the flux at the Earth is only

$$F_G = 3\times 10^{-15} \text{ Wm}^{-2} \tag{1.110}$$

This energy will be radiated predominantly at twice the fundamental frequency of the binary with higher harmonics becoming important as the orbital eccentricity increases. Even for a nearby close binary such as ι Boo (distance 23 pc, $M_1 = 2.7 \times 10^{30}$ kg, $M_2 = 1.4 \times 10^{30}$ kg, $e = 0$, $P = 2.3 \times 10^4$ s), the flux only rises to

$$F_G = 5\times 10^{-14} \text{ Wm}^{-2} \text{ (at a frequency near 40 μHz)} \tag{1.111}$$

[172] The '+' polarisation state changes the shape of a circular test object that is oriented perpendicular to the direction of propagation of the gravitational wave in the cycle: [circle] → [oval squeezed horizontally] → [circle] – [oval squeezed vertically] → [circle]. The 'x' polarisation state has similar effects but oriented at 45° to the effects of the '+' polarisation state.

[173] For comparison with these calculations, the e-m energy flux in the visual region for an $m_v = 0$ star is $F_v = 2.48 \times 10^{-8}$ W m^{-2}.

[174] The observed intensity varies with the angle to the plane of its orbit at which the binary is observed. The intensity when a binary with zero eccentricity is observed from above the plane of the orbit is about one-third of the intensity compared with when it is observed along the plane of its orbit.

Rotating elliptical objects radiate quadrupole radiation with an intensity given approximately by

$$L_G \approx \frac{GM^2\omega^6 r^4 (A+1)^6 (A-1)^2}{64c^5} \text{ W} \tag{1.112}$$

where M is the mass, ω is the angular velocity, r is the minor axis radius and A is the ratio of the major and minor axis radii. So that for a pulsar with

$$\omega = 100 \text{ rad s}^{-1} \tag{1.113}$$

$$r = 15 \text{ km} \tag{1.114}$$

$$A = 0.99998 \tag{1.115}$$

we obtain

$$L_G = 1.5 \times 10^{26}\ W \tag{1.116}$$

which for a distance of 1000 pc leads to an estimate of the flux at the Earth of:

$$F_G = 10^{-14} \text{ W m}^{-2} \left(\text{at a frequency near 60 mHz}\right) \tag{1.117}$$

Binaries formed from black holes (Section 1.6.1) have much higher gravitational luminosities than these for 'ordinary' binaries and for rotating pulsars, but their generally much greater distances mean that their fluxes at the Earth are not that much larger. Thus, for GW150914, when the black holes' separation was (say) around 1,000 times their Schwarzschild radii (~100,000 km),[175] L_G would have been somewhere in the region of 10^{37} W with a flux near the Earth of ~10^{-8} W m^{-2} at a frequency near 10 mHz. While for the quasar PKS 1302-102, thought, perhaps, to have two super-massive black holes in a binary system at its centre, the possible parameters are around $M_1 \approx M_2 \approx 10^{11}\ M_\odot$ for the black holes' masses, $P \approx 5$ years for their period. Then with e assumed to be zero, we get L_G somewhere in the region of 10^{47} W. But the quasar's distance of ~1.1 Gpc gives its flux at the Earth as ~10^{-5} W m^{-2} at a frequency near 1 μHz.

Now these last two examples give gravitational wave fluxes that are quite respectable when compared with the visible light flux of a zero apparent magnitude star (see previous footnote). However, AdvLIGO and AdvVirgo are sensitive to gravitational wave frequencies in about the 1- to 1,000-Hz region, not the milli- and micro-hertz emissions from the binary and pulsar examples given previously. Also and more significantly, GWDs such as AdvLIGO and AdvVirgo operate in effect using the *tides* of the gravitational waves and *not* their absolute fluxes (Section 1.6.3). Thus, although the flux from GW150914 peaked[176] at 2×10^{-2} W m^{-2}, with an apparent gravitational brightness at the Earth of about 1.5 mW m^{-2} (about the brightness of the full Moon in visual terms), the measurement *actually* made by the GWDs was of a change in the 4-km separation of two test masses by about 4×10^{-18} m (= 4 am ≈ 1/400 of the diameter of a proton).[177]

Supernovae within the Milky Way galaxy may produce strains at the Earth of ~10^{-18} (higher if close to us). But although the general frequency of supernovae in other galaxies suggests that such

[175] This would have occurred about half a million years ago.

[176] Its luminosity was then, for a millisecond or so, around a few times 10^{49} W – some ten times the total energy emissions of all the other stars in the visible universe put together (depending a *lot* on how you estimate this last figure!).

[177] This is based upon the proton's charge radius of ~900 am. Multiple reflections between the mirrors multiply the effective path length up to 1,120 km (Section 1.6.3) and the shift to nearly 1 fm.

events could occur several times a century, in practice the Milky Way seems to host a supernova less than once a century. Supernovae out to some 100 Mpc might produce strains of ~10^{-21} and be expected to occur a few times a year – although again their actual rate of occurrence seems to be quite a bit lower than this.

At the time of writing, both of the AdvLIGO instruments and AdvVirgo are in the midst of their third observing run (O3). When they are joined by Kamioka Gravitational wave detector (KAGRA) and Indian Initiative in Gravitational wave Observations (INDIGO; see Section 1.6.3), detection rates for black hole and/or neutron star mergers of several per week may be achieved. Five operating detectors will also enable the polarisation of the waves to be determined completely, providing a strong test of GR as well as further information of astrophysical interest. KAGRA will initially be operating at less than its design sensitivity, but when that is achieved, the four detectors should be able to pin down gravitational wave source to areas of the sky of less than 10 deg^2.

Given the difficulties of observing gravitational disturbances coming from the rest of the Universe, one might be tempted to set about generating them in a laboratory, under controlled conditions and so on; this, after all, would be the usual scientific approach. However, the GW luminosity produced by two (say) 1,000-kg masses rotating around each other once a second and separated by 1 m is ~10^{-53} W. The flux for a distance of 4 km (about as near to AdvLIGO as we could get such a GW generator) would thus be ~5×10^{-62} W m^{-2} – nearly 10^{60} times fainter than GW150914 at its peak. So clearly, laboratory-based gravitational wave science is still some way off.

1.6.3 Detectors

It was mentioned in Section 1.6.1, that those GWDs so far constructed or planned, rely upon the tidal effects of the wave for their operations (a detection method potentially looking for the direct effects of gravitational waves is discussed in Section 1.6.3.4).

In fact, this is a Newtonian gravitational view of what is happening. Using this approach, however, may give the reader a 'feel' for what is happening before tackling the GR explanation. In any case, the practical differences between the Newtonian and GR viewpoints when we are dealing with weak gravitational fields is usually small (the angle by which a light beam skimming the solar surface is deflected is predicted to be 0.875″ by Newtonian physics and to be 1.75″ by GR, for example).

Tides, such as those of the Moon upon the Earth, arise because the force of gravity varies with distance. The mean distance between centres of the Earth and of the Moon is 384,400 km and the Earth's diameter is 12,756 km. A kilogram block of matter on the surface of the Earth where the Moon is at the zenith is thus at a distance of 378,022 km from the Moon. Hence, a similar block of matter on the far side of the Earth is 390,778 km from the Moon. The gravitational force of the Moon on those two blocks of matter is thus 34.3 and 32.1 μN, respectively. The difference between those two forces, 2.2 μN, is the tidal force (note how much less in magnitude is the tidal force than the actual attractive force towards the Moon).[178]

Now in this situation, the two blocks of matter will be held in their places by the Earth's gravity and so will not move with respect to each other. However, in the absence of the Earth, two such blocks of matter would experience the Moon's force and be free to respond to it. Whilst both blocks will be attracted towards the Moon, the nearer block will be attracted more and so the difference between the two forces (the tidal force) acts to increase the distance between the blocks. When the blocks of matter are free to move, then their separation will increase. In the example just discussed, the separation will have increased from 12,756 km to more than 85,000 km by the time the first

---

[178] The actual ebb and flow of the sea that we see on the sea shore and call the 'tide' arises from the differing orbital motions that the two objects would have if independent of each other, the fact that they do not have these motions because the Earth is a solid body and the facts that the solid parts of the Earth move by only small amounts under the tidal forces, whilst the liquid parts (seas, oceans) can flow and move large distances under the same forces. Additionally, topography makes a major contribution, leading to 16-m tides in the Bay of Fundy and 1-m tides at Venice etc.

block of matter hits the lunar surface, assuming that both blocks were initially stationary with respect to the Moon and started their falls simultaneously.

The GR viewpoint of the situation is that the mass of the Moon distorts the space-time continuum in which the blocks of matter are moving. Thus, the fabric of space itself is stretched as the blocks move deeper into the distortion arising from the presence of the Moon. The separation of the blocks thus increases because space which they occupy has expanded.

The situation when a gravitational wave passes by is largely the same as that of the lunar tides. The Newtonian view is that blocks of material (say, for example, the test mass mirrors in an Michelson Interferometer-GWD [MI-GWD]; see Section 1.6.3.2.1) will experience a change in the gravitational forces acting upon them. If the distance between the two blocks is reasonably comparable with the wavelength of the gravitational wave, then at a given instant the blocks are likely to be experiencing differing forces. They will, therefore, move in differing directions and/or by differing amounts in response to those forces and so change their separation. The GR view is that as the gravitational wave passes by it stretches or compresses the fabric of space and so the blocks' separation changes because the fabric of space between then has changed.

One significant difference from Newtonian gravity, however is that in GR, gravitational waves have two states of polarisation (Section 1.6.2). The MI-GWDs (Section 1.6.3.2.1) operate using two pairs of blocks oriented perpendicularly to each other. In most circumstances, the gravitational wave will cause one separation to expand whilst the other contracts and vice-versa. This would not occur, except by chance, on the Newtonian tidal analogy of the interaction.

The effect of the gravitational tide is measured by the change in the distance between two points separated in space: opposite sides of a large block of material for the resonant GWDs (Section 1.6.3.1) and the distance between two test masses for the non-resonant GWDs (Section 1.6.3.2.1). The changes are most usually recorded as the strain induced by the tide, that is, $\delta x/x$ where x is the size of the test object or separation of the test masses and δx is its change. The strain is conventionally denoted by the symbol h in gravitational wave astronomy. It is a dimensionless quantity. Another parameter, the strain sensitivity, symbolised by $\tilde{h}$ (pronounced: H-tilde[179]) is also in use. It is defined as the square root of the input noise that is required to produce the noise observed at the output. It takes account of the bandwidth of the signal and is useful when comparing the performances of various GWDs. It has units of $(\text{seconds})^{0.5}$, but customarily $\text{Hz}^{-0.5}$ are used. For burst gravitational wave sources, such as supernovae and binary mergers, h and $\tilde{h}$ are related by

$$\tilde{h} = h\,\Delta t\,\sqrt{\frac{\pi B}{2}}\quad Hz^{-0.5} \tag{1.118}$$

where Δt is the duration of the burst and B is the GWD bandwidth.

For GW150914, the value of h peaked at about 10^{-21}. With x having a value of 1,120 km for AdvLIGO and 120 km for AdvVirgo (Section 1.6.3.2) this strain corresponds to values of δx of ~10^{-15} m and 1.2×10^{-16} m, respectively. The resonant GWDs, with $x \approx 1$ m would have changed in size by ~10^{-21} m.

[179] Not to be confused with the symbol for the reduced Planck constant, $\hbar$. Other symbols and names for $\tilde{h}$ may be encountered such as h_{rss} (h-root-sum-square), strain amplitude, root-sum-square amplitude, spectral strain sensitivity, strain noise power spectral density, etc. A third parameter that is sometimes used is the horizon distance. This is the maximum distance at which the coalescence of two 1.4-solar mass neutron stars could be detected. The basic definition assumes that the coalescence is optimally oriented for detection. To allow for non-optimum situations the basic horizon distance needs dividing by a factor of about 2¼. Unfortunately, it is not always apparent which of all these figures is being quoted. In this book, therefore, we shall stick with the clearer h and $\tilde{h}$ as the detector criteria.

1.6.3.1 Direct Resonant Detectors

All the early gravity wave telescopes fall into this category. The first such was constructed and operated by Joseph Weber in the 1960s. Subsequent similar instruments are essentially based upon the same design. Weber's GWD used a massive (>1 tonne) aluminium cylinder which was isolated by all possible means from any external disturbances and whose shape was monitored by piezo-electric crystals attached around its 'equator'. Two such cylinders separated by up to 1000 km were used and only coincident events were regarded as significant to eliminate any remaining external interference. With this system, Weber could detect a strain of 10^{-16} (cf. 10^{-21} for the detection of GW150914[180] The cylinders had a natural vibration frequency of 1.6 kHz, this was also the frequency of the gravitational radiation that they would detect most efficiently.

Weber detected about three pulses per day at 1.6 kHz with an apparent sidereal correlation and a direction corresponding to the galactic centre (or to a point 180° away from the centre). If originating in the galactic centre, some 500 solar masses would have to be totally converted into gravitational radiation each year to provide the energy for the pulses. Unfortunately (or perhaps fortunately for the continuing existence of the galaxy), the results were not confirmed even by workers using identical equipment to Weber's, and it is now generally agreed that to date there have been no definite detections of gravitational waves by direct resonant instruments.

There have been a few follow-up experiments using Weber's bar design (sometimes with spherical test masses), but the best sensitivities achieved to date have been around $\tilde{h} \approx 10^{-20}$ $\text{Hz}^{-0.5}$ at kHz frequencies (h = a few times 10^{-19}). One Weber-bar type experiment is still in operation and now using some techniques developed for the double beam interferometers (Section 1.6.3.2.1). This is the Opto-acoustical Gravitational Antenna (OGRAN) instrument housed underground at the Baksan Neutrino Observatory. It is based around a 1-tonne cryogenic bar whose vibrations are detected by the changes in length of a Fabry-Perot cavity (Section 1.6.3.2.1.2.1). It (probably) has the ability to detect the gravitational waves from a supernova occurring within our own galaxy, that is, one detection every 50 years or so. We shall therefore not consider these instruments further here.[181]

1.6.3.2 Direct, Non-Resonant Detectors

These designs are of two main types and have the potential advantage of being capable of detecting gravity waves with a wide range of frequencies. The first design (Section 1.6.3.2.1) is based upon a Michelson type interferometer (Section 2.5), while the second design (Section 1.6.3.2.2) would detect gravitational wave bursts by the bursts' individual effects upon a radio transmitter separated from a receiver by more than the burst's physical length (i.e., >30,000 km for a 100-ms bursts). Only the first type has been actually implemented in practice.

1.6.3.2.1 Double Beam Interferometers

The first design uses a Michelson type two-armed interferometer (Section 2.5) operating in a vacuum to detect the relative changes in the positions of four test masses. The basic layout for the system is shown in Figure 1.110. In operation, when no gravitational disturbance is present, the system is adjusted so that there is destructive interference between the two beams at the beam splitter, i.e., in anti-phase with each other and so there is no output signal, sometimes called 'dark fringe operation' (or slightly away from this point so that the detector always has a very low output). The light, or most of it, can then be returned to the arms of the interferometer, thus increasing the power available – a technique called 'power recycling' (Section 1.6.3.2.1.2.3).

---

[180] Although these resonant detectors are physically small in size when compared with those considered in subsequent sections, when 'kicked' by a gravitational wave, they continue to ring like a bell for many seconds or even minutes at their resonant frequencies, thus significantly increasing the chances of detecting the signal.

[181] Readers interested in these experiments may find researching the following projects, in addition to OGRAN, useful: ALLEGRO, AURIGA, EXPLORER, NIOBE, MiniGRAIL and NAUTILUS as well as Weber's own work.

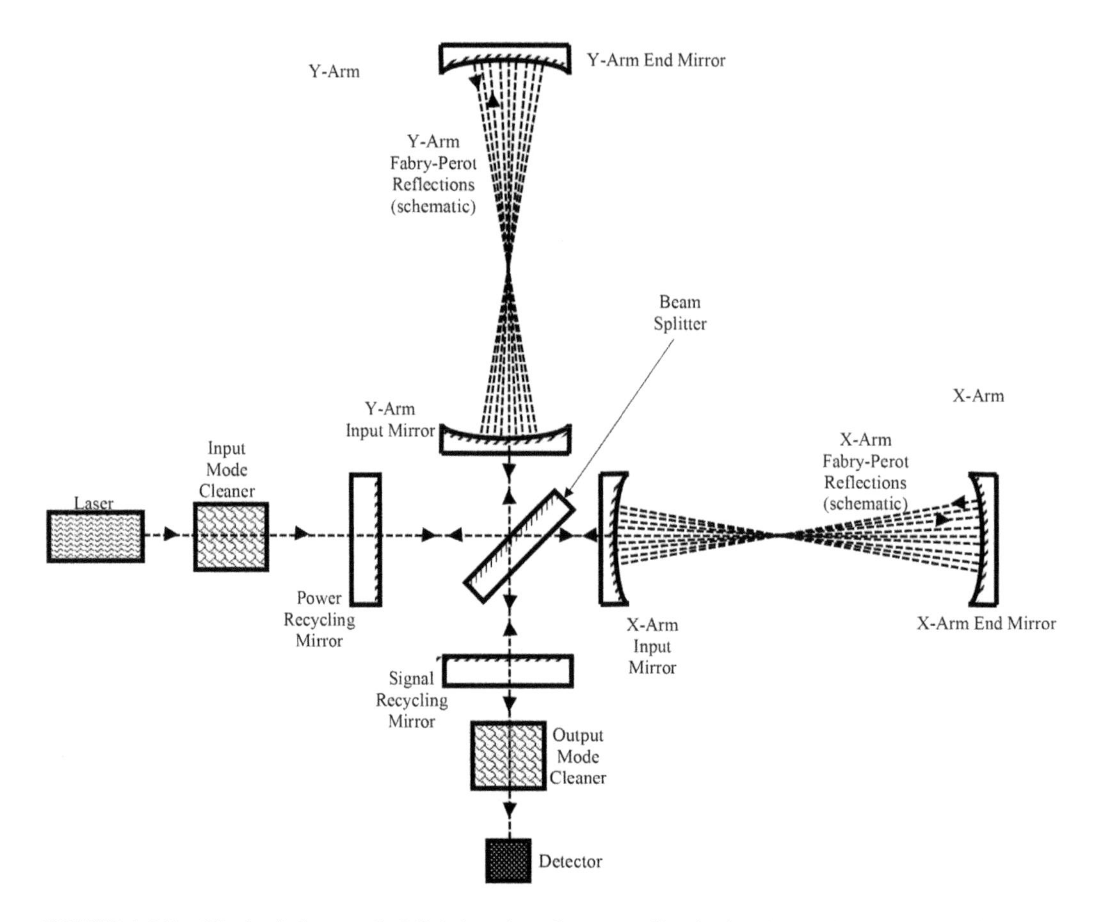

FIGURE 1.110 The basic layout of a Michelson interferometer Gravitational Wave Detector (MI-GWD).

When the interferometer is affected by a gravitational disturbance, then the difference between the forces (the tide – see above) on the two test masses in one of its arms will cause their separation to change. This results in a phase change, of around 10^{-11} radians, in the light wave on its return to the light detector when compared with quiescent situation. A similar process occurs in the interferometer's second arm but the phase change will generally differ from that in the first arm. Thus, the two beams will now no longer be in exact anti-phase.[182] The interference between the two beams will thus be at least partially constructive, and there will be an output signal from the detector (Section 1.6.3.2.1.2.2).

Unfortunately, many other processes can change the separations of the test masses or otherwise change the relative phases of the light beams resulting in false signals (i.e., noise). Furthermore, as already noted, even for a strong gravitational wave source, such as GW150914, the change in the separations of the test masses is *very* small – just 4 am for the 4-km-long arms of AdvLIGO. Thus, to detect gravitational disturbances, the Michelson type interferometers have both to reduce

[182] Except in the unlikely situation (because the polarisations of the wave tend to cause one arm to lengthen whilst the other shortens), that the changes in test mass separations for the two arms are identical. Then, however, with several instruments in operation at different sites on Earth (or even in space), the wave will be detected by the other instruments because these will have different differently orientations towards the source of the gravitational disturbance.

noise as much as possible (Section 1.6.3.2.1.1) and to strengthen the signal as much as possible (Section 1.6.3.2.1.2).[183]

1.6.3.2.1.1 Michelson Interferometer Gravitational Wave Detectors – Noise For best sensitivity to incoming signals, the noise in a GWD's output should be Gaussian in nature. However, this is not always achievable, though it was the case for the LIGO instruments when GW150914 was detected. Individual sources of noise are discussed in the next few subsections.

1.6.3.2.1.1.1 Shot Noise and Radiation Pressure Shot noise (also called 'Poisson noise', 'quantum noise' or 'photon noise'; see Section 1.1.17.1) arises from the photon (quantum) nature of light. When a light beam is very faint, the number of photons within it will be small and their arrival at a detector (say) can vary significantly on a short timescale from just the random fluctuations in their numbers. Thus, in a light beam with an intensity of (say) one photon per second (about 10^{-18} W on average for visible light), a fast detector will actually measure the intensity as being about 10^{-9} W for 1 ns[184] and zero for the remaining 999,999,999 ns.

The fluctuation in the number of photons in a beam of laser light[185] with an otherwise uniform intensity obeys Poisson statistics and so has a standard deviation of $\sqrt{N}$, where N is the number of photons arriving at a detector within a specified time. The signal-to-noise ratio thus increases as $\sqrt{(1/N)}$ and so the relative effect of shot noise may be reduced by increasing the light beam's intensity. For example, for AdvVirgo, the 700-kW final beam[186] (Section 1.6.3.2.1.1.4) has less than 2% of the relative shot noise of the 200-W input beam and less than 0.2% that of the initial 2-W starter beam. 'Squeezing' the photons (Section 1.6.3.2.1.2.4) may also help to reduce the shot noise.

Not only does shot noise affect the intensities of each beam independently, but acting at the beam splitter it may result in the relative intensities of the two beams varying.

Additionally, a perfectly reflected light beam (Section 1.6.3.2.1.1.4) exerts a radiation force of

$$F_R = \frac{2\,E_R}{c}\,N \qquad (1.119)$$

where F_R is the radiation force and E_R is the energy of the radiation beam.

A 700-kW beam (AdvLIGO – Section 1.6.3.2.1.1.4) thus exerts a force of 4.7 mN on the test mass mirrors – sufficient to give a 40-kg mass an acceleration of 1.2×10^{-4} m s^{-2}. Such acceleration is almost sufficient to damage the test masses' suspensions (see KAGRA, Section 1.6.3.2.1.1.2). Fortunately, it is the variations in the intensity of the 700-kW light beam that are the problem, not its

[183] Despite the number of items that *are* considered in the following subsections, they are only a small fraction of the total number of measures, adjustments, precautions and so on required before an MI-GWD can operate successfully. These vital, but perhaps less central and/or more detailed and/or more technical aspects of the instruments have to be left for the reader to pursue in more specialist publications than this one. For example, lock acquisition (getting all the light beams into their correct relative phases and mutual resonances and the physical components free of relative motions) is one of the major and lengthy preparatory steps and requires multiple adjustments of all the instrument's components which it would need hundreds of pages to describe. Shielding and baffling the optics from scattered and otherwise unwanted radiation is another major concern not discussed here. There is also the physical security of the instrument to be considered – the vacuum systems and other constructs must be protected from excessive heat or cold, from water and from other damage caused by, for example, the hoists, fork-lift trucks, cranes, cars, vans and lorries that are needed to move people and equipment around the instrument's site. The clean rooms for the assembly of components are essential, but details of their construction and operation are far beyond the scope of this book. Then there are the specially written data-processing packages, dedicated computers, vacuum pumping stations, atomic clocks, power supplies, cryogenics and so on and so-on and, … .

[184] Based on the typical response time f 1 ns for a PMT (a quantum-type detector; see Section 1.1.2).

[185] Thermal photons have a Bose-Einstein distribution.

[186] The energy eventually reaching the detector, though, is typically a few tens of milliwatts.

total intensity, i.e., the shot noise. Now the relative direct shot noise decreases as the power of the light beam increases whilst the variation in the radiation power effects increase with increasing light beam power. There is thus an optimum value for the light beam's power which is called the Standard Quantum Limit (SQL) when the two effects are uncorrelated. When there is some correlation, such as through the use of squeezed light (Section 1.6.3.2.1.2.4), then the performance of the instrument may be improved beyond the SQL.

1.6.3.2.1.1.2 Seismic Noise In the context of Michelson interferometer-type GWDs, 'seismic disturbance' refers to any type of ground tremor – not just those originating from earthquakes. Thus, road and rail traffic, wind-induced vibrations, construction and demolition activities, quarrying, mining and so on can cause problems as severe as those of 'traditional' seismic activity. Even at a good (low seismic noise) terrestrial site, the seismic noise still requires reducing by a factor of 10^9–10^{10} to ensure that the optics are kept to their correct positions with the required picometre accuracy.

The early GWD designs used passive measures to reduce seismic effects. The test masses in the interferometer arms are mirrors which reflect the light beam between themselves (see Section 1.6.3.2.1.2.1). Increasing the mass of a mirror increases its inertia (and also its thermal inertia; see Section 1.6.3.2.1.1.3) and so reduces the magnitude of the movement in response to any impulse that it may receive (except that from gravitational waves which affect all masses equally). Thus, the mirrors in AdvLIGO now have masses of 40 kg, whilst in its precursor, LIGO, they were 10.7 kg.

The mirrors need to be (as far as possible) 'free-falling' masses, i.e., able to move under the influence of gravitational changes without constraint. In space (see 'drag-free spacecraft' Section 1.6.3.2.2), it is possible to achieve this to high levels of accuracy in all three dimensions, but on the Earth this is not possible. It has been known since 1851[187] though, that a mass suitably suspended by a thin wire from a fixed point – a pendulum – acts as a close approximation to a 'free-falling' mass in the two horizontal directions. Only if the mass tries to move vertically will it encounter constraints. The mirrors are therefore mounted as pendulums within the GWD. Springs and/or cantilever mountings can help to reduce vertical disturbances.

For LIGO, a single pendulum design was used with 310-μm diameter steel wires on each side of the mirror as the supports (so that it was only truly free to move in one direction – along the line of the interferometer arm).

For AdvLIGO, four pendulums are used in succession and employ 400-μm diameter silica fibres[188] as the supports. To begin with, a massive metal frame is suspended from the top of the mounting by four fibres (allowing unconstrained movement only along the line of the interferometer arm). This frame then suspends a second metal frame, using four fibres. The second frame carries a 40-kg mass silica 'penultimate' mass of similar shape to the mirror, again using four fibres. Finally, the 40-kg mirror (= the test mass) is suspended from the penultimate mass by eight fibres. Gravitational European Observatory (GEO600) uses a triple pendulum suspension system, KAGRA, a quintuple pendulum and Virgo's 'Super-Attenuators' have seven pendulum stages, each 1 m in length which reduce the incident seismic noise amplitude by a factor of 10^{10} (i.e., to 0.00000001%). AdvVirgo's five attenuator stages reduce seismic noise to a similar level.

All the optical components – other mirrors, beam splitters, spacers, etc., of an MI-GWD need to be similarly shielded from seismic noise. Suspensions and supports analogous to those for the test mass mirrors are thus used. Generally, though, these supports can be of somewhat simpler designs. The input mode cleaner's mirrors for AdvVirgo, for example, use single pendulum supports.

---

[187] When Léon Foucault first demonstrated 'his' pendulum.

[188] Termed 'monolithic supports' when the fibres are made from the same material as the mirror's substrate, as is the case here.

AdvLIGO also uses an active seismic noise reduction system. This comprises mountings for the optical and other components of the instrument that incorporate sets of vibration sensors. The sensors drive actuators within the mountings that move to counteract the vibrations' disturbances. Since the sensors cover a wide range of vibration frequencies, the actuators are driven to counteract the integrated effects of all the vibrations at each instant.

Seismic noise can, of course, cause rather more severe problems than vibrations by tiny fractions of metres. In April 2016, a magnitude 7.3 earthquake and several strong aftershocks occurred about 700 km away from the KAGRA site. Accelerations at KAGRA reached 0.0003 g (0.003 m s^{-2}) and caused damage to mirror suspensions. Another earthquake in 2011 delayed the start of the instrument's construction. Normally, however, by operating KAGRA some 200-m underground the seismic noise will be reduced by up to a factor of 100.

It seems likely that, despite scientists' best endeavours, ground-based detectors will be limited by the seismic problems to operating at frequencies above a few hertz and that for the lower frequency detections, space-based detectors (like LISA; see Table 1.8) or Doppler tracking (Section 1.6.3.2.2) or pulsar timing (Section 1.6.3.3) will be needed.

1.6.3.2.1.1.3 Thermal Noise and Mirror Thermal Control Vibrations in the test mass mirrors due to the thermal motions of their component atoms and those of their suspensions and reflecting layers have significant effects at the low frequency ends of current MI-GWDs' detection ranges, particularly if they excite the resonances of the masses.

With light beam intensities within the interferometer arms approaching 1 MW (Section 1.6.3.2.1.1.4), even with the very low absorptions of the mirror coatings (Section 1.6.3.2.1.2.5), the mirrors will experience some heating. Although the Fabry-Perot reflections are spread over an area of the mirror (Section 1.6.3.2.1.2.1), there will still be some thermal gradient across the mirror. This may lead to stresses in the mirror's substrate and so to distortions of the optical surface.

Somewhat counterintuitively, therefore, active heating measures are used reduce these thermal stresses, both for the main test mass mirrors and the subsidiary mirrors and beam splitter. For AdvLIGO, the test mass mirrors have electrical ring heaters which can also be used to fine-tune the mirrors' radii of curvature (Section 1.6.3.2.1.2.1). The instrument's subsidiary mirrors have their temperatures regulated using carbon dioxide lasers whose emissions are strongly absorbed by the silica substrate of the mirrors. AdvVirgo and the early GEO600 similarly use ring heaters and with KAGRA the test masses of the instrument are all cooled to 20 K with the test mass mirrors' excess thermal energies being conducted away along their suspension fibres. GEO600 now uses 128 individually controlled heaters for thermal control which are separated from the optics and then imaged onto the beam splitter.

Apart from cooling the test masses (as with KAGRA), and as much of the rest of the instrument as possible, this thermal noise can only be controlled by designing the test mass mirrors' shapes to ensure the resonant frequencies are as high as possible and by using low-mechanical-loss materials, such as silica and sapphire (Section 1.6.3.2.1.2.5), for their construction.

1.6.3.2.1.1.4 Laser While just about every component of an MI-GWD, from the humble nuts and bolts upwards, has to do its job to perfection if the whole instrument is to work successfully, the source providing the light beams that go down the interferometer's arms and return to produce the interference pattern is a little more crucial than most parts.

The light sources are Nd:YAG[189] lasers for all current instruments. The lasers' outputs must be extremely stable. The lengths of the two interferometer arms are *not* identical (except perhaps occasionally by chance). They are adjusted to have extremely constant lengths that are fairly similar to each other. However, differences of several millimetres between those lengths may occur. Since the detection mechanism of the gravitational wave depends on a changing interference pattern, the arm

[189] Neodymium-yttrium-aluminium-garnet.

length difference does not matter, provided that the laser beam is coherent[190] over distances greater than the arm length differences. Nd: YAG lasers can be adjusted to have coherence lengths of many tens of kilometres, so they are ideal for this application.

The powers of the light beams within the interferometer arms are measured in hundreds of kW (e.g., 700 kW for AdvVirgo). Given that the initial laser power is just a few watts and this has to be divided between the two arms, clearly there must be several amplification and/or accumulation stages before the operational light beams are available.

As an example, the arm light beams in AdvLIGO start at a 4-W diode laser[191] emitting at 808 nm. This output is fed into a small Nd:YAG crystal (called Non-Planar Ring Oscillator [NPRO]). The 808-nm radiation acts as the pump for stimulated emission from the crystal and after undergoing several total internal reflections within the crystal, a 2-W laser beam emerges with a wavelength of 1.064 μm.[192] The power of that beam is then amplified twice. The amplifiers are high power oscillators (HPOs) which are essentially single pass lasers in which the output from the NPRO is fed into another Nd:YAG crystal[193] whose electrons are being pumped up into the laser's metastable state by a powerful external light source. The passage of the 1.064-μm radiation from the NPRO triggers stimulated emission within the crystal. The stimulated photons are also at 1.064 μm and are added in phase with those in the NPRO beam. The beam finally emerges with a gain of ~17.5 at 35 W. A second HPO then boosts the beam to 200 W. The power of the amplified light beam is stabilised by comparison with a constant voltage source. In practice, the input beam is not always used at full power so that, for example, for AdvLIGO's first run it was only 22 W.

Now to reach the 700 kW or so of the final beam, it might be expected that another two or three HPOs would be used. Instead the 200 W is not further amplified in power but is reused many times (Section 1.6.3.2.1.2.3) and multiply-reflected between the mirrors comprising the test masses at each end of an arm (Section 1.6.3.2.1.2.1) until the single beam in an interferometer is made up of many thousands of 200-W component beams.

1.6.3.2.1.1.5 Residual Gas Noise The whole of the optical systems of MI-GWDs operate in a vacuum. There are '2½' reasons for this. The first is that the atoms and molecules comprising the gas can collide with the mirrors inducing movements (cf. Brownian motion), so operating within a hard vacuum (typically 0.1 to 10 μPa) reduces this problem. Secondly, even a rarefied gas has a refractive index which can affect the effective path length of the interferometer arm. The '½' reason is that it allows any dust (whose presence is minimised anyway) to settle out. Dust can cause problems by scattering the photons in the light beams thus adding to the background photon noise of the system. More unexpectedly, dust particles can cause the silica fibre mirror suspensions to shatter by inducing micro-cracks into the fibres' surfaces when they settle there (AdvVirgo has suffered from this problem).

1.6.3.2.1.1.6 Control Interference Operation of the instrument, especially such actions as adjusting components' positions can introduce noise. This, however, can largely be avoided by careful design and appropriate operating practices.

1.6.3.2.1.1.7 Data Processing Processing the data from MI-GWDs is mostly relatively normal. One less-than-usual feature though, is the need to look for gravitational wave signatures within the data. Thus, for example, the final stages of a coalescence, such as GW150914, produce a signal that

---

[190] That is, the sine wave nature of the light beam is smoothly continuous and does not have any discontinuities.

[191] For comparison, the diode laser pointers widely used by teachers and lecturers are a few milliwatts in power.

[192] Frequency doubling to 532 nm would have some advantages, but at present such systems do not provide adequate power or stability for this application.

[193] Nd:YVO_4 for Virgo.

strengthens rapidly and also rises in frequency. In audio terms, it is something that we would hear as a 'chirp', with the highest frequencies around 1 kHz. Binary systems would be 'seen' as a steady sine wave, with harmonics if the objects' orbits were significantly elliptical. Their highest frequencies could be in the kHz region, but most examples will have much lower frequencies – μHz or smaller (i.e., the orbital periods for the binaries upwards of days or weeks) – and be undetectable at least until space-based detectors become available. The collapse phase of a supernova may generate waves with frequencies rising from less than 100 Hz to 1 kHz and with the rate of rise decreasing somewhat over a second of time or so. Other events probably not have simple characteristic signatures and will need to be found just as strong transient events coincident between many detectors.

Anyone with a computer and an internet link can help with gravitational wave data processing through the Einstein@home project. This currently has more than 500,000 volunteers whose computers process data for the project when they would otherwise be idle. An internet search for 'Einstein@home' will take anyone interested in the project straight to its home page.

1.6.3.2.1.2 Michelson Interferometer Gravitational Wave Detectors – Operation

1.6.3.2.1.2.1 Beam Length The light beam in an MI-GWD arm is reflected back and forth many times between the input and end mirrors (test masses). There are several reasons for this approach but the two main ones are to increase the effective length of the arm and to increase the light beam's intensity. With AdvLIGO, the strain of 10^{-21} caused by GW150914 changed the 4-km[194] separation of its test masses by about 4×10^{-18} m (the often quoted 1/400 of the size of a proton). However, the light beams in AdvLIGO's arms are multiply-reflected some 280 times, making the effective arm length (also called the 'optical path') about 1,120 km. Once the light present in the arms at time of the gravitational wave's occurrence has cycled through, the effective arm length will have changed by $280 \times 4 \times 10^{-18}$ m or just over 10^{-15} m (about two-third of the diameter of a proton). The latter figure, of course, is still small and only about 10^{-9} of the 1.064-μm wavelength of the light in the beams, but this amplification is one of the major contributors to the instrument's ability to detect the actually much smaller physical shifts due to the gravity wave.

Multiple reflections such as those in AdvLIGO's arms can be realised in practice in two main ways: delay lines and Fabry-Perot etalons. Delay lines generally require larger mirrors than etalons and so it is the latter that have been used to-date by the existing and planned MI-GWDs, except for GEO600, which uses a three-reflection delay line (i.e., it simply folds the beam in half).

In a delay line the mirrors are (ideally) *perfectly* reflecting and the light beam enters through one hole in the mirror and exits through another hole. In this situation, the number of light beams in the cavity is clear-cut. The total number of light beams travelling both up and down the cavity is simply the number of reflections plus one. Slightly confusingly, this is called the 'bounce number', and so GEO600 has three reflections and a bounce number of four.[195] With its physical arm length of 600 m, it thus has an effective arm length of 2,400 m.

The theoretical basis of Fabry-Perot etalons is covered in Section 4.1.4.1 in the physically much smaller context of spectroscopic and filter applications. The large constructs used in MI-GWDs are usually termed Fabry-Perot cavities (FPCs). The essential point in etalons and FPCs is the use of opposed, *partially* reflecting mirrors which transmit light through the whole system only when rays with differing numbers of reflections are in phase with each other.

In an FPC, the input mirror is deliberately made to be only partially reflecting (the end mirror is made as close to 100% reflectivity as is practically possible). The input beam will thus bounce back and forth between the mirrors, losing a small part of its intensity each time it returns to the input mirror by transmission through that mirror. In AdvVirgo, the input mirror has a transmissivity of 1.4% and so (assuming an end mirror with 100% reflectivity and ignoring any other losses)

---

[194] Actually 3.996 km.

[195] Readers interested in further information on these devices should research 'folded optical delay lines' and/or 'Herriott delay lines'.

only 98.6% of a returning beam will be reflected again. If the intensity of the input beam as it first emerges from the input mirror into the FPC is put at 1, it will be 0.986 after 1 return journey, 0.972 after 2 return journeys, 0.959 after 3 return journeys, … 0.868 after 10 return journeys, … 0.244 after 100 return journeys, … 0.029 after 500 return journeys … and so on. Mathematically at least, it will require an infinite number of reflections to reach zero intensity. Unless we say (not very usefully) that AdvVirgo's FPCs have bounce numbers of ∞, it is not clear what the effective length of the instrument's arms should be. Therefore, for FPCs, the weighted average bounce number (ABN) is used and this is based upon the finesse (Equation 4.68) of the cavity:

$$\text{ABN} = \frac{2 \times \text{Finesse}}{\pi} = \frac{2\sqrt{R}}{1-R} \tag{1.120}$$

where R is the mirror's reflectivity.

Since R is close to unity, the $(1 - R)$ denominator in Equation 1.120 means that the ABN value is sensitive to small changes in the reflectivity or to other loss mechanisms. The effective arm lengths in FPC-based MI-GWOs are thus not necessarily of fixed values. For AdvLIGO, the effective arm lengths are ~1,120 km, whilst for AdvVirgo they are ~120 km.

We are accustomed (especially in every-day life) to regarding light as moving around practically instantaneously. However astronomers, familiar with using units of light-years, should know better and many people will have noticed when watching live TV reports and the like from distant news spots, that there are often quite noticeable breaks in the conversation between a locally based news-caster and the distant reporter – half a second or more if the broadcast is routed through a couple of geostationary communications spacecraft. For gravity wave detections, where we are concerned with timescales of nanoseconds and less, the 'slow' speed of light[196] becomes a major concern.

Thus, a photon of light entering one of AdvLIGO's 1,120-km effective arms will be trapped in there for about 3.7 ms, whilst AdvVirgo will hold onto a photon for about 400 μs. This is called the 'storage time of the detector', and it imposes limits on the usable frequency range of an MI-GWD. A 1-kHz gravity wave has a wavelength of 300 km and so AdvLIGO's effective arms will span nearly four such wavelengths. Now the sampling theorem (Section 2.1.1) tells us that to obtain all the information in a signal with a certain frequency, it suffices to sample it at twice that frequency (i.e., at physical intervals of half the signal's wavelength). For AdvLIGO, the ideal gravitational wave frequency would thus be in the region of 135 Hz (wavelength 2,200 km) and its sensitivity will be lower for 1 kHz signals

At higher frequencies (shorter wavelengths), there will be many cycles of the gravitational wave acting along the MI-GWD's arms simultaneously. The output from the instrument will thus become smoothed out (blurred) as the effects from the many cycles start to cancel each other out.

These two limitations mean that AdvLIGO should provide good measurements from about 100 Hz (wavelength 3,000 km) to about 800 Hz (wavelength 400 km) and useable ones over ~10 Hz to a few kHz. AdvVirgo, with its shorter effective arm lengths, should be best at about 1 kHz and provide good measurements from about 500 Hz (wavelength 600 km) to about 5 kHz (wavelength 60 km) and useable ones over ~100 Hz to a few tens of kHz.

FPCs in their spectroscopic application (etalons; see Section 4.1.4.1) use mirrors that are flat, usually parallel to each other and have separations of up to a few tens of millimetres. When a beam of light of mixed wavelengths illuminates the device at a slight angle to its normal, only those wavelengths whose paths after 0, 2, 4, … reflections differ by a whole number of wavelengths emerge from it. Scanning the angle of incidence then allows the whole spectrum to be built up.

[196] In this context, a useful, if non-SI, conversion is that the speed of light is about 1 ft/ns (actually 0.9836 feet/ns).

Narrow-band filters are similar in their principles but are illuminated normally to the surface and have much smaller mirror separations. Only the desired wavelengths emerge from the device.

By contrast, the FPCs in GI-GWDs usually use concave mirrors. This ensures that the beams are kept centred within the cavity (any slight mutual inclination of *flat* mirrors would result in the beams progressively moving sideways across those mirrors), it spreads the beams over larger areas of the mirrors (thus reducing thermal gradients), refocuses the individual parts of the overall beam and spreads out the radiation pressure on the mirrors so that any unbalanced twisting forces are minimised. The input mirror has a radius of curvature slightly less than half the test mirror separation and the end mirror slightly more than half so that the beam is more concentrated at the input mirror. AdvVirgo, for example, with a separation of 3,000 m has mirror radii of 1,420 and 1,683 m. The thermal control heaters (Section 1.6.3.2.1.1.3) can be used to fine tune the mirrors' curvatures and so adjust their radii.

1.6.3.2.1.2.2 Optical Detectors The MI-GWDs themselves are detectors of gravitational waves, but they operate using 1.064-μm light beams. Therefore, they need optical detectors within them and these are generally relatively simple InGaAs p-i-n photodiodes (Section 1.1.9.1). There are two main approaches to the light detection – the radio frequency (RF)[197] and the direct current (DC)[198] schemes. The RF scheme was used for the earlier instruments and the DC scheme for today's second-generation instruments.

The problem with the detection is that the frequency of the radiation in the light beams is 2.82 THz, while the photodiodes' fastest response is about 100 MHz. Thus, the diodes cannot detect the beams' phase changes directly. In the RF scheme, a low-frequency local oscillator signal is added to the laser signal producing sidebands that differ from the laser's frequency by multiples of the local oscillator's frequency. When the arm lengths are adjusted for dark fringe operation at the laser's frequency (sometimes called the 'carrier frequency'), it is not at the dark fringe position for the sidebands and these emerge towards the optical detector. When there is a gravitational wave affecting the MI-GWD, then this will also be at a low frequency (10–1,000 Hz) and will also produce sidebands to the laser's main frequency. The output to the optical detector will then contain beat frequencies between the sidebands to which it can respond.

In the DC scheme, there is no separate local oscillator. Instead the instrument is operated slightly away from the dark fringe position. Some laser light thus emerges towards the optical detector and this acts as the local oscillator.

1.6.3.2.1.2.3 Power Recycling As we have seen previously, in the absence of a gravitational disturbance, the two returning beams from the arms of the interferometer are adjusted to be in anti-phase when at the beam splitter (dark fringe operation or nearly so, see Sections 1.6.3.2.1 and 1.6.3.2.1.2.2). This means that the output from the detector is then zero. It also means that beams directed back towards the laser by the beam splitter contain all the energy from the arms.[199] A partially reflecting mirror (95% reflectivity for AdvLIGO for example) placed between the laser and the beam splitter then returns most of this energy to the instrument. When operating at the dark fringe, the beam splitter and the arm FPCs are equivalent to a single highly reflective mirror and so with the additional (recycling) mirror they form a second FPC. This procedure is called 'power recycling' and its effect is to multiply the input power by the ABN of the recycling FPC; in AdvLIGO, it is around ×40, whilst GEO600 has a power gain of ×1,000.

---

[197] Or heterodyne.

[198] Essentially a heterodyne receiver in which the LO is at the same frequency as the signal and so there are no beat frequencies (IFs). Also called a Homodyne system.

[199] If it seems that both the 'detector beam' and the 'laser beam' should have their two component beams in anti-phase, then remember that the internal reflection involves a phase change of 180°.

Together with the multiple reflections in the main arms' FPCs (Section 1.6.3.2.1.2.1), the combined effect is to boost the beam energy in the MI-GWD's arms up to the required hundreds of kW (Sections 1.6.3.2.1.1.1 and 1.6.3.2.1.1.4). It takes a little while though, for this boost to reach its maximum effect – without any losses at all, it would take, for example, the 200-W output from AdvLIGO's amplified laser about an hour to reach 700 kW.

1.6.3.2.1.2.4 Signal Recycling When a gravitational disturbance is received by an MI-GWD, the normal zero output from the detector changes to a real output. There is thus energy in the beams of radiation sidebands emerging from the beam splitter. Like power recycling (Section 1.6.3.2.1.2.3), this can be amplified by a partially reflecting mirror before the detector. Such amplification is called 'signal recycling' and can improve the instrument's sensitivity by up to a factor of ten or more.

Adjustments to the signal recycling mirror can be used to fine-tune the instrument as a whole. Thus, for AdvVirgo, small changes in the mirror's position change the instrument's most sensitive frequency, while changing the reflectivity/transmissivity of the partially reflecting mirror changes its bandwidth.

Power recycling (Section 1.6.3.2.1.2.3) and signal recycling when both are used in the same MI-GWD are often called dual recycling.

1.6.3.2.1.2.5 Squeezing Squeezed light[200] is a phenomenon closely related to quantum entanglement. In squeezed light, the photons in a light beam are arranged in a more uniform order than normal, so that the shot noise (Section 1.6.3.2.1.1.1) is reduced (by up to a factor of ten). The technique uses a doubly refracting non-linear crystal[201] (Section 5.2.2.1, Figure 5.11) illuminated by light at half the GWD's laser's operating wavelength (i.e., 532-nm green light for current GWD instruments). The excited crystal electrons are then able to store photons at the laser's wavelength of 1.064 μm. When the laser light passes though the crystal some of its photons are stored in this manner and some are re-emitted back into the beam, until an equilibrium is reached. When a shot noise fluctuation increases the laser beam's intensity, more of the photons are removed from the beam and stored, whilst when the fluctuation reduces the beam intensity, stored photons are released back into the beam, thus smoothing out the fluctuations.

The GEO600[202] started using squeezed light in 2010, improving its sensitivity by 50% at kHz frequencies and AdvLIGO and AdvVirgo adopted the technique in 2015 giving a similar improvement. More recently, with improved optics and tuning GEO600 has almost doubled its sensitivity through further improvements in the squeezing system and at the time of writing, both AdvLIGO and AdvVirgo are expected to be using state-of-the-art squeezed light systems shortly.

1.6.3.2.1.2.6 Mirrors (Test Masses) The test masses in an MI-GWD are mirrors and their properties are as crucial to the instrument's operation as those of the laser light source (Section 1.6.3.2.1.1.4). The manner in which the mirrors are held in place and their protection from seismic noise sources are discussed in Section 1.6.3.2.1.1.2, here we are concerned with their optical properties.

In Section 1.6.3.2.1.2.1, we saw that to increase the interferometer's optical arm length where the light beam is reflected many times between the two end mirrors before emerging. This requires the mirror to have an extremely high reflectivity so that absorption losses are minimised and to have an extremely accurate surface so that the beam keeps to its required track.

---

[200] Also known as squeezed vacuum, squeezed states.

[201] Such as $MgO{:}LiNbO_3$ or $KTiOPO_4$.

[202] The second-generation instruments, AdvLIGO and AdvVirgo are essentially complete re-builds of their prior versions, taking several years to complete and with the instruments then being out of use. KAGRA is a new-build instrument yet to start operations. GEO600 has adopted an incremental upgrading programme called GEO-high frequency (GEO-HF) which has enabled it to continue operating without long-term unavailabilities.

The mirrors are formed from fused silica (SiO_2) or, in KAGRA, artificial sapphire (Al_2O_3), with multilayer reflective coatings. For AdvVirgo, for example, the coating layers are silica and titanium-doped tantalum oxide ($Ti{:}Ta_2O_5$) giving absorptions levels of around 300–400 parts per billion, The instrument's input mirrors' transmissions are then 1.4%, while the end mirrors have transmissions very close to zero (<0.0001%) and so reflectivities very close to 100%. The KAGRA's mirrors similarly have reflectivities of 99.6% and 99.9945%.

The surfaces of the mirrors are figured to standards around 50 times better than those of a typical optical reflecting telescope. Overall, the mirror surfaces are within a few nanometres of theoretical perfection and the roughness (small-scale irregularities) is less than a few tens of picometres.

1.6.3.2.1.2.7 Mode Cleaner Two additional components of some MI-GWDs are the mode cleaners: the input mode cleaner placed between the laser and the power recycling mirror and the output mode cleaner placed between the signal recycling mirror and the detector. These help to suppress beam jitter and remove unwanted higher order modes of the beam arising from mirror imperfections by using three-mirror ring cavities resonant with the fundamental mode of the laser.

1.6.3.2.1.3 Michelson Interferometer Gravitational Wave Detectors – Instruments The MI-GWDs currently operating have all been mentioned in the previous sections. Here their details (where known) are summarised (Table 1.8) alongside those of some future possibilities. Early versions of some detectors, whose current details are given in full, are omitted from this table.

1.6.3.2.2 Doppler Tracking

An ingenious idea underlies the second method in the direct, non-resonant GWD class. An artificial satellite and the Earth will be independently influenced by gravity waves whose burst length is less than the physical separation of the two objects. If an accurate frequency source on the Earth is broadcast to the satellite and then returned to the Earth (Figure 1.111), its frequency will be changed by the Doppler shift as the Earth or satellite is moved by a gravity wave. Each gravity wave pulse will be observed three times enabling the reliability of the detection to be improved. The three detections will correspond to the effect of the wave on the transmitter, the satellite and the receiver, although not necessarily in that order.

A drag-free spacecraft (Figure 1.112) would be required for this approach. Such a spacecraft comprises a test mass positioned at the centre of a hollow satellite (the shield). After arriving at its desired orbit, all physical links between the test mass and the shield are removed. The position of the test mass within the shield is monitored (capacitatively or by lidar, etc. – Figure 1.112), and the shield is then driven by thrusters to keep the test mass centred. Since the test mass will follow an orbit dictated only by the gravitational fields that it encounters (including the gravitational wave), then so also will the shield. The position and/or velocity of the shield can then be measured just as for any other spacecraft, but it will not have been influenced by external non-gravitational perturbations arising from the solar wind, radiation pressure, atmospheric drag, etc.

Although a mission specifically designed to detect gravitational waves using a drag-free spacecraft has not yet taken place, spacecraft on other missions have had their transmissions monitored for the purpose. These missions include Ulysses, Galileo and Cassini-Huygens, but so far this has not resulted in any detections (strains of about 10^{-15} or a little smaller could have been picked up).

1.6.3.3 Pulsar Timing Arrays

The third direct(ish) approach to gravitational wave detection, after resonant and non-resonant GWDs, is to detect the waves in the microhertz frequency region from their effects upon the arrival times of pulses from pulsars. A gravitational wave contracts space in some directions and expands it in others as it passes through the solar system. The pulses should therefore arrive earlier or later than normal by a few nanoseconds. By monitoring the arrival times of the pulses from several tens of pulsars, distributed around the whole sky, the passage of the wave should be recordable.

TABLE 1.8
Interferometric Gravitational Wave Observatories

Name	Location Operating Agency or Countries	Start of Operations Current Status	Mirror Separation (*m*)	Effective Path Length (*m*)	Sensitivity	Notes
A+ LIGO	Hanford & Livingston NSF plus others	2020–2026? Planned improvements for AdvLIGO	4,000	1.12×10^6	$\tilde{h} = \sim 2 \times 10^{-24}$ Hz$^{-0.5}$ at ~300 Hz $h = 10^{-23}$ at ~300 Hz	G-B GWD Squeezed light
AGIS	?	?	1,000	?	$\tilde{h} = \sim 10^{-19}$ Hz$^{-0.5}$ at a few Hz	S-B GWD Possible future new design based upon atom interferometers (instruments in which the wave nature of atoms enables them to act like radiation-based interferometers). A space-based version might be comparable with LISA.
AIGSO	Space	2030s? 2040? Concept			$\tilde{h} < 10^{-20}$ Hz$^{-0.5}$ at mHz frequencies	S-B GWD Possible future new design based upon atom ring interferometers and the Sagnac effect.[a] Three drag-free spacecraft (Section 1.6.3.2.2) in Earth orbit.
AdvLIGO	Hanford (2 instruments) and Livingston NSF plus others	2015 Operating periodically	4,000	1.12×10^6	$\tilde{h} = \sim 4 \times 10^{-24}$ Hz$^{-0.5}$ at ~300 Hz $h = \sim 3 \times 10^{-23}$ at ~300 Hz	Second-generation G-B GWD First detection – GW150914 Power recycling Signal recycling Squeezed light Storage time – 3.7 ms Possibly one Hanford Instrument may be moved to India to form INDIGO
ASTROD-GW	Space China	2020s? Design study	2.6×10^8	2.6×10^8	$\tilde{h} \approx 10^{-22}$ at mHz frequencies	S-B GWD 100-nHz to 1-mHz coverage Three drag-free spacecraft near Sun-Earth L3, L4 and L5 points
AdvVirgo	Cascina EGO	2017 Operating periodically	3,000	1.2×10^5	$\tilde{h} = \sim 3 \times 10^{-24}$ Hz$^{-0.5}$ at ~100 Hz	Second-generation G-B GWD First detection – GW170814 Power recycling Signal recycling Squeezed light Storage time – 300 μs

(Continued)

TABLE 1.8 (*Continued*)
Interferometric Gravitational Wave Observatories

Name	Location Operating Agency or Countries	Start of Operations Current Status	Mirror Separation (*m*)	Effective Path Length (*m*)	Sensitivity	Notes
BBO	Space ESA	2030s? 2040s? Concept	5×10^7	5×10^7	?	Third- (or fourth-)generation S-B GWD Three to 12 drag-free spacecraft in one to four triplets in heliocentric orbits
Cosmic Explorer	Hanford & Livingston? NSF plus others	2030s? 2040s? Concept	4×10^7?	?	?	Third-generation G-B GWD Possible larger replacement for LIGO Voyager. Cryogenic, with 40-km arms and a factor of ten improvement on AdvLIGO. Possibly three double arms (cf. ET).
DECIGO	Space Japan	2027? Planned	10^6		$h = 10^{-24}$ at ~1 Hz 100-mHz to 10-Hz coverage	S-B GWD Three drag-free spacecraft Possibly may involve up to two precursor test-bed missions – DECIGO-Pathfinder and/or Pre-DECIGO
ET	? Europe	Design study 2020s? 2030s?	10^4		$\tilde{h} = \sim 3$ to $\sim 6 \times 10^{-25}$ $Hz^{-0.5}$ at a few hundred Hz	Third-generation G-B GWD Underground Three arms in an equilateral triangle with two detectors at each corner Frequency bands 1–250 Hz and 10 Hz–10 kHz with beam powers of 18 kW and 3 MW, respectively Polarisation sensitive
GEO600 -HF	Sarstedt Germany & UK	2002 Operating periodically	600	2400	$\tilde{h} = \sim 2 \times 10^{-22}$ $Hz^{-0.5}$ at ~600 Hz $h = 10^{-21}$ at ~600 Hz	First- and second-generation G-B GWD GEO-HF is an up-grade to the instrument Three-reflection delay lines Power recycling Signal re-cycling Squeezed light

(*Continued*)

TABLE 1.8 (*Continued*)
Interferometric Gravitational Wave Observatories

Name	Location Operating Agency or Countries	Start of Operations Current Status	Mirror Separation (*m*)	Effective Path Length (*m*)	Sensitivity	Notes
INDIGO	Hingoli India	2020s?	4,000 (?)	1.12×10^6(?)	$\tilde{h} = \sim 4 \times 10^{-24}$ Hz$^{-0.5}$ at ~300 Hz (?) $h = \sim 3 \times 10^{-23}$ at ~300 Hz (?)	Second-generation G-B GWD Power recycling Signal recycling Squeezed light Storage time – 3.7 ms (?) Possibly one Hanford AdvLIGO Instrument to be moved to India to form INDIGO
KAGRA	Kamioka ICRR	December 2019 Being commissioned	3,000	?	$\tilde{h} = \sim 3 \times 10^{-24}$ Hz$^{-0.5}$ at ~100 Hz $h = \sim 3 \times 10^{-23}$ at ~100 Hz	Second-generation G-B GWD Underground Cooled to 20 K Power recycling Signal recycling
LIGO Voyager	Hanford & Livingston NSF plus others	2030s? Concept	4,000	1.6×10^6	?	Second- and third-generation G-B GWD
LISA	Space ESA (+ NASA?)	Early 2030s Planned – LISA pathfinder has tested the technology required	2.5×10^9	2.5×10^9	$\tilde{h} \approx 4 \times 10^{-21}$ at mHz frequencies $h < 10^{-23}$ at mHz) frequencies	S-B GWD Three drag-free spacecraft separated by 2.5×10^6 km in a 1 AU orbit 20° behind the Earth, gLISA is a variant using spacecraft in geosynchronous orbits

(*Continued*)

TABLE 1.8 (*Continued*)
Interferometric Gravitational Wave Observatories

Name	Location Operating Agency or Countries	Start of Operations Current Status	Mirror Separation (*m*)	Effective Path Length (*m*)	Sensitivity	Notes
LISA pathfinder	Space ESA	2015 Decommissioned 2017	0.38	0.38		S-B GWD Drag-free spacecraft in an Earth L1 orbit Test bed for LISA – two test masses' positions measured to an accuracy better than 10 pm.
MAGIS-100	Space	2030s? 2040s? Concept				S-B GWD Interferometer using ultracold strontium atom clouds falling 100 m vertically in a vacuum as the test masses
MFB	Space	2030s Concept	7×10^7	7×10^7	10 mHz to 1 Hz	SW-B GWD Drag-free spacecraft in a geosynchronous orbit
Tian Quin	Space China	2025–2030 Construction started	10^8	10^8		S-B GWD Three drag-free spacecraft in an Earth orbit

[a] An effect whereby in a rotating ring interferometer, the clockwise light beam's transit time differs from the anticlockwise light beam's transit time. Further details are beyond the scope of this book and are left for interested readers to research for themselves.

Abbreviations: AdvLIGO, Advanced Laser Interferometer Gravitational-wave Observatory; AdvVirgo, Advanced Virgo; AIGSO, Atom Interferometric Gravitational wave Space Observatory; AGIS, Atomic Gravitational wave Interferometric Sensor; ASTROD-GW, Astronomical Space Test of Relativity using Optical Devices-Gravity Waves; AU, astronomical unit; BBO, Big Bang Observer; DECIGO, Deci-Hertz Interferometer Gravitational wave Observatory; EGO, European Gravitational Observatory; ESA, European Space Agency; ET, Einstein gravitational wave Telescope; G-B GWD, Ground-Based Gravitational Wave Detector; GEO600-HF, GEO600-High Frequency; ICRR, Institute for Cosmic Ray Research (University of Tokyo); INDIGO, Indian Initiative in Gravitational wave Observations; KAGRA, Kamioka Gravitational wave detector; LIGO, Laser Interferometer Gravitational-wave Observatory; LISA, Laser Interferometer Space Antenna; MAGIS, Matter-wave Atomic Gradiometer Interferometric Sensor; MFB, Mid-Frequency Band; NASA, National Aeronautics and Space Administration; NSF = National Science Foundation; S-B GWD, Space-Bassed; Gravitational Wave Detector; UK, United Kingdom.

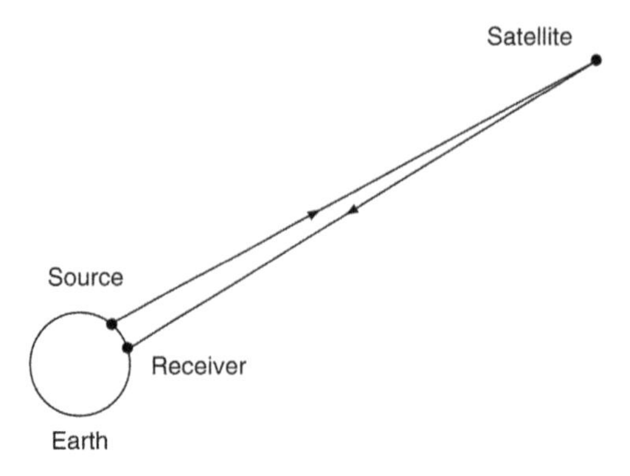

FIGURE 1.111 Arrangement for a satellite-based Doppler tracking Gravitational Wave Detector (GWD).

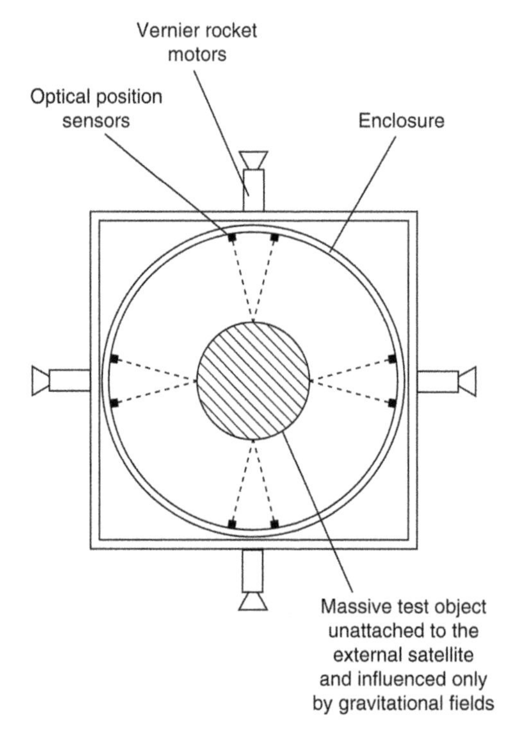

FIGURE 1.112 Schematic cross section through a drag-free spacecraft. The position of the massive test object is sensed optically and the external satellite driven so that it keeps the test object centred.

Several projects such as the European Pulsar Timing Array (EPTA),[203] North American Nanohertz Observatory for Gravitational Waves (NANOGrav)[204] and the Parkes Pulsar Timing Array (PPTA)[205] are presently searching for gravitational waves in this manner from supermassive black hole binaries, so far without success, although upper limits for h of less than a few times 10^{-15} for the background intensity of these waves have been set. The blazar, Ohio Radio Survey, object 287 (OJ 287),[206] varies in its optical brightness quasi-periodically every 11 to 12 years. This may arise from the effects of a ~$1.8 \times 10^9 M_\odot$ central binary black hole pair. If correct, it is predicted to have a peak GW strain

203 Lovell, Westerbork, Effelsberg and Nançay radio telescopes, see Section 1.2.
204 Green Bank and Arecibo radio telescopes.
205 Parkes radio telescope.
206 The 'J' indicates an RA of 9^h.

amplitude of nearly 10^{-15} and so potentially be detectable at the time of writing, but it is likely to fade for some years after 2021. When the SKA is in full operation, far more pulsars will be able to be included in these monitoring programmes.

1.6.3.4 Indirect Detectors

Proposals for these so far only involve binary star systems. The principle of the method is to attempt to observe the effect on the period of the binary of the loss of angular momentum as a result of the gravitational radiation. A loss of angular momentum by some process is required to explain the present separations of dwarf novae and other close binary systems with evolved components. These separations are now so small that the white dwarf would have completely absorbed its companion during its previous giant stage, thus the separation must have been greater in the past. Gravitational radiation can provide an adequate orbital angular momentum loss to explain the observations, but it is not the only possible mechanism. Stellar winds, turbulent mass transfer and tides may also operate to reduce the separation, so that these systems may not provide unequivocal evidence for gravitational radiation.

The prime candidate for studying the evidence of low-frequency gravitational radiation is the binary pulsar (PSR 1913 +16). Its orbital period is 2.79×10^4 s, and it is decreasing at a rate of 7.56×10^{-5} s/y. If the decrease in the binary's period were the result of gravitational radiation, then GR would predict a value of 7.57×10^{-5} s/y for the decrease. The agreement between these two figures is striking, and prior to GW150914, they provided the strongest evidence by far for the actual existence of gravitational waves. Thus, the 1993 Nobel Prize for Physics was awarded to Russell Hulse and Joseph Taylor for their discovery of the pulsar and their interpretation of its orbital decay.

However, it has been suggested that a helium star may be a part of the PSR 1913 +16 system and so tides or other interactions might also explain the observations. GW150914's data though now give the strong support to the gravitational wave interpretation of PSR 1913 +16's binary period changes.

Even if correctly interpreted, PSR 1913 +16's binary gravitational wave emissions are currently far too weak at a strain of $h \approx 10^{-26}$ for direct detection by the current or any envisaged MI-GWDs. Since the radio emissions from pulsars are tightly beamed and the PSR1913 +16 system is processing, it will soon no longer be visible from the Earth; long-lived gravitational wave astronomers, however, can look forward to observing an intense gravitational wave burst in about 300 million years, when the two neutron stars coalesce.

To date, the only gravitational wave event to have been clearly associated with electromagnetic outbursts is GW170817, the double neutron star merger. This is no coincidence. Although black hole mergers are extremely violent events, they generally take place in a region of space that has been thoroughly cleared of 'ordinary' material by the black holes before their merger. There is, thus, no ordinary matter left to emit the electromagnetic radiation. However, some black hole mergers may leave the resulting more massive single black hole with a high velocity with respect to its immediate neighbourhood – perhaps up to 100 km/s. That single, more massive, black hole's gravitational field could then pull distant gaseous material with it as the black hole moves through space. The interactions of the disturbed gas with other, still unaffected, gas would then lead to electromagnetic emissions. Such a process might thus provide a means of detecting binary black hole mergers long after their occurrence, but some way would need to be found for detecting this type of turbulent gas electromagnetic emission from all other similar, but non-black-hole-merger, processes.

1.6.3.5 The Future

Space-based MI-GWDs such as LISA (Table 1.8) are currently being planned for operation in the 2020s, 2030s, and after. LISA pathfinder has demonstrated the feasibility of much of the technology that will be required. These missions use single beam interferometers with path lengths ranging from 10^3 to 10^9 km. Most designs employ three drag-free spacecraft (Section 1.6.3.2.2) at the corners of isosceles triangles. In many ways, they are simpler than the ground-based instruments but do have their own complications, such as having to aim their interferometer beams at the points in space where the receiving spacecraft will be when the beams get to them tens of seconds or minutes after they have been emitted.

The gravitational wave strains that the space-based GWDs are expected to be able to detect are comparable with the present-day AdvLIGO and AdvVirgo sensitivities, but the space-based systems will operate at mHz frequencies, where noise from variations in the Earth's gravitational field masks extra-terrestrial emissions for ground-based GWDs, and so the space-based systems should be able to detect many binary systems, super-massive black hole mergers and stellar mass objects falling into super-massive black holes directly. LISA and similar instruments may be able to self-calibrate their phases using the gravitational wave emissions from ultra-compact binaries. These binaries, like pulsars, will emit their signals at almost constant frequencies and should be detectable by LISA with about a week of observations. Thus, they may help to improve the instruments' stabilities as well as being observed for their own interest. In what might seem a surprising suggestion, given that, to date, only black holes and neutron stars with masses many times that of the Sun have been detected, it may be that LISA will be able to detect exoplanets with masses above 50 times that of the Earth (~0.000 15 $M_{\odot}$). Such exoplanets, though, would need to be in orbit around binary white dwarf systems and the detection would be of the Doppler shift produced in the gravitational wave signals from the binaries by the exoplanets. LISA and other space-based instruments, as well as the proposed terrestrial Einstein telescope, should be able to detect the gravitational waves from core-collapse supernovae throughout most of the Milky Way galaxy, if not beyond. It is perhaps possible that LISA or even AdvLIGO or AdvVirgo could detect dark matter if the latter interacts sufficiently with their test mass mirrors to cause them to move (Section 1.7). Microhertz gravitational waves might be detectable using a space-based interferometer using two beams of light from a bright star.

Another suggestion which might be cheaper than LISA but perhaps have a similar level of sensitivity could be to use a ground-based laser feeding several geostationary spacecraft visible from its site.

One suggested approach to detecting low-frequency gravitational waves is to use the Gaia spacecraft's (or some similar future mission's) data to look for changes in star's positions (i.e., their proper motions). The star positions would change as the gravitational wave passed through the Earth. Although the change for an individual star would be small and lost in the noise, the wave would cause all the observed stars' positions to change in a regular pattern depending upon the nature of the gravitational disturbance and so might be detectable. Since Gaia would have to measure many star positions at least twice, only slow changes would be detectable; Gaia's first and second data releases took three and five years, respectively, to appear after its launch. The method would thus not detect rapid events like GW150914 but might detect components of the cosmic gravitational wave background.

Another gravitational wave detection method might be via monitoring regular variable stars' brightness. A variable star with the same frequency as a passing gravitational wave would absorb energy from the wave and brighten slightly. Like the previous suggestion, this would be best suited to detecting low-frequency disturbances. Similarly, when stellar surface velocities can be measured to be better than a millimetre per second, then these might also show evidence of gravitational disturbances.

The levels of precision now reached by atomic clocks (better than one part in 10^{18}) is probably good enough that low-frequency gravitational waves would be detectable via their time dilation effects upon an array of such clocks in heliocentric orbits. The clocks onboard global positioning spacecraft, with precisions of about one part in 10^{15}, have in this way set an upper limit to the intensity of the background gravitational radiation in the mHz region.

Low-frequency gravitational waves may also be detectable by test masses comprising long bars. Suspended perpendicularly at their centres of rotation, a gravitational wave would cause one bar to move through an angle in one direction while the other moves in the opposite direction. Angle changes between the bars as small as 10^{-17} of a degree would need to be measured.

MHz frequency gravitational wave may be detectable using a double Michelson interferometer system. The Fermilab's holometer instrument was designed to study the nature of space-time at the attometre scale. It was essentially two MI-GWDs with 40-m arms, and it operated by correlating the outputs from the interferometers when they were orientated in 'nested' or 'back-to-back' arrangements. Fermilab's holometer failed to detect gravitational waves, but a more sensitive instrument might do so. Other high frequency (HF)-GWDs based upon small FPCs and other resonant cavities have set upper limits on $\tilde{h}$ of around 10^{-16} $Hz^{-0.5}$ for the 100-MHz gravitational wave background and these may be improved in the future.

Significant improvements to MI-GWDs may be brought about by the use of FPCs and delay lines in combination. Also known as 'folded FPCs', these devices might reduce noise by illuminating the test mass mirrors more uniformly. Detectors physically larger than LIGO (up to 40-km long arms) are also under study. Ring interferometer systems using fibre-optic delay lines may provide yet another means of detecting gravitational waves.

The possibility of using ultra-cold clouds of atoms as the test masses has already been mentioned (Table 1.8). A rather similar method would be to use small test masses optically levitated inside a cavity. A gravitational wave would set such a mass oscillating at its resonant frequency, thus combining the resonant (Section 1.6.3.1) and non-resonant (Section 1.6.3.2) approaches to the waves' detection.

1.7 DARK MATTER AND DARK ENERGY DETECTION

1.7.1 Introduction

We know that an object is in our Universe because it emits, reflects or scatters e-m radiation, gravitational radiation, cosmic rays or neutrinos (or has done so in the past) towards us and which we then detect. Thus, we generally regard the Universe as made up from planets, stars, nebulae, gas clouds, galaxies, etc. Such objects themselves are formed from atoms, molecules, ions and nuclei, which in turn are built up from protons, neutrons and electrons. Now protons and neutrons are subatomic particles classed, along with many lesser known particles, as baryons.[207] Hence, the matter that we encounter in our everyday lives, the Earth, the Sun and all the other 'normal' components of the Universe is called 'baryonic matter'.

Main stream cosmology, though, currently suggests that baryonic matter makes up only about 5% of the total mass-energy content of the Universe.[208] This prediction is based upon two main lines of evidence. First, from the abundances of nuclei of light elements and isotopes, such as deuterium, helium and lithium, which were mostly built up from protons and neutrons in the first 20 minutes or so following the Big Bang. These primordial element abundances require the Universe's content of baryonic matter to be close to 5%. Secondly, irregularities in the CMB suggest a similar total abundance of baryonic matter.[209]

---

[207] Baryons are composite subatomic particles formed from three quarks. Electrons, along with neutrinos (Section 1.5), are called 'leptons' and are, as far as we can tell at the moment, fundamental particles (i.e., they are not constructed from even more basic units). Further details of elementary particle physics are beyond the scope of this book and are left for interested readers to research for themselves.

[208] In many branches of physics, mass and energy are regarded as just different manifestations of the same entity. Mass and energy may be inter-converted via Einstein's equation (see Section 1.5.1). The total amount of mass and energy and expressed in mass or energy units as may be convenient is a quantity called mass-energy. The total mass-energy content of the Universe is thought to be just sufficient for gravity to halt the expansion of the Universe after an infinite length of time and it sometimes called the 'Critical Mass'. It implies that the average density of the Universe today (the critical density) must be about 10^{-26} kg m^{-3} – if the Universe were composed exclusively of hydrogen atoms, this would give a density of about 6 hydrogen atoms per cubic metre (about 10^{-30} times the density of a human body).

[209] It is worth noting though, that recently, two separate lines of evidence (light element abundances in old stars and the possible changes in the maximum brightness of type 1-a supernovae) have cast some doubt upon the existence of both dark matter and dark energy.

Most of the Universe is, therefore, *not* in the form of planets, stars, nebulae, gas clouds or galaxies. Furthermore, if we add up the masses of all the planets, stars, nebulae, gas clouds, galaxies, and so on which we currently *can* observe, then they amount to only about 2.5% of that total mass-energy and not to the required 5%.

Hence of the missing ~97.5% of the Universe, missing baryonic matter may form another 2% to 3% (and that may now no longer be 'missing' – see below), but some ~95% of the Universe's content is still presently unobserved by ourselves. That missing content is thought to take two forms – matter (dark matter or missing mass/matter) and energy (dark energy). The dark matter comprises perhaps ~25% of the Universe's total mass-energy and the dark energy, ~70%.[210] Explaining what dark mass and dark energy is and detecting them directly are often called the 'missing mass problem'.

1.7.1.1 Dark Matter

That some of the material making up galaxies might be missing was first suggested Jan Oort in 1932, and it was followed by Fritz Zwicky in 1933. Oort found that more mass was needed to account for the motions of nearby stars than was visible in the Milky Way's disk. Zwicky's observations of the velocities of the galaxies within the Coma galaxy cluster showed that they were too high for the galaxies to be retained by the cluster's gravitational field. Over a few hundred million years the cluster would 'evaporate' and the galaxies move away as independent entities. However, the existence of many clusters of galaxies suggests that they *are* stable over much longer periods of time. Zwicky, therefore, theorised that there must be additional material, within the cluster or the galaxies or both, whose presence meant that the gravitational field of the cluster was sufficient for the cluster to be stable. Since this matter was not directly visible, it became known as 'dark matter'.

Zwicky found that the amount dark matter that was needed to stabilise the Coma cluster was 400 times the amount of visible matter. However, subsequent measurements have brought this figure down to the quantity of dark matter required generally being 5 to 6 times that of the visible matter. Recently though, at least one example has been found of a galaxy that does not seem to contain any significant amount of dark matter. For this galaxy, NGC1052-DF2, a faint ultra-diffuse galaxy some 20 Mpc away from us, the velocities of its globular clusters have the values to be expected from the gravitational effects of just the visible matter. Other galaxies, including the Milky Way, also seem to have less than the expected amounts of dark matter in the regions inside their halos. These observations not only go to confirm the need for the presence of 'something' in other galaxies to explain their velocities but also shows that normal and dark matter can be separated out from each other.

Nearly a century has passed since Oort's and Zwicky's observations and although the *existence* of dark matter has been confirmed by many other circumstantial observations – such as the rotation curves of galaxies – the *nature*[211] of dark matter remains a mystery:

> We live in a vast sea of dark matter, but decades of intensive research have mostly established what it is not.[212]

However, there are two parts to the dark matter part of the overall problem; the *baryonic* missing mass[213] and the *non-baryonic* missing mass – and the former may now have been solved. For some time, one candidate proposed for the missing baryonic matter was that it could be hot gas lying in the voids between galaxies. The gas might have a temperature between 10^5 K and 10^7 K but would

[210] The proportions used here are deliberately approximate. Various other values, often given to three or even four significant figures, are currently to be found in the literature. These apparently more exact values however, are for specific cosmological models and are, therefore, likely to change with developments in those models.

[211] A similar situation once existed for the neutrino; its existence was predicted by Wolfgang Pauli in 1930, but it was not found experimentally until 1956.

[212] Laura Baudis, DARWIN Project Coordinator, *European Review*, 26, 2018, 70.

[213] Aka dim matter, the cosmic web or the inter-galactic medium.

be of such low density that any direct X-ray emissions (Section 1.3) would be unobservable by current X-ray space observatories. Indications of the presence of such hot filamentary material were found in 2010 using the Cosmic Web Imager (CWI; see Section 4.2.4) on the 5-m Hale telescope. The material was observed via the light that it scattered from bright quasars and galaxies in the same region of space.

More evidence recently came from the Planck spacecraft's (Section 1.2) large archive of observations of the CMB. In 2016, two teams used that archive to check for the presence of excess microwave radiation between pairs of galaxies. Such an excess could result through inverse Compton scattering by high energy electrons[214] of radiation left over from the Big Bang. After analysing 1.25 million pairs of galaxies, the teams did find evidence of baryonic material between the galaxies. This material was three to six times denser than normal outside galaxies and probably in the form of megaparsec-long filamentary clouds. If the presence of this filamentary baryonic material is confirmed, then in combination with the directly visible baryonic matter (stars, etc.), the total amount of baryonic material now observed in the Universe is close to 5% and in good agreement with the predictions.

The missing *non-baryonic* material however remains as a problem. One property though, is thought to be known: the dark matter constituents, whatever they may be, are moving slowly – (i.e., it is Cold Dark Matter [CDM]). If dark matter is cold, then after the Big Bang, the smaller objects would have formed first, and then these combined to build up the larger objects, whilst with hot dark matter–relativistic particles (HDM) large objects would form first and then fragment to form the smaller ones. The actually observed structure of the Universe favours the distribution of the sizes of its large-scale objects predicted by CDM models.

Suggestions for what dark matter might comprise is divided into three main groups – Axions, Massive Compact Halo Objects (MACHOs), and WIMPs. MACHOS, however, are objects such as mini black holes, brown dwarfs, large planets, and small stars. Such objects were once thought of as possible forms of dark matter, but recent gravitational lensing searches show that they are probably too rare to be of any significance. Also, they are baryonic matter and, as we have just seen, the Universe's 'quota' of baryonic matter is probably filled. Thus, we are left with axions and WIMPs.[215]

Axions are hypothetical subatomic particles whose existence is required to solve a problem with our current theory of the forces holding atomic nuclei together.[216] If they exist, then they are predicted to have low masses ($\sim 10^{-12}$ to $\sim 5 \times 10^{-9}$ times the mass of an electron is the current estimate). They could have been produced in large numbers in the early stages of the Big Bang and would along with related particles known as axion-like particles, be classed as CDM.

The WIMPs are also hypothetical particles and the four words making up their acronym pretty well summarise all we know about them. There is a huge choice of possible candidates for the actual dark matter WIMP – it's a bit like saying that the dark matter particle is a plant – it could then be a Californian redwood tree or a rosebush or a carrot or a liverwort or....

One possible WIMP candidate that is in favour at the moment is the lowest mass neutralino.[217] Neutralinos have yet to be found but are predicted by the particle physicists' supersymmetry theory to be produced in large numbers at the energies prevalent during the early stages of the Big Bang. There are expected to be four types of neutralinos with the lightest one possibly being stable and possibly having a mass in the range from 10 to 10,000 times the mass of the proton.

[214] The resulting distortion of the CMB is knowns as the Sunyaev-Zel'dovitch effect.

[215] An alternative explanation for the missing mass problems has been that it is not missing at all, but the apparent discrepancy arises because at large distances the gravitational force varies slightly from the Newtonian prediction. Theories in this class are usually called 'Modified Newtonian Dynamics' (MOND). MOND-based theories however, failed to describe the properties of galaxy clusters well. They also, generally, required different velocities for e-m and gravitational waves. The recent observations that the two velocities are almost identical (Section 1.6) means that most of these theories are now discounted.

[216] The strong charge-parity (CP) problem in quantum chromodynamics (QCD).

[217] A currently hypothetical particle. Not to be confused with the neutrino (Section 1.5).

1.7.1.2 Dark Energy

The existence of dark energy, like that of dark matter, has yet to be proven. There are two main circumstantial lines of evidence for its existence – the mean density of the Universe and the brightness of Type Ia supernovae. The critical density of the Universe is the density that would enable gravity to slow down the expansion of the Universe to zero after an infinite length of time (see above). A density lower than the critical value and the Universe will expand forever, a higher density and its expansion will eventually halt and it will collapse back again, perhaps ending in the 'Big Crunch'. As we have seen, the combined density of the matter and dark matter in the Universe amounts to about 30% of the critical density. For reasons beyond the scope of this book,[218] if the apparent mean density is as close to the critical density as it appears to be, then the true mean density must almost certainly be exactly equal to the critical density. Thus, 'something' other than matter must comprise 70% of the Universe and that 'something' is called 'Dark Energy'.

The second line of evidence for dark energy emerged in the late 1990s. Observations of distant (≥2,000 Mpc) Type Ia supernovae seemed to suggest that after an initial period during which the expansion of the Universe was decelerating, it is now accelerating. One way of causing such acceleration would be the presence within the Universe of a large amount of dark energy. If the density of the Universe is assumed to be the critical density, then the amount of dark energy needed to provide the observed acceleration turns out to be around 70% of the total mass-energy of the Universe.

1.7.2 Dark Matter and Dark Energy Detectors

Now that gravitational waves have been detected (Section 1.6), this section is the only one left in the book which discusses instruments that have yet to make any successful direct detections of their quarry. Indeed, given the uncertainty about what should be looked for (Section 1.7.1), it is perhaps surprising that there are efforts being made at all. In fact, almost no new detection principles will be described in this section (save phonons and nanoexplosions – see below); the instruments used in the searches for dark matter and dark energy are mostly amongst those that we have seen in the previous sections of this chapter and in later chapters, though some may be optimised for this purpose. Many of the searches use the existing results and capabilities of those instruments to look for data with the signature of some aspect(s) of dark matter or dark energy. For dark matter, the search concentrates on trying to detect the particles which form it via detectors such as those used for γ rays and neutrinos, whilst evidence is sought for dark energy via its influence on the CMB radiation or on galaxy formation and clustering or via improved observations of distant supernovae. There are now a large number of individual experiments in this area, so the details of only a few examples from each different approach to dark matter particle (DMP) detection are discussed.

Dark matter detectors divide into direct and indirect types. The direct types look for interactions of the DMPs with atoms within the test apparatus. Those interactions are primarily scintillations or ionisations arising from a high-energy normal matter particle produced by a collision with a DMP or the acoustic pulses arising from the heat deposited into the material surrounding such a collision. Indirect detectors look for decay products of DMPs or for their influences on aspects of the Universe that can be observed (Oort's and Zwicky's observations [see Section 1.7.1] are thus indirect DMP detectors).

1.7.2.1 Non-Baryonic Dark Matter – Direct Detectors

If DMPs are WIMPs, then their interactions with potential detectors would be similar to those of neutrinos. Thus, despite the elusiveness of neutrinos (Section 1.6), solar neutrinos are a source of background noise limiting the potential sensitivities of these types of detectors. It is possible that DMP WIMP detectors will start to reach this limit – called the 'neutrino floor' – by the early to mid-2020s.

[218] See *Dark Side of the Universe* by I. Nicolson (2007) for an excellent explanation for all of this from a layman's point of view.

1.7.2.1.1 Scintillation-Based Detectors

These detectors are designed to detect the high energy nuclei resulting from an elastic scattering collision between a DMP and at constituent atom within the detector. The nucleus itself produces a scintillation flash (Section 1.3.2) after the collision. The pulses resulting from recoiling nuclei have longer durations than those produced by γ rays or high-energy electrons, enabling discrimination to be made between potential DMP detections and those of other events.

The Argon Dark Matter (ArDM) experiment is designed to detect WIMPs via their elastic scattering collisions in 850 kg of liquid argon. It operates as a TPC (Section 1.5) with PMTs (Section 1.1.11) detecting the optical scintillations from the recoiling argon nuclei and from electrons produced in ionisations. The shapes of the scintillation pulses are used to help separate any desired signals from the background. It started operations in 2015 and is located in the Spanish Laboratorio Subterraneo de Canfranc some 800-m underground below Mt Tobazo. The Dark matter Experiment using Argon Pulse shape (DEAP) detector series has progressed from using 7 kg of liquid argon to 3600 kg today, and there are plans for a 50-tonne machine in the future. The current instrument's detection principle is the same as that of ArDM. The DEAP3600 uses 1,000 kg of its argon for DMP detection and the remainder for shielding and background noise reduction, and it has some 255 PMTs as its optical detectors. It started operations in 2015 and is housed in SNO Laboratory (SNOLAB) 2-km below ground in the same mine in Ontario as the SNO neutrino detector (Section 1.5.2).

Liquid xenon is also used in a somewhat similar way to liquid argon for DMP detectors. The Xenon 1 Tonne (XENON1T) TPC experiment is located in the Gran Sasso tunnel so that the Apennines provide some 1.4 km of overlying rock to shield the experiment from unwanted interactions. It uses 3,200 kg of liquid xenon with 2,000 kg of this forming the active part of the detector (and there are plans for a future instrument with 6,000 kg of liquid Xenon). A water jacket also acts as a passive shield and 248 PMTs are the optical detectors. Electrons produced in the interactions are driven through the liquid by an electric field to emerge at its surface and there be detected separately so that there is a recognisable double pulse (double-phase detection). PandaX at the Jinping underground laboratory[219] in Sichuan is a similar instrument using 580 kg of xenon. Recently, XENON1T has been able to determine an upper limit to the possible interaction cross-sectional area (σ) of WIMP particles of $\leq 4 \times 10^{-51}$ m^2.[220]

The Large Underground Xenon experiment (LUX) experiment operated until 2016. It comprised a 370-kg liquid xenon TPC detector that was located in the Sanford lab in South Dakota, 1.5-km underground. It is currently being replaced by LUX-ZEPLIN (LZ)[221] experiment using 10 tonnes of xenon and with an expected completion date of 2020. The proposed DARWIN project will use 50 tonnes of liquid xenon. It will be a double phase TPC design with an expected completion date of 2023.

The DMP detectors may also be based on solid scintillators with NaI(Tl) and plastic being used to date. The Dark Matter (DAMA)/Libra experiment is housed in the Gran Sasso tunnel using 250 kg of NaI(Tl). Similarly, the Annual modulation with NaI Scintillators (ANAIS) detector in the Laboratorio Subterraneo de Canfranc uses 112 kg of NaI(Tl) and has, as a main objective, searching for any annual modulation rate in its detection that might arise through the Earth's motion through the dark matter. The COSINE100[222] at the Yangyang laboratory located 1.1-km

---

[219] This is currently the best shielded underground laboratory available with the absorption of its overlying layers equivalent to 6.7 km of water.

[220] The measured 'size' of a subatomic particle varies with its energy and the interaction being used to determine its value. However, for comparison, for proton-proton collisions in the LHC, intended for the production of Higgs particles, σ is around 10^{-32} m^2 at 1 TeV and around 10^{-30} m^2 at 10 TeV. The non-SI unit of 'barn' (symbol – 'b') is sometimes used as a measure of interactional cross-sectional area of atoms, but at 1 b = 10^{-28} m^2, it has few advantages over the standard SI unit at these scales.

[221] LUX – Zoned Proportional scintillation in Liquid Noble gases experiment – a DMP detector using 12 kg of liquid xenon and located in the Boulby potash mine in North Yorkshire. It operated from 2006 to 2011.

[222] Derivation unknown.

underground in South Korea also aims to search for annual modulations using 106 kg of NaI(Tl) crystals to detect muons and started its operations in 2016; its early results have not confirmed any modulations. The possibility of detecting electron scintillations in doped GaAs is also being investigated at the time of writing.

1.7.2.1.2 Calorimeters and Bolometers

The DMP detectors whose operating principles are based upon the heating effects of absorbing energy from DMPs are discussed here. Sections 1.1.15, 1.3.2.7 and 1.5.2.4 give further details of the basic operating principles of the detectors.

The Cryogenic Rare Event Search with Superconducting Thermometers (CRESST) II experiment which was in the Gran Sasso tunnel used calcium tungstate cryogenic calorimeters and scintillators. In 2011, it found 67 signals that could have been due to WIMPs. The expected background of such events was only about half this figure. The balance could, therefore, have arisen from WIMPs with masses in the region of 10–25 GeV. Reanalysis of the data and later measurements with an upgraded instrument, however, have not confirmed that first result. The CRESST III instrument is currently operating with 250 g of $CaWO_4$.

The series of Expèrience pour Dètecter les WIMPs en Site Souterrain (EDELWEISS) experiments use both ionisation and heat deposition in germanium bolometers cooled to 20 mK and are housed in the Laboratoire Souterrain de Modane near the French-Italian border. The European Underground Rare Event Calorimeter Array (EURECA) instrument will be the successor to the EDELWEISS instruments. EURECA is planned have between 100 and 1,000 kg of cryogenic calorimeters made from a variety of materials.

1.7.2.1.3 Bubble Chambers

The recoiling nucleus from a DMP-atom collision can be detected using bubble chambers (Sections 1.2.3 and 1.4.2). Thus, Chicagoland Observatory for Underground Particle Physics (COUPP) which was based in SNOLAB used 60 kg of CF31, whereas PICO-60[223] also at SNOLAB uses 57 kg of C_3F_8. The latter instrument has acoustic detectors as well as cameras to monitor the bubbles and improve its discrimination between possible DMP events and other interactions. There are plans to develop instruments with up to 500 kg of active material.

1.7.2.1.4 Ionisations

The Coherent Germanium Neutrino Technology (CoGeNT) experiment in the 700-m deep Soudan underground laboratory in Minnesota uses a 440 g of germanium crystal cooled by liquid nitrogen (Section 1.3.2.7). It senses ionisations arising from nuclei hit by DMPs. Several years' worth of its data seem to show an excess of detections over the expected background level but at less than the 3σ level (Section 1.1.17.4). The instrument may be upgraded to several kg of germanium at some time in the future. A similar detector, China Dark Matter Experiment (CDEX), has been operating in the Jinping underground laboratory since 2013.

The Dark Matter in CCDs (DAMIC100) instrument in SNOLAB uses 18 three-phase, buried-channel CCDs (Section 1.1.8), with an active mass of 100 g, as its detecting elements to collect possible nuclear recoils from DMP events. The CCDs' low noise levels mean that its WIMP mass detection threshold is around 10^{-4} electron masses.

1.7.2.1.5 Phonons

Phonons are the quanta of sound (i.e., of the vibrational energy in a material). Like photons, their energies are related to their frequencies by Planck's constant (h). Normal sound waves (like normal light beams) are formed from huge numbers of individual phonons – a 100-dB level sound has an

[223] Previously COUPP60. The PICO name derives from the merger of PICASSO (Project in Canada to Search for Super Symmetric Objects – a previous experiment) and COUPP.

intensity of 10 mW m^{-2} or about 1.5×10^{29} phonons s^{-1} m^{-2} at 100 Hz. However, atomic-level events, such as the elastic scattering of an atom's nucleus by a DMP can excite individual phonons, and these in turn can be used in dark matter detectors.

The Cryogenic Dark Matter Search (CDMS) instrument used germanium and silicon disks cooled to mK temperatures to detect both the phonons and ions from interactions. It used TES detectors (Section 1.1.15.3) for the phonons and the two pulses enabled the many electron-scattering events to be separated from the far fewer nuclear (and thus possible WIMP) events. It was housed in the Soudan laboratory. A few events were identified as possibly being due to WIMPs but only at the 3σ level (Section 1.1.17.4). SuperCDMS has been operating since 2012 with 225 kg of active material. An upgraded version of that instrument (SuperCDMS SNOLAB) is currently being developed to be sited in SNOLAB. The latter instrument will have more and thicker detector disks and better shielding from background radiation and may be operational by the early 2020s.

A recently proposed detector concept based upon superfluid helium would detect DMPs via phonons. The phonons would propagate through the liquid helium bath and upon reaching the surface release individual helium atoms. The helium atoms would be ionised and then accelerated by intense electrostatic fields. The final detection of the helium ions would use conventional calorimeters (Section 1.3.2.7).

1.7.2.1.6 Haloscope

For axion detection, there are several possible approaches and one of these is to convert the axion into a photon. The instrument is called a haloscope[224] or Sikivie-type detector[225] and comprises a microwave resonator embedded in a strong magnetic field. An axion passing through the instrument can interact with the field to produce a photon with the axion's energy[226] when the microwave frequency is tuned to that same energy. The Axion Dark Matter Experiment (ADMX) instrument at the University of Washington has a cryogenic microwave cavity, which is 1-m long and 0.42-m in diameter and embedded in a 7.6-T magnetic field. It recently started searching for axions with masses near to 10^{-11} times that of the electron. A higher-frequency version of the instrument (ADMX-HF) at Yale University is looking for axions with masses 5 to 6 times higher than those of ADMX.

1.7.2.1.7 Helioscope

Rather confusingly, axions are also being sought by a rather similarly named but different instrument called a 'helioscope'. Even more confusingly, this instrument is quite different from the spectrohelioscope discussed in Section 5.3.3.

Axion helioscopes, like haloscopes, depend upon the Primakoff process (see footnote) for their operation, although the transformation does not occur only within the instrument. The centre of the Sun has an enormous abundance of X-ray photons plus the ions' and nuclei's electric fields and the larger scale magnetic fields from the plasma. We may, therefore, expect the Sun to be a strong source of axions converted via the Primakoff process from the 200 pm X-rays which predominate at the solar centre. These axions would have masses around a millionth of that of an electron – about 100,000 times more massive than those of concern to the haloscope.

The CAST instrument (see also Section 1.3.2.2), located near Geneva and which has undergone several modifications since it started operating in 2003, uses a prototype magnet from the Large Hadron Collider (LHC) to produce a 9-T magnetic field over a 10-m-long volume. It uses a

---

[224] Because its aim is to detect axions from the galactic dark matter *halo*.

[225] After Peter Sikivie who invented the instrument in 1983.

[226] Via the Primakoff process. In the presence of strong electric or magnetic fields, photons can transform into axions and axions into photons. Further details are beyond the scope of this book and are left for interested readers to research.

focussing X-ray camera (Section 1.3.4) with CCD detectors (Section 1.1.8) to pick up any axions that have reconverted back into high-energy photons in the instrument's magnetic field. Like most astronomical instruments, it has to compensate for the Earth's motion, and so it is mounted on a moveable platform that allows it to track the Sun for about 1.5 hours close to sunrise and again at sunset.

IAXO is expected to improve upon CAST's sensitivity by a factor of ten by using a custom-built magnet to provide a 2.5-T magnetic field over a 25-m length. It is also expected to use micromega detectors (Section 1.3.2.2), Wolter-1–type optics (Section 1.3.4.4) and to track the Sun for up to 12 hours a day.

1.7.2.1.8 Other Possible Direct Detectors

Some possible, though perhaps more speculative, suggestions for direct non-baryonic dark matter detectors include nanoexplosions and superconducting metals. The former would be based on nanometre-sized metal particles embedded in an oxide. The thermal energy deposited by a DMP interaction would ignite one or more of the metal particles into an exothermic explosion which would then be detected. Metallic superconductors could be used for detecting DMP particles with masses down to 10^{-9} times that of an electron by splitting the Cooper pairs (Section 1.1.12) of electrons. Ancient (500 million-year-old) minerals such as halite (NaCl) and Olivine ($[Mg, Fe_2]SiO_4$) may retain tracks of DMP particles that could be revealed by etching or ablating (with lasers) the surfaces and scanning by electron, helion or atomic force microscopes or via X-ray scattering.

Axions passing through a magnetic field might generate small electric fields, and these could be used to generate oscillations in cold plasma. The plasma oscillations would then be detected to form an axion detector. It is also possible that DMPs might interact with atomic nuclei, altering their nuclear magnetic resonance (NMR) spectra and so be detectable from those changes.

It may be possible to investigate whether or not the gravitational effects of dark matter follow GR's predictions by measuring precisely the orbital motions of millisecond pulsars located near possible dark matter concentrations. So far one such attempt has been undertaken (using PSR J1713 + 0747) without showing any discrepancies from GR. Somewhat similarly, if DMPs interact at all with ordinary matter, then GWDs (Section 1.6) might be able to detect the Brownian motion induced in their test masses (mirrors) by such interactions.

A recent and radically different suggestion is that the evaporation of small black holes via their Hawking radiation halts when their Schwarzschild radii are equal to their Planck lengths. This would be at a size of a few times 10^{-35} m and at a mass of a few times 10^{-8} kg. Now 10^{-8} kg equals about 10^{28} eV or about 10^{19} protons, so a small number of such Planck black holes could easily make up the whole of dark matter and be undetectable by any of the experiments discussed previously. Experimental searches for magnetic monopoles (all of which, to date, have failed), might however be able to detect these objects instead.

1.7.2.2 Non-Baryonic Dark Matter – Indirect Detectors

1.7.2.2.1 Galaxy Rotation Curves

Oort's measurements of stellar velocities in the Milky Way was the first indication of dark matter's possible existence by this or any other method. There are now many galaxies whose material (gas clouds, nebulae as well as stars) has had its velocity measured over a range of distances from the centre of the galaxy. The resulting velocity/distance plot is known as the galaxy's rotation curve. For material well outside the central regions of the galaxy (so that most of the galaxy's mass is inside its orbit), we might expect that the velocity be given by Kepler's third law and so be inversely proportional to the square of its distance from the galaxy's centre (lower curve in Figure 1.113). However, for disk-type galaxies (spiral and lenticular galaxies), the plots show a consistent discrepancy; the velocity far out from the centre of the galaxy is higher than that predicted from the observed distribution of matter in the galaxy (upper curve in Figure 1.113). The discrepancy can be

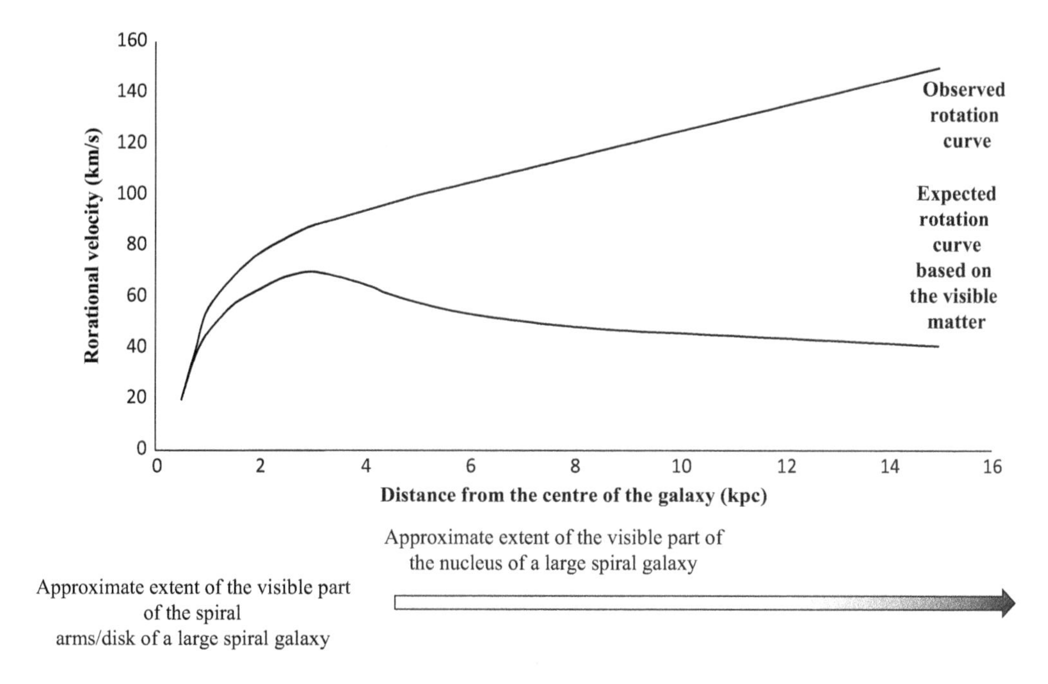

FIGURE 1.113 Schematic rotation curve for a spiral galaxy.

explained if there is additional matter extending out to many times the visible size of the galaxy and which we are not seeing (i.e., a dark matter halo).

1.7.2.2.2 Cosmic Rays

Indirect dark matter detectors search for possible collision or decay products from DMPs, such as γ rays, positrons, anti-protons or anti-deuterons (although because it is not known what DMPs are, it is also not known what such decay products might be). The searches attempt to find excesses of particles apparently coming from regions where dark matter might be expected to be concentrated, such as the centres of galaxies. Since γ-rays are unaffected by interstellar or intergalactic magnetic fields, they are the best candidates for the searches but could also be preferentially produced in the centres of galaxies by other processes. The detectors are identical to the Čerenkov and other γ-ray detectors as discussed in Sections 1.3 and 1.4.

No significant results have been found to date, although the Fermi spacecraft may have detected a γ-ray emission line originating from near the centre of the Milky Way at an energy of 135 GeV. Such emission might result from the mutual annihilation of DMPs; however, there are several other possible explanations such as pulsars or multiple bursts. Some years ago, a search for anti-deuterons from DMP decays was carried out by the balloon-borne General Anti-Particle Spectrometer (GAPS) instrument, using ToF silicon detectors but without success.

Although such dark matter indications may be found in the results from standard cosmic ray detectors, the High Energy cosmic Radiation Detection (HERD) calorimeter will be optimised for a dark matter search and is planned to be mounted on China's future space station, Tianhe-1,[227] in the early 2020s.

[227] Which translates to Harmony of the Heavens.

1.7.2.2.3 Neutrinos

Neutrinos (Section 1.5) were once candidates for dark matter themselves, but they would be hot and so are probably ruled out by being unable to reproduce the observed large-scale structure of the Universe (see above). A hypothetical particle called the 'sterile neutrino' though remains a possibility for a DMP. It would be a neutrino whose only interaction with the rest of the Universe was via gravity. A faint X-ray emission line at 3.55 keV in the spectra of some galaxies has no means of origin via currently known atomic reactions and so has been attributed to the decay of sterile neutrinos. That interpretation remains to be confirmed.

It is possible that DMPs could accumulate towards the centre of the Sun or even the centre of the Earth. Their destruction or decay there might produce neutrinos which could be detected by neutrino telescopes (Section 1.5) on the Earth's surface. In 2010 and 2011, IceCube conducted a search for such neutrinos without success, although a lower limit to the lifetimes of 10 TeV and higher mass dark matter particles of 10^{28} s was determined.

1.7.2.2.4 Gravitational Lenses

Clumps of dark matter with masses equivalent to those of tens or hundreds of galaxies will act as weak gravitational lenses affecting the images of objects behind them. A recent deep sky survey by the Subaru telescope's Hyper Suprime-Cam (an 870 megapixel optical camera; see Section 1.1) has detected nine features that could be such clumps of dark matter.

1.7.2.2.5 Gravitational Waves

It has been suggested that when sufficient gravitational wave events from binary black hole and/or neutron star mergers have been detected (Section 1.6), there should be a statistical link between the merger rates within individual dwarf galaxies and the quantity of dark matter in that galaxy.

1.7.2.2.6 Direct Gravitational Effects

Another recent suggestion is based on the fact that the only guaranteed interaction between dark and normal matter is via gravity. It is suggested that impulse sensors based upon quantum effects are now so sensitive that an array of them could pick up the gravitational force of DMPs passing through the array directly. The particles though, would need to be some 10^{19} times the mass of a proton[228] before being detectable currently by this approach.

1.7.2.2.7 Anti-Matter

The assumption is usually made that dark matter will interact with anti-matter in a similar fashion to the way that it interacts with matter (symmetric interactions). However, this may not be the case and that possibility has just recently started to be investigated by measuring the precession frequency of anti-protons caught in a Penning trap[229] at CERN. Passing dark matter, especially if axion-like, might cause periodic variations in the precession frequency. To date, though, no such variations have been found.

1.7.2.3 Non-Baryonic Dark Matter – Making Your Own

1.7.2.3.1 Large Hadron Collider

A quite different approach to confirming the existence of dark matter would be to make some ourselves – and that is exactly what the LHC may be able to do. When its two beams of protons collide head on at a combined energy of 1.4×10^{13} eV (14 TeV), amongst the many by-products of the

[228] This is HUGE and is equivalent to about 10,000,000 e-coli cells.

[229] A device that traps a charged particle in two dimensions using magnetic fields and in the third dimension by using electric fields.

collisions there may be DMPs. These DMPs would, of course, be no easier to detect than those from any other source, but their presence might become apparent if there is something missing when all the other (detectable) products of the interaction have been taken into account.

1.7.2.3.2 Light-through-a-wall Detectors

The basic principle of this type of detector is similar to that of the helioscope (Section 1.7.2.1.6): they have a source of trapped non-DMPs, converted to DMPs which *can* escape from the trap , then convert those escaped DMPs back into the non-DMPs and detect the non-DMPs.

In the case of the helioscope, the non-DMPs are the X-rays trapped at the centre of the Sun by its overlying layers (the 'wall'). The DMPs converted from the X-rays can escape from the Sun and some then caught in laboratory instruments on Earth (CAST, IAXO, etc.), converted back to X-rays and detected.

Deutsches Elektronen-Synchrotron (DESY), with sites near Hamburg and Berlin in Germany, operates the Any Light Particle Search (ALPS) detectors. These work in the same way as the helioscope except that the solar X-rays are replaced by photons produced by a laser and the Sun's overlying layers are replaced by a barrier opaque to the laser light. ALPS I (2007–2010) used a 5-T tubular magnet some 8.8-m long and a 1.064-μm Nd:YAG laser photon source frequency doubled to 532 nm. To increase the laser's power, its photons were fed into an optical resonator (cf. FPCs – Section 1.6.3.2.1.2) which occupied one-half of the magnet. An opaque wall sealed the light source from the second half of the magnet's central hole where the DMPs were to be regenerated. Apart from demonstrating the principle of the instrument, ALPS I's set an upper limit on the existence of a type of DMP called a weakly interacting slow particle (WISP).[230]

ALPS II (2013 to the present) is an upgraded version of ALPS I with about ten times the latter's sensitivity. It operates at the laser's fundamental frequency (1.064 μm) and uses 10-m-long resonant cavities to amplify both the laser photons and those regenerated back from DMPs that have passed through the wall. It uses TES sensors (Section 1.1.15.3) as the photon detectors, and its first scientific operations are expected in 2020. A further upgrade to 100-m-long resonant cavities is planned for the future.

1.7.2.4 Dark Energy Detectors

Dark energy has two main lines of evidence for its existence: the total mass of the Universe and the Universe's large-scale structure and behaviour. The first of these has been mentioned in the introduction and is based upon the critical density of the Universe. Once that has been set by a cosmological theory, there is currently little in the way of further tests of its validity that are possible. Thus, the active measures to investigate dark energy are mostly concerned with observing the large-scale structure of the Universe.

Vesto Slipher, starting in 1912, found that the lines in the spectra of many spiral nebulae (then thought to lie within what we now call the Milky Way galaxy) had wavelengths longer than those same lines produced in the laboratory. Since he was working at visible wavelengths, this meant that his observed lines were moved towards the red end of the visible spectrum. Recognition that the spiral nebulae were in fact not within the Milky Way but were galaxies in their own right at distances of megaparsecs and Edwin Hubble's systematic spectroscopic observations showed that the magnitudes of these 'red shifts' were (roughly) proportional to the distances to the galaxies (i.e., the further away the galaxy was from us, the faster it was moving away). Now we and the Milky Way do not occupy some especially undesirable part of the Universe from which every other entity is fleeing; in fact, an identical picture of objects in the Universe appearing to move away at ever-increasing speeds would be obtained by observers within any of Slipher's and Hubble's external galaxies: it is the Universe as a whole that is expanding,[231] not just our small part of it.

[230] Also known as a Weakly Interacting Slim Particle or Weakly Interacting Sub-eV Particle.

[231] This result is not difficult to prove but is well beyond the remit of this book; interested readers are challenged to prove it for themselves (or search amongst history of cosmology sources).

Cosmological theories based upon GR had been developed from 1915 onwards and Einstein had expected the solution to be a static universe. Since his initial models showed the Universe to be expanding or contracting, in 1917, he added a new term to the cosmological equation called the cosmological constant and which enabled a static solution to be found. Had observers discovered the general recession of the galaxies (= expansion of the Universe) a few years earlier, there would never have been any need for the cosmological constant. A couple of decades later, Einstein called the invention of the cosmological constant his 'greatest blunder'. But perhaps Einstein blundered twice – once in introducing the cosmological constant and then later in discarding it – because one of the (several) suggestions for the nature of dark energy is that it acts in opposition to the normal forces of gravity (i.e., as a type of 'anti-gravity') and that is exactly what the effect of the cosmological constant can be.

In 1998, observations of type 1a supernovae[232] showed that the expansion of the Universe, after first slowing down as one would expect from the effects of gravity had started to speed up[233] some 5,000 million years ago. Despite some contrary results (or at least contrary interpretations of results) the reality of this acceleration is now generally regarded as confirmed. The main explanations for the acceleration of the Universe invoke a new entity called 'dark energy'. Whilst dark energy itself is not understood, the properties it needs to have to explain the Universe's expansion are known; they are that it acts as an outward pressure (or in opposition to gravity), it reduces more slowly than matter as the Universe expands and it is more uniformly distributed than matter. As remarked in the previous paragraph, the use of a cosmological constant[234] can fulfil these requirements and the resulting cosmological model for the Universe is the one that is currently most widely accepted. Since the cosmological constant usually has the symbol Λ (lambda) and the Universe's matter is thought to be cold (CDM), the model goes by the name of the Lambda-CDM model.

Additional investigations of the nature and properties of dark energy are currently sought via detailed studies of the large-scale distribution of galaxies, baryonic matter and of the CMB. Several surveys of large numbers of galaxies and their redshifts have been undertaken in recent years. Thus, the Deep Extragalactic Evolutionary Probe (DEEP2) Galaxy Redshift Survey which was completed in 2013 using the Keck telescopes has an archive of about 50,000 galaxy redshifts, the Galaxy and Mass Assembly (GAMA) survey used the AAO telescope and was completed in 2014 with an archive of about 300,000 galaxy redshifts, the Four Star Galaxy Evolution (ZFOURGE) survey used the Magellan telescopes to obtain some 60,000 galaxy redshifts with its last data release in 2016. Meanwhile, at the time of writing, the Sloan Digital Sky Survey (SDSS) has accumulated more than three million galaxy images and spectra covering a third of the sky and the SKA (Section 1.2.3.2) should soon provide longer wavelength data on at least a billion galaxies.

For the future, the construction of the VRO has recently begun and its first light is expected around 2021. This will be an 8.4-m telescope designed to have a 10 deg^2 field of view and using a 3,000 megapixel camera. It will be able to image all the available sky with 15-s exposures every three nights, covering three-quarters of the whole sky over a year. The resulting survey should provide a detailed 3D map of large-scale structure of the Universe. Dark Energy Spectroscopic Instrument (DESI), for the Mayall 4-m telescope, achieved its first light in October 2019. Over the next five years, it is expected to obtain spectra and redshifts of up to 30 million galaxies with up to 5,000 images on each exposure using fibre-optic cables positioned robotically. The ESA's Euclid spacecraft, scheduled for a launch in 2022, will use a 1.2-m Korsch (Section 1.1.19.2) type telescope with the aim of mapping galaxies to investigate the acceleration of the Universe. The future WFIRST spacecraft may carry the Wide-filed Opto-Mechanical Assembly (WOMA) instrument for dark energy and exoplanet surveys.

---

[232] Which are all thought to have about the same maximum absolute visual magnitude of about -19.3^m and so can act as 'standard candles' for distance measurements.

[233] The 2011 Nobel Prize in Physics was awarded to Saul Perlmutter, Adam Reiss and Brian Schmidt for this discovery.

[234] Sometimes called the 'vacuum energy'.

As previously mentioned, studies of distant supernovae provide a history of the acceleration/deceleration of the Universe over time and, this in turn, can put constraints on the nature of dark energy. Such data has already come from surveys like the SDSS and forms the current observational basis for the existence of dark energy. For example, the Dark Energy Camera (DECam) has been operating on the 4-m Victor M. Blanco telescope since 2012, conducting the Dark Energy Survey (DES). It uses 62 CCDs as its detectors (Section 1.1.8) with a field of view 2.2° across and its main purpose to search for distant supernovae. The Hobby-Eberly telescope (Section 1.2.3.2) has recently been upgraded to have a field of view about a third of a degree across and is currently conducting the Hobby-Eberly Telescope Dark Energy Experiment (HETDEX).

Confirmation of the existence of dark energy and of its nature may also come through its effect upon the large-scale structure of ordinary (baryonic) matter throughout the Universe. Galaxies and clusters of galaxies started to form soon after the Big Bang, and their presence produces ripples in the intensity of the CMB (Section 1.2.3). Many other processes also produce ripples in the CMB. The results from the WMAP and Planck spacecraft have so far proved to be inconclusive, whilst a claimed discovery of the signature of gravitational waves (Section 1.6) in the CMB in data from the BICEP-2 telescope in the Antarctic was later withdrawn. Regular variations in ordinary matter (Baryonic Acoustic Oscillations [BAOs]), if they can be found, would also be likely to provide evidence of the natures of both dark matter and dark energy. Such oscillations are currently being sought in the optical region by Baryon acoustic Oscillations Spectroscopic Survey (BOSS; see Section 4.2.5) and in the future by the currently under construction, HIRAX radio array.

Clues to the large-scale structure of the Universe and so to dark energy may also come in the future from studies of the X-ray emission from clusters of galaxies and from weak gravitational lensing.

2 Imaging

2.1 THE INVERSE PROBLEM

A problem that occurs throughout much of astronomy and other remote sensing applications is how best to interpret noisy data so that the resulting deduced quantities are real and not artefacts of the noise. This problem is termed the 'inverse problem'.

For example, stellar magnetic fields may be deduced from the polarisation of the wings of spectrum lines (Section 5.2). The noise (e.g., errors, uncertainties) in the observations, however, will mean that a range of field strengths and orientations will fit the data equally well. The three-dimensional (3D) distribution of stars in a globular cluster must be found from observations of its two-dimensional (2D) projection onto the plane of the sky. Errors in the measurements will lead to a variety of distributions being equally good fits to the data. Similarly, the images from radio and other interferometers (Section 2.5) contain spurious features because of the side lobes of the beams. These spurious features may be removed from the image if the effects of the side lobes are known. But because there will be uncertainty in both the data and the measurements of the side lobes, there can remain the possibility that features in the final image are artefacts of the noise or are incompletely removed side lobe effects, etc.

The latter illustration is an instance of the general problem of instrumental degradation of data. Such degradation occurs from all measurements because no instrument is perfect – even a faultlessly constructed telescope will spread the image of a true point source into the Airy diffraction pattern (Figure 1.35). If the effect of the instrument and other sources of blurring on a point source or its equivalent are known (the point spread function [PSF] or instrumental profile) then an attempt may be made to remove its effect from the data. The process of removing instrumental effects from data can be necessary for any type of measurement, but is perhaps best studied in relation to imaging, when the process is generally known as 'deconvolution'.

2.1.1 Deconvolution

This form of the inverse problem is known as deconvolution because the true image *convolves* with the PSF to give the observed (or dirty) image. Inversion of its effect is thus *deconvolution*. Convolution is most easily illustrated in the one-dimensional (1D) case (such as the image of a spectrum or the output from a Mills Cross radio array) but applies equally well to 2D images.

A 1D image, such as a spectrum, may be completely represented by a graph of its intensity against the distance along the image (Figure 2.1). The PSF may similarly be plotted and may be found, in the case of a spectrum, by observing the effect of the spectroscope on a monochromatic source.

If we regard the true spectrum as a collection of adjoining monochromatic intensities, then the effect of the spectroscope will be to broaden each monochromatic intensity into the PSF. At a given point (wavelength – λ_1, say) in the observed spectrum, therefore, some of the original energy will have been displaced out to nearby wavelengths, while energy will have been added from the spreading out from nearby wavelengths (λ_2, say – Figure 2.2). The process may be expressed mathematically by the convolution integral:

$$O(\lambda_1) = \int_0^\infty T(\lambda_2) I(\lambda_1 - \lambda_2) d\lambda_2 \tag{2.1}$$

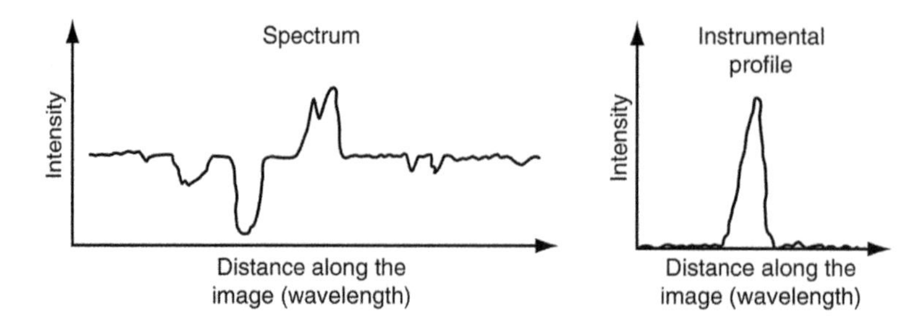

FIGURE 2.1 Representation of a one-dimensional image and the instrumental profile (point spread function [PSF]) plotted as intensity versus distance along image.

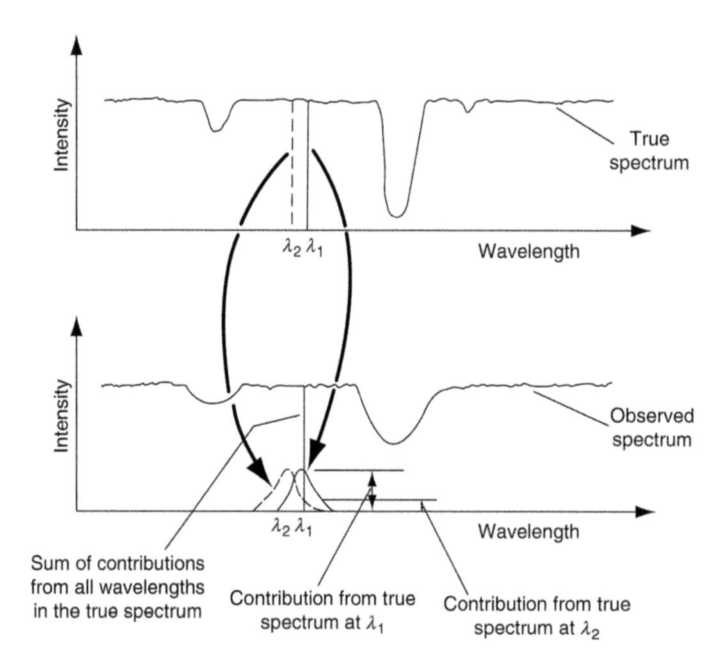

FIGURE 2.2 Convolution of the true spectrum with the point spread function (PSF) to produce the observed spectrum.

where $O(\lambda_1)$ is the intensity in the observed spectrum at wavelength λ_1, $T(\lambda_2)$ is the intensity in the true spectrum at wavelength λ_2 and $I(\lambda_1 - \lambda_2)$ is the response of the instrument (spectroscope) at a distance $(\lambda_1 - \lambda_2)$ from its centre.

Equation 2.1 is normally abbreviated to

$$O = T * I \tag{2.2}$$

where * is the convolution symbol.

The inversion of Equation 2.1 to give the true spectrum cannot be accomplished directly but involves the use of Fourier transforms.

With the Fourier transform and its inverse in the form

$$F(s) = F\left(f(x)\right) = \int_{-\infty}^{\infty} f(x) e^{-2\pi i x s} dx \tag{2.3}$$

$$f(x) = F^{-1}(F(s)) = \int_{-\infty}^{\infty} F(s) e^{2\pi i x s} ds \tag{2.4}$$

then the convolution theorem states that:

Convolution of two functions corresponds to the *multiplication* of their Fourier transforms.

Thus, taking Fourier transforms of Equation 2.2 we have

$$F(O) = F(T * I) \tag{2.5}$$

$$= F(T) \times F(I) \tag{2.6}$$

and so the true spectrum (etc.) may be found from inverting Equation 2.6 and taking its inverse Fourier transform:

$$T = F^{-1}\left[\frac{F(O)}{F(I)}\right] \tag{2.7}$$

In practice, obtaining the true data (or source function) via Equation 2.7 is complicated by two factors. First, data is sampled at discrete intervals so is not the continuous function required by Equations 2.3 and 2.4, and also it is not available over the complete range from $-\infty$ to $+\infty$. Secondly, the presence of noise will produce ambiguities in the calculated values of T.

The first problem may be overcome by using the discrete versions of the Fourier transform and inverse transform;

$$F_D(s_n) = F_D(f(x))_n = \sum_{k=0}^{N-1} f(x_k) e^{-2\pi i k n/N} \Delta \tag{2.8}$$

$$f(x_k) = F_D^{-1}(F_D(s_n)) = \sum_{n=0}^{N-1} F_D(s_n) e^{2\pi i k n/N} \frac{1}{N} \tag{2.9}$$

where $F_D(s_n)$ is the nth value of the discrete Fourier transform of $f(x)$, N is the total number of measurements, $f(x_k)$ is the kth measurement and Δ is the step length between measurements and setting the functions to zero outside the measured range.

Now a function that has a maximum frequency of f is completely determined by sampling at $2f$ (the sampling theorem). Thus, the use of the discrete Fourier transform involves no loss of information providing the sampling frequency ($1/\Delta$) is twice the highest frequency in the source function. The highest frequency that can be determined for a given sampling interval ($1/2\Delta$) is known as the 'Nyquist frequency' or 'critical frequency'. If the source function contains frequencies higher than the Nyquist frequency, then these will not be determined by the measurements, and the finer detail in the source function will be lost when it is reconstituted via Equation 2.7. Rather more seriously, however, the higher frequency components may beat with the measuring frequency to produce spurious components at frequencies lower than the Nyquist frequency. This phenomenon is known as 'aliasing' and can give rise to major problems in finding the true source function.

The actual evaluation of the transforms and inverse transforms may now be relatively easily accomplished using the fast Fourier transform algorithm on even quite small computers. The details of this algorithm are outside the scope of this book but may be found from books on numerical computing.

The 1D case just considered may be directly extended to two or more dimensions, though the number of calculations involved then increases dramatically. Thus, for example, the 2D Fourier transform Equations are

$$F(s_1, s_2) = F\left(f\left(x_1, x_2\right)\right) = \int_{-\infty}^{\infty}\int_{-\infty}^{\infty} f\left(x_1, x_2\right) e^{-2\pi i x_1 s_1} e^{-2\pi i x_2 s_2}\, dx_1\, dx_2 \tag{2.10}$$

$$f(x_1, x_2) = F^{-1}\left(F\left(s_1, s_2\right)\right) = \int_{-\infty}^{\infty}\int_{-\infty}^{\infty} F\left(s_1, s_2\right) e^{2\pi i x_1 s_1} e^{2\pi i x_2 s_2}\, ds_1 ds_2 \tag{2.11}$$

Some reduction in the noise in the data may be achieved by operating on its Fourier transform. In particular, cutting back on or removing the corresponding frequencies in the transform domain may reduce cyclic noise such as 50 or 60 Hz mains hum or the stripes on scanned images. Random noise may be reduced by using the optimal (or Wiener) filter defined by

$$W = \frac{[F(O)]^2}{\left[F(O)\right]^2 + \left[F(N)\right]^2} \tag{2.12}$$

where $F(O)$ is the Fourier transform of the observations, without the effect of the random noise and $F(N)$ is the Fourier transform of the random noise. (The noise and the noise-free signal are separated by assuming the high-frequency tail of the power spectrum to be just due to noise and then extrapolating linearly back to the lower frequencies.) Equation 2.7 then becomes

$$T = F^{-1}\left[\frac{F(O)W}{F(I)}\right] \tag{2.13}$$

Although processes such as those outlined may reduce noise in the data, it can never be totally eliminated. The effect of the residual noise, as previously mentioned, is to cause uncertainties in the deduced quantities. The problem is often ill-conditioned; that is to say, the uncertainties in the deduced quantities may, proportionally, be much greater than those in the original data.

A widely used technique, especially for reducing the instrumental broadening for images obtained via radio telescope is the RL[1] algorithm. This is an iterative procedure in which the $n + 1$th approximation to the true image is related to the PSF and nth approximation by

$$T_{n+1} = T_n \int \frac{O}{T_n * I} I \tag{2.14}$$

The first approximation to start the iterative process is usually taken as the observed data. The RL algorithm has the advantages compared with some other iterative techniques of producing a normalised approximation to the true data without any negative values, and it also usually converges quite rapidly.

Recently, several methods have been developed to aid choosing the 'best' version of the deduced quantities from the range of possibilities. Termed 'non-classical' or 'Bayesian methods', they aim to stabilise the problem by introducing additional information, not inherently present in the data, as constraints and, thus, to arrive at a unique solution. The best known of these methods is the maximum entropy method (MEM). The MEM introduces the external constraint that the intensity cannot be negative and finds the solution that has the least structure in it that is consistent with the data.

[1] After William Richardson and Leon Lucy.

The name derives from the concept of entropy as the inverse of the structure (or information content) of a system. The maximum entropy solution is thus the one with the least structure (the least information content or the smoothest solution) that is consistent with the data. The commonest measure of the entropy is

$$s = -\sum p_i \ln p_i \tag{2.15}$$

where:

$$p_i = \frac{d_i}{\sum_j d_j} \tag{2.16}$$

and d_i is the ith datum value, but other measures can be used. A solution obtained by a MEM has the advantage that any features in it must be real and not artefacts of the noise. However, it also has the disadvantages of perhaps throwing away information and that the resolution in the solution is variable, being highest at the strongest values.

Other approaches can be used either separately or in conjunction with MEM to try and improve the solution. The CLEAN[2] method, much used on interferometer images, is discussed in Section 2.5. The method of regularisation stabilises the solution by minimising the norm of the second derivative of the solution as a constraint on the smoothness of the solution. The non-negative least squares (NNLS) approach solves for the data algebraically, but subject to the added constraint that there are no negative elements. Myopic deconvolution attempts to correct the image when the PSF is poorly known by determining the PSF as well as the corrected image from the observed image. While blind deconvolution is used when the PSF is unknown and requires several images of the same object, preferably with dissimilar PSFs, such as occurs with adaptive optics images (Section 1.1). Burger-Van Cittert deconvolution first convolves the PSF with the observed image and then adds the difference between the convolved and observed images from the observed image. The process is repeated with the new image and iterations continue until the result of convolving the image with the PSF matches the observed image. The $n + 1$th approximation is thus

$$T_{n+1} = T_n + \left(O - T_n * I\right) \tag{2.17}$$

Compared with the RL approach, the Burger-van Cittert algorithm has the disadvantage of producing spurious negative values. The recently developed technique of spectral deconvolution is based upon the relative positions of two real objects (such as a bright star with a much fainter companion) being the same at different wavelengths, while artefacts, such as internal reflections and so on will generally change their positions with wavelength, thus enabling real objects to be separated from false ones.

2.2 PHOTOGRAPHY

2.2.1 Requiem for a Well-Loved Friend

In the first edition of this book, photography occupied 24 pages – *three* times as much as required for all forms, then, of astronomical electronic imaging.

Even in the sixth edition, just describing photography's idiosyncrasies sufficiently adequately for someone to be able to work with photographic archives, took up 13 pages. However, most photographic archives are now digitised, and it is two decades since any professional observatories used the technique (see also Section 1.1.10). Most amateur astronomers now use charge-coupled device

[2] A name.

(CCD) imagers and even in Wikipedia's definition of 'Photography', electronic imaging is mentioned twice as frequently as the use of photographic emulsions. It is thus time to close the door on this topic and to say farewell to one of astronomy's oldest and best servants. It is due to take its exit and to go to its long-deserved retirement.

If any readers *do* need to read up on the topic then older editions of this book (if any still lie around) may help, and there are other books and internet sources that may be hunted out – though, now, they are mostly for special art and niche craft work.

This author, at least, though would like to acknowledge, with great gratitude, the enormous contribution photography has made, but, with Thomas Gray[3]

No farther seek his merits to disclose
Or draw his frailties from their dread abode.

2.3 ELECTRONIC IMAGING

2.3.1 Introduction

The alternatives to photography for recording images directly are all electronic in nature (unless you are into sketching things yourself). They have two great advantages over the photographic emulsion. First, the image is usually produced as an electrical signal and can therefore be relayed or transmitted to a remote observer, which is vital for satellite-borne instrumentation and useful in many other circumstances and also enables the data to be fed directly into a computer. Secondly, with many of the systems, the quantum efficiency is up to a factor of 100 or so higher than that of the photographic plate. Subsidiary advantages can include intrinsic amplification of the signal, linear response and long or short wavelength sensitivity.

The most basic form of electronic imaging simply consists of an array of point-source detecting elements. The method is most appropriate for the intrinsically small detectors such as photoconductive cells (Section 1.1). Hundreds, thousands or even more of the individual elements (usually termed 'pixels') can then be used to give high spatial resolution. The array is simply placed at the focus of the telescope or spectroscope, etc. Other arrays such as CCDs, infrared arrays and superconducting tunnel junctions (STJs) were reviewed in detail in Section 1.1. Millimetre-wave and radio arrays were also considered in Section 1.1 and in Section 1.2. Here, therefore we are concerned with other electronic approaches to imaging, most of which have been superseded by array detectors but which have historical and archival interest.

2.3.2 Television and Related Systems

Low-light-level television systems, combined with image intensifiers and perhaps using a slower scan rate than normal have been used in the past. They were particularly found on the guide systems of large telescopes where they enable the operator at a remote-control console to have a viewing, finding, or guiding display. Also, many of the early planetary probes and some other spacecraft such as the International Ultraviolet Observer used TV cameras of various designs. These systems have now been replaced by solid-state arrays.

2.3.3 Image Intensifiers

Microchannel plates (MCPs) are a form of image intensifier (Section 1.3). That term, though, is usually reserved for the types of devices that produce an intensified image visible to the eye. Although used in the past for astronomical purposes, they are now rarely encountered and are mostly found as night sights for the military and for wildlife observers. Details may be found in previous editions of this book, if required.

Although it is not an *image* intensifier as such, a large area detector (commercially known by the name, ABALONE) has recently been devised using an avalanche photodiode (APD) as its actual

[3] 'Elegy Written in a Country Churchyard'.

detector and is perhaps best mentioned here. This uses a glass hemisphere with a photocathode on its inner surface. The photoelectrons are accelerated and focussed by electric fields and travel through the vacuum inside the hemisphere to the APD at its centre. The output signal is thus preamplified by the accelerations of the photoelectrons and then further amplified by the APD. The hemisphere can be made physically large, relatively cheaply, so that the device may be used to replace large photomultipliers (PMTs; see Section 1.1.11) in future astronomical applications.

2.3.4 Photon Counting Imaging Systems

A combination of a high-gain image intensifier and TV camera led in the 1970s to the development by Alec Boksenberg and others of the image photo counting system (IPCS). In the original IPCS, the image intensifier was placed at the telescope's (or spectroscope's, etc.) focal plane and produced a blip of some 10^7 photons at its output stage for each incoming photon in the original image. A relatively conventional TV camera then viewed this output and the video signal fed to a computer for storage. Again, this type of device is now superseded by solid-state detectors.

More recently a number of variants on Boksenberg's original IPCS have been developed. These include MCP image intensifiers linked to CCDs and other varieties of image intensifiers linked to CCDs or other array-type detectors. The principles of the devices remain unchanged however.

2.4 SCANNING

This is such an obvious technique as scarcely to appear to deserve a separate section and indeed at its simplest it hardly does so. A 2D image may be built up using any point source detector, if that detector is scanned over the image or vice versa. Many images at all wavelengths are obtained in just this way. Sometimes the detector moves and sometimes the whole telescope or satellite, etc. Occasionally it may be the movement of the secondary mirror of the telescope or of some ad hoc component in the apparatus which allows the image to be scanned. Scanning patterns are normally raster or spiral (Figure 2.3). Other patterns may be encountered, however, such as the continuously nutating roll employed by some of the early artificial satellites. The only precautions advisable are the matching of the scan rate to the response of the detector or to the integration time being used and the matching of the separation of the scanning lines to the resolution of the system.

Scanning by radio telescopes (Section 1.2) used to be by observing individual points within the scan pattern for a set interval of time. More recently the scanning has been continuous, or 'on the fly' because this enables slow changes due to the Earth's atmosphere or to the instrument to be eliminated. Specialised applications of scanning occur in the spectrohelioscope (Section 5.3) and the scanning spectrometer (Section 4.2). Many Earth observation satellites use push-broom scanning. In this system, a linear array of detectors is aligned at right angles to the spacecraft's ground track. The image is then built up as the spacecraft's motion moves the array to look at successive slices of the swathe of ground over which the satellite is passing.

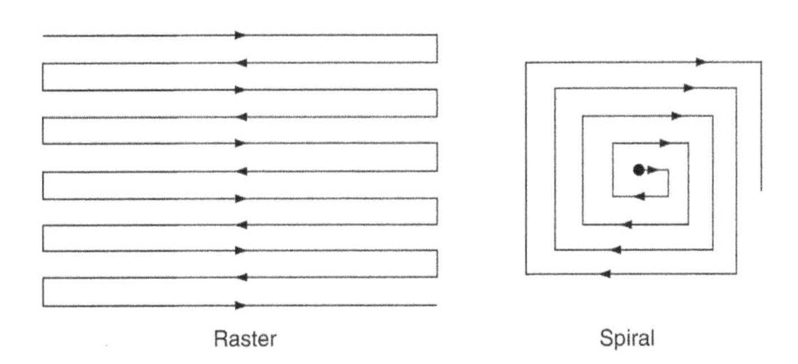

FIGURE 2.3 Scanning patterns.

A more sophisticated approach to scanning is to modulate the output of the detector by interposing a mask of some type in the light beam. Examples of this technique are discussed elsewhere and include the modulation collimator and the coded array mask used in X-ray imaging (Section 1.3.4.1). A great improvement over either these methods or the basic approach discussed previously may be made by using a series of differing masks or by scanning a single mask through the light beam so that differing portions of it are used. This improved method is known as 'Hadamard mask imaging'. We may best illustrate its principles by considering 1D images such as might be required for spectrophotometry. The optical arrangement is shown in Figure 2.4. The mask is placed in the image plane of the telescope and the Fabry lens directs all the light that passes through the mask onto the single detector. Thus, the output from the system consists of a simple measurement of the total intensity passed by the mask. If a different mask is now substituted for the first, then a new and, in general, different intensity reading will be obtained. If the image is to be resolved into N elements, then N such different masks must be used and N intensity readings determined. If **D** is the vector formed from the detector output readings, **I** is the vector of the intensities of the elements of the image and **M** the $N \times N$ matrix whose columns each represent one of the masks, with the individual elements of the matrix corresponding to the transmissions of the individual segments of the mask, that is,

$$D = \left[D_1, D_2, D_3, \ldots D_N\right] \tag{2.18}$$

$$I = \left[I_1, I_2, I_3, \ldots I_N\right] \tag{2.19}$$

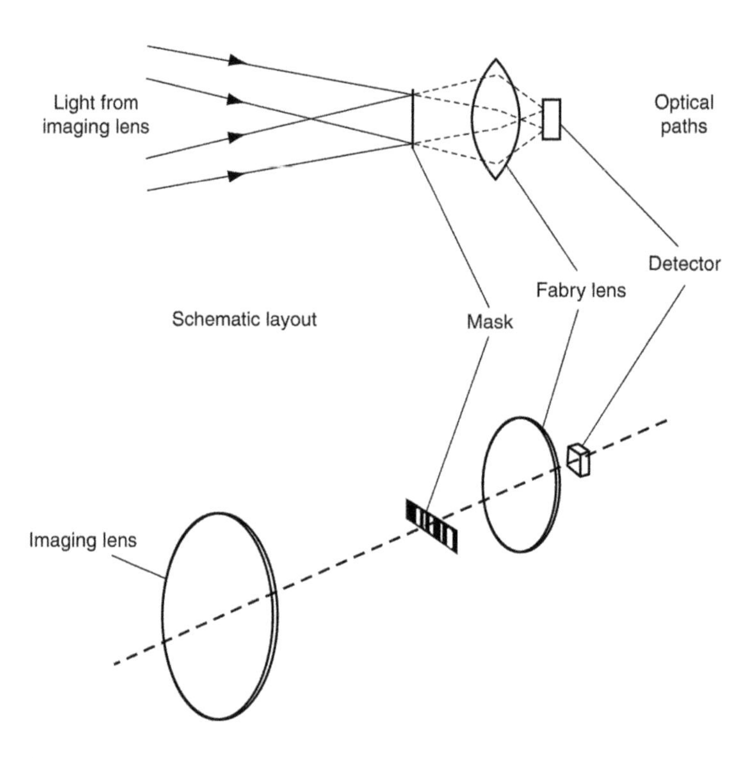

FIGURE 2.4 Schematic arrangement for a Hadamard mask imaging system.

$$M = \begin{bmatrix} m_{11} & m_{12} & m_{13} & \cdots & m_{1N} \\ m_{21} & & & & \\ m_{31} & & & & \\ \cdot & & & & \\ \cdot & & & & \\ \cdot & & & & \\ m_{N1} & & & & m_{NN} \end{bmatrix} \tag{2.20}$$

Then, ignoring any contribution arising from noise, we have

$$\mathbf{D} = \mathbf{I}\,\mathbf{M} \tag{2.21}$$

and so

$$\mathbf{I} = \mathbf{D}\,\mathbf{M}^{-1} \tag{2.22}$$

Thus, the original image is simply obtained by inverting the matrix representing the masks. The improvement of the method over a simple scan lies in its multiplex advantage (cf. the Fourier transform spectrometer, Section 4.1). The masks usually comprise segments that either transmit or obscure the radiation completely, that is

$$m_{ij} = 0 \text{ or } 1 \tag{2.23}$$

and on average, about half of the total image is obscured by a mask. Thus $N/2$ image segments contribute to the intensity falling onto the detector at any one time. Hence, if a given signal-to-noise ratio is reached in a time, T, when the detector observes a single image element, then the total time required to detect the whole image with a simple scan is

$$N \times T \tag{2.24}$$

and is approximately

$$\sqrt{2N}\,T \tag{2.25}$$

for the Hadamard masking system. Thus, the multiplex advantage is approximately a factor of

$$\sqrt{\frac{N}{2}} \tag{2.26}$$

improvement in the exposure length.

In practice, moving a larger mask across the image generates the different masks. The matrix representing the masks must then be cyclic, or in other words, each successive column is related to the previous one by being moved down a row. The use of a single mask in this way can lead to a considerable saving in the construction costs because $2N–1$ segments are needed for it compared with N^2 if separate masks are used. An additional constraint on the matrix **M** arises from the presence of noise in the output. The errors introduced by this are minimised when

$$\mathrm{Tr}\left[\mathbf{M}^{-1}\left(\mathbf{M}^{-1}\right)^{\mathrm{T}}\right] \tag{2.27}$$

is minimised. There are many possible matrices that will satisfy these two constraints, but there is no completely general method for their generation. One method of finding suitable matrices and so of specifying the mask is based upon the group of matrices known as the 'Hadamard matrices' (hence, the name for this scanning method). These are matrices whose elements are ±1's and which have the property

$$\mathbf{H}\,\mathbf{H}^{\mathrm{T}} = \mathbf{NI} \tag{2.28}$$

where **H** is the Hadamard matrix and **I** is the identity matrix. A typical example of the mask matrix, **M**, obtained in this way might be

$$M = \begin{bmatrix} 0 & 1 & 0 & 1 & 1 & 1 & 0 \\ 0 & 0 & 1 & 0 & 1 & 1 & 1 \\ 1 & 0 & 0 & 1 & 0 & 1 & 1 \\ 1 & 1 & 0 & 0 & 1 & 0 & 1 \\ 1 & 1 & 1 & 0 & 0 & 1 & 0 \\ 0 & 1 & 1 & 1 & 0 & 0 & 1 \\ 1 & 0 & 1 & 1 & 1 & 0 & 0 \end{bmatrix} \tag{2.29}$$

for N having a value of 7.

Variations to this scheme can include making the obscuring segments out of mirrors and using the reflected signal as well as the transmitted signal or arranging for the opaque segments to have some standard intensity rather than zero intensity. The latter device is particularly useful for infrared work. The scheme can also easily be extended to more than one dimension, although it then rapidly becomes complex. A 2D mask may be used in a straightforward extension of the 1D system to provide 2D imaging. A 2D mask combined suitably with another 1D mask can provide data on three independent variables – for example, spectrophotometry of 2D images. Two 2D masks could then add (for example) polarimetry to this and so on. Except in the 1D case, the use of a computer to unravel the data is obviously essential, and even in the 1D case, it will be helpful as soon as the number of resolution elements rises above five or six.

2.5 INTERFEROMETRY

2.5.1 Introduction

Interferometry is the technique of using constructive and destructive addition of radiation beams to determine information about the source of that radiation.[4] The double beam Michelson interferometer as used by Michelson and Morley in their attempts to detect the aether, is now, of course, the basis of all current gravitational wave detectors (GWDs) and its details are discussed in Section 1.6. The theory of the GWDs' interferometers overlaps extensively with the theory of the instruments discussed in this section, but the physical realisation of the instruments differs enormously. Interferometers are also used to obtain high precision positions for optical and radio sources and this application is considered in Section 5.1.

Two principal types of interferometer exist: the Michelson stellar interferometer, first proposed by Armand Fizeau in 1868 and the intensity interferometer proposed by Robert Hanbury Brown in 1949. There are also a number of subsidiary techniques such as amplitude interferometry, nulling interferometry, speckle interferometry, Fourier spectroscopy, intensity interferometry, etc. The latter until recently has only been of historical interest; however, intensity interferometry may soon be revived to map stellar surfaces using γ-ray air Čerenkov telescopes.

[4] Astronomical interferometers generally have their two (or more) beams with the same frequency – known as 'homodyne operation'. Other applications may use beams with differing frequencies – heterodyne operation (c.f. radio receivers, Section 1.2.2). The interferometers carried by the Laser Interferometer Space Antenna (LISA) Pathfinder spacecraft (Table 1.8) were of the heterodyne type.

2.5.2 Michelson Optical Stellar Interferometer

The Michelson *stellar* interferometer is so called to distinguish it from the type of interferometer used in the 'Michelson-Morley' experiment, gravitational wave interferometers and in Fourier spectroscopy. Because the latter type of interferometer is discussed primarily in Sections 1.6 and 4.1, the 'stellar' qualification for this section will be dropped when referring to the first of the two main types of interferometer.

In practice, many interferometers use the outputs from numerous telescopes; however, this is just to reduce the time taken for the observations, and it is actually the outputs from pairs of telescopes that are combined to produce the interference effects. We may start, therefore, by considering the interference effects arising from two beams of radiation.

The complete optical system that we shall study is shown in Figure 2.5. The telescope objective, which for simplicity is shown as a lens (but exactly the same considerations apply to mirrors), is covered by an opaque screen with two small apertures and is illuminated by two monochromatic equally bright point sources at infinity. The objective diameter is D, its focal length F, the separation

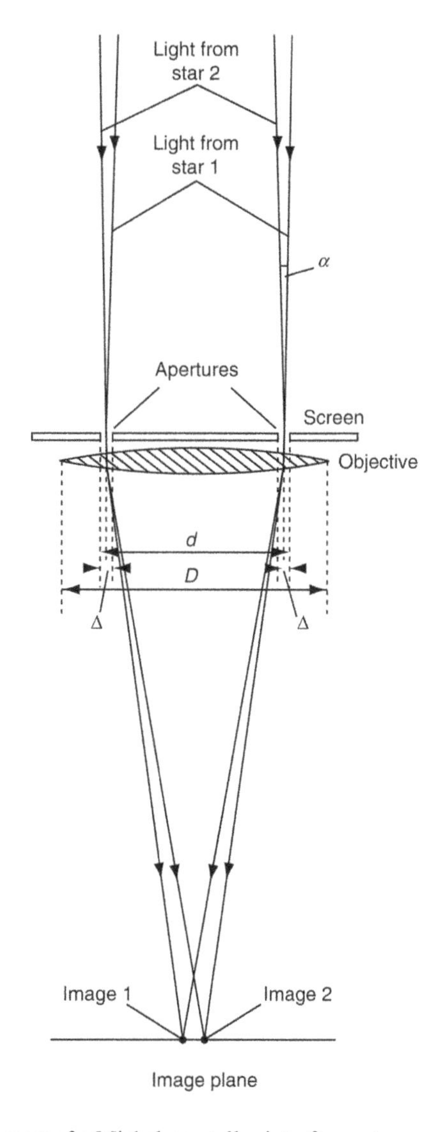

FIGURE 2.5 Optical arrangement of a Michelson stellar interferometer.

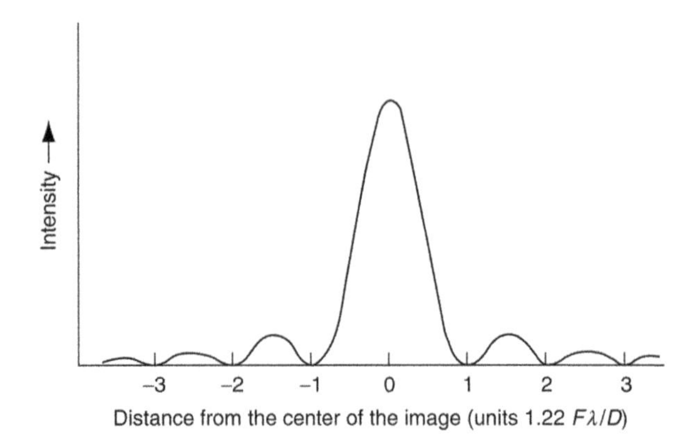

FIGURE 2.6 Image structure for one point source and the whole objective.

of the apertures is d, the width of the apertures is Δ, the angular separation of the sources is α, and their emitted wavelength is λ.

Let us first consider the objective by itself, without the screen and with just one of the sources. The image structure is then just the well-known diffraction pattern of a circular lens (Figure 2.6). With both sources viewed by the whole objective, two such patterns are superimposed. There are no interference effects between these two images because their radiation is not mutually coherent. When the main maxima are superimposed upon the first minimum of the other pattern, we have Rayleigh's criterion for the resolution of a lens (Figures 1.35 and 2.7). The Rayleigh criterion for the minimum angle between two separable sources α', as we saw in Section 1.1 is

$$\alpha' = \frac{1.22\lambda}{D} \tag{2.30}$$

Now let us consider the situation with the screen in front of the objective and first consider just one aperture looking at just one of the sources. The situation is then the same as that illustrated in Figure 2.6, but the total intensity and the resolution are reduced because of the smaller size of the aperture compared with the objective (Figure 2.8). Although the image for the whole objective is also shown for comparison in Figure 2.8, if it were truly to scale for the situation illustrated in Figure 2.5, then it would be one-seventh of the width that is shown and 1,800 times higher.

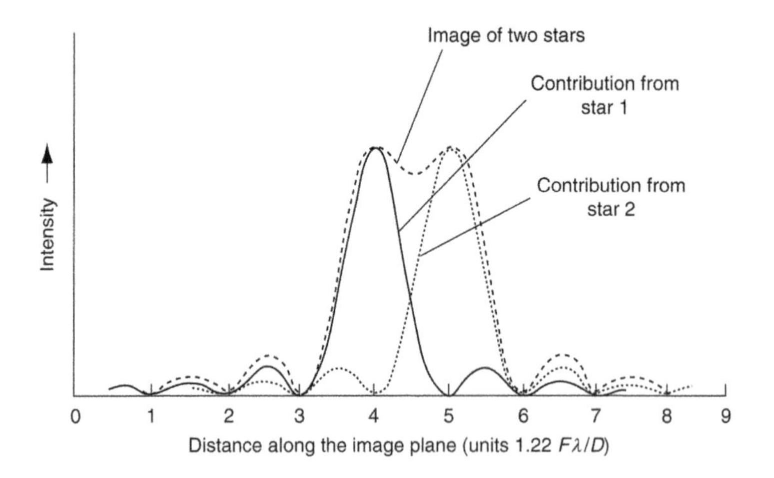

FIGURE 2.7 Image structure for two point sources and the whole objective.

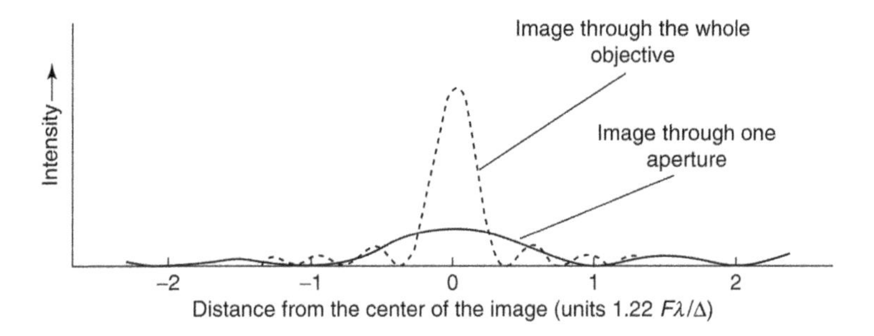

FIGURE 2.8 Image structure for one point source and one aperture.

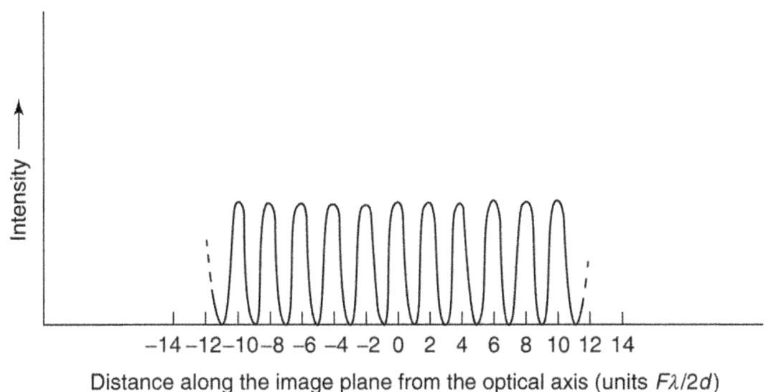

FIGURE 2.9 Image structure for a single point source viewed through two infinitely small apertures.

Now consider what happens when one of the sources is viewed simultaneously through both small apertures. If the apertures were infinitely small, then ignoring the inconvenient fact that no light would get through anyway, we would obtain a simple interference pattern (Figure 2.9). The effect of the finite width of the apertures is to modulate the straightforward variation of Figure 2.9 by the shape of the image for a single aperture (Figure 2.8). Because two apertures are now contributing to the intensity and the energy 'lost' at the minima reappears at the maxima, the overall envelope of the image peaks at four times the intensity for a single aperture (Figure 2.10). Again, for the actual situation shown in Figure 2.5, there should be a total of 33 fringes inside the major maximum of the envelope.

Finally, let us consider the case of two equally bright point sources viewed through two apertures. Each source has an image whose structure is that shown in Figure 2.10, and these simply add together in the same manner as the case illustrated in Figure 2.7 to form the combined image.

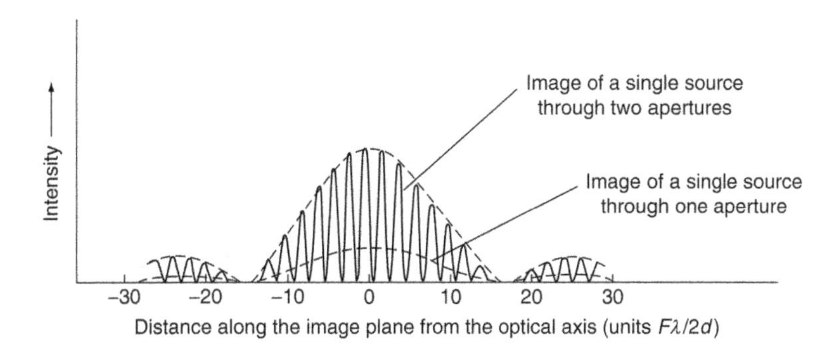

FIGURE 2.10 Image structure for one point source viewed through two apertures.

The structure of this combined image will depend upon the separation of the two sources. When the sources are almost superimposed, the image structure will be identical with that shown in Figure 2.10, except that all the intensities will have doubled. As the sources move apart, the two fringe patterns will also separate, until when the sources are separated by an angle α'', given by

$$\alpha'' = \frac{\lambda}{2d} \tag{2.31}$$

the maxima of one fringe pattern will be superimposed upon the minima of the other and vice versa. The fringes will then disappear and the image will be given by their envelope (Figure 2.11). There may still be a slight ripple on this image structure because the incomplete filling of the minima by the maxima, but it is unlikely to be noticeable. The fringes will reappear as the sources continue to separate until the pattern is almost double that of Figure 2.10 once again. The image patterns are then separated by a whole fringe width and the sources by $2\alpha''$. The fringes disappear again for a source separation of $3\alpha''$ and reach yet another maximum for a source separation of $4\alpha''$ and so on. Thus, the fringes are most clearly visible when the source's angular separation is given by $2n\alpha''$ (where n is an integer) and they disappear or reach a minimum in clarity for separations of $(2n + 1)\alpha''$. Applying the Rayleigh criterion for resolution, we see that the resolution of two apertures is given by the separation of the sources for which the two fringe patterns are mutually displaced by half a fringe width. This as we have seen is simply the angle α'' and so we have

$$\frac{\text{Resolution through two apertures}}{\text{Resolution of the objective}} = \frac{\alpha''}{\alpha'} \tag{2.32}$$

$$= \frac{2.44d}{D} \tag{2.33}$$

Imagining the two apertures placed at the edge of the objective (i.e., $d = D$), we see the quite remarkable result that the resolution of an objective may be increased by almost a factor of $2^{1/2}$ by screening it down to two small apertures at opposite ends of one of its diameters.[5] The improvement in the resolution is only along the relevant axis – perpendicular to this, the resolution is just that of one of the apertures. The resolution in both axes may, however, be improved by using a central occulting disc which leaves the rim of the objective clear. This also enables us to get a feeling for the physical basis for this improvement in resolution. We may regard the objective as composed of a series of concentric narrow rings, each of which has a resolution given by Equation 2.31, with d

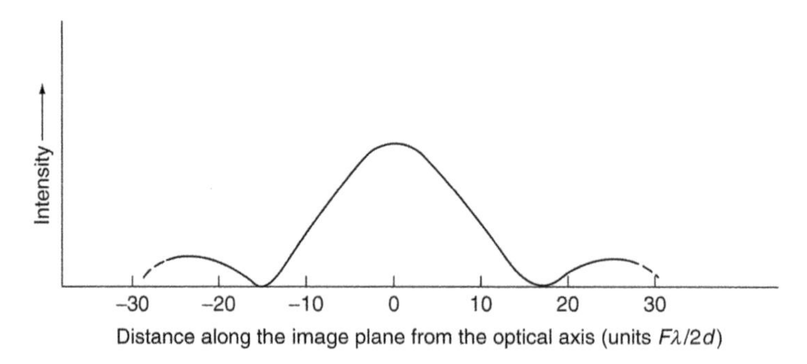

FIGURE 2.11 Image structures for two point sources separated by an angle, $\lambda/2d$, viewed through two apertures.

[5] Remember, resolution is *improved* when the resolved angle is *smaller*. See James Webb Space Telescope's (JWST's) Near Infrared Imager and Slitless Spectrograph (NIRISS) instrument for an application (Section 2.6).

being the diameter of the ring. Because the objective's resolution is the average of all these individual resolutions, it is naturally less than their maximum. In practice, some blurring of the image may be detected for separations of the sources that are smaller than the Rayleigh limit (Section 1.1). This blurring is easier to detect for the fringes produced by the two apertures, than it is for the images through the whole objective. The effective improvement in the resolution may therefore be even larger than that given by Equation 2.33.

Now for the situation illustrated in Figure 2.5, the path difference between the two light beams on arriving at the apertures is likely to be small because the screen will be perpendicular to the line of sight to the objects. To produce fringes with the maximum clarity, however, that path difference must be close to zero. This is because radiation is never completely monochromatic; there is always a certain range of wavelengths present known as the 'bandwidth of the signal'. Now for a zero-path difference at the apertures, all wavelengths will be in phase when they are combined and will interfere constructively. However, if the path difference is not zero, some wavelengths will be in phase but others will be out of phase to a greater or lesser extent because the path difference will equal different numbers of cycles or fractions of cycles at the different wavelengths. There will thus be a mix of constructive and destructive interference and the observed fringes will have reduced contrast. The path difference for which the contrast in the fringes reduces to zero (i.e., no fringes are seen) is called the 'coherence length', l and is given by

$$l = \frac{c}{\Delta\nu} = \frac{\lambda^2}{\Delta\lambda} \tag{2.34}$$

where $\Delta\nu$ and $\Delta\lambda$ are the frequency and wavelength bandwidths of the radiation. So that for $\lambda = 500$ nm and $\Delta\lambda = 1$ nm, we have a coherence length of 0.25 mm; for white light ($\Delta\lambda \approx 300$ nm) this reduces to less than a micron, but in the radio region it can be large; 30 m, for example, at $\nu = 1.5$ GHz and $\Delta\nu = 10$ MHz, for example. However, as shall seen, the main output from an interferometer is the fringe contrast (usually known as the 'fringe visibility', V, – Equation 2.35), so that for this not to be degraded, the path difference to the apertures, or their equivalent, must be kept to a small fraction of the coherence length. For interferometers such as Michelson's 1921 stellar interferometer (Figure 2.14), the correction to zero path difference is small and in that case was accomplished through the use of adjustable glass wedges. However, most interferometers, whether operating in the optical or radio regions, now use separate telescopes on the ground (Figures 2.27 and 2.35, for example), so the path difference to the telescopes can be more than 100 m for optical interferometers and up to 1,000 km for very-long-base-line radio interferometry. These path differences have to corrected either during the observation by hardware, or afterwards during the data processing.

Michelson's original interferometer was essentially identical to that shown in Figure 2.5, except that the apertures were replaced by two movable parallel slits to increase the amount of light available and a narrow band filter was included to give nearly monochromatic light (previously, it had assumed that the *sources* were monochromatic). To use the instrument to measure the separation of a double star with equally bright components, the slits are aligned perpendicularly to the line joining the stars and moved close together, so that the image fringe patterns are undisplaced, and the image structure is that of Figure 2.10. The actual appearance of the image in the eyepiece at this stage is shown in Figure 2.12. The slits are then moved apart until the fringes disappear (Figures 2.11 and 2.13). The distance between the slits is then such that the fringe pattern from one star is filling-in the fringe pattern from that other, and their separation is given by α'' (Equation 2.31 – with d given by the distance between the slits). If the two stars are of differing brightness, then the fringes will not disappear completely, but the fringe visibility, V, given by

$$V = \frac{(I_{\max} - I_{\min})}{(I_{\max} + I_{\min})} \tag{2.35}$$

FIGURE 2.12 Appearance of the image of a source seen in a Michelson interferometer at the instant of maximum fringe visibility.

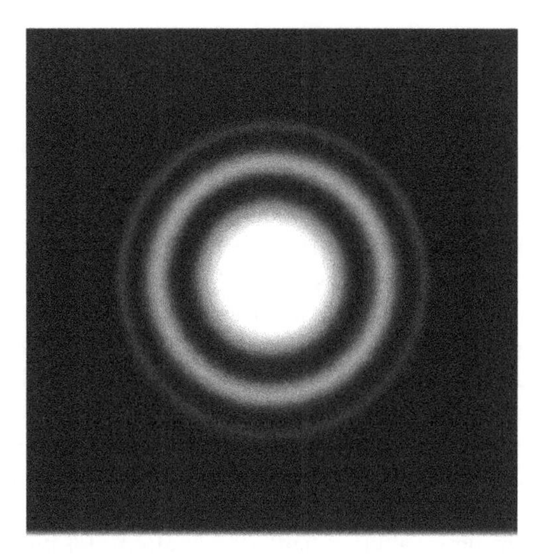

FIGURE 2.13 Appearance of the image of a source seen in a Michelson interferometer at the time of zero fringe visibility.

where I_{max} is the intensity of a fringe maximum and I_{min} is the intensity of a fringe minimum, will reach a minimum at the same slit distance.

To measure the diameter of a symmetrical object like a star or the satellite of a planet, the same procedure is used. However, the two sources are now the two halves of the star's disk, represented by point sources at the optical centres[6] of the two semicircles. The diameter of the star is then given by 2.44 α'', so that for determining the diameters of objects, an interferometer has the same resolution as a conventional telescope with an objective diameter equal to the separation of the slits.

Stringent requirements on the stability and accuracy of the apparatus are required for the success of this technique. As we have seen, the path difference to the slits must not be greater than a small fraction

[6] The same as their centres of gravity if they are thin flat uniform semicircles.

of the coherence length of the radiation (Equation 2.34). Furthermore, that path difference must remain constant to considerably better than the wavelength of the radiation that is being used as the slits are separated. Vibrations and scintillation are additional limiting factors. In general, the paths will not be identical when the slits are close together and the fringes are at their most visible. But the path difference will then be some integral number of wavelengths and provided that it is significantly less than the coherence length the interferometer may still be used. Thus, in practice the value that is required for d is the difference between the slit separations at minimum and maximum fringe visibilities. A system such as has just been described was used by Michelson in 1891 to measure the diameters of the Galilean satellites. Their diameters are in the region of 1″, so that d has a value of a few tens of millimetres (Figure 2.13).

Now the angular diameters of stellar disks are much smaller than those of the Galilean satellites; 0.047″ for α Orionis (Betelgeuse), for example, one of the largest angular diameter stars in the sky. A separation for the apertures of several metres is thus needed if stellar diameters are to be measured. This led to the improved version of the system that is rather better known than the one used for the Galilean satellites and which is the type of interferometer that is usually intended when the Michelson stellar interferometer is mentioned. It was used by Albert Michelson and Francis Pease in 1921 to measure the diameters of half a dozen or so of the larger stars. The slits were replaced by movable mirrors on a rigid 6-m-long bar mounted on top of the 2.5-m Hooker telescope (Figure 2.14). By this means, d could be increased beyond the diameter of the telescope. A system of mirrors then reflected the light into the telescope. The practical difficulties with the system were great, and although larger systems have since been attempted, none has been successful; so the design is now largely of historical interest only.

Modern optical interferometers are based upon separate telescopes rather than two apertures feeding a single instrument. They are still primarily used to measure stellar diameters, although some aperture synthesis systems (see below) are now working and others are under construction.

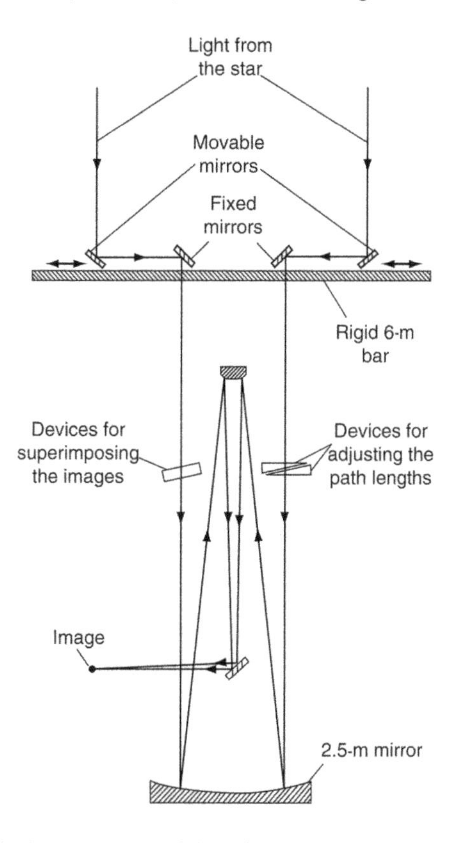

FIGURE 2.14 Schematic optical arrangement of the Michelson stellar interferometer.

FIGURE 2.15 One of the retro-reflectors making up a delay line for the VLTI. The retro-reflector is essentially a small Cassegrain telescope in which the incoming light beam is reflected from the secondary mirror back to be reflected from the primary mirror a second time. The light beam is thus returned back to the other end of the delay line. The retro-reflector is mounted on a wheeled carriage which can move on rails over a 60-m distance and so provide a maximum delay line length of 120 m. The required positioning accuracy of the reflector is about 50 nm, and this is achieved by a combination of movement of the carriage and piezoelectric transducers monitored by a laser measuring system. (Courtesy of ESO.)

There is no difference in the principle of operation of these interferometers from that of Michelson, but there are many practical changes. The main change arises because the path difference to the two telescopes can now be many metres. A system for compensating for the path difference is thus required. In most designs, the path compensation is via a delay line; the light from one telescope is sent on a longer path than that from the other one before they are mixed together (Figure 2.15). The extra path length is arranged to be equal to the path difference between the telescopes. Because the path difference will change as the object moves across the sky, the delay must also change. This is usually accomplished by having one or more of the mirrors directing the light into the longer path mounted on a moveable carriage. Even using narrow bandwidths, the small value of the coherence length (Equation 2.34) at optical wavelengths means that the carriage must be capable of being positioned to submicron accuracy and yet be able to travel tens of metres.

The second major difference is that the telescopes are usually fixed in position or moveable only slowly between observations. It is thus not possible to change their separation to observe maximum and minimum fringe visibilities. A change in the effective separation occurs as the angle between the baseline and the source in the sky alters, but this is generally insufficient to range from maximum to minimum fringe visibility. Instead, therefore, a measurement of the fringe visibility is made for just one, or a small number, of separations. This measurement is then fitted to a theoretical plot of the variation of the fringe visibility with the telescope separation and the size of the source to determine the latter.

The third difference arises from the atmospheric turbulence. We have seen that there is an upper limit to the size of a telescope aperture, given by Fried's coherence length (Equation 1.60). In the visible region, this is only about 120 mm, though in the infrared it can rise to 300 mm. Telescopes with apertures larger than Fried's limit will thus be receiving radiation through several, perhaps many, atmospheric cells, each of which will have different phase shifts and wavefront distortions (Figure 1.61). When the beams are combined, the desired fringes will be washed out by all these differing contributions. Furthermore, the atmosphere changes on a time scale of around 5 milliseconds. So that even if small apertures are used, exposures have to be short enough to freeze the atmospheric motion (see also speckle interferometry, Section 2.6). Thus, many separate observations have to be added together to get a value with a sufficiently high signal-to-noise ratio. To use large telescopes as a part of an interferometer, their images have to be corrected through adaptive optics (Section 1.1). Even a simple tip-tilt correction will allow the useable aperture to be increased by a factor of three. However, to use 8-m and 10-m telescopes, such as European Southern Observatory's (ESO's) Very Large Telescope (VLT) and the Keck telescope, full atmospheric correction is needed.

When only two telescopes form the interferometer, the interruption of the phase of the signal by the atmosphere results in the loss of that information. Full reconstruction of the morphology of the original object is not then possible, only the fringe visibility can be determined (see below for a discussion of multi-element interferometers, closure phase and aperture synthesis). However, the fringe visibility still enables stellar diameters and double-star separations to be found, and so several two-element interferometers either have been until recently or are currently in operation.

Nulling interferometry can be undertaken with two-element interferometers as well as those with more elements. It has recently come into prominence as a way of detecting exoplanets. The interferometer is adjusted so that the stellar fringe pattern is at a maximum (Figure 2.12). The exoplanet's fringe pattern will not then, in general, simultaneously be at a fringe maximum because the exoplanet is separated by a few micro- to milliarcseconds from the star in the sky. The exoplanet's fringe pattern can therefore be studied separately from that of the star. Operating in this fashion, their purposes overlap with those of stellar coronagraphs (Section 2.7.4).

The Large Binocular Telescope Interferometer (LBTI) operates as a nulling instrument from 3 to 14 μm with an effective 14.4-m baseline giving it an angular resolution of better than 100 milliarcseconds. It also has several other modes of operation including other types of interferometry, coronography, spectroscopy and direct imaging. Recently, a concept proposal for the future, envisages a space-based nulling interferometer with four 0.75-m mirrors and a minimum baseline of 40 m. Its primary mission would be to study the (possible) exoplanet Proxima Centauri-b between 5 and 20 μm. The instrument might also act as a demonstrator for a later larger mission.

Recently, a nulling interferometer based upon photonic components has been demonstrated in the laboratory and tested with the Subaru telescope using SCExAO, on a number of stars. Guided Light Interferometric Nulling Technology (GLINT) subdivides the telescope's image into two segments and feeds the segments into separate 9-μm diameter single mode[7] waveguides within a single block of glass. The waveguides are produced within the glass (i.e., the photonic chip) by increasing the material's refractive index along a 10.75-μm wide track inside the block by a momentary heating with a finely focussed laser beam. The beam splittings, recombinations and so on required for the interferometer, all take place photonically within the chip and the four output beams are then fed by fibre-optic cables to the detectors.

A variation on the system that also has some analogy with intensity interferometry (Section 2.5.6) is known as 'amplitude interferometry'. This has been used recently in a successful attempt to measure stellar diameters. At its heart is a device known as a 'Köster prism' that splits and combines the two separate light beams (Figure 2.16). The interfering beams are detected in a straightforward manner using PMTs. Their outputs may then be compared. The fringe visibility of the Michelson interferometer appears as the anticorrelated component of the two signals, while the atmospheric scintillation and so on affect the correlated component. Thus, the atmospheric effects can be screened out to a large extent and the stability of the system is vastly improved.

---

[7] Photonics, in various forms, is becoming increasingly important within astronomical instruments and references to it are made in several places in this book. There is not space to discuss it here properly and so (for this edition at least) it still has to be left to readers to research for themselves. Simplistically though, a *multimode* waveguide can allow the simultaneous propagation down its length of several light beams, each inclined at a different small angle to its longitudinal axis. A *single* mode waveguide is one that is so narrow that there is only 'room' for a light beam at a single angle to the longitudinal axis to travel along it. Thus, a telecommunications fibre-optic using doped silicon for its core, pure silicon for its cladding and transmitting radiation at 1.5 μm would be single mode for core diameters up to about 10 μm and multi-mode at larger sizes.

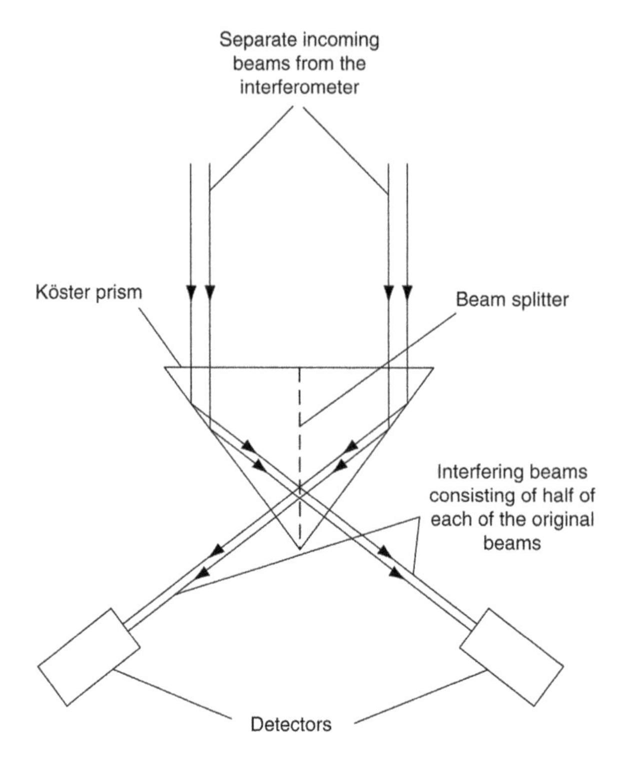

FIGURE 2.16 The Köster prism.

2.5.3 Michelson Radio Interferometer

Radio aerials are generally only a small number of wavelengths in diameter (Section 1.2), and so their intrinsic resolution (Equation 1.15) is poor. The use of interferometers to obtain higher resolution therefore dates back to soon after the start of radio astronomy. There is no difference in principle between a radio and an optical interferometer when they are used to measure the separation of double sources, or the diameters of uniform objects. However, the output from a radio telescope contains both the amplitude and phase of the signal, so that complete imaging of the source is possible. Radio interferometers generally use more than two antennae, and these may be in a 2D array. The basic principle, however, is the same as for just two antennae, although the calculations become considerably more involved.

The individual elements of a radio interferometer are usually fairly conventional radio telescopes (Section 1.2), and the electrical signal that they output varies in phase with the received signal. In fact, with most radio interferometers, it is the electrical signals that are mixed to produce the interference effect rather than the radio signals themselves. Their signals may then be combined in two quite different ways to provide the interferometer output. In the simplest version, the signals are simply added together before the square law detector (Figure 1.72), and the output will then vary with the path difference between the two signals. Such an arrangement, however, suffers from instability problems, particularly because of variations in the voltage gain.

A system that is now preferred to the simple adding interferometer is the correlation or multiplying interferometer. In this process, as the name suggests, the intermediate frequency (IF) signals from the receivers are multiplied together. The output of an ideal correlation interferometer will contain only the signals from the source (which are correlated); the other components of the outputs of the telescopes will be zero as shown below.

If we take the output voltages from the two elements of the interferometer to be $(V_1 + V_1')$ and $(V_2 + V_2')$ where V_1 and V_2 are the correlated components (i.e., from the source in the sky) and V_1' and V_2' are the uncorrelated components, the product of the signals is then

$$V = \left(V_1 + V_1'\right) \times \left(V_2 + V_2'\right) \tag{2.36}$$

$$= V_1V_2 + V_1'\,V_2 + V_1V_2' + V_1'\,V_2' \tag{2.37}$$

If we average the output over time, then any component of Equation 2.37 containing an uncorrelated component will tend to zero. Thus,

$$\bar{V} = \overline{V_1V_2} \tag{2.38}$$

In other words, the time-averaged output of a correlation interferometer is the product of the correlated voltages. Because most noise sources contribute primarily to the uncorrelated components, the correlation interferometer is inherently much more stable than the adding interferometer. The phase-switched interferometer (see below) is an early example of a correlation interferometer, though direct multiplying interferometers are now more common.

The systematics of radio astronomy differ from those of optical work (Section 1.2), so that some translation is required to relate optical interferometers to radio interferometers. If we take the polar diagram of a single radio aerial (Figure 1.78 for example), then this is the radio analogue of our image structure for a single source and a single aperture (Figures 2.6 and 2.8). The detector in a radio telescope accepts energy from only a small fraction of this image at any given instant (though array detectors with small numbers of pixels are now coming into use; see Section 1.2). Thus, scanning the radio telescope across the sky corresponds to scanning this energy-accepting region through the optical image structure. The main lobe of the polar diagram is therefore the equivalent of the central maximum of the optical image and the side lobes are the equivalent of the diffraction fringes. Because an aerial is normally directed towards a source, the signal from it corresponds to a measurement of the central peak intensity of the optical image (Figures 2.6 and 2.8).

If two stationary aerials are now considered and their outputs combined, then when the signals arrive without any path differences, the final output from the radio system as a whole corresponds to the central peak intensity of Figure 2.10. If a path difference does exist however, then provided that it is less than the coherence length, the final output will correspond to some other point within that image. In particular, when the path difference is a whole number of wavelengths, the output will correspond to the peak intensity of one of the fringes, and when it is a whole number plus half a wavelength, it will correspond to one of the minima. Now the path differences arise in two main ways: from the angle of inclination of the source to the line joining the two antennae and from delays in the electronics and cables between the antennae and the central processing station (Figure 2.17). The latter will normally be small and constant and may be ignored or corrected. The former will alter as the rotation of the Earth changes the angle of inclination. Thus, the output of the interferometer will vary with time as the value of P changes. The output over a period of time, however, will not follow precisely the shape of Figure 2.10 because the rate of change of path difference varies throughout the day, as we may see from the equation for P

$$P = s\cos\phi\cos(\psi - E) \tag{2.39}$$

where μ is the altitude of the object and is given by

$$\mu = \sin^{-1}[\sin\delta\sin\phi + \cos\delta\cos\phi\cos(T - \alpha)] \tag{2.40}$$

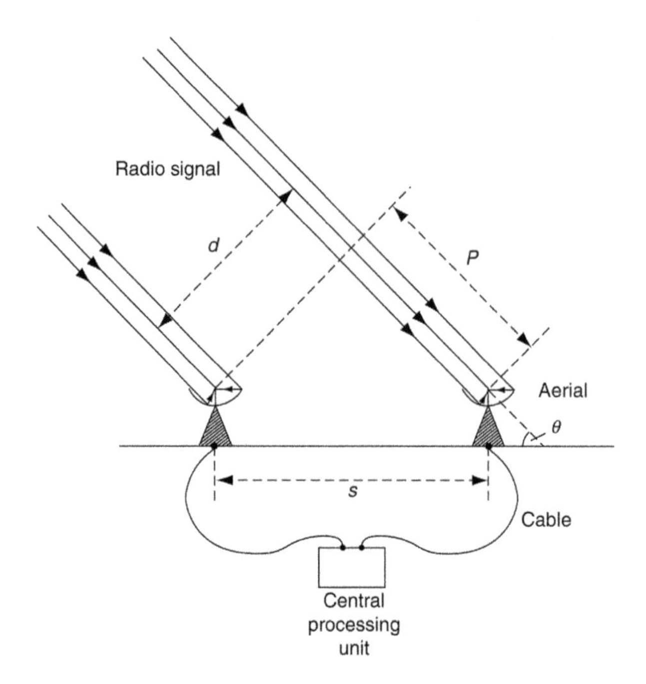

FIGURE 2.17 Schematic arrangement of a radio interferometer with fixed aerials.

ψ is the azimuth of the object and is given by

$$\psi = \cot^{-1}[\sin\phi\cot(T-\alpha) - \cos\phi\tan\delta\,\text{cosec}\,(T-\alpha)] \tag{2.41}$$

In these Equations E is the azimuth of the line joining the aerials, α and δ are the right ascension and declination of the object, T is the local sidereal time at the instant of observation and ϕ is the latitude of the interferometer. The path difference also varies because the effective separation of the aerials, d, where

$$d = s[\sin^2\mu + \cos^2\mu\sin^2(\psi - E)]^{1/2} \tag{2.42}$$

also changes with time; thus, the resolution and fringe spacing (Equation 2.31) are altered. Hence, the output over a period of time from a radio interferometer with fixed antennae is a set of fringes whose spacing varies, with maxima and minima corresponding to those of Figure 2.10 for the instantaneous values of the path difference and effective aerial separation (Figure 2.18).

An improved type of interferometer, which has increased sensitivity and stability, is the phase-switched interferometer (see previous discussion and phased arrays, Section 1.2.3). The phase of the signal from one aerial is periodically changed by 180° by, for example, switching an extra piece of cable half a wavelength long into or out of the circuit. This has the effect of oscillating the beam pattern of the interferometer through half a fringe width. The difference in the signal for the two

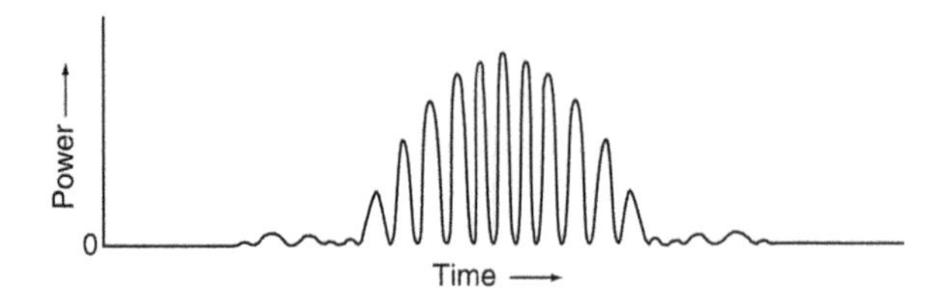

FIGURE 2.18 Output from a radio interferometer with stationary aerials, viewing a single source.

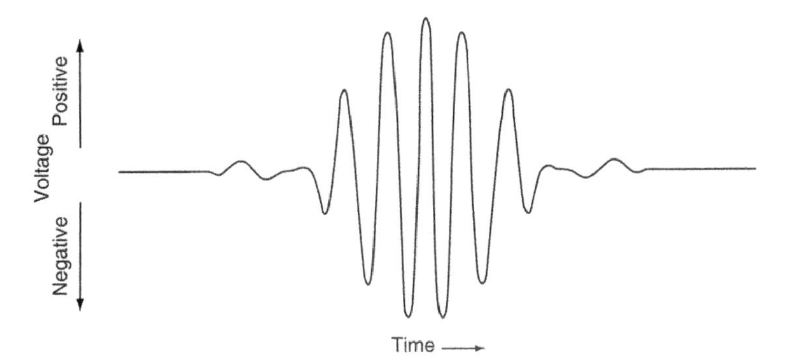

FIGURE 2.19 Output from a radio interferometer with stationary aerials and phase switching, viewing a single point source.

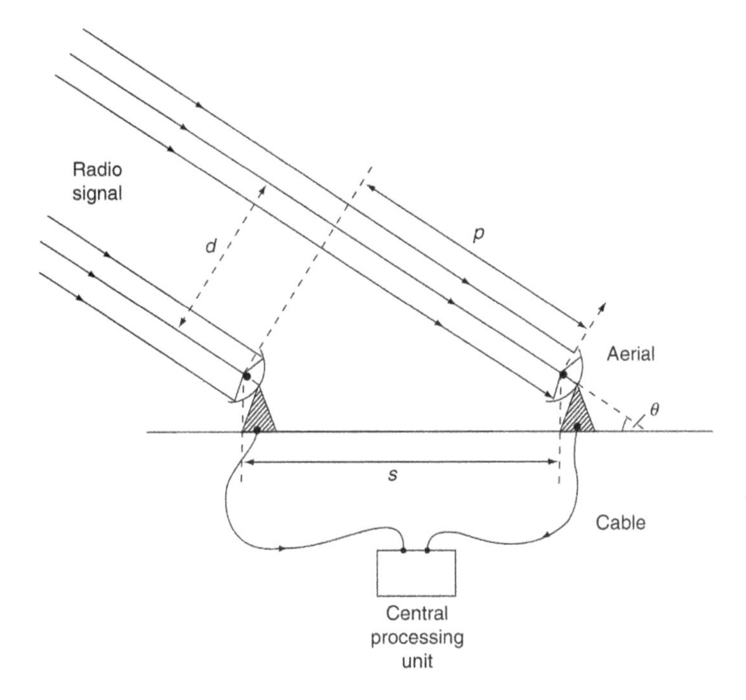

FIGURE 2.20 Schematic arrangement of a radio interferometer with tracking aerials.

positions is then recorded. The phase switching is generally undertaken in the latter stages of the receiver (Section 1.2), so that any transient effects of the switching are not amplified. The output fluctuates either side of the zero position as the object moves across the sky (Figure 2.19).

When the aerials are driven so that they track the object across the sky (Figure 2.20), as would usually be the case, then the output of each aerial corresponds to the central peak intensity of each image (Figures 2.16 and 2.18). The path difference then causes a simple interference pattern (Figure 2.21 and cf. Figure 2.9) whose fringe spacing alters due to the varying rate of change of path difference and aerial effective spacing as in the previous case. The maxima are now of constant amplitude because the aerials' projected effective areas are constant. In reality, many more fringes would occur than are shown in Figure 2.21. As with the optical interferometer however, the path differences at the aerials arising from the inclination of the source to the baseline have to be compensated to a fraction of the coherence length. However, this is much easier at the longer wavelengths because the coherence length is large – 30 m for a 10-MHz bandwidth and 300 m for a

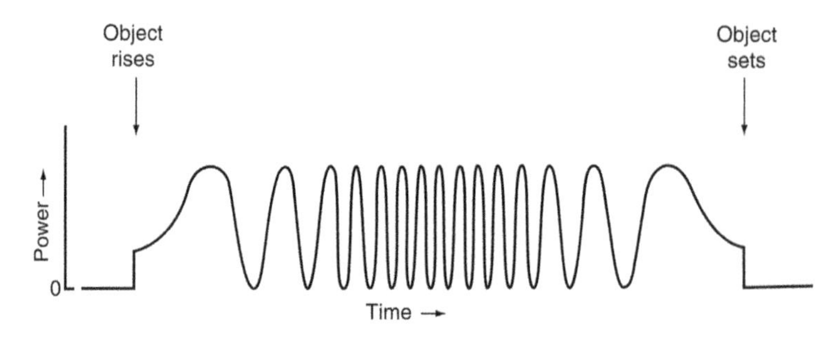

FIGURE 2.21 Output from a radio interferometer with tracking aerials viewing a single point source.

1-MHz bandwidth. The path difference can therefore be corrected by switching in extra lengths of cable or by shifting the recorded signals with respect to each other during data processing.

2.5.4 Aperture Synthesis

An interferometer works most efficiently, in the sense of returning the most information, for sources whose separation is comparable with its resolution. For objects substantially larger than the resolution little or no useful information may be obtained. We may, however, obtain information about a *larger* source by using an interferometer with a *smaller* separation of its elements. The resolution is thereby degraded until it is comparable with the angular size of the source. By combining the results of two interferometers of differing separations, one might thus obtain information on both the large- and small-scale structure of the source. Following this idea to its logical conclusion led Sir Martin Ryle in the early 1960s to the invention and development of the technique of aperture synthesis. He was awarded the Nobel Prize in Physics in 1974 for this work.

By this technique, which also goes under the name of Earth-rotation synthesis and is closely related to synthetic aperture radar (Section 2.8), observations of a stable source by a number of interferometers are combined to give the effect of an observation using a single large telescope. The simplest way to understand how aperture synthesis works is to take an alternative view of the operation of an interferometer. The output from a two-aperture interferometer viewing a monochromatic point source is shown in Figure 2.9 and is a simple sine wave. This output function is just the Fourier transform (Equation 2.3) of the (point) source. Such a relationship between source and interferometer output is no coincidence but is just a special case of the van Cittert-Zernicke theorem.

The instantaneous output of a two-element interferometer is a measure of one component of the 2D Fourier transform (Equation 2.10) of the objects in the field of view of the telescopes.

Thus, if a large number of two-element interferometers were available, so that all the components of the Fourier transform could be measured, then the inverse 2D Fourier transform (Equation 2.11) would immediately give an image of the portion of the sky under observation.

Now the complete determination of the Fourier transform of even the tiniest fields of view would require an infinite number of interferometers. In practice, therefore, the technique of aperture synthesis is modified in several ways. The most important of these is relaxing the requirement to measure all the Fourier components *at the same instant.* However, once the measurements are spread over time, the source(s) being observed must remain unvarying over the length of time required for those measurements.

Given, then, a source that is stable at least over the measurement time, we may use one or more interferometer pairs to measure the Fourier components using different separations and angles. Of course, it is still not possible to make an infinite number of such measurements, but we may use the discrete versions of Equations 2.10 and 2.11 to bring the required measurements down to a finite number (cf. the 1D analogues, Equations 2.3, 2.4, 2.8 and 2.9) though at the expense of losing the high-frequency components of the transform and, hence, the finer details of the image (Section 2.1).

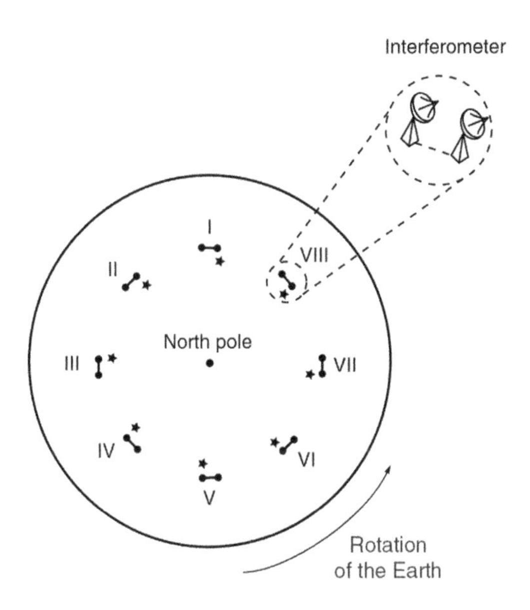

FIGURE 2.22 Changing orientation of an interferometer. The Earth is viewed from above the North Pole and successive positions of the interferometer at 3-hours intervals are shown. Notice how the orientation of the starred aerial changes through 360° with respect to the other aerial during a day.

The problem of observing with many different baselines is eased because we are observing from the rotating Earth. Thus, if a single pair of telescopes observes a source over 24 hours, the orientation of the baseline revolves through 360° (Figure 2.22). If the object is not at the North (or South) pole then the projected spacing of the interferometer will also vary, and they will seem to trace out an ellipse. The continuous output of such an interferometer is a complex function whose amplitude is proportional to the amplitude of the Fourier transform and whose phase is the phase shift in the fringe pattern.

The requirement for 24 hours of observation would limit the technique to circumpolar objects. Fortunately, however, only 12 hours are actually required; the other 12 hours can then be calculated from the conjugates of the first set of observations. Hence, aperture synthesis can be applied to *any* object in the *same* hemisphere as the interferometer.

Two elements arranged upon an East-West line follow a circular track perpendicular to the Earth's rotational axis. It is thus convenient to choose the plane perpendicular to the Earth's axis to work in and this is usually termed the u–v plane. If the interferometer is not aligned East-West, then the paths of its elements will occupy a volume in the u–v–w space, and additional delays will have to be incorporated into the signals to reduce them to the u–v plane. The paths of the elements of an interferometer in the u–v plane range from circles for an object at a declination of $\pm 90°$ through increasingly narrower ellipses to a straight line for an object with a declination of 0° (Figure 2.23).

A single 12-hours observation by a two-element interferometer samples all the components of the Fourier transform of the field of view covered by its track in the u–v plane. We require, however, the whole of the u–v plane to be sampled. Thus, a series of 12-hours observations must be made, with the interferometer baseline changed by the diameter of one of its elements each time (Figure 2.24). The u–v plane is then sampled completely out to the maximum baseline possible for the interferometer, and the inverse Fourier transform will give an image equivalent to that from a single telescope with a diameter equal to the maximum baseline.

A two-element radio interferometer with (say) 20-m diameter aerials and a maximum baseline of 1 km would need to make fifty 12-hours observations to synthesise a 1-km diameter telescope. By using more than two aerials, however, the time required for the observations can be much reduced.

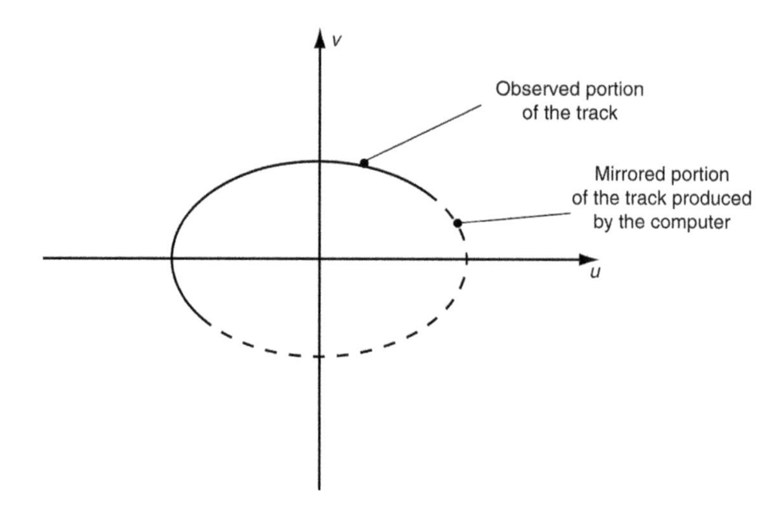

FIGURE 2.23 Track of one element of an interferometer with respect to the other in the u–v plane for an object at a declination of ±35°.

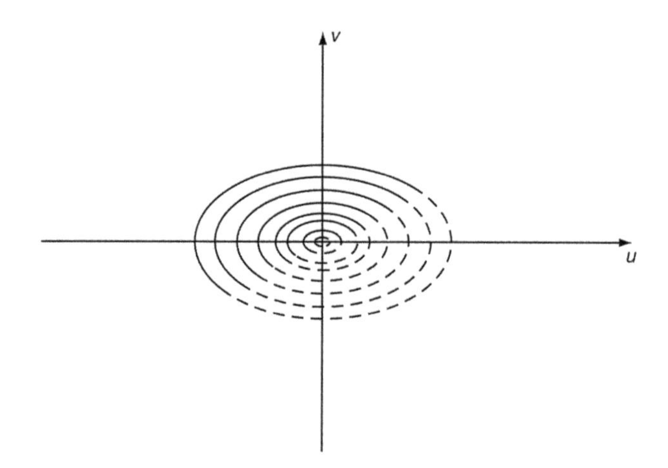

FIGURE 2.24 Successive tracks in the u–v plane of a two-element interferometer as its baseline is varied.

Thus with six elements, there are 15 different pairings.[8] If the spacings of these pairs of elements are all different (non-redundant spacing– something that it is not always possible to achieve) then the 50 visibility functions required by the previous example can be obtained in just four 12-hours observing sessions. In 1997, Eric Keto showed that the optimum (in the sense of sampling the Fourier transform best) layout for the elements of an interferometer was in the shape of a curve of constant width. The circle is the best known such curve but is the least satisfactory choice for this purpose. Other curves, known as Reuleaux polygons, are better. A Reuleaux polygon is just a straight-sided polygon with an odd number of sides in which the straight sides are replaced by arcs of circles whose centres are the vertices opposite to each of the sides.[9] Of these curves, the Reuleaux triangle is the best of all for an interferometer.

If the source has a reasonably smooth spectrum, the u–v plane may be sampled even more rapidly by observing simultaneously at several nearby frequencies (multi-frequency synthesis). Because the

[8] For N elements there are $[N(N-1)/2]$ possible pairs.

[9] The Reuleaux heptagon will be familiar to some readers, at least, because it is the shape of the United Kingdom's 20 and 50 p and the United Arab Emirates' 50 fils coins.

path differences between the aerials are dependent upon the operating wavelength, this effectively multiplies the number of interferometer pairs with different separations by the number of frequencies being observed. The analysis of the results from such observations, however, will be considerably complicated by any variability of the source over the frequencies being used.

A radio aperture synthesis system such as we have just been considering is a filled-aperture system. That is, the observations synthesise the sensitivity *and* resolution of a telescope with a diameter equal to the maximum available baseline. Although some aperture synthesis systems are of this type, it becomes increasingly impractical and expensive to sample the whole u–v plane as the baseline extends beyond a kilometre or so. For many observations, even arrays that *are* capable of synthesising a filled aperture may not *need* to do so.

Most aperture synthesis arrays, in practice, therefore, synthesise sparse apertures in which only selected annuli within the whole aperture are completed. Sparse apertures can range from systems which are almost filled to Very Long Base-line Interferometry (VLBI – see below) where a few to a few tens of annuli, each only 10–100 m in width may be completed within a baseline 10,000 km long. In such cases, the Fourier transform of the source is not fully sampled. Special techniques, known as 'hybrid mapping' (see below) are then required to produce the final maps of radio sources.

The Westerbork Synthesis Radio Telescope (WSRT) in the Netherlands is an example of an array that can come close to synthesising a filled aperture. It uses ten fixed and four movable 25-m dishes over a maximum baseline of 2.7 km. The WSRT is due to be upgraded shortly to using array detectors, so increasing its field of view by a factor of 25.

In the United Kingdom (UK), enhanced-Multi-Element Remotely Linked Interferometer Network (e-MERLIN) is an array of seven fixed radio telescopes with a maximum baseline of 217 km sited across central England. It has been upgraded (from radio) to using super-fast fibre-optics to link the telescopes. It observes down to the μJ level from 1.2 to 25 GHz and can reach a resolution of 50 milliarcseconds at 5 GHz.

The Karl G. Jansky Very Large Array (VLA) in New Mexico (Figure 2.25) uses twenty-seven[10] 25-m dishes arranged in a 'Y' pattern with a maximum baseline of 36 km.

Details, insofar as they are determined at the time of writing, of the planned and partially constructed Square Kilometer Array (SKA) were discussed in Section 1.2. It is possible that minor upgrades to provide phased array outputs for some of the instrument, could add VLTI capabilities to the SKA.

FIGURE 2.25 The central part of the Very Large Array (VLA). (Reproduced by courtesy of National Radio Astronomy /Associated Universities Inc. [NRAO/AUI] and NRAO).

[10] Twenty-seven dishes imply 351 baselines – and this is sufficient for the VLA to be able to obtain images in the time it takes to make a single set of measurements of the source and so not have to wait for the Earth's rotation to 'do any of the work' for it. This is known as the VLA's snapshot mode of operation. The snapshot images though are much noisier than those obtained by conventional aperture synthesis.

Beyond e-MERLIN, the VLA and similar radio systems, we have VLBI. For VLBI, the elements of the interferometer may be separated by thousands of kilometres, range over several continents and provide milliarcsecond resolutions or better. In VLBI, the signals from each element are separately recorded along with timing pulses from an atomic clock (c.f. Event Horizon Telescope [EHT] – Section 1.2.3). The recordings are then physically brought together and processed by a computer which uses the time signals to ensure the correct registration of one radio signal with respect to another. Real-time analysis of the data from a VLBI system comprising the Arecibo radio dish and telescopes in the UK, Sweden, the Netherlands and Poland became possible a decade ago when they were linked via internet research networks – earning the system the name of e-VLBI.

The Very Long Baseline Array (VLBA) VLBI system uses ten 25-m telescopes spread out over the United States from Hawaii to the US Virgin Islands. Its maximum baseline is 8,600 km and around 50 GHz, its resolution can be 150 microarcseconds. It has, for example, recently been able to measure the parallax for the Orion nebula to an accuracy of 100 microarcseconds, reducing the previously accepted distance for the nebula from 475 to 385 pc. It has also recently had new receivers installed and with upgraded computers it is now improved by a factor of around 5,000 compared with its performance when it first started operating in 1993. Along with the Arecibo, Effelsberg and Green Bank Telescope (GBT) dishes and the VLA it also forms the High Sensitivity Array (HSA). This can improve upon the sensitivity of the VLBA by a factor of ten and has a maximum baseline of 10,300 km.

For even longer baselines than the HSA, it is necessary to put one or more receivers into space. The Japanese Halca spacecraft which carried an 8-m radio telescope was linked with up to 40 ground-based dishes, to give baselines up to 30,000 km in length. The spacecraft lost attitude control in 2003 however and has now ceased to operate. Also, as mentioned in Section 1.2, the Russian Radio Astron spacecraft carrying a 10-m dish operated until 2019 and, by linking with ground-based instruments, could offer baselines up to 390,000 km and angular resolutions of 8 microarcseconds.

The fields of view in VLBI have generally been only a few arc seconds or less. However, the computational problems previously encountered arising from the available computer capacities are now being overcome as computer power increases. It seems likely that fields of view of up to half a degree may become possible in the near future, greatly increasing the utility of VLBI. The increasing availability of array feeds (radio cameras; e.g., Large APEX Bolometer Camera [LABOCA], WSRT, etc. – Section 1.2.3) will also aid in increasing VLBI's fields of view.

VLBI is also potentially extendable to submillimetre wavelengths (see below) with more-or-less the existing equipment and telescopes. Already APEX, Submillimeter Array (SMA) and the 10-m Submillimeter Telescope in Arizona have conducted VLBI at a wavelength of 1.3 mm (230 GHz) and achieved 28 microarcsecond resolutions. Some of the EHT instruments with Atacama Large Millimetre Array (ALMA) in its compact configuration have been conducting VLBI observations at 300 GHz since about 2017. It seems likely that submillimetre observations will be made in the fairly near future.

At millimetre and submillimetre wavelengths, the Owens valley radio observatory in California operated the Combined Array for Research in Millimeter-wave Astronomy (CARMA) array until 2015. This was actually two independent arrays. One had 15 dishes with 6.1-m or 10.4-m diameters. The second had eight 3.5-m dishes. The former observed 85–115 GHz and 215–270 GHz bands (3.5–2.6 mm and 1.4–1.1 mm). The latter covered 26–36 GHz and 80–115 GHz (11.5–8.3 mm and 3.8–2.6 mm). The maximum baseline was 2 km and the resolutions were 150 milliarcseconds and 1′, respectively.

On the Plateau de Bure in France, Institut de Radioastronomie Millimétrique (IRAM) operates a system with six 15-m telescopes with a maximum baseline of 760 m. It observes in the 100–300 GHz band (3–1 mm) with a best resolution of 500 milliarcseconds.

The SMA on Mauna Kea observes between 180 and 700 GHz (1.7 mm and 430 μm) with eight 6-m dishes. It can also link with the James Clerk Maxwell Telescope (JCMT) and the 10.4-m Caltech Submillimeter Observatory telescope (both also on Mauna Kea) via fibre-optic cables as

FIGURE 2.26 Left, Some of the individual radio telescopes making up ALMA. Seen at night. (Courtesy of Atacama Large Millimetre Array [ALMA]) (European Southern Observatory/National Astronomical Observatory of Japan/National Radio Astronomy Observatory [ESO/NAOJ/NRAO], C. Padilla.) Right, Moving one of ALMA's radio telescopes to a new site. (Courtesy of ALMA [ESO/NAOJ/NRAO], S. Rossi [ESO].)

a ten-element interferometer (when it is called the 'extended SMA' or eSMA). Baselines range up to 500 m for SMA and to 780 m for eSMA. It can achieve a resolution of a few tenths of an arc. The SMA has four configurations that are based upon Reuleaux triangles insofar as this is practicable.

ALMA (see also Sections 1.2.2 and 1.2.3) has fifty 12-m antennas plus a central smaller array of twelve 7-m and four 12-m dishes for observing extended sources (Figure 2.26 left). It operates between 30 GHz and 1 THz, has angular resolutions down to 20 milliarcseconds at its maximum aerial separation of 16 km and can resolve velocities down to 50 m s^{-1}. The individual radio telescopes weigh around 100 tonnes each but can be moved in one piece from one position to another within the array so that the configuration of the array can be changed (Figure 2.26 right). There are a total of 196 available positions for the 66 antennas. Correction for the seeing at millimetre wavelengths is used (see also optical real-time atmospheric compensation – Section 1.1.22) using radiometers to measure the atmospheric properties (especially water vapour) along the line of sight and by observing point sources (quasars) near to the field of view.

At visible and near infrared (NIR) wavelengths aperture synthesis has only recently been successfully attempted because of the stringent requirements on the stability and accuracy of construction of the instruments imposed by the small wavelengths involved.

The Centre for High Angular Resolution Astronomy (CHARA) array on Mount Wilson can reach a resolution of 200 microarcseconds in the visible and NIR using six 1-m telescopes in a Y-shaped array with a maximum baseline of 330 m. The six light beams are fed to the central combining laboratory via vacuum light pipes.

The four 8.2-m telescopes of ESO's VLTI (Figure 2.15) can be combined with four 1.8-m auxiliary telescopes for aperture synthesis and the latter can be moved to 30 different positions offering baselines of up to 200 m (see Figure 2.15). The system operates in the NIR and mid-infrared (MIR) with angular resolutions down to 2 milliarcseconds for imaging and 10 microarcseconds for astrometry. There are several second-generation instruments either just commissioned or about to be commissioned for the VLTI at the time of writing. These include Gravity,[11] Multi AperTure mid-Infrared SpectroScopic Experiment (MATISSE) and Precision Integrated-Optics Near-infrared Imaging ExpeRiment (PIONIER).

Gravity is a group of NIR instruments designed for imaging (2 milliarcsecond resolution), spectroscopy (R up to 4,000) and astrometry (10 microarcsecond accuracy) using the four 8.2-m unit telescopes. It achieved first light in 2016 and operates from 18 to 27 GHz (K band). MATISSE had

[11] Gravity; a name.

first light in 2018 and operates in the MIR combining the outputs from up to four of the telescopes as a spectro-interferometer. PIONIER has been operating as an imaging interferometer in the NIR since 2010. It is a visitor instrument now converted to a VLTI facility, built in France and which combines the outputs from the four 8.2-m telescopes or the four 1.8-m telescopes.

The Kenneth J. Johnston Navy Precision Optical Interferometer (NPOI) has been operational since 1994. It has six movable telescopes with effective apertures of 0.12 m in a Y-shaped distribution and with baselines up to 98 m. The light beams are sent to the central processing laboratory through vacuum light pipes. It operates in the visible (600–850 nm) for imaging and low-resolution spectroscopy, with a best angular resolution of about 1.5 milliarcseconds. It is currently being upgraded with three 1-m adaptive-optics telescopes and with Electron Multiplying Charge Coupled Device (EMCCD) detectors and will soon be able to use baselines from 8 to 430 m. Further upgrades have been proposed within the Astro 2020 decadal review.

The Magdalena Ridge Observatory Interferometer (MROI) in New Mexico is currently under construction. It will operate from 600 nm to 2.4 μm with ten 1.4-m telescopes in a Y-shaped configuration and with baselines ranging from 7.8 to 340 m. It is hoped eventually to achieve a resolution of 600 microarcseconds at a wavelength of 1 μm. It also uses vacuum light pipes to direct the light beams to the beam combiner and has delay lines to bring them into phase. First light occurred in 2019.

There have been a number of proposals for space-based interferometer systems. Most of these have been cancelled at some stage. The Halca and RadioAstron spacecraft-based radio interferometers, though, have made it into space and have been discussed previously. The Space Millimetre-Wavelength VLBI Array (SMVA) proposal (Section 1.2.3) would use two spacecraft to improve the longer baseline coverage.

One future ground-based advance, which had a prototype called Dragonfly[12] built in 2012, was based upon photonic optical components (Section 1.1). It has been tried out successfully on the Anglo-Australian Telescope (AAT). It remains to be seen, however, whether this approach can improve upon existing instruments.

2.5.5 Data Processing

The KERN[13] computing package has been developed from the earlierAstronomical Image Processing System (AIPS), AIPS++ and Common Astronomy Software Applications (CASA)[14] packages produced by the National Radio Astronomy Observatory (NRAO). It is used at many interferometric observatories and provides for many of the processing steps outlined here. Its details change biannually, if not faster, and so seriously interested readers should consult the current version (to be found at https://casa.nrao.edu/).

The extraction of the image of the sky from an aperture synthesis system is complicated by the presence of noise and errors. Overcoming the effects of these adds additional stages to the data reduction process. In summary, it becomes:

a. Data calibration
b. Inverse Fourier transform
c. Deconvolution of instrumental effects
d. Self-calibration

Data calibration is required to compensate for problems such as errors in the locations of the elements of the interferometer and variations in the atmosphere. It is carried out by comparing the theoretical instrumental response function with the observed response to an isolated

[12] Not to be confused with the Dragonfly instrument discussed in Section 1.1.23.
[13] 'Core' in Dutch and Afrikaans.
[14] Still in use for some instruments.

stable point source. These responses should be the same, and if there is any difference, then the data calibration process attempts to correct it by adjusting the amplitudes and phases of the Fourier components. Often ideal calibration sources cannot be found close to the observed field. Then, calibration can be attempted using special calibration signals, but these are insensitive to the atmospheric variations, and the result is much inferior to that obtained using a celestial source.

Applying the inverse Fourier transform to the data has already been discussed and further details are given in Section 2.1.1.

After the inverse Fourier transformation has been completed, we are left with the 'dirty' map of the sky. This is the map contaminated by artefacts introduced by the PSF of the interferometer. The primary components of the PSF, apart from the central response, are the side lobes. These appear on the dirty map as a series of rings that extend widely over the sky and which are centred on the central response of the PSF. The PSF can be calculated from interferometry theory to a high degree of precision. The deconvolution can then proceed as outlined in Section 2.1.

The maximum entropy method discussed in Section 2.1 has recently become widely used for determining the best source function to fit the dirty map. Another method of deconvolving the PSF, however, has long been in use and is still used by many workers and that is the method known as CLEAN.

The CLEAN algorithm was introduced by Jan Högbom in 1974. It involves the following stages:

a. Normalise the PSF (instrumental profile or 'dirty' beam) to (gI_{MAX}), where I_{MAX} is the intensity of the point of maximum intensity in the dirty map and g is the 'loop gain' with a value between 0 and 1.
b. Subtract the normalised PSF from the dirty map.
c. Find the point of maximum intensity in the new map – this may or may not be the same point as before – and repeat the first two steps.
d. Continue the process iteratively until I_{MAX} is comparable with the noise level.
e. Produce a final clear map by returning all the components removed in the previous stages in the form of 'clean beams' with appropriate positions and amplitudes. The clean beams are typically chosen to be Gaussian with similar widths to the central response of the dirty beam.

CLEAN has proved to be a useful method for images made up of point sources despite its lack of a substantial theoretical basis. For extended sources MEMs are better because CLEAN may then require thousands of iterations. As pointed out in Section 2.1.1 though, MEMs also suffer from problems, especially the variation of resolution over the image.

The final stage of self-calibration is required for optical systems and for radio systems when the baselines become more than a few kilometres in length because the atmospheric effects then differ from one telescope to another. Under such circumstances with three or more elements, we may use the closure phase that is independent of the atmospheric phase delays. The closure phase is defined as the sum of the observed phases for the three baselines made by three elements of the interferometer. It is independent of the atmospheric phase delays as we may see by defining the phases for the three baselines, it in the absence of an atmosphere to be ϕ_{12}, ϕ_{23} and ϕ_{31} and the atmospheric phase delays at each element as a_1, a_2 and a_3. The observed phases are then

$$\phi_{12} + a_1 - a_2$$

$$\phi_{23} + a_2 - a_3$$

$$\phi_{31} + a_3 - a_1$$

The closure phase is then given by the sum of the phases around the triangle of the baselines:

$$\phi_{123} = \phi_{12} + a_1 - a_2 + \phi_{23} + a_2 - a_3 + \phi_{31} + a_3 - a_1 \tag{2.43}$$

$$= \phi_{12} + \phi_{23} + \phi_{31} \tag{2.44}$$

From Equation 2.44 we may see that the closure phase is independent of the atmospheric effects and is equal to the sum of the phases in the absence of an atmosphere. In a similar way, an atmosphere-independent closure amplitude can be defined for four elements:

$$G_{1234} = \frac{A_{12}A_{34}}{A_{13}A_{24}} \tag{2.45}$$

where A_{12} is the amplitude for the baseline between elements 1 and 2, etc. Neither of these closure quantities are actually used to form the image, but they are used to reduce the number of unknowns in the procedure. At millimetre wavelengths, the principal atmospheric effects are the result of water vapour. Because this absorbs the radiation, as well as leading to the phase delays, monitoring the sky brightness can provide information on the amount of water vapour along the line of sight and so provide additional information for correcting the phase delays.

For VLBI, 'hybrid mapping' is required because there is insufficient information in the visibility functions to produce a map directly. Hybrid mapping is an iterative technique that uses a mixture of measurements and guesswork. An initial guess is made at the form of the required map. This may well actually be a lower resolution map from a smaller interferometer. From this map the visibility functions are predicted and the true phases estimated. A new map is then generated by Fourier inversion. This map is then improved and used to provide a new start to the cycle. The iteration is continued until the hybrid map is in satisfactory agreement with the observations.

2.5.6 Intensity Interferometer

Until the last couple of decades, the technical difficulties of an optical Michelson interferometer severely limited its usefulness. Most of these problems, however, may be reduced in a device that correlates intensity fluctuations. Robert Hanbury-Brown originally invented the device in 1949 as a radio interferometer, but it has found its main application in the optical region. The disadvantage of the system compared with the Michelson interferometer is that phase information is lost and so the structure of a complex source cannot be reconstituted. The far greater ease of operation[15] of the intensity interferometer led to it being able to measure some 30 stellar diameters between 1965 and 1972. Hanbury-Brown's instrument has long been decommissioned, but there is now a revival of interest in the technique since existing γ-ray Čerenkov arrays (Sections 1.3 and 1.4) have the potential to use the approach to image stellar surfaces, although their normal *direct* images are of low resolution (typically 1°). Recently, the Very Energetic Radiation Imaging Telescope Array System (VERITAS) instruments have been used to produce the first such trial measurements (of Bellatrix – γ Ori). Other, similar instruments, including the future Čerenkov Telescope Array (CTA) could soon follow suit.

The principle of the operation of the interferometer relies upon phase differences in the low-frequency beat signals from different mutually incoherent sources at each aerial,[16] combined with electrical filters to reject the high-frequency components of the signals. The schematic arrangement of the system is shown in Figure 2.27. We may imagine the signal from a source resolved into its

[15] Essentially the two telescopes observe a non-point source and compare their outputs. The telescopes are gradually moved apart and when the two signals start to differ significantly (become decorrelated), the telescopes' separation provides the measure of source's size.

[16] That is, there are *two* detectors needed instead of the *single* detector for a Michelson interferometer.

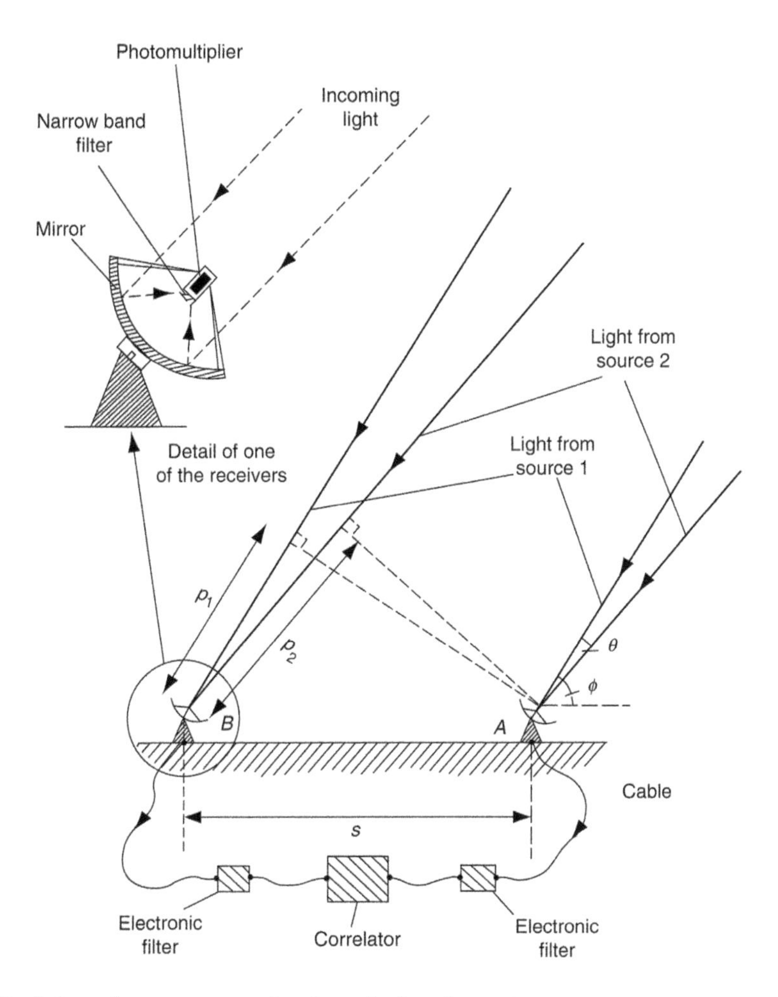

FIGURE 2.27 Schematic arrangement of an intensity interferometer.

Fourier components and consider the interaction of one such component from one source with another component from the other source. Let the frequencies of these two components be ν_1 and ν_2, then when they are mixed, there will be two additional frequencies – the upper- and lower-beat frequencies, $(\nu_1 + \nu_2)$ and $(\nu_1 - \nu_2)$ involved. For light waves, the lower beat frequency will be in the radio region (typically 10–100 MHz), and this component may easily be filtered out within the electronics from the much higher original and upper beat frequencies. The low-frequency outputs from the two telescopes are multiplied and integrated over a short time interval and the signal bandwidth to produce the correlation function, K. Hanbury-Brown and Richard Twiss were able to show that K was simply the square of the fringe visibility (Equation 2.35) for the Michelson interferometer.

The intensity interferometer was used to measure stellar diameters and $K(d)$ reaches its first zero when the angular stellar diameter, θ', is given by

$$\theta' = \frac{1.22\lambda}{d} \tag{2.46}$$

where d is the separation of the receivers. So, the resolution of an intensity interferometer (and also of a Michelson interferometer) for stellar discs is the same as that of a telescope (Equation 2.30) whose diameter is equal to the separation of the receivers (Figure 2.28).

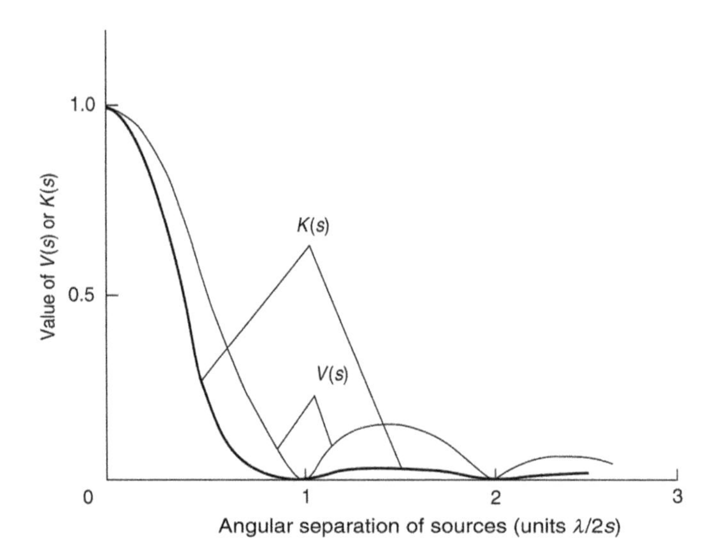

FIGURE 2.28 Comparison of the fringe visibility of a Michelson interferometer with the correlation function of an intensity interferometer.

The greater ease of construction and operation of the intensity interferometer over the Michelson interferometer arises from its dependence upon the beat frequency of two light beams of similar wavelengths, rather than upon the actual frequency of the light. A typical value of the lower-beat frequency is 100 MHz, which corresponds to a wavelength of 3 m. Thus, the path differences for the two receivers may vary by up to about 0.3 m during an observing sequence without ill effects. Scintillation, by the same argument, is also negligible.

Only one purpose-built working intensity interferometer has been constructed. It was built by Hanbury-Brown at Narrabri in Australia. It has now been decommissioned. It used two 6.5-m reflectors that were formed from several hundred smaller mirrors. There was no need for high optical quality because the reflectors simply acted as light buckets and only the brightest stars could be observed. The reflectors were mounted on trolleys on a circular track 94 m in radius. The line between the reflectors could therefore always be kept perpendicular to the line of sight to the source and their separation could be varied from 0 to 188 m (Figure 2.29). It operated at a wavelength of 433 nm, giving it a maximum resolution of about 500 microarcseconds.

High optical quality is not needed for the telescopes involved in intensity interferometry.[17] The telescopes in Hanbury-Brown's instrument for example had individual resolutions of 6′. Furthermore, as already mentioned, the detectors operated only in an analogue fashion. γ-ray air Čerenkov arrays (Sections 1.3 and 1.4) are therefore well suited to being used as intensity interferometers. Because they can have up to five telescopes (and may have more in the future) with sizes up to 28 m and because photon-counting detectors can be used, the performances of such arrays will be far better than that of the original instrument. Computer modelling suggests that at a wavelength of 400 nm resolutions of 60 microarcseconds could be obtainable. This would be sufficient for maps of the angularly larger stars to be produced with several tens or even a few hundred resolution elements – more than adequate to monitor the emissions from larger active regions.

Recently, two high-speed photometers using Single Photon Avalanche Photo-diode (SPAD) detectors, Asiago Quantum Eye (Aqueye+), and Italian Quantum Eye (Iqueye) built for pulsar timing have been able to function as an intensity interferometer with a 3.9-km baseline. The photometers were linked to the 1.8-m Copernicus and the 1.2-m Galileo telescopes at Asiago (Italy) by

[17] Of course, better resolution does not do any harm, so there is no reason why (say) the VLT or the Keck instruments could not be used this way as well.

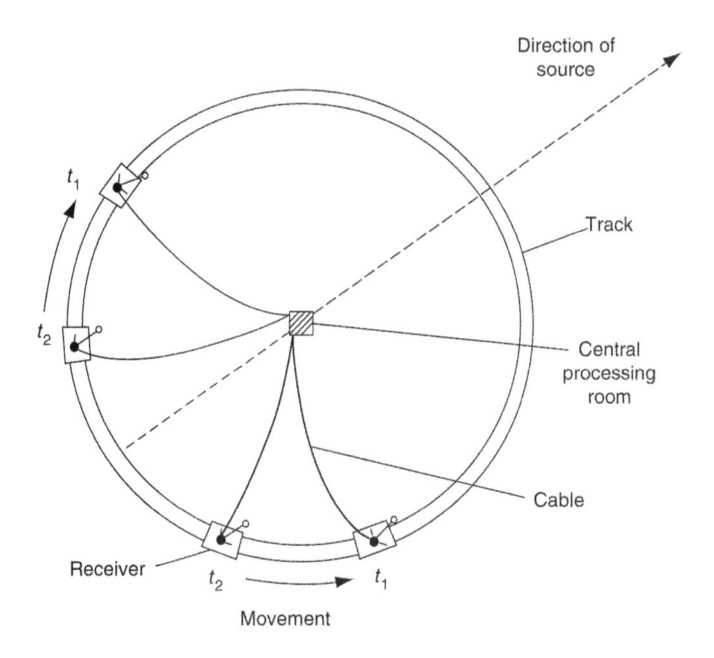

FIGURE 2.29 Schematic layout of the intensity interferometer at Narrabri, showing the positions of the receivers for observing in the same direction with different baselines. Note there were only *two* receivers; the diagram shows the arrangement at two separate baselines, superimposed.

fibre-optic cables. The effective baseline length was changed by the Earth's rotation (c.f. Aperture Synthesis, Section 2.5.4). Successful operation of the system as an intensity interferometer was demonstrated (using Deneb).

2.6 SPECKLE INTERFEROMETRY

This technique is a kind of poor man's space telescope because it provides near diffraction-limited performance from Earth-based telescopes. It works by obtaining images of the object sufficiently rapidly to freeze the blurring of the image that arises from atmospheric scintillation. The total image then consists of a large number of small dots or speckles, each of which is a diffraction-limited interference fringe for some objective diameter up to and including the diameter of the actual objective (Figure 1.61). An alternative to this technique is the adaptive optics telescope (Section 1.1), where adjusting the telescope optics to compensate for the atmospheric effects recombines the speckles. Adaptive optics, especially with artificial laser guide stars, has replaced speckle interferometry for most of the largest instruments, but the technique can be encountered in use on smaller telescopes – such as, for example, the US Naval Observatory's 0.66-m refractor which is still used to make up to 150 observations per night using intensified charge-coupled device (ICCD) detectors. Also using ICCD detectors, speckle interferometry is used by the 2.5-m Hooker telescope on Mount Wilson. The 4.1-m Southern Astrophysical Research (SOAR) telescope has an active programme of studying close binary stars with an EMCCD-based camera. A modern adaptation of the technique is used on the VLT and will be available on the James Webb Space Telescope (JWST; see below).

We may see how this speckled image structure arises by considering the effect of scintillation upon the incoming wavefront from the object. If we assume that above the atmosphere the wavefront is planar and coherent, then the main effect of scintillation is to introduce differential phase delays across it. The delays arise because the atmosphere is non-uniform with different cells within it having slightly different refractive indices. A typical cell size is 0.1 m and the scintillation frequencies usually lie in the range of 1–100 Hz. Thus, some 100 atmospheric cells will affect an average image from a 1-m telescope at any given instant. These will be rapidly changing, and over a normal exposure time, which can range from seconds to hours, they will form an integrated image that will be large and blurred compared with the diffraction-limited image. Even under the best seeing conditions, the image is rarely less than 1″ across. An exposure of a few milliseconds, however, is sufficiently rapid to freeze the image motion and the observed image is then just the resultant of the contributions from the atmospheric cells across the telescope objective at that moment.

Now the large number of these cells renders it highly probable that some of the phase delays will be similar to each other, and so some of the contributions to the image will be in phase with each other. These particular contributions will have been distributed over the objective in a random manner. Considering two such contributions, we have in fact a simple interferometer and the two beams of radiation will combine in the image plane to produce results identical with those of an interferometer whose baseline is equal to the separation of the contributions on the objective. We have already seen what this image structure might be (Figure 2.10). If several collinear contributions are in phase, then the image structure will approach that shown in Figure 4.2 modulated by the intensity variation due to the aperture of a single cell. The resolution of the images is then given by the maximum separation of the cells. When the in-phase cells are distributed in two dimensions over the objective, the images have resolutions in both axes given by the maximum separations along those axes at the objective. The smallest speckles in the total image, therefore, have the diffraction-limited resolution of the whole objective, always assuming, of course, that the optical quality of the telescope is sufficient to reach this limit. Similar results will be obtained for those contributions to the image that are delayed by an integral number of wavelengths with respect to each other.

Intermediate phase delays will cause destructive interference to a greater or lesser extent at the point in the image plane that is symmetrical between the two contributing beams of radiation but will, again, interfere constructively at other points to produce an interference pattern that has been shifted with respect to its normal position. Thus, all pairs of contributions to the final image interfere with each other to produce one or more speckles.

The true image is the Fourier transform of the interference pattern, just as for any other type of interference pattern. To obtain the true image from a speckled image, it must thus be Fourier analysed and the power spectrum obtained. This is the square of the modulus of the Fourier transform of the image intensity (Equation 2.3). Today, the Fourier transform is mostly obtained directly using quite small computers. However, when the technique was first developed in the 1970s and 1980s even the largest computers were inadequate for that task. The Fourier transform was, therefore, obtained optically by illuminating the image with collimated coherent light (Figure 2.30). The image (then a photographic negative) was placed at one focus of an objective and its Fourier transform imaged at the back focus of the objective. A spatial filter at this point can be used to remove unwanted frequencies if required and then the Fourier transform re-imaged.

Howsoever it may be obtained, the power spectrum can then be inverted to give centrosymmetric information such as diameters, limb darkening, oblateness, and binarity. Non-centrosymmetric structure can only be obtained if there is a point source close enough for its image to be obtained simultaneously with that of the object and for it to be affected by the atmosphere in the same manner as the image of the object (i.e., the point source is within the same isoplanatic patch of sky as the object of interest – Figure 1.62). Then deconvolution (Section 2.1) can be used to retrieve the image structure.

An alternative way of processing speckle images that requires little in the way of computing power is called 'Shift and Add'. It is the same process that is mentioned in Section 1.1 as a means of (somewhat) correcting atmospheric image degradation for small telescopes. With a number of speckle images of the same object, the brightest speckle in each image is found and all the images moved until those brightest speckles are aligned with each other. The images are then just added together to produce a generally much less noisy and a sharper image.

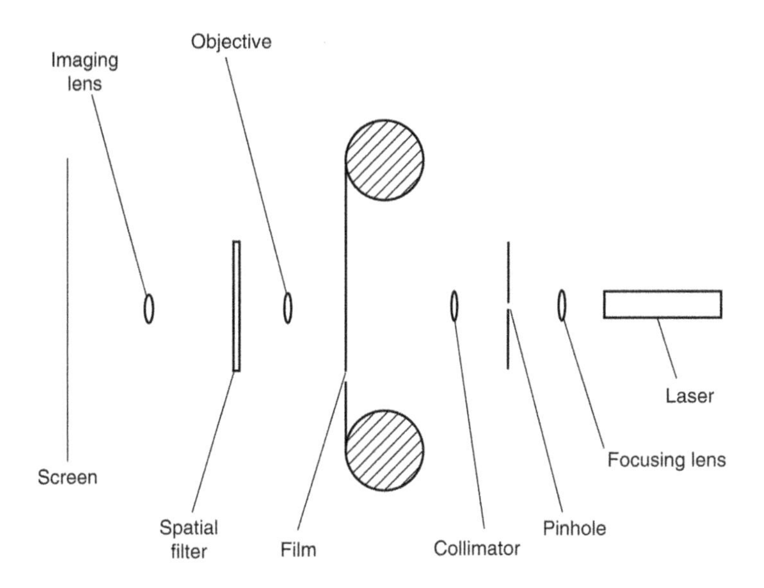

FIGURE 2.30 Arrangement for obtaining the Fourier transform of a speckle photograph by optical means.

The practical application of speckle interferometry requires the use of large telescopes because not only short exposures are required (0.001–0.1 seconds), but large image scales (0.1–1″/mm) are also needed to separate the individual speckles. Furthermore, the wavelength range must be restricted by a narrow band filter of 20–30 nm. Even with such a restricted wavelength range, it may still be necessary to correct any remaining atmospheric dispersion using a low-power direct vision spectroscope (Chapter 4). Thus, the limiting magnitude of the technique is currently about $+18^{m}$, and it has found its major applications in the study of red supergiants, the orbits of close binary stars, Seyfert galaxies, asteroids and fine details of solar structure.

A recent improvement to the technique is obtained, rather counterintuitively, by masking the whole telescope objective down to a few, much smaller, holes. The sizes of the holes are typically similar to those of an individual atmospheric cell (100–300 mm depending upon the wavelength at which the observations are being made). The mask may be placed directly over the telescope objective. More conveniently however, a lens after the telescope's focal point may be used to produce a parallel beam of light with a much smaller diameter than that of the main mirror. A suitably scaled mask is then placed across the beam of parallel light and the light is re-imaged. The holes are usually arranged over the mask so that the distances between them are not duplicated along either the x- or y-axes of a plane perpendicular to the light beam – although the x separations *should* be duplicated along the y direction to give equal resolutions in both. In effect the original single telescope mirror is now an interferometer array (Section 2.5) directly analogous to systems like the VLTI and the VLA.

Masking the main telescope mirror in this way is called 'speckle masking' or 'sparse aperture masking'. When, as described, the separations of the holes (baselines in interferometric parlance) are not duplicated, it is termed a 'non-redundant mask'. Sometimes, to increase the amount of light available, separations may be duplicated and then it is termed a 'partially redundant mask'.

The image obtained through a sparse aperture mask is a collection of speckles from each pair of holes. Because the holes are comparable in size with the atmospheric cells, many will only be affected by one such cell (i.e., no wavefront distortion for that particular hole). The speckle from two such holes will have no noise contribution from the atmosphere – just a displacement across the field of view if there is a phase difference between the light beams. The speckles will thus be much cleaner than those obtained with an un-masked mirror. Analysis of the data is based upon the calculation of an average closure phase (Equations 2.53 and 2.54; the closure phase is sometimes called the 'bispectrum in this context') which is inverted to obtain the image.

Sparse aperture masking is currently being undertaken on several telescopes – for example with the Nasmyth Adaptive Optics System-Coudé Near Infrared Camera (NAOS-CONICA) camera of the VLT where 340 milliarcsecond resolutions have been reached. It will also be an option on the JWST where Near InfraRed Imager and Slitless Spectrography (NIRISS) will provide the telescope's highest resolution imaging at around 75 milliarcseconds for a 4.6-μm wavelength. It will be a 7-aperture non-redundant mask and used, not to compensate for the Earth's atmosphere but to gain the ×2.4 interferometer improvement in angular resolution (Equation 2.33).

2.7 OCCULTATIONS

2.7.1 BACKGROUND

Three astronomical objects whose movements through space have put them into a straight line are a phenomenon termed as 'syzygy'. With the Earth as one of the three objects involved, there are three commonly encountered such alignments: Eclipses (when the two objects seen from the Earth have comparable angular sizes[18]), transits (when the nearer of the two objects has a much smaller angular size as seen from the Earth compared with the more distant one) and occultations (when the nearer of the two objects has a much larger angular size as seen from the Earth compared with the more distant one).

Occultations of more distant objects by the Moon occur frequently because of the large angular size of the Moon. Observation of such occultations has a long history, with records of lunar occultations stretching back some two and a half millennia. Occultations by planets, their satellites and asteroids occur much less often and are usually only to be seen from a restricted part of the Earth's surface. Recently, interest in the events has been revived for their ability to give precise positions for objects observed at low angular resolution in, say, the X-ray region and for their ability to give structural information about objects at better than the normal diffraction-limited resolution of a telescope. To see how the latter effect is possible, we must consider what happens during an occultation.

First, let us consider Fresnel diffraction at a knife-edge (Figure 2.31) for radiation from a monochromatic point source. The phase difference, δ, at a point P between the direct and the diffracted rays is then

$$\delta = \frac{2\pi}{\lambda}\left\{d_1 + \left[d_2^2 + (d_2 \tan\theta)^2\right]^{1/2} - \left[(d_1 + d_2)^2 + (d_2 \tan\theta)^2\right]^{1/2}\right\} \tag{2.47}$$

which, since θ is small, simplifies to

$$\delta = \frac{\pi d_1 d_2}{\lambda\left(d_1 + d_2\right)}\theta^2 \tag{2.48}$$

The intensity at a point in a diffraction pattern is obtainable from the Cornu spiral (Figure 2.32), by the square of the length of the vector, **A**. Here, P' is the point whose distance along the curve from the origin, l, is given by

$$l = \left(\frac{2 d_1 d_2}{\lambda\left(d_1 + d_2\right)}\right)^{1/2}\theta \tag{2.49}$$

[18] A lunar eclipse is not, therefore, a true eclipse, although it is a syzygy. If observed from the Moon by an astronaut it would properly be called a' transit of the Sun by the Earth'. It also seems most appropriate to consider stellar coronagraphs under this heading because they operate by producing artificial eclipses (see the end of this section). Solar coronagraphs are considered in Section 5.3.

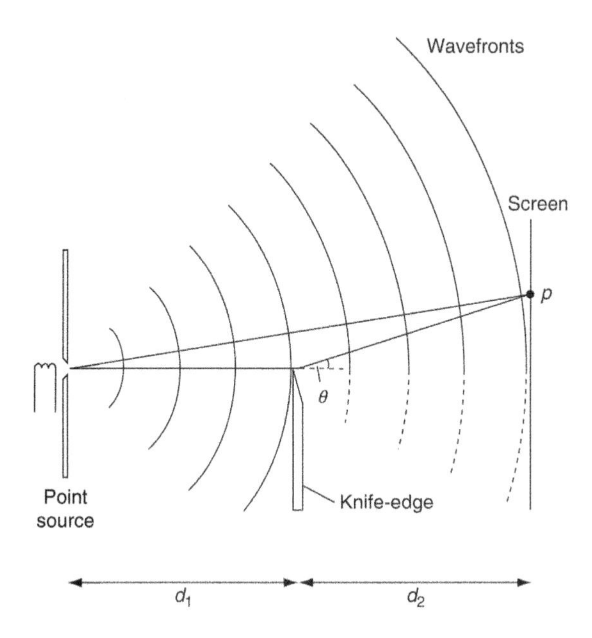

FIGURE 2.31 Fresnel diffraction at a knife-edge.

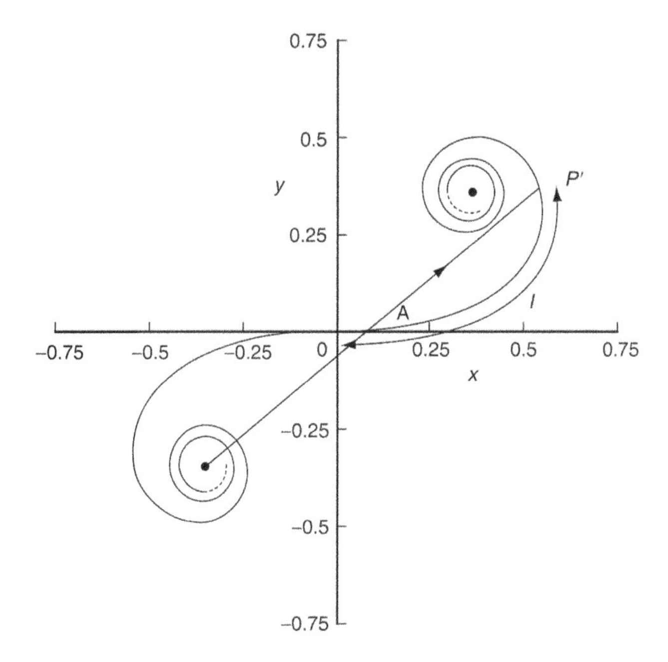

FIGURE 2.32 The Cornu spiral.

and the phase difference at P' is the angle that the tangent to the curve at that point makes with the x-axis, or from Equation 2.48

$$\delta = \frac{1}{2}\pi l^2 \tag{2.50}$$

The coordinates of P', which is the point on the Cornu spiral giving the intensity at P, x and y, are obtainable from the Fresnel integrals

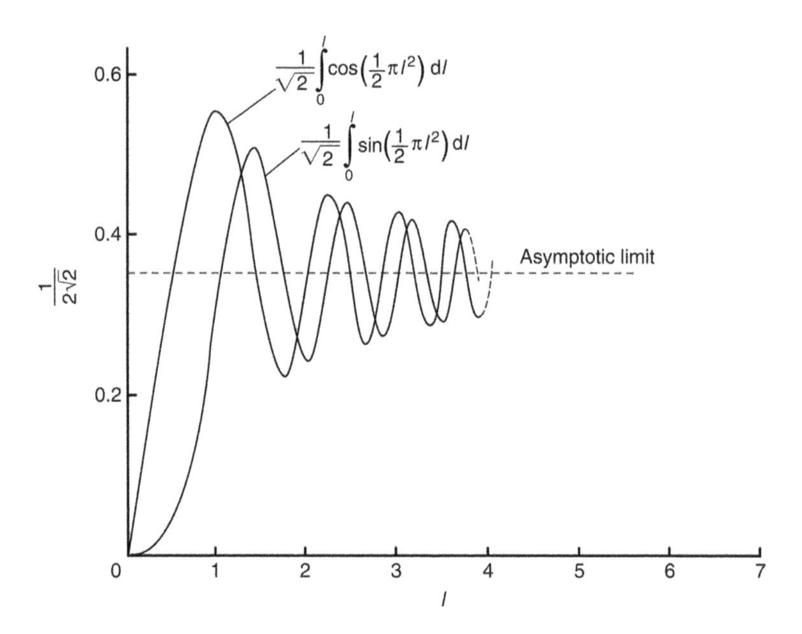

FIGURE 2.33 The Fresnel integrals.

$$x = \frac{1}{\sqrt{2}} \int_0^l \cos\left(\frac{1}{2}\pi l^2\right) dl \tag{2.51}$$

$$y = \frac{1}{\sqrt{2}} \int_0^l \sin\left(\frac{1}{2}\pi l^2\right) dl \tag{2.52}$$

whose pattern of behaviour is shown in Figure 2.33.

If we now consider a star occulted by the Moon, then we have

$$d_1 >> d_2 \tag{2.53}$$

so that

$$l \approx \left(\frac{2d_2}{\lambda}\right)\theta \tag{2.54}$$

but otherwise the situation is unchanged from the one we have just discussed. The edge of the Moon of course is not a sharp knife-edge, but because even a sharp knife-edge is many wavelengths thick, the two situations are not in practice any different. The shadow of the Moon cast by the star onto the Earth, therefore, has a standard set of diffraction fringes around its edge. The intensities of the fringes are obtainable from the Cornu spiral and Equations 2.51 and 2.52 and are shown in Figure 2.34. The first minimum occurs for

$$l = 1.22 \tag{2.55}$$

so that for the mean Earth-Moon distance

$$d_2 = 3.84 \times 10^8 \quad \text{m} \tag{2.56}$$

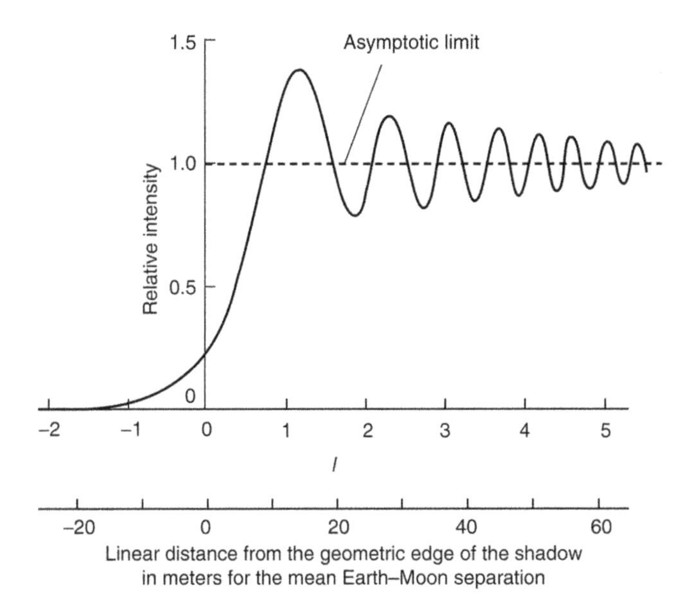

FIGURE 2.34 Fringes at the edge of the lunar shadow.

we obtain

$$\theta = 6.4 \text{ milliarcseconds} \tag{2.57}$$

at a wavelength of 500 nm. The fringes have a linear width of about 12 m, therefore, when the shadow is projected onto a part of the Earth's surface that is perpendicular to the line of sight. The mean rate of angular motion of the Moon, $\dot{\theta}$, is

$$\dot{\theta} = 0.55'' \text{ s}^{-1} \tag{2.58}$$

so that at a given spot on the Earth, the fringes will be observed as intensity variations of the star as it is occulted, with a basic frequency of up to 85 Hz. Usually, the basic frequency is lower than this because the Earth's rotation can partially offset the Moon's motion and because the section of the lunar limb that occults the star will generally be inclined to the direction of the motion.

If the star is not a point source, then the fringe pattern shown in Figure 2.34 becomes modified. We may see how this happens by imagining two point sources separated by an angle of about 3.7 milliarcseconds in a direction perpendicular to the limb of the Moon. At a wavelength of 500 nm, the first maximum of one star is then superimposed upon the first minimum of the other star and the amplitude of the resultant fringes is much reduced compared with the single-point source case. The separation of the sources parallel to the lunar limb is, within reason, unimportant in terms of its effect on the fringes. Thus, an extended source can be divided into strips parallel to the lunar limb (Figure 2.35) and each strip then behaves during the occultation as though it were a centred point source of the relevant intensity. The fringe patterns from all these point sources are then superimposed in the final tracing of the intensity variations. The precise nature of the alteration from the point source case will depend upon the size of the star and upon its surface intensity variations through such factors as limb darkening and gravity darkening and so on as discussed later in this section. A rough guide, however, is that changes from the point source pattern are just detectable for stellar diameters of 2 milliarcseconds, while the fringes disappear completely for object diameters

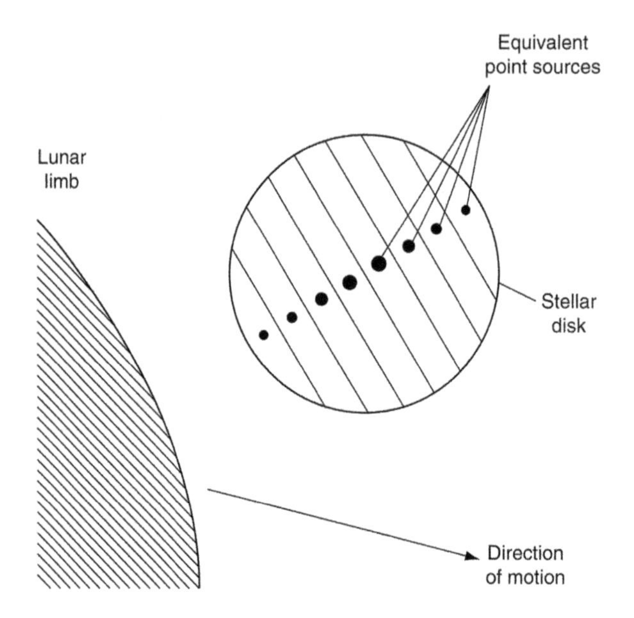

FIGURE 2.35 Schematic decomposition of a non-point source into equivalent point sources for occultation purposes.

of about 200 milliarcseconds. In the latter situation, the diameter may be recoverable from the total length of time that is required for the object to fade from sight. Double stars may also be distinguished from point sources when their separation exceeds about 2 milliarcseconds.

2.7.2 Techniques

In complete contrast to almost all other areas of astronomy, the detection of occultations in the optical region is best undertaken using comparatively small telescopes. This arises from the linear size of the fringes, which we have seen is about 12 m. Telescopes larger than about 1 m will, therefore, simultaneously sample widely differing parts of the fringe so that the detected luminosity variations become smeared. Because the signal-to-noise ratio decreases with increasing size of telescope, the optimum size for observing occultations usually lies between 0.5 and 2 m.

Most of the photometers described in Section 3.2 can be used to detect the star. However, we have seen that the observed intensity variations can have basic frequencies of 85 Hz with harmonics of several hundred hertz. The photometer and its associated electronics must therefore be capable of responding sufficiently rapidly to pick up these frequencies. The photometers used to detect occultations thus require response times of one to a few milliseconds.

Fringes are also blurred by the waveband over which the photometer is operating. We have seen from Equation 2.54 that the monochromatic fringe pattern is wavelength dependent. A bichromatic source with wavelengths differing by a factor of 2.37 would have the first maximum for one wavelength superimposed upon the first minimum for the other, so that the fringes would almost disappear. Smaller bandwidths will still degrade the visibility of the fringes although to a lesser extent. Using the standard UBV filters (Section 3.1) only five or six fringes will be detectable even in the absence of all other noise sources. A bandwidth of about 20 nm must be used if the fringes found using a medium sized telescope are not to deteriorate through this effect.

CCDs can also be used to determine the diffraction pattern of an occulted star. The diffraction pattern moves over the CCD at a calculable velocity and the charges in the pixels are moved through the device at the same rate (cf. image tracking for liquid mirrors, Section 1.1). By this process (time delayed integration [TDI]), higher signal-to-noise ratios can be reached because each portion of the diffraction pattern is observed for a longer time than when a single element detector is used.

Because the observations are obviously always carried out within a few arcminutes of the brightly illuminated portion of the Moon, the scattered background light is a serious problem. It can be minimised by using clean, dust-free optics, precisely made light baffles and a small entrance aperture for the photometer, but it can never be completely eliminated. The total intensity of the background can easily be many times that of the star even when all these precautions have been taken. Thus, electronic methods must be used to compensate for its effects. The simplest system is to use a differential amplifier that subtracts a pre-set voltage from the photometer's signal before amplification. It must be set manually a few seconds before the occultation since the background is continually changing. A more sophisticated approach uses a feedback to the differential amplifier that continually adjusts the pre-set voltage so that a zero output results. The time constant of this adjustment, however, is arranged to be long, so that while the slowly changing background is counteracted, the rapid changes during the occultation are unaffected. The latter system is particularly important for observing disappearances because it does not require last minute movements of the telescope onto and off the star.

An occultation can be observed at the leading edge of the Moon (a disappearance event) or at the trailing edge (when it is a reappearance) providing that the limb near the star is not illuminated in each case. Disappearances are by far the easiest events to observe because the star is visible and can be accurately tracked right up to the instant of the occultation. Reappearances can be observed only if a precise offset is available for the photometer or telescope, or if the telescope can be set onto the star before its disappearance and then kept tracking precisely on its position for the tens of minutes required before its reappearance. The same information content is available, however, from either event. The Moon often has quite a large motion in declination, so that it is occasionally possible for both a disappearance and the following reappearance to occur at a dark limb. This is a particularly important situation because the limb angles will be different between the two events and a complete 2D map of the object can be produced.

Scintillation is another major problem in observing an occultation. The frequencies involved in scintillation are similar to those of the occultation so that they cannot be filtered out. The noise level is therefore largely due to the effects of scintillation and it is rare for more than four fringes to be detectable in practice. Little can be done to reduce the problems caused by scintillation because the usual precautions of observing only near the zenith and on the steadiest nights cannot be followed; occultations have to be observed when and where they occur.

Occultations can be used to provide information other than whether the star is a double or not or to determine its diameter. Precise timing of the event can give the lunar position to 50 milliarcseconds or better and can be used to calibrate ephemeris time. Until GPS satellites became available, lunar occultations were used by surveyors to determine the positions on the Earth of isolated remote sites, such as oceanic islands.

Occultations of stars by planets, asteroids and the like are quite rare but are of considerable interest. Their main use is as a probe for the upper atmosphere of the planet because it may be observed through the upper layers for some time before it is completely obscured. There has also been the serendipitous discovery of the rings of Uranus in 1977 through their occultation of a star. In spectral regions other than the visible, lunar occultations have in the past found application in the determination of precise positions of objects.

The rapid recent improvement in the angular resolution of the observational techniques available at most wavelengths (see previous sections) has somewhat reduced the interest in occultation measurements. However, there are research occultation programmes still being undertaken. Thus, the International Occultation Timing Association (http://occultations.org/) currently provides predictions of occultations by the Moon, planets and their satellites and asteroids together with guidance on how to observe occultations and then process the data. The High-speed imaging camera for occultations (HIPO) is still listed as being available on the Stratospheric Observatory for Infrared Astronomy (SOFIA) telescope, though as an investigator class instrument, actual access to it is probably limited.

2.7.3 Analysis

The analysis of the observations to determine the diameters and/or duplicity of the sources is simple to describe, but time-consuming to carry out. It involves the comparison of the observed curve with synthesised curves for various models. The synthetic curves are obtained in the manner indicated in Figure 2.35, taking account also of the bandwidth and the size of the telescope. One additional factor to be taken into account is the inclination of the lunar limb to the direction of motion of the Moon. If the Moon were a smooth sphere this would be a simple calculation, but the smooth limb is distorted by lunar surface features. Sometimes the actual inclination at the point of contact may be determinable from artificial satellite photographs of the Moon. On other occasions, the lunar slope must appear as an additional unknown to be determined from the analysis. Rarely, a steep slope may be encountered and the star may reappear briefly before its final extinction. More often the surface may be too rough on a scale of tens of metres to provide a good 'straight-edge'. Both these cases can usually be recognised from the data and have to be discarded, at least for the purposes of determination of diameters or duplicity.

The analysis for positions of objects simply comprises determining the precise position of the edge of the Moon at the instant of occultation. One observation limits the position of the source to a semicircle corresponding to the leading (or trailing) edge of the Moon. Two observations from separate sites or from the same site at different lunations fix the object's position to the one or two intersections of two such semicircles. This will usually be sufficient to allow an unambiguous optical identification for the object, but if not, further observations can continue to be added to reduce the ambiguity in its position and to increase the accuracy for as long as the occultations continue. Generally, a source that is occulted one month, as observed from the Earth or from a close Earth-orbiting satellite, will continue to be occulted for several succeeding lunations. Eventually, however, the rotation of the lunar orbit in the Saros cycle of 18 years will move the Moon away from the object's position and no further occultations will occur for several years.

2.7.4 Stellar Coronagraphs

A coronagraph is an instrument that is designed to enable a faint object to be studied when that object is so close to a much brighter object that it would normally be swamped by the scattered light from that brighter object. The solar corona is typically less than 0.0001% as bright as the solar photosphere and the first coronagraphs were built to enable the solar corona to be studied outside those occasions when there was a total solar eclipse; hence, the name given to the instruments. Solar coronagraphs are discussed in Section 5.3; here we are concerned with high contrast situations involving other objects.

There are many high contrast situations – such as stars with circumstellar gas and dust envelopes and faint stellar companions to brighter stars – that are of interest, but much of the recent work in this area has been motivated by the desire to study and/or to image exoplanets directly. Although exoplanets can be ten times (or more) the mass of Jupiter, the latter is close to the maximum *physical* size attained by exoplanets.[19] To put the problem of directly observing exoplanets in perspective, if alien astronomers were to observe the solar system from a distance of (say) ~100 pc, then Jupiter's maximum brightness would be just 0.0000002% that of the Sun and its image would be separated from that of the Sun by just 50 milliarcseconds – and most exoplanets are physically *much* closer to their host stars than Jupiter is to the Sun *and* a lot further away from us than 100 pc.

Exoplanets hosted by cool, faint stars or brown dwarfs and in large orbits may, however, reduce the contrast problem considerably – the brown dwarf 2 M1207[20] in the TW Hydrae association is

[19] The material towards the centre of exoplanets which are higher in mass than Jupiter, is much more compressed than that at the centre of Jupiter, due to the more massive planets' higher central pressures. So their sizes are not much more than that of Jupiter.

[20] Its full name is 2MASSWJ1207334-393254.

just 100 times brighter than its ≥4 $M_{Jupiter}$ exoplanet.[21] The exoplanet, 2 M1207 b, is 45 astronomical units (AU) out from its host brown dwarf (800 milliarcminutes as seen from Earth) and was first imaged in 2004 using the VLT's NaCo[22] instrument but without using the coronagraph option. Because many exoplanets are intrinsically hot (whether because they are still condensing, have internal tidal heating or are heated by their host star) observations in the NIR or MIR may also reduce contrast since the exoplanet's own emissions will be added to its reflected light.

We have already seen that the nulling interferometer (Section 2.5) can be used to reduce the luminosity of a bright object whilst leaving the fainter object unaffected. This is also essentially what a coronagraph does, except that the latter does so by producing an artificial eclipse. Stellar coronagraphs can be ground- or space-based and the eclipse can be external or internal to the telescope. In all of the variants, careful precautions must be taken to minimise all forms of scattered light and it is in this respect that the coronagraph principally differs from a normal telescope. For the ground-based varieties, the use of real-time atmospheric compensation is essential (Section 1.1) and quasi-static speckles (below) need suppressing by differential observations of some type.

Quasi-static speckles have a similar appearance to the speckles produced by atmospheric disturbances (Figure 1.61) but change on a timescale of tens of minutes to hours. They may easily be mistaken for faint companions to the main star. These speckles arise from diffraction and interference effects caused by the mirror mountings and supports and other parts of the structure of the telescope and possibly also slight imperfections and misalignments of the mirrors. They alter because of the changing gravitational loading on the telescope structure and consequent sagging as it tracks an object across the sky and the expansion or contraction of the structure as a result of temperature changes. The quasi-static speckles however can often be distinguished from faint companions by obtaining two or more images under circumstances where the speckles change but the companion's image does not. The three current approaches to undertaking this process are called angular differential imaging (ADI), polarimetric differential imaging (PDI), and spectral differential imaging (SDI).

Angular differential imaging relies upon the, usually unwanted, phenomenon of field rotation (Section 1.1) for images at Nasmyth foci. With a telescope on an alt-az mounting, the telescope structure maintains a fixed orientation to the vertical. The speckles, therefore, also maintain a fixed orientation to the vertical. However, without the field de-rotator in action the field of view will rotate as the telescope tracks the image across the sky. Several exposures obtained at intervals of an hour or so (the image de-rotator will need to operate *during* each exposure) will thus change the mutual orientation of the real objects in the sky and the quasi-static speckles. Subtracting the average frame from each of the raw frames will leave just the planet's image on each frame, plus any uncompensated noise. De-rotating the processed frames and then adding them together will reinforce the exoplanet's image while further averaging the background noise. With PDI, two images of the field of view are obtained simultaneously through orthogonally oriented linear polarizers. The star light and the quasi-static speckle light will be almost unpolarized, but the reflected light from the exoplanet will be polarized to some degree or other. Subtracting the two images will then reduce the brightness of the stellar and speckle images by proportionally more than the reduction in the brightness of the exoplanet's image, so improving the contrast of the latter.

SDI is essentially similar to PDI except that the images are obtained through narrow band filters whose wavelengths are closely adjacent to each other. SDI also differs from PDI in that more than two images can be used by employing more filters, so improving on the level of noise reduction. For example, the image may be split into four and passed through narrow band filters centred on and just outside the strong methane absorption band in the near infrared. Jovian-type exoplanets contain methane and will be fainter when seen through the filters within the absorption region than when

[21] The mass of an exoplanet is usually only a lower limit obtained by assuming that the plane of the exoplanet's orbit lies along the line of sight. For other orientations of the orbit, the exoplanetary mass will have a larger value.

[22] NAOS and CONICA; now decommissioned.

compared with the image through filters outside that region. The host star, though, will have the same brightness in all images. Subtracting one image from another will eliminate the star's image but leave that of the planet. Exoplanets have yet to be observed using the SDI, but brown dwarfs have been detected.

The Lyot coronagraph is the basic instrument for producing the artificial eclipse internally (shown in Figure 5.27 and described in Section 5.3). The first image of an exoplanet obtained with a coronagraph of this design was that of Fomalhaut b[23] announced in 2008. The Hubble Space Telescope's (HST's) Advanced Camera for Surveys (ACS) operating in its coronagraphic mode had obtained the images in 2004 and 2006. More recently HiCIAO on the 8.2-m Subaru telescope and which is also uses a Lyot-pattern coronagraph has detected a companion to Gliese 878, a sixth-magnitude star in Lyra. The companion has a mass estimated to lie between 10 and 40 $M_{Jupiter}$ and so could be an exoplanet but is more likely to be a brown dwarf. The now decommissioned Project 1640,[24] was a 5-m Hale telescope instrument. Its Lyot-type coronagraph used an apodized image (Section 5.3) and was combined with an integral field spectrograph and the Hale's Palm-3000 adaptive optics system. Its observations of Vega were able to place an upper limit on the brightness of any companions. The Zurich Polarimeter (ZIMPOL; see Section 5.2.3.2) which forms a part of the Spectro-Polarimetric High contrast Exoplanet Research (SPHERE) instrument on the VLT can incorporate Lyot-Type coronagraphs into its system.

The Gemini Planet Imager (GPI), currently available on the 8.1-m Gemini South telescope, operates in the NIR using a coronagraph based upon apodised masks, plus a spectrograph and adaptive optics and obtained its first image of an exoplanet, 51 Eri b, in 2014.

In space, The JWST will carry three coronagraphs operating from 2.1 μm to longer than 10 μm. While for Wide Field Infrared Survey Telescope (WFIRST), it is planned to incorporate a coronagraph with two interchangeable masks and EMCCD detectors. It would also be designed to be able to operate with an external occulter (see below), should one be launched at some future time.

A long-discussed, but not yet implemented, concept, is for a space-based stellar coronagraph using an external occulter (aka; a starshade – see also Section 1.1.23). It would comprise two spacecraft flying in formation with a separation of thousands of kilometres. One spacecraft would carry the telescope and the second an opaque disk (the occulter). Just like the Moon obscures the Sun during a solar eclipse and allows the faint corona to be seen, so the disk would obscure the star from the telescope and allow much fainter exoplanets or other companions to be seen. The two spacecraft's positions would need to be tightly controlled, but that is now a possibility. WFIRST has been mentioned in this connection in the preceding paragraph and there is a pre-Astro 2020 proposal for just such a star shade (Exo-Starshade [Exo-S]) to be launched around the mid-2020s as a probe class mission. European Space Agency's (ESA's) Project for On-Board Autonomy (PROBA-3), currently scheduled for a 2021 launch date, will be a two-spacecraft technology-demonstration mission comprising a solar coronagraph with 150-m separation between the 1.4-m diameter occulter and the 0.05-m telescope. PROBA-3's primary mission, though, is to test out techniques for the precision formation flying of two spacecraft; the coronagraph is 'guest payload' on the mission. Other similar, though currently rather vague, concepts range up to systems with 1-km diameter occulters and 40,000-km separations.

A recent development for NIR coronagraphs uses a phase plate. The phase plate is a disk of zinc selenide with annular zones of differing thicknesses. Interference effects arising from the different times that it takes the infrared light to pass through the different thicknesses of the material again lead to the suppression of the star's light. A phase plate commissioned for the VLT's NaCo instrument was used to obtain the first image of β Pic b.

---

[23] 2M1207b was the first exoplanet (if it is a planet) to be imaged in any fashion – by the VLT in 2004 using just normal direct NIR imaging – see earlier discussion.

[24] Named for its optimum operating wavelength of 1,640 nm.

Another recent development is to apodise the coronagraph using two aspherical mirrors in place of the apodising masks. The technique, known as Phase-Induced Amplitude Apodisation (PIAA), retains the full throughput of the instrument and its angular resolution and is achromatic. It can observe contrasts between two objects at the 10^{10} level and promises to enable exoplanet imaging using 1- to 2-m-class telescopes. A PIAA coronagraph was under consideration for the planned Exo Circumstellar Environments and Disk Explorer (EXCEDE) spacecraft, but this mission was not funded.

A completely different approach to producing a stellar coronagraph has recently been suggested and this is to use an optical vortex. The optical vortex looks like a 360° turn of the steps of a spiral staircase. It is a helical mask with steps of progressively increasing thickness constructed from a transparent material. When the phase delay of the radiation passing through the thinnest step is two wavelengths less than that through the thickest step, destructive interference occurs such that the radiation passing through the centre of the mask is eliminated. The operating wavelength of the optical vortex depends upon the step heights and so it is essentially a monochromatic device; nonetheless it may provide the possibility for future stellar coronagraphs that are much more compact than existing devices. A vortex coronagraph fabricated from synthetic diamond was recently added to the Keck's NIRC2 instrument, enabling the imaging of the brown dwarf HIP79124 B.

2.8 RADAR

2.8.1 Introduction

Radar astronomy and radio astronomy are closely linked because the same equipment is often used for both purposes. Radar astronomy is less familiar to the astrophysicist however, because only one star, the Sun, has ever been studied with it, and this is likely to remain the case until other stars are visited by spacecraft (which will be a while yet[25]). Other aspects of solar system astronomy benefit from the use of radar to a much greater extent, so the technique is included, despite being almost outside the limits of this book, for completeness and because its results may find applications in some areas of astrophysics.

Radar, as used in astronomy, can be fairly conventional equipment for use onboard spacecraft, or for studying meteor trails, or it can be of highly specialised design and construction for use in detecting the Moon, planets, Sun and so on from the Earth. Its results potentially contain information on three aspects of the object being observed – distance, surface fine scale structure and relative velocity.

The name 'Radar' is derived from 'Radio Detection and Ranging' however e-m radiation of any wavelength could theoretically be used in place of the radio waves. To date, only light has been so used; resulting in light detecting and ranging (LiDAR) systems. These have found occasional use for studying solar system objects and much wider use as accurate ranging systems during spacecraft rendezvous missions with asteroids and similar objects.

[25] Suggestions recently imply this comment might be (slightly) pessimistic. Several studies have shown that sending a spacecraft to a nearby star is, as a conservative estimate, likely in 2012 to cost between US$100 billion for a minimal non-return fly-by mission, with results transmitted back to the Earth within a century, and $200 trillion for an unmanned sampling and return mission (see the author's *Exoplanets – Finding, Exploring and Understanding Alien Worlds* [2012] for a more detailed discussion of interstellar mission possibilities). The new studies suggest that a radar system with a receiver/transmitter area of 10,000 km^2 might detect Earth-sized planets up to 6 pc (20 ly) away at a cost of $20 trillion. The results would be received within 9 years (α Centauri) and 40 years (stars and planets 20 ly away). To put these figures in perspective, the current US gross domestic product is around $20 trillion and in 2020 dollars the Apollo Moon landing programme cost ~$220 billion.

2.8.2 Theoretical Principles

2.8.2.1 Basic Radar Systems

Active radar systems may use the same antenna both for transmission and reception (a mono-static radar) or two separate antennae (bi-static radar – usually with a significant spatial separation between the two antennae). Passive radar is a form of bi-static radar using some adventitious source as the radio signal transmitter. Mostly, to date, the sources have been terrestrial radio or TV broadcasting stations and they have been used to observe meteor trails. However, there seems to be no prohibitive reason why strong single astronomical radio point (or nearly so) sources (e.g., Cas A – ~3 kJy at 1 GHz or the Crab nebula – ~1 kJy at 1 GHz) should not be used similarly for observing other astronomical objects.

The theoretical principles are much the same for both of the active systems. Consider a radar system as shown in Figure 2.36. The transmitter has a power P and is an isotropic emitter. The radar cross section of the object is α, defined as the cross-sectional area of a perfectly isotropically scattering sphere which would return the same amount of energy to the receiver as the object. Then the flux at the receiver, f is given by

$$f = \frac{P\alpha}{4\pi R_1^2 4\pi R_2^2} \tag{2.59}$$

Normally we have the transmitter and receiver close together, if they are not actually the same antenna, so that

$$R_1 = R_2 = R. \tag{2.60}$$

Furthermore, the transmitter would not in practice be isotropic but would have a gain, g (see Section 1.2). If the receiver has an effective collecting area of A_e then the received signal, F, is given by

$$F = \frac{A_e \alpha P g}{16\pi^2 R^4} \tag{2.61}$$

For an antenna, we have the gain from Equation 1.70

$$g = \frac{4\pi \nu^2 A_e'}{c^2} \tag{2.62}$$

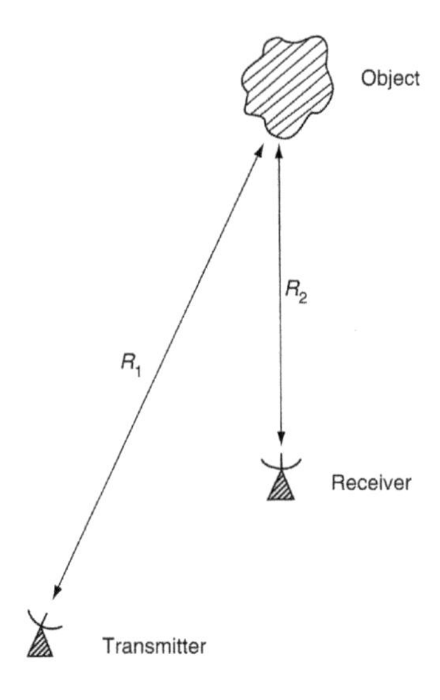

FIGURE 2.36 Schematic radar system.

where A_e' is the effective area of the transmitting antenna. Much of the time in radar astronomy, the transmitting and receiving dishes are the same, so that

$$A_e' = A_e \tag{2.63}$$

and

$$F = \frac{P\alpha A_e^2 \nu^2}{4\pi c^2 R^4} \tag{2.64}$$

This last equation is valid for objects that are not angularly resolved by the radar. For targets that are comparable with the beam width, or larger, the radar cross section, α, must be replaced by an appropriate integral. Thus, for a spherical target which has the radar beam directed towards its centre, we have

$$F \approx \frac{PA_e^2 \nu^2}{4\pi c^2 R^4} \int_0^{\pi/2} 2\pi r \alpha(\phi) \sin\phi \left[s\left(\frac{r \sin\phi}{R} \right) \right]^2 d\phi \tag{2.65}$$

where r is the radius of the target and is assumed to be small compared with the distance, R, ϕ is the angle at the centre of the target to a point on its surface illuminated by the radar beam, $\alpha(\phi)$ is the radar cross section for the surface of the target when the incident and returned beams make an angle ϕ to the normal to the surface. The function is normalised so that the integral tends to α as the beam width increases. $s(\theta)$ is the sensitivity of the transmitter/receiver at an angle θ to its optical axis (cf. Equation 1.71).

The amount of flux that is received is not the only criterion for the detection of objects as we saw in Section 1.2 (Equation 1.64). The flux must also be sufficiently stronger than the noise level of the whole system if the returned pulse is to be distinguishable. Now we have seen in Section 1.2 that the receiver noise may be characterised by comparing it with the noise generated in a resistor at some temperature. We may similarly characterise all the other noise sources and obtain a temperature, T_s, for the whole system, which includes the effects of the target and of the transmission paths as well as the transmitter and receiver. Then from Equation 1.65, we have the noise in power terms, N

$$N = 4\,k\,T_s \Delta\nu \tag{2.66}$$

where $\Delta\nu$ is the bandwidth of the receiver in frequency terms. The signal-to-noise ratio is, therefore,

$$\frac{F}{N} = \frac{P\alpha A_e^2 \nu^2}{16\pi k c^2 R^4 T_s \Delta\nu} \tag{2.67}$$

and this must be unity or larger if a single pulse is to be detected. Because the target which is selected fixes α and R, the signal-to-noise ratio may be increased by *increasing* the power of the transmitter, the effective area of the antenna, or the frequency, or by *decreasing* the system temperature or the bandwidth. Only over the last of these items does the experimenter have any real control, all the others will be fixed by the choice of antenna. However, the bandwidth is related to the length of the radar pulse. Even if the transmitter emits monochromatic radiation, the pulse will have a spread of frequencies given by the Fourier transform of the pulse shape. To a first approximation

$$\Delta\nu \approx \frac{1}{\tau} \quad \text{Hz} \tag{2.68}$$

where τ is the length of the transmitted pulse in seconds. Thus, increasing the pulse length can increase the signal-to-noise ratio. Unfortunately for accurate ranging, the pulse needs to be as short as possible and so an optimum value which minimises the conflict between these two requirements must be sought and this will vary from radar system to radar system, from target to target and from purpose to purpose.

When ranging is not required, so that information is only being sought on the surface structure and the velocity, the pulse length may be increased considerably. Such a system is then known as a continuous wave (CW) radar, and its useful pulse length is only limited by the stability of the transmitter frequency and by the spread in frequencies introduced into the returned pulse through Doppler shifts arising from the movement and rotation of the target. In practice, CW radars work continuously even though the signal-to-noise ratio remains that for the optimum pulse. The alternative to CW radar is pulsed radar. Astronomical pulsed radar systems have a pulse length typically between 10 μs and 10 ms, with peak powers of several tens of megawatts. For both CW and pulsed systems, the receiver pass band must be matched to the returned pulse in both frequency and bandwidth. The receiver must, therefore, be tunable over a short range to allow for the Doppler shifting of the emitted frequency.

The signal-to-noise ratio may be improved by integration. With CW radar, samples are taken at intervals given by the optimum pulse length. For pulsed radar, the pulses are simply combined. In either case the noise is reduced by a factor of $N^{1/2}$, where N is the number of samples or pulses that are combined. If the radiation in separate pulses from a pulsed radar system is coherent, then the signal-to-noise ratio improves directly with N. Coherence, however, may only be retained over the time interval given by the optimum pulse length for the system operating in the CW mode. Thus, if integration is continued over several times the coherence interval, the signal-to-noise ratio for a pulsed system decreases as $(NN^{*1/2})$, where N is the number of pulses in a coherence interval and N^* the number of coherence intervals over which the integration extends.

For most astronomical targets, whether they are angularly resolved or not, there is some depth resolution with pulsed systems. The physical length of a pulse is $(c\tau)$ and this ranges from about 3–3,000 km for most practical astronomical radar systems. Thus, if the depth of the target in the sense of the distance along the line of sight between the nearest and furthest points returning detectable echoes, is larger than the physical pulse length, the echo will be spread over a greater time interval then the original pulse. This has both advantages and disadvantages. First, it provides information on the structure of the target with depth. If the depth information can be combined with Doppler shifts arising through the target's rotation, then the whole surface may be mappable, although there may still remain a two-fold ambiguity about the equator. Secondly and on the debit side, the signal-to-noise ratio is reduced in proportion approximately to the ratio of the lengths of the emitted and returned pulses.

Equation 2.67 and its variants which take account of resolved targets and so on is often called the 'radar equation'. Of great practical importance is its R^{-4} dependence. Thus, if other things are equal, the power required to detect an object increases by a factor of 16 when the distance to the object doubles. Thus, radar detection of Venus, at greatest elongation, needs 37 times the power required for its detection at inferior conjunction. The current state-of-the-art for Earth-based planetary radar studies is given by the Arecibo radio telescope's detection of Saturn's rings and its satellite, Titan.

The high dependence upon R can be made to work in our favour, however, when using spacecraft-borne radar systems. The value of R can then be reduced from tens of millions of kilometres to a few hundred. The enormous gain in the sensitivity that results by R being reduced in this way, when a spacecraft orbits or flies-by its target, means that small low power radar systems, which are suitable for spacecraft use, are sufficient to map the target in detail.

The prime example of this bonus to date is, of course, Venus. Most of the data we have on the surface features have come from radars carried by Magellan and earlier spacecraft. Magellan (1989–1994) though, was the last spacecraft to study Venus' surface using radar.[26] Two Venus

[26] Venus Express did study Venus' atmosphere with a bi-static radar sounder from 2006 onwards. The spacecraft broadcast the transmitted signal and it was received by ground-based radio telescopes on the Earth. Also, Cassini tested its radar on Venus on its way to Saturn (1998).

missions are currently in the planning stages and may continue Magellan's radar work. These are the Indian Shukrayaan-1[27] Venus orbiter mission (perhaps with a mid-2020s launch date and with a radar instrument currently short-listed to be onboard) and the Russian Venera-D orbiter and lander (a possibility for the late 2020s, with the orbiter's primary mission being radar studies of Venus' atmosphere).

All the planets out to Saturn, plus some other inner solar system objects, have currently been studied by ground-based radars – often bi-static with (say) the GBT as the transmitter and Arecibo as the receiver. With the arrival of BepiColombo at Mercury in 2025, all the inner planets and some asteroids and comets will also have been studied by spacecraft-borne radars of one sort or another (BepiColombo's orbiter carries a LiDAR for altitude measurement). Thus, many space missions to Mars (e.g., Mars Express, Mars Odyssey, etc.) have carried radars, including ground-penetrating systems (Section 2.8.5) and they have also been mounted on some ground-based instruments on board landers and rovers, Rosetta studied Comet 67P/Churyumov–Gerasimenko with a type of bi-static radar and, at the time of writing, the Origins, Spectral Interpretation. Resource Identification, Security, Regolith Explorer (OSIRIS-REx) is investigating asteroid Bennu with a LiDAR system.

Cassini has operated a radar at Saturn and the planned Europa Clipper (launch possible mid to late 2020s) and Jupiter Icy Moons Explorer (JUICE; launch currently expected in 2022) spacecraft are expected to carry radars and LiDAR to the Galilean satellites. Finally (for now), New Horizons used a type of bi-static radar to study the interplanetary medium between Pluto and the Earth (a radio transmitter on the spacecraft and ground-based radio telescopes on the Earth).

2.8.2.2 Synthetic Aperture Radar Systems

The radar on board Magellan was of a different type from previously space-based and from Earth-based systems. It is known as synthetic aperture radar (SAR) and has much in common with the technique of aperture synthesis (Section 2.5). Such radars have also been used onboard many remote sensing satellites studying the Earth, ranging from Seasat, launched a decade before Magellan, to the recent NovaSAR-1.

SAR uses a single antenna which, because it is onboard a spacecraft, is moving with respect to the surface of the planet below. The single antenna, therefore, successively occupies the positions of the multiple antennae of an array (Figure 2.37). To simulate the observation of such an array, all that is necessary, therefore, is to record the output from the radar when it occupies the position of one of the elements of the array and then to add the successive recordings with appropriate shifts to counteract their various time delays. In this way the radar can build up an image whose resolution,

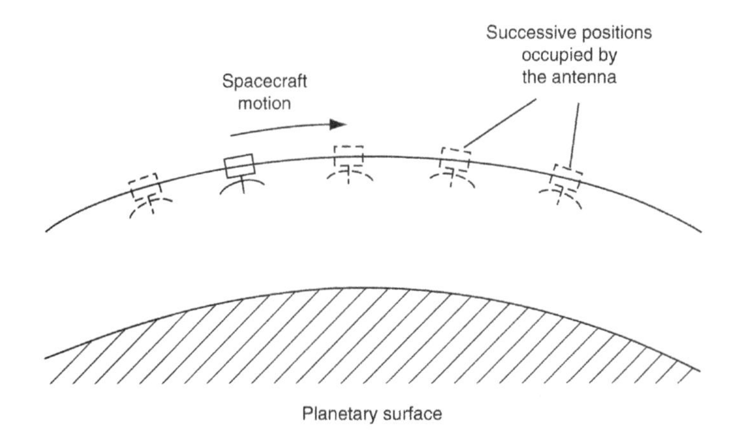

FIGURE 2.37 Synthesis of an interferometer array by a moving single antenna.

[27] Venus craft.

along the line of flight of the spacecraft, is many times better than that which the simple antenna would provide by itself. In practice, of course, the SAR operates continually and not just at the positions of the elements of the synthesised array.

The maximum array length that can be synthesised in this way is limited by the period over which a single point on the planet's surface can be kept under observation. The resolution of a parabolic dish used as a radar is approximately λ/D. If the spacecraft is at a height, h, the radar 'footprint' of the planetary surface is thus about

$$L = \frac{2\lambda h}{D} \tag{2.69}$$

in diameter (Figure 2.38). Since h, in general, will be small compared with the size of the planet, we may ignore the curvature of the orbit. We may then easily see (Figure 2.39) that a given point on the surface only remains observable while the spacecraft moves a distance, L.

Now, the resolution of an interferometer (Section 2.5) is given by Equation 2.31. Thus, the angular resolution of an SAR of length L is

$$\text{Resolution} = \frac{\lambda}{2L} = \frac{D}{4H} \quad \text{radians} \tag{2.70}$$

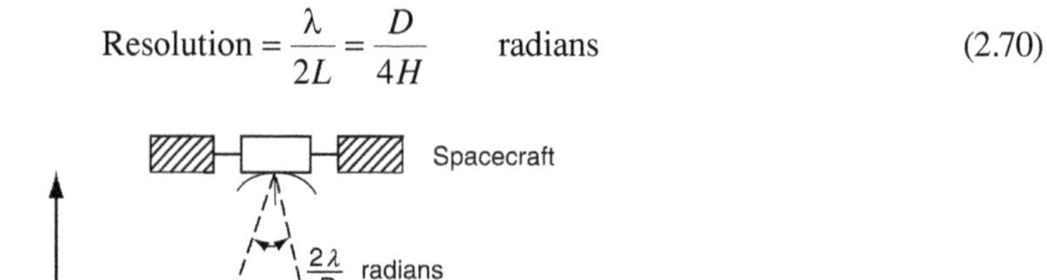

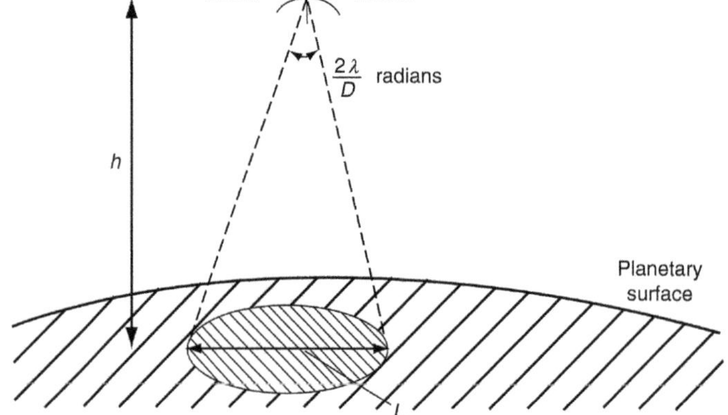

FIGURE 2.38 Radar footprint.

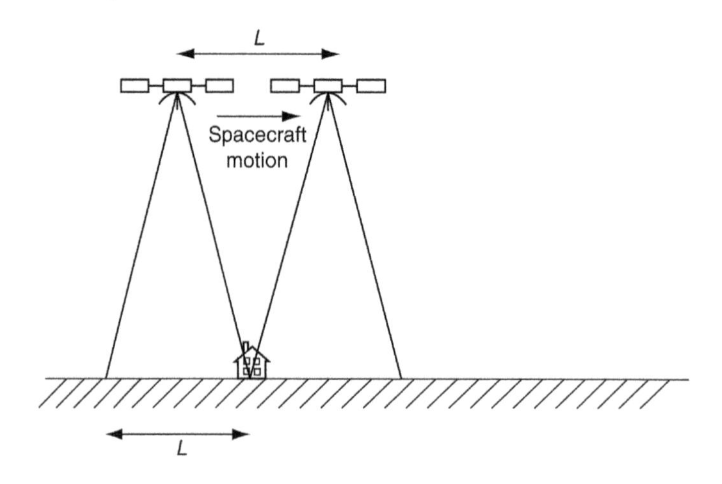

FIGURE 2.39 Maximum length of asynthetic aperture radar (SAR).

The diameter of the synthesised radar footprint is then

$$\text{Footprint diameter} = 2\left(\frac{D}{4h}\right)h = \frac{D}{2} \tag{2.71}$$

Equation 2.71 shows the remarkable result that the linear resolution of an SAR is improved by *decreasing* the size of the radar dish! The linear resolution is just half the dish diameter. Other considerations, such as signal-to-noise ratio (Equation 2.67) require the dish to be as large as possible. Thus, an actual SAR system has a dish size that is a compromise and existing systems typically have 5–10-m diameter dishes.

This analysis applies to a focused SAR, that is to say, an SAR in which the changing distance of the spacecraft from the point of observation is compensated by phase shifting the echoes appropriately. In an unfocused SAR, the point at which the uncorrelated phase shifts due to the changing distance to the observed object reach about $\lambda/4$ limits the synthesised array length. The diameter of the array footprint is then about

$$\text{Unfocused SAR footprint} = \sqrt{2\lambda h} \tag{2.72}$$

2.8.3 Equipment

Radar systems for use in studying planets and so on have much in common with radio telescopes. Almost any filled aperture radio antenna is usable as a radar antenna. Steerable paraboloids are particularly favoured because of their convenience in use. Unfilled apertures, such as collinear arrays (Section 1.2) are not used because most of the transmitted power is wasted.

The receiver is also generally similar to those in use in radio astronomy, except that it must be accurately tunable to provide compensation for Doppler shifts and the stability of its frequency must be high. Its band pass should also be adjustable so that it may be matched accurately to the expected profile of the returned pulse to minimise noise.

Radar systems, however, do require three additional components that are not to be found in radio telescopes. First, and obviously, there must be a transmitter, secondly a high stability master oscillator and thirdly an accurate timing system. The transmitter is usually second only to the antenna as a proportion of the cost of the whole system. Different targets or purposes will generally require different powers, pulse lengths, frequencies and so on from the transmitter. Thus, it must be sufficiently flexible to cope with demands that might typically range from pulse emission at several megawatts for a millisecond burst to continuous wave generation at several hundred kilowatts. Separate pulses must be coherent, at least over a time interval equal to the optimum pulse length for the CW mode, if full advantage of integration to improve signal-to-noise ratio is to be gained. The master oscillator is in many ways the heart of the system. In a typical observing situation, the returned pulse must be resolved to 0.1 Hz for a frequency in the region of 1 GHz, and this must be accomplished after a delay between the emitted pulse and the echo that can range from minutes to hours. Thus, the frequency must be stable to one part in 10^{12} or 10^{13}, and it is the master oscillator that provides this stability. It may be a temperature-controlled crystal oscillator, or an atomic clock. In the former case, corrections for aging will be necessary. The frequency of the master oscillator is then stepped up or down as required and used to drive the exciter for the transmitter, the local oscillators for the heterodyne receivers, the Doppler compensation system and the timing system. The final item, the timing system, is essential if the range of the target is to be found.

It is also essential to know with quite a high accuracy *when* to expect the echo. This arises because the pulses are not broadcast at constant intervals. If this were to be the case then there would be no certain way of relating a given echo to a given emitted pulse. Thus, instead, the emitted pulses are sent out at varying and coded intervals and the same pattern sought amongst the echoes. Because the distance between the Earth and the target is also varying, an accurate timing system is vital to allow the resulting echoes with their changing intervals to be found amongst all the noise.

2.8.4 Data Analysis

The analysis of the data to obtain the distance to the target is simple in principle, being merely half the delay time multiplied by the speed of light. In practice, the process is much more complicated. Corrections for the atmospheres of the Earth and the target (should it have one) are of particular importance. The radar pulse will be refracted and delayed by the Earth's atmosphere, and there will be similar effects in the target's atmosphere with, in addition, the possibility of reflection from an ionosphere as well as, or instead of, reflection from the surface. This applies especially to solar detection. Then reflection can often occur from layers high above the visible surface; 10,000 km above the photosphere, for example, for a radar frequency of 1 GHz. Furthermore, if the target is a deep one, that is, it is resolved in depth, then the returned pulse will be spread over a greater time interval than the emitted one, so that the delay length becomes ambiguous. Other effects such as refraction or delay in the interplanetary medium may also need to be taken into account.

For a deep, rotating target, the pulse is spread out in time and frequency, and this can be used to plot maps of the surface of the object (Figure 2.40). A cross section through the returned pulse at a given instant of time will be formed from echoes from all the points over an annular region such as that shown in Figure 2.41. For a rotating target, which for simplicity we assume has its rotational axis perpendicular to the line of sight, points such as A and A' will have the same approach velocity,

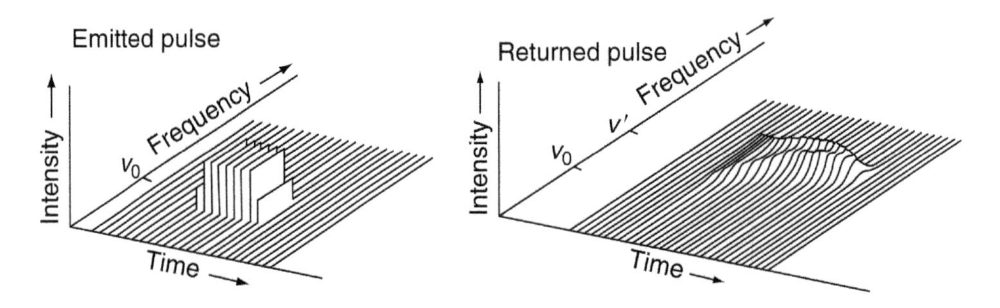

FIGURE 2.40 Schematic change in pulse shape for radar pulses reflected from a deep rotating target. The emitted pulse would normally have a Gaussian profile in frequency, but it is shown here as a square wave to enhance the clarity of the changes between it and the echo.

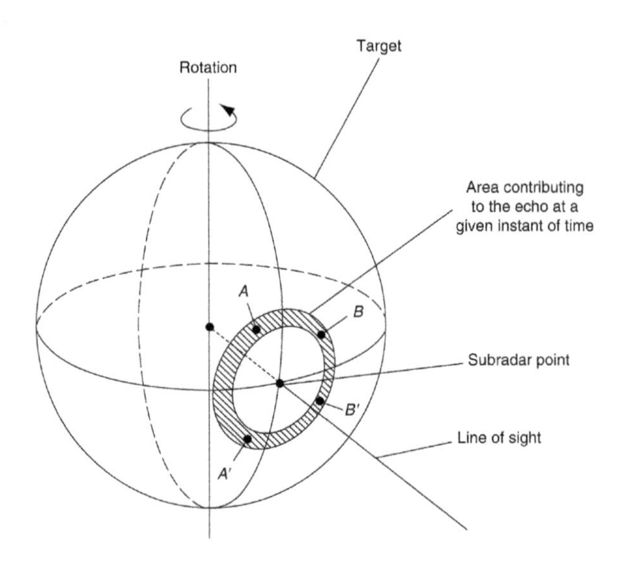

FIGURE 2.41 Region of a deep radar target contributing to the echo at a particular instant of time.

and their echoes will be shifted to the same higher frequency. Similarly, the echoes from points such as *B* and *B′* will be shifted to the same lower frequency. Thus, the intensity of the echo at a particular instant and frequency is related to the radar properties of two points that are equidistant from the equator and are North and South of it. Hence, a map of the radar reflectivity of the target's surface may be produced, although it will have a two-fold ambiguity about the equator. If the rotational axis is not perpendicular to the line of sight, then the principle is similar, but the calculations are more complex. However, if such a target can be observed at two *different* angles of inclination of its rotational axis, then the two-fold ambiguity may be removed and a genuine map of the surface produced. Again, in practice the process needs many corrections and must have the frequency and time profiles of the emitted pulse removed from the echo by deconvolution (Section 2.1). The features on radar maps are often difficult to relate to more normal surface features because they are a conflation of the effects of surface reflectivity, small scale structure, height, gradient, etc.

The relative velocity of the Earth and the target is much simpler to obtain. Provided that the transmitter is sufficiently stable in frequency (see the previous discussion), then it may be found from the shift in the mean frequency of the echo compared with the mean frequency of the emitted pulse, by the application of the Doppler formula. Corrections for the Earth's orbital and rotational velocities are required. The former is given by Equation 4.40, while the latter may be obtained from

$$\mathrm{v}_c = \mathrm{v}_o + 462\cos\delta\sin(\alpha - \mathrm{LST})\sin\left(\mathrm{LAT}\right) \quad \mathrm{m\ s^{-1}} \tag{2.73}$$

where v_c is the corrected velocity (m s^{-1}), v_0 is the observed velocity (m s^{-1}), α and δ are the right ascension and declination of the target, LST is the local sidereal time at the radar station when the observation is made and LAT is the Earth latitude of the radar station.

The combination of accurate knowledge of both distance and velocity for a planet on several occasions enables its orbit to be calculated to a high degree of precision. This then provides information on basic quantities such as the AU and enables tests of metrical gravitational theories such as general relativity to be made and so is of considerable interest and importance even outside the area of astronomy. Producing a precise orbit for a planet by this method requires, however, that the Earth's and the planet's orbits both be known precisely so that the various corrections and so on can be applied. Thus, the procedure is a 'bootstrap' one, with the initial data leading to improved orbits, and these improved orbits then allowing better interpretation of the data and so on over many iterations.

The radar reflection contains information on the surface structure of the target. Changes in the phase, polarisation and of the manner in which the radar reflectivity changes with angle, can in principle be used to determine rock types, roughness of the surface and so on. In practice, the data are rarely sufficient to allow anything more than vague possibilities to be indicated. The range of unfixed parameters describing the surface is so large, that many different models for the surface can fit the same data. This aspect of radar astronomy has therefore been of little use to date.

2.8.5 Ground Penetrating Radar

Ground penetrating radar (also known as 'microwave tomography') is frequently used on the Earth as a non-destructive means of studying the top 100 m of so of the Earth's crust. The radar operates in the 10-MHz to 1-GHz region (30 m to 300 mm) with the radar antenna usually in contact with the ground, although air-borne instruments are also used. It seems likely that such radars will be installed on planetary landers in the future. In the mean time, ESA's Mars Express carries Mars Advanced Radar for Subsurface and Ionosphere Sounding (MARSIS) to study the top few kilometres of Mars' crust from orbit. It operates in the 2- to 5-MHz region (150–60 m) looking, in particular, for water/ice interfaces. The instruments and data interpretation for ground penetrating radar are little different from those of other radars except for the low frequencies at which they operate.

2.8.6 Meteors

The ionised vapour along the track of a meteor reflects radar transmissions admirably. Over the last few decades, therefore, numerous stations have been set up to detect meteors by such means. However, meteors can be observed equally well during the day as at night, many previously unknown daytime meteor showers have been discovered. Either pulsed or CW radar is suitable and a wavelength of a few metres is usually used.

The echoes from a meteor trail are modulated into a Fresnel diffraction pattern (Figure 2.34), with the echoed power expressible in terms of the Fresnel integrals (Equations 2.61 and 2.62) as

$$F(t) = F\left(x^2 + y^2\right) \tag{2.74}$$

where F is the echo strength for the fully formed trail, $F(t)$ is the echo strength at time t during the formation of the trail. The parameter l of the Fresnel integrals, in this case, is the distance of the meteor from the minimum range point. The modulation of the echo arises because of the phase differences between echoes returning from different parts of the trail.

The distance of the meteor and, hence, its height may be found from the delay time as usual. In addition to the distance however, the cross-range velocity can also be obtained from the echo modulation. Several stations can combine, or a single transmitter can have several well-separated receivers, to provide further information such as the position of the radiant, atmospheric deceleration of the meteor, etc.

2.9 ELECTRONIC IMAGES

2.9.1 Image Formats

There are various ways of storing electronic images. One currently in widespread use for astronomical images is Flexible Image Transport System (FITS). There are various versions of FITS, but all have a header, the image in binary form and an end section. The header must be a multiple of 2,880 bytes.[28] The header contains information on the number of bits representing the data in the image, the number of dimensions of the image (e.g., one for a spectrum, two for a monochromatic image, three for a colour image, etc.), the number of elements along each dimension, the object observed, the telescope used, details of the observation and other comments. The header is then padded out to a multiple of 2,880 with zeros. The image section of a FITS file is simply the image data in the form specified by the header. The end section then pads the image section with zeros until it too is an integer multiple of 2,880 bytes. FITS version 4.0 was released in 2016.

Other formats may be encountered such as Joint Photographic Experts Group (JPEG), Graphic Interchange Format (GIF), and Tagged Image File Format (TIFF) and which are generally more useful for processed images than raw data.

2.9.2 Image Compression

Most images, howsoever they may be produced and at whatever wavelength they were obtained, are now stored in electronic form. This is even true of photographic images that may be scanned and fed into computers. Electronic images may be stored by simply recording the precise value of the intensity at each pixel in an appropriate 2D matrix. However, this uses large amounts of memory; for example, 400 megabytes for a 32-bit 10,000 × 10,000-pixel image. Means whereby storage space for electronic images may be used most efficiently are therefore at a premium. The most widely used approach to the problem is to compress the images.

[28] This number originated in the days of punched card input to computers and represents 36 cards each of 80 bytes.

Image compression relies for its success upon some of the information in the image being redundant. That information may therefore be omitted or stored in shortened form without losing information from the image as a whole. Image compression that conserves all the information in the original image is known as 'lossless compression'. For some purposes, a reduction in the amount of information in the image may be acceptable (for example, if one is only interested in certain objects or intensity ranges within the image) and this is the basis of lossy compression.

There are three main approaches to lossless image compression. The first is differential compression. This stores not the intensity values for each pixel, but the differences between the values for adjacent pixels. For an image with smooth variations, the differences will normally require fewer bits for their storage than the absolute intensity values. The second approach, known as 'run compression', may be used when many sequential pixels have the same intensity. This is often true for the sky background in astronomical images. Each run of constant intensity values may be stored as a single intensity combined with the number of pixels in the run, rather than storing the value for every pixel. The third approach, which is widely used on the internet and elsewhere, where it goes by various proprietary names such as ZIP and DoubleSpace, is similar to run compression except that it stores repeated sequences whatever their nature as a single code, not just constant runs. For some images, lossless compressions by a factor of five or ten are possible.

Sometimes not all the information in the original images is required. Lossy compression can then be by factors of 20, 50 or more. For example, a 32-bit image to be displayed only on a computer monitor which has 256 brightness levels need only be stored to 8-bit accuracy, or only the (say) 64 × 64 pixels covering the image of a planet, galaxy, and so on need be stored from a 1,000 × 1,000 pixel image if that is all that is of interest within the image. Most workers, however, will prefer to store the original data in a lossless form.

Lossless compression techniques do not work effectively on noisy images because the noise is incompressible. Then, a lossy compression may be of use since it will primarily be the noise that is lost. An increasingly widely used lossy compression for such images is based on the wavelet transform (c.f. Fourier transforms – Equations 2.3 and 2.4). The basic function used in a wavelet transform is a short, often cyclical, variation sometimes called the 'Mother wavelet' (Figure 2.19 shows an example of this type of variation, though it is in a totally different context – here it is called a 'Morlet wavelet'). This compares with the (theoretically) infinitely long sine and cosine variations of Fourier transforms. The shortness of the wavelet means that a wavelet transform deals much better than the Fourier transform with abrupt changes in the initial function (e.g., sharp edges to the image of an object, for example). The wavelet transform, T, is given by

$$T = \frac{1}{\sqrt{|a|}} \int f(t) \psi\left(\frac{t-b}{a}\right) dt \tag{2.75}$$

where f is the function (image) to be compressed, ψ is the mother wavelet, t is time, a is a scale factor that helps match the mother wavelet to the variation(s) within the function and b is a translational shift to align the wavelet appropriately with the function. The wavelet transform can give high compression ratios for noisy images. Further details of the technique are, however, beyond the scope of this book.

2.9.3 Image Processing

Image processing is the means whereby images are produced in an optimum form to provide the information that is required from them.

Image processing divides into data reduction and image optimisation. Data reduction is the relatively mechanical process of correcting for known problems and faults. Many aspects of it are discussed elsewhere, such as dark frame subtraction, flat fielding and cosmic ray spike elimination on CCD images (Section 1.1), deconvolution and maximum entropy processing (Section 2.1) and CLEANing aperture synthesis images (Section 2.5), etc. Much, if not all, of the data reduction needed may be done automatically with suitable computer programs.

Image optimisation is more of an art than a science. It is the further processing of an image with the aim of displaying those parts of the image that are of interest to their best effect for the objective for which they are required. The same image may therefore be optimised in quite different ways if it is to be used for different purposes. For example, an image of Jupiter and its satellites would be processed quite differently if the surface features on Jupiter were to be studied, compared with if the positions of the satellites with respect to the planet were of interest. There are dozens of techniques used in this aspect of image processing, and there are few rules other than experience to suggest which will work best for a particular application. The main techniques are outlined below, but interested readers will need to look to other sources and to obtain practical experience before being able to apply them with consistent success.

2.9.3.1 Grey Scaling[29]

This is probably the technique in most widespread use. It arises because many detectors have dynamic ranges far greater than those of the computer monitors or hard copy devices that are used to display their images. In many cases the interesting part of the image will cover only a restricted part of the dynamic range of the detector. Grey scaling then consists of stretching that section of the dynamic range over the available intensity levels of the display device.

For example, the spiral arms of a galaxy on a 16-bit CCD image (with a dynamic range from 0 to 65,535) might have CCD level values ranging from 20,100 to 20,862. On a computer monitor with 256 intensity levels, this would be grey scaled so that CCD levels 0 to 20,099 corresponded to intensity 0 (i.e., black) on the monitor, levels 20,100 to 20,103 to intensity 1, levels 20,104 to 20,106 to intensity 2 and so on up to levels 20,859 to 20,862 corresponding to intensity 254. CCD levels 20,863 to 65,535 would then all be lumped together into intensity 255 (i.e., white) on the monitor.

Other types of grey scaling may aim at producing the most visually acceptable version of an image. There are many variations of this technique whose operation may best be envisaged by its effect on a histogram of the display pixel intensities. In histogram equalisation, for example, the image pixels are mapped to the display pixels in such a way that there are equal numbers of image pixels in each of the display levels. For many astronomical images, a mapping which results in an exponentially decreasing histogram for the numbers of image pixels in each display level often gives a pleasing image because it produces a good dark background for the image. The number of other mappings available is limited only by the imagination of the operator.

2.9.3.2 Image Combination

Several images of the same object may be added together to provide a resultant image with lower noise. More powerful combination techniques, such as the subtraction of two images to highlight changes, or false-colour displays, are, however, probably in more widespread use.

2.9.3.3 Spatial Filtering

This technique has many variations that can have quite different effects upon the image. The basic process is to combine the intensity of one pixel with those of its surrounding pixels in some manner. The filter may best be displayed by a 3×3 matrix (or 5×5, 7×7 matrices, etc.) with the pixel of interest in the centre. The elements of the matrix are the weightings given to each pixel intensity when they are combined and used to replace the original value for the central pixel (Figure 2.42). Some commonly used filters are shown in Figure 2.42.

[29] This term is also used for the conversion of a colour image to a monochromatic image.

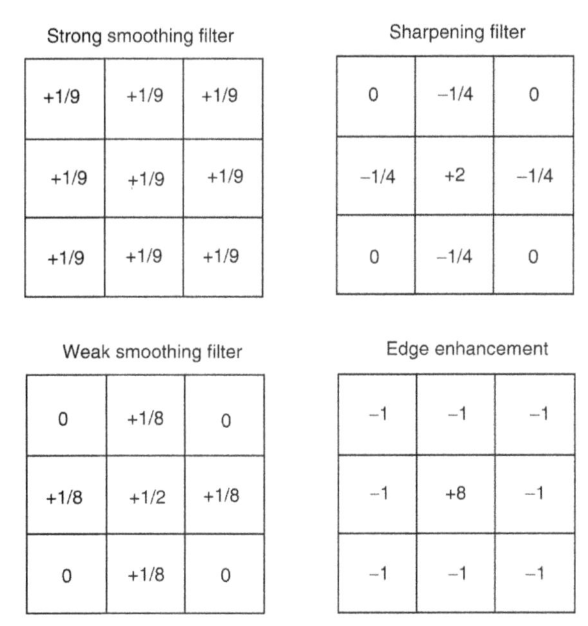

FIGURE 2.42 Examples of commonly used spatial filters.

2.9.3.4 Ready-Made Computer Packages

Various commercial image processing packages are available and enable anyone to undertake extensive image processing. Much of this though is biased towards commercial artwork and the production of images for advertising, etc. Adobe's Photoshop™ and/or its rivals are probably already known to many readers (and are useful). The (free) image processing and graphic program GNU Image Manipulation Program (GIMP), may be found at www.gimp.org/. Microsoft's Excel™ spreadsheet (usually standard in most Microsoft Office packages) is widely used for dealing with large amounts of data in catalogue and other forms and it is easy to custom-write in it small (or even quite large) algorithms for ad hoc, bespoke, or specialised data-processing needs.

There are, however, computer packages specifically designed for astronomical image processing and many are available free of charge. Until recently, the most widespread of these packages for optical images was Image Reduction and Analysis Facility (IRAF), produced and maintained by National Optical Astronomy Observatories (NOAO). However, at the time of writing NOAO has given notice that support for the package is being withdrawn. It is not clear what, if anything, will replace IRAF, although the UK's open source Starlink package[30] is a good substitute. The Astropy project is a community-developed package of programmes for many aspects of astronomy and astrophysics, written in the Python computer language and in widespread use. At the time of writing, Python 3.8.1 may be downloaded from www.python.org/ and Astropy from www.astropy.org/. Custom-written software for processing astronomical catalogues is available at www.star.bris.ac.uk/~mbt/topcat/ and called Tool for Operations on Catalogues and Tables (TOPCAT). While Spectroscopy Made Easy (SME) may be useful for spectroscopists.

If you have purchased a CCD camera designed for amateur astronomical use, then the manufacturer will almost certainly have provided at least a basic image-processing package as a part of the

[30] Now supported by the East Asian Observatory.

deal. Most such manufacturers also have internet sites where other software produced by themselves and by users of the cameras may be found.

For radio astronomical data and interferometric data, KERN (and the previous CASA) is the most widely used data-reduction package (Section 2.5). Virtual Observatories (Section 5.5.3) all have their own programs and software to operate with their data bases. Many major astronomical instruments, of all types, have their own specific data-processing packages already available. These are often called 'pipelines' and their details will be made available to anyone using those particular instruments (indeed, their use may be a condition of being awarded time on the instrument and/or the instrument's raw output may be automatically processed before the investigator gets any of the data).

3 Photometry

3.1 PHOTOMETRY

3.1.1 BACKGROUND

3.1.1.1 Introduction

The measurement of the intensity of electromagnetic energy coming to us from an astronomical object has traditionally been called 'photometry'. However, physics distinguishes between 'photometry' and 'radiometry'. The former is the brightness of an object as perceived by the *eye*, and the latter the brightness measured in *absolute terms*.[1] Thus, when the eye was the only means of determining the brightness of stars, nebulae and so on then astronomers were, indeed, making photometric measurements. By definition, however, photometry can only be undertaken within the visible part of the spectrum – typically from 390 to 700 nm (Section 1.1). Thus, in 1800 when Sir William Herschel discovered infrared radiation from the Sun and in 1801 when Johann Ritter discovered solar ultraviolet radiation, they were making the first radiometric measurements.

Today, no one would consider making eye estimates of the brightness of stars for any work that had the least pretension to being serious. Basic photometers based upon pin photodiodes are available cheaply and can be used with the smallest of astronomical telescopes. The charge-coupled devices (CCDs) are also used extensively by amateur as well as professional astronomers, and both CCDs and pin diodes fundamentally measure brightness in absolute terms. Even off-the-shelf 'every-day' digital single lens reflex (DSLR) cameras have been used (with some ingenuity) for stellar photometry.

Thus, on a strict definition, when astronomers now measure the brightness of objects in the sky, they are undertaking radiometry and not photometry. However, the use of the term 'photometry' in this context is so well established amongst astronomers that a change to the accurate terminology seems unlikely to occur – and so we will continue to use 'photometry' in the astronomers' sense in this book.

Photometry is the measurement of the energy coming from an astronomical object within a restricted range of wavelengths. On this definition just about every observation that is made is a part of photometry. It is customary and convenient, however, to regard spectroscopy and imaging as separate from photometry although the differences are blurred in the overlap regions and outside the optical part of the spectrum.

Spectroscopy (Chapter 4) is the measurement of the energy coming from an object within several (usually hundreds or thousands) of adjacent wavebands and with the widths of those wavebands a hundredth of the operating wavelength or less (usually a thousandth to a hundred-thousandth of the operating wavelength). However, some narrow band photometric systems on this definition *could* be regarded as low-resolution spectroscopy.

It is harder to separate imaging (Chapters 1 and 2) from photometry, especially with array detectors, because much of photometry is now just imaging through an appropriate filter or filters.

[1] The difference is reflected in the units used. For radiometry, *Radiant Flux* or *Radiant Power* is the total power emitted as e-m radiation by a source in all directions and over all wavelengths. It is measured in watts. For photometry, the equivalent quantity is the *Luminous Flux* or *Luminous Power* and is the total visible light power emitted by an e-m source in all directions as *perceived* by the *eye*. It is measured in lumens. At a wavelength of 555 nm, 1 lm_{555} = 0.001464 W = 4.1 × 10^{15} 555 nm photons but the relationship varies with wavelength, so that at 500 nm, 1 lm_{500} = 0.004043 W = 1.0 × 10^{16} 500 nm photons and so on.

So, it is perhaps best to regard photometry as imaging with the *primary* purpose of measuring the *brightness* of some or all of the objects within the image and imaging as having the *primary* purpose of determining the *spatial structure* or *appearance* of the object. The situation is further confused by the recent development of integral field spectroscopy (Section 4.2) where spectra are simultaneously obtained for some or all of the objects within an image.

Most radio receivers are narrow band instruments, and many of the detectors used for far infrared (FIR) and microwave observations operate over restricted wavebands. Essentially, therefore, almost all their observations come under photometry and so have been covered in Sections 1.1 and 1.2. Likewise, extreme ultraviolet (EUV), X-ray and gamma-ray detectors often have some intrinsic sensitivity to wavelength (i.e., to photon energy) and so automatically operate as photometers or low-resolution spectroscopes. Their properties have been covered in Section 1.3.

In this section we are thus concerned with photometry as practiced over the near ultraviolet (UV), visible, near infrared (NIR) and mid-infrared (MIR) regions of the spectrum (roughly 100 nm to 100 μm).

3.1.1.2 Magnitudes

The system used by astronomers to measure the visible brightness of stars is an ancient one and for that reason is an awkward one. It originated with the earliest known astronomical catalogue, that one due to Hipparchus in the late second century BCE. The stars in that catalogue were divided into six classes of brightness, the brightest being of the first class and the faintest of the sixth class. With the invention and development of the telescope, such a system proved to be inadequate, and it had to be refined and extended. William Herschel initially undertook this, but our present-day system is based upon the work of Norman Pogson in 1856. He suggested a logarithmic scale which approximately agreed with the previous measures. As we have seen in Section 1.1, the response of the eye is nearly logarithmic, so that the ancient system attributed to Hipparchus was actually also roughly logarithmic in terms of intensity. Hipparchus' class five stars were about 2.5 times brighter than his class six stars and were about 2.5 times fainter than class four stars and so on. Pogson expressed this as a precise mathematical law in which the magnitude was related to the logarithm of the diameter of the minimum size of telescope required to see the star (i.e., the limiting magnitude – Equation 3.3). Soon after this, the proposal was adapted to express the magnitude directly in terms of the relative energy coming from the star and the equation is now usually used in the form:

$$m_1 - m_2 = -2.5\log_{10}\left(\frac{E_1}{E_2}\right) \tag{3.1}$$

where m_1 and m_2 are the magnitudes of stars 1 and 2 and E_1 and E_2 are the energies per unit area at the surface of the Earth for stars 1 and 2.

The scale is awkward to use because of its peculiar base, leading to the ratio of the energies of two stars whose brightness differ by one magnitude being 2.512.[2] Also, against usual practice, the *brighter* stars have magnitudes whose actual values are *smaller* than those of the *fainter* stars. However (like 'photometry'), it seems unlikely that astronomers will change the system, so the student must perforce become familiar with it as it stands.

One aspect of the scale is immediately apparent from Equation 3.1 and that is that the scale is a relative one: the magnitude of one star is expressed in terms of that of another, and there is no direct reference to absolute energy or intensity units. The zero of the scale is therefore fixed by reference to standard stars. On Hipparchus' scale, those stars that were just visible to the naked eye were of class six. Pogson's scale is thus arranged so that stars of *magnitude six* (6^m) are just visible to a normally

[2] $= 10^{0.4}$. For a difference of two magnitudes, the energies differ by ×6.31 (2.512^2), for three magnitudes by ×15.85 (2.512^3), four magnitudes by ×39.81 (2.512^4) and by five magnitudes by exactly ×100 (2.512^5).

sensitive eye from a good observing site on a clear, moonless night. The standard stars are chosen to be non-variables and their magnitudes are assigned so that the above criterion is fulfilled. The primary stars are known as the 'North Polar Sequence'[3] and comprise stars within 2° of the pole star, so that they may be observed from almost all Northern Hemisphere observatories throughout the year. Secondary standards are set up from these primary standards to allow observations to be made in any part of the sky. A commonly used standard for the Johnson and Morgan UBV filter system (see below) is the star Vega (α Lyrae) which has a magnitude of $+0.03^m$ in the visible region. Several recent large photometric surveys have started measuring the brightness of objects in terms of absolute units (see below). The data however is still presented as a magnitude scale and is made as consistent with previous work as possible.

The faintest stars visible to the naked eye, from the definition of the scale, are of magnitude six; this is termed the 'limiting magnitude of the eye'. For point sources, the brightness is increased by the use of a telescope by a factor, G, which is called the 'light grasp of the telescope' (Section 1.1). Because the dark-adapted human eye has a pupil diameter of about 7 mm, G is given by

$$G \approx 2 \times 10^4 d^2 \tag{3.2}$$

where d is the telescope's objective diameter in metres. Thus, the limiting magnitude through a visually used telescope, m_L, is

$$m_L = 16.8 + 5 \log_{10} d \tag{3.3}$$

CCDs and other detection techniques will generally improve on this by some 5–10 stellar magnitudes.

The magnitudes so far discussed are all Apparent Magnitudes and so result from a combination of the intrinsic brightness of the object *and* its distance. This is obviously an essential measurement for an observer because it is the criterion determining exposure times, etc. However, it is of little intrinsic significance for the object in question. A second type of magnitude is therefore defined which is related to the actual brightness of the object. This is called the 'Absolute Magnitude' and it is 'the apparent magnitude of the object if its distance from the Earth were to be ten parsecs'.

It is usually denoted by M, while apparent magnitude uses the lower case, m. The relation between apparent and absolute magnitudes may easily be obtained from Equation 3.1. Imagine the object moved from its real distance to ten parsecs from the Earth. Its energy per unit area at the surface of the Earth will then change by a factor $(D/10)^2$, where D is the object's actual distance in parsecs. Thus,

$$M - m = -2.5 \log_{10}\left(\frac{D}{10}\right)^2 \tag{3.4}$$

$$M = m + 5 - 5 \log_{10} D \tag{3.5}$$

The difference between apparent and absolute magnitudes is called the 'distance modulus' and is occasionally used in place of the distance itself

$$\text{Distance modulus} = m - M = 5 \log_{10} D - 5 \tag{3.6}$$

[3] The stars of the North Polar Sequence were listed in first to fourth editions of this book. If anyone should need this information, then it may be found in those prior editions of this book, or in the primary source; Leavitt H. S. *Annals of the Harvard College Observatory* **71** (3), 49–52.

Equations 3.5 and 3.6 are valid so long as the only factors affecting the apparent magnitude are distance and intrinsic brightness. However, light from the object often has to pass through interstellar gas and dust clouds, where it may be absorbed. A more complete form of Equation 3.5 is therefore

$$M = m + 5 - 5\log_{10} D - AD \tag{3.7}$$

where A is the interstellar absorption in magnitudes per parsec. A typical value for A for lines of sight within the galactic plane is 0.002 mag pc^{-1}. Thus, we may determine the absolute magnitude of an object via Equations 3.5 or 3.7 once its distance is known. More frequently, however, the equations are used in the reverse sense to determine the distance. Often the absolute magnitude may be estimated by some independent method and then

$$D = 10^{\left[(m-M+5)/5\right]} \quad \text{pc} \tag{3.8}$$

Such methods for determining distance are known as 'standard candle methods' because object is in effect acting as a standard of some known luminosity. The best-known examples are the classical Cepheids with their period-luminosity relationship.

$$M = -1.9 - 2.8 \log_{10} P \tag{3.9}$$

where P is the period of the variable in days. Many other types of stars such as dwarf Cepheids, *RR* Lyrae stars, *W* Virginis stars and β Cepheids are also suitable. Also, Type Ia supernovae, or the brightest novae, or the brightest globular clusters around a galaxy, or even the brightest galaxy in a cluster of galaxies, can provide rough standard candles. Yet another method is due to Wilson and Bappu who found a relationship between the width of the emission core of the ionised calcium line at 393.3 nm (the Ca II K line) in late-type stars and their absolute magnitudes. The luminosity is then proportional to the sixth power of the line width.

Both absolute and apparent magnitudes are normally measured over some well-defined spectral region. Although this discussion is quite general, the equations only have validity within a given spectral region. Because the spectra of two objects may be quite different, their relationship at one wavelength may be different from that at another. The next section discusses the definitions of these spectral regions and their interrelationships.

3.1.2 Filter Systems

Numerous filter systems and filter-detector combinations have been devised. They may be grouped into wide, intermediate and narrow band systems according to the bandwidths of their transmission curves. In the visible, wide band filters typically have bandwidths of around 100 nm, intermediate band filters range from 10 to 50 nm, while narrow band filters range from 0.05 to 10 nm. The division is convenient for the purposes of discussion here, but it is not of any real physical significance.

The filters used in photometry are of two main types based upon either absorption/transmission or on interference. The absorption/transmission filters use salts such as nickel or cobalt oxides dissolved in glass or gelatine, or as a suspension of colloid particles. These filters typically transmit over a 100-nm wide region. They are thus mostly used for the wide band photometric systems. Many of these filters will also transmit in the red and infrared and so must be used with a red blocking filter made with copper sulphate. Short wave blocking filters may be made using cadmium sulphide or selenide, sulphur or gold. Although not normally used for photometry, two other types of filters may usefully be mentioned here: dichroic mirrors and neutral density filters. Neutral density filters absorb by a constant(ish) amount over a wide range of wavelengths and may be needed when observing bright objects like the Sun (Section 5.3). They are normally a thin deposit of a reflecting metal such as aluminium or stainless steel on a glass or thin plastic substrate. Dichroic mirrors reflect over one wavelength range and transmit over another. For example, a mirror thinly plated with a gold coating is transparent to visual radiation but reflects the infrared. They may also be produced using multi-layer interference coatings. Dichroic mirrors may be used in spectroscopy to feed two spectroscopes operating over different wavelength ranges (Section 4.2).

Interference filters are mostly Fabry-Perot etalons (Section 4.1 – see also Section 1.6) with a small separation of the mirrors. The transmission wavelength and the bandwidth of the filter can be tuned by changing the mirror separation and/or their reflectivities (see also tunable filters, Section 4.1 and solar H-α filters, Section 5.3). Combining several Fabry-Perot cavities can alter the shape of the transmission band. Interference filters are usually used for the narrower filters of a photometric system because they can be made with bandwidths ranging from a few tens of nanometres to a hundredth of a nanometre. The recently developed rugate filter can transmit or reflect simultaneously at several wavelengths. It uses interference coatings in which the refractive index varies continuously throughout the depth of the layer.

The earliest 'filter' system was given by the response of the human eye (Figure 1.3) and this peaks around 510 nm, with a bandwidth of 400 nm or so. Because all normal (i.e., not colour-blind, etc.) eyes have roughly comparable responses, the importance of the spectral region within which a magnitude was measured was not realised until the application of the photographic plate to astronomical detection. Then, it was discovered that the magnitudes of stars that had been determined from such plates often differed significantly from those found visually. The discrepancy arose from the differing sensitivities of the eye and the photographic emulsion. Early emulsions had a response that peaks near 450 nm; little contribution is made to the image by radiation of a wavelength longer than about 500 nm. The magnitudes determined with the two methods became known as 'visual magnitude' and 'photographic magnitude', respectively, and are denoted by m_v and m_p or M_v and M_p. These magnitudes may be encountered when reading older texts or working on archive material but have now largely been replaced by more precise filter systems.

With the development of photoelectric methods of detection of light and their application to astronomy by Joel Stebbins and others in the twentieth century, other spectral regions became available and more precise definitions of filters were then required. There are now a great number of different photometric systems. Although it is probably not quite true that there is one system for every observer, there is certainly at least one separate system for every slightly differing purpose to which photometric measurements may be put. Many of these are developed for highly specialised reasons and so are unlikely to be encountered outside their particular area of application. Anyone working in photometry will find details of the filter systems for the photometer which they are using in that instrument's technical guides, including (probably) how to process and analyse the data. Just a few of the more general systems are, therefore, reviewed here as an introduction to the topic.

For a long while, the most widespread of these more general systems was the ultraviolet, blue, visible (UBV) system, defined by Harold Johnson and William Morgan in 1953. This is a wide band system with the B and V regions corresponding approximately to the photographic and visual responses and with the U region in the violet and UV. The precise definition requires a reflecting telescope with aluminised mirrors and uses an RCA 1P21 photomultiplier (PMT).

The filters are:

Corning 3384 for the V region
Corning 5030 plus Schott GG 13 for the B region
Corning 9863 for the U region

Filter name	U	B	V
Central wavelength (nm)	365	440	550
Bandwidth (nm)	70	100	90

The response curves for the filter-PMT combination are shown in Figure 3.1. The absorption in the Earth's atmosphere normally cuts off the short wavelength edge of the U response. At sea level

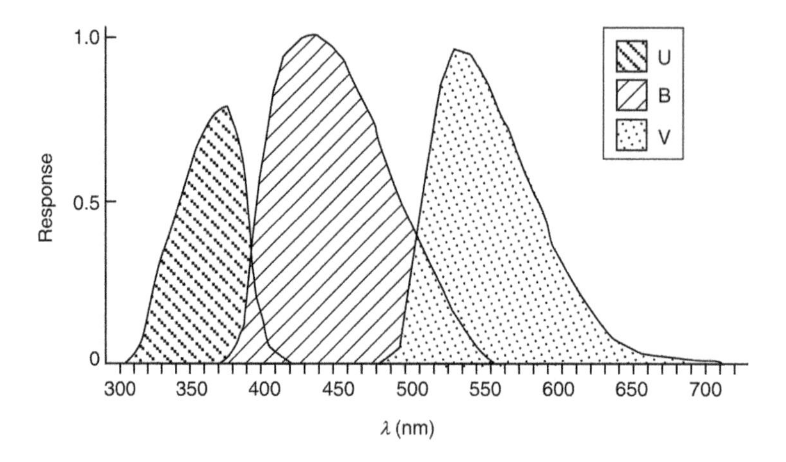

FIGURE 3.1 Ultraviolet, Blue, Visible (UBV) response curves, excluding the effects of atmospheric absorption.

roughly half of the incident energy at a wavelength of 360 nm is absorbed, even for small zenith angles. More importantly, the absorption can vary with changes in the atmosphere. Ideal U response curves can therefore only be approached at high altitude, high-quality observing sites. The effect of the atmosphere on the U response curve is shown in Figure 3.2. The B and V response curves are comparatively unaffected by differential atmospheric absorption across their wavebands; there is only a total reduction of the flux through these filters (see Section 3.2). The scales are arranged so that the magnitudes through all three filters are equal to each other for A0 V stars such as α Lyrae (Vega). The choice of filters for the UBV system was made on the basis of what was easily available at the time. The advent of CCDs has led to several new photometric systems being devised, with filters more related to the requirements of astrophysics, but the UBV system and its extensions are still in use, if only to enable comparisons to be made with measurements taken in the past.

The relationship between UBV magnitudes and absolute physical quantities has been determined using laboratory sources such as platinum furnaces. It is sometimes called the 'Vega system'. It is defined by the flux for a zero-magnitude star of spectral type, A0, at 550 nm wavelength (V filter) being

$$F_{550\,\mathrm{nm}} = 3.60 \times 10^{-11}\ \mathrm{W\,m^{-2}nm^{-1}} \tag{3.10}$$

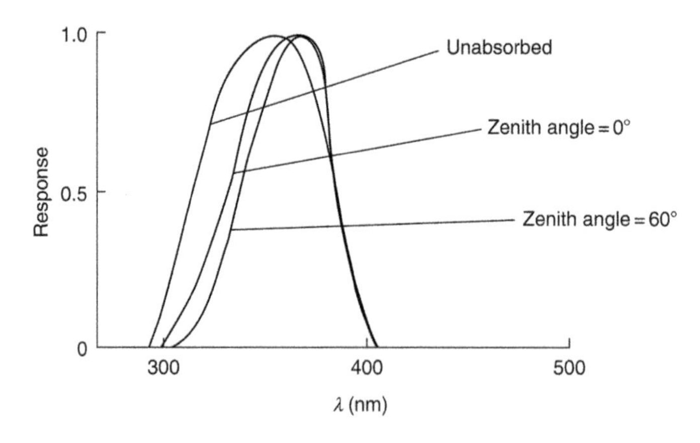

FIGURE 3.2 Effect of atmospheric absorption at sea level upon the U response curve (normalised).

or using frequency[4]

$$F_{545\,\text{THz}} = 3.64 \times 10^{-23}\ \text{W m}^{-2}\text{Hz}^{-1} \tag{3.11}$$

Any measurement of an object's brightness expressed in the units used in Equation 3.10 can be converted into its V magnitude through the equation

$$V = -2.5 \log_{10} \frac{\int R(\lambda)\, F_{\text{Object}}(\lambda)\, d\lambda}{\int R(\lambda)\, F_{\text{Vega}}(\lambda)\, d\lambda} + 0.03 \tag{3.12}$$

(or its equivalent in frequency terms) where $R(\lambda)$ is the V filter response at wavelength λ and the 0.03 is Vega's V magnitude. The relationships for the other filters are based upon the colour indices (see below) being zero for A0 stars (i.e., $B - V = U - B = 0$).

There are several extensions to the UBV system for use at longer wavelengths. A widely used system was introduced by Johnson in the 1960s for lead sulphide infrared detectors. With the photographic R and I bands this adds another eight regions:

Filter Name	*R*	*I*	*J*	*K*	*L*	*M*	*N*	*Q*
Central wavelength (nm)	700	900	1,250	2,200	3,400	4,900	10,200	20,000
Bandwidth (nm)	220	240	380	380	700	300	5,000	5,000

The longer wavelength regions in this system are defined by atmospheric transmission and so are variable. Recently, therefore, filters have been used to define all the pass bands and the system has been adapted to make better use of the CCD's response and the improvements in infrared detectors. This has resulted in slight differences to the transmission regions in what is sometimes called the 'JCG system'.[5]

Filter Name	*U*	*B*	*V*	*R*	*I*	*Z*[6]	*J*
Central wavelength (nm)[7]	367	436	545	638	797	908	1,220
Bandwidth (nm)	66	94	85	160	160	96	213

Filter Name	*H*	*K*	*L*	*M*
Central wavelength (nm)	1,630	2,190	3,450	4,750
Bandwidth (nm)	307	390	472	460

Two other wide band filter systems are now used, because of the large data sets that are available. These are from the Hubble space telescope (HST) and the Sloan Digital Sky Survey (SDSS). The HST Wide Field and Planetary Camera 2 (WFPC2) system had six filters:

[4] The similarity of the significant figures in these two equations is happenstance – arising from the numerical value of λ^2 (at 550 nm) being almost exactly 10^{21} times the numerical value of the speed of light. The corresponding values at 360 nm (U filter) are $F_{360\,\text{nm}} = 4.18 \times 10^{-11}$ W m^{-2} nm^{-1} and $F_{833\,\text{THz}} = 1.81 \times 10^{-23}$ W m^{-2} Hz^{-1}.

[5] For Johnson, Alan Cousins and Ian Glass.

[6] Added for use with CCDs.

[7] Data from *The Encyclopaedia of Astronomy and Astrophysics*, Ed. P. Murdin (IoP Press, 2001), 1642.

Central wavelength (nm)	336	439	450	555	675	814
Bandwidth (nm)	47	71	107	147	127	147

Whereas the Wide Field and Planetary Camera 3 (WFC3) has 12 filters:

Central wavelength (nm)	222.4	235.9	270.4	335.5	392.1	432.5
Bandwidth (nm)	32.2	46.7	39.8	51.1	89.6	61.8

Central wavelength (nm)	477.3	530.8	588.7	624.2	764.7	802.4
Bandwidth (nm)	134.4	156.2	218.2	146.3	117.1	153.6

The magnitudes obtained using the HST (known as STMag) are related to the flux from the object (cf. Equation 3.10) by

$$m(\lambda) = -2.5 \log_{10} F(\lambda) - 21.1 \tag{3.13}$$

where the constant normalises the magnitudes to that for Vega in the Johnson and Morgan V band.

The SDSS is based upon measurements made using the 2.5-m telescope at the Apache Point Observatory in New Mexico and uses five filters that cover the whole range of sensitivity of CCDs:

Filter Name	u'	g'	r'	i'	z'
Central wavelength (nm)	358	490	626	767	907
Bandwidth (nm)	64	135	137	154	147

The SDSS magnitudes are based upon the brightness of spectral type, *F*, sub-dwarf stars (known as the Gunn system after James Gunn who proposed the idea along with Trinh Thuan in 1976). The choice of these stars is due to their spectra being much smoother than those of Vega and similar A0 stars. The star BD + 17 4708 (spectral type sdF8, $U = 9.724$, $B = 9.886$, $V = 9.45$) is the main standard and has all its SDSS magnitudes equal to 9.50 (thus again more-or-less normalising the system to the Johnson and Morgan V band). The SDSS measurements are based upon the flux per unit frequency interval (cf. Equation 3.11) and the magnitudes[8] obtained are related to the flux from the object by

$$m(\nu) = -2.5 \log_{10} F(\nu) - 48.6 \tag{3.14}$$

Amongst the intermediate pass band systems, the most widespread is the *uvby* or Strömgren system. This was proposed by Bengt Strömgren in the late 1960s and is now in fairly common use. Its transmission curves are shown in Figure 3.3. It is often used with two additional filters centred on the $H\beta$ line, 3 nm and 15 nm wide, respectively, to provide better temperature discrimination for the hot stars.

[8] Magnitudes based upon the flux per unit frequency interval are sometimes called AB magnitudes, the AB being derived from ABsolute. Unfortunately, the possibility of confusion with absolute magnitude (Equation 3.5) is great and so the term should be avoided.

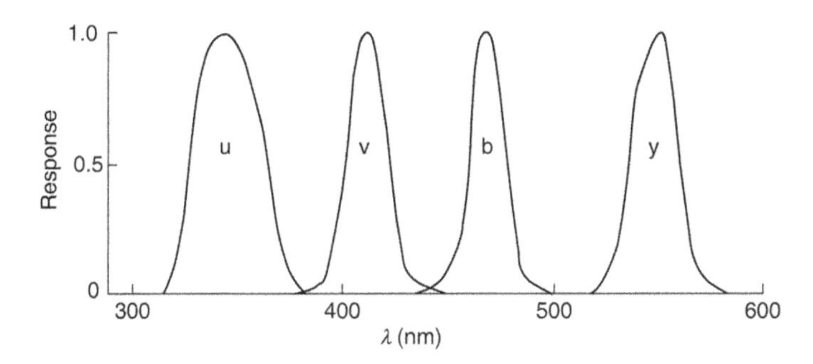

FIGURE 3.3 Normalised transmission curves for the *uvby* system, not including the effects of atmospheric absorption.

Filter Name	*u*	*v*	*b*	*y*	β_n	β_w
Central wavelength (nm)	349	411	467	547	486	486
Bandwidth (nm)	30	19	18	23	3	15

Narrow band work mostly concentrates on isolating spectral lines. $H\alpha$ and $H\beta$ are common choices, and their variations can be determined by measurements through a pair of filters which are centred on the line but have different bandwidths. No single system is in general use, so that a more detailed discussion is not profitable. Other spectral features can be studied with narrow band systems where one of the filters isolates the spectral feature and the other is centred on a nearby section of the continuum. The reason for the invention of these and other photometric systems lies in the information that may be obtained from the comparison of the brightness of the object at different wavelengths. Many of the processes contributing to the final spectrum of an object as it is received on Earth preferentially affect one or more spectral regions. Thus, estimates of their importance may be obtained simply and rapidly by measurements through a few filters. Some of these processes are discussed in more detail in the next subsection and in Section 3.2. The most usual features studied in this way are the Balmer discontinuity and other ionisation edges, interstellar absorption and strong emission or absorption lines or bands.

There is one additional photometric 'system' that has not yet been mentioned. This is the bolometric system, or rather bolometric magnitude, because it has only one pass band. The bolometric magnitude is based upon the total energy emitted by the object at all wavelengths. Since it is not possible to observe this in practice, its value is determined by modelling calculations based upon the object's intensity through one or more of the filters of a photometric system. Although X-ray, UV, infrared and radio data are now available to assist these calculations, many uncertainties remain and the bolometric magnitude is still rather imprecise, especially for high-temperature stars. The calculations are expressed as the difference between the bolometric magnitude and an observed magnitude. Any filter of any photometric system could be chosen as the observational basis, but in practice the V filter of the standard UBV system is normally used. The difference is then known as the bolometric correction (BC),

$$\mathrm{BC} = m_{\mathrm{bol}} - V \tag{3.15}$$

$$= M_{\mathrm{bol}} - M_V \tag{3.16}$$

and its scale is chosen so that it is zero for main sequence stars with a temperature of about 6,500 K (i.e., about spectral class F5 V). The luminosity of a star is directly related to its absolute bolometric magnitude

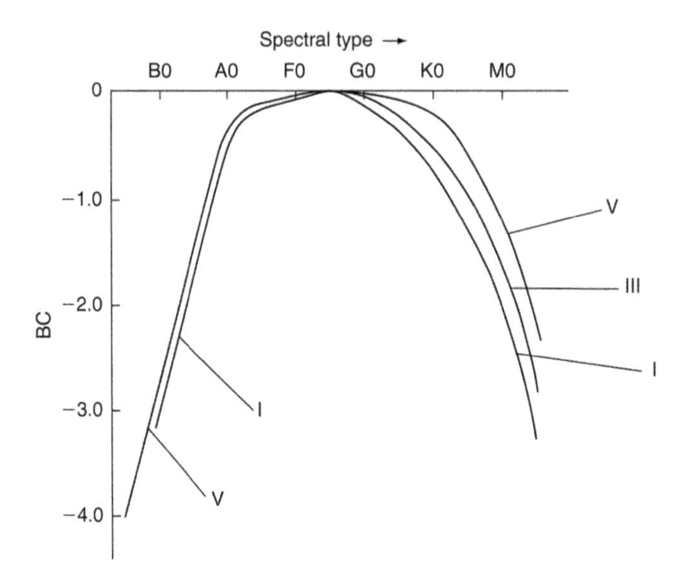

FIGURE 3.4 Bolometric corrections for main sequence stars (type V), giants (type III), and supergiants (type I).

$$L_* = 3 \times 10^{28} \times 10^{-0.4 M_{bol}} \quad \text{W} \tag{3.17}$$

Similarly, the flux of the stellar radiation just above the Earth's atmosphere, f_*, is related to the apparent bolometric magnitude

$$f_* = 2.5 \times 10^{-8} \times 10^{-0.4 m_{bol}} \quad \text{Wm}^{-2} \tag{3.18}$$

The BCs are plotted in Figure 3.4.

Measurements in one photometric system can sometimes be converted to another system. This must be based upon extensive observational calibrations or upon detailed calculations using specimen spectral distributions. In either case, the procedure is second best to obtaining the data directly in the required photometric system and requires great care in its application.

3.1.3 Stellar Parameters

The usual purpose of making measurements of stars in the various photometric systems is to determine some aspect of the star's spectral behaviour by a simpler and more rapid method than that of actually obtaining the spectrum. The simplest approach to highlight the desired information is to calculate one or more colour indices. The colour index is just the difference between the star's magnitudes through two different filters, such as the B and V filters of the standard UBV system. The colour index, C, is then just

$$C = B - V \tag{3.19}$$

where B and V are the magnitudes through the B and V filters, respectively. Similar colour indices may be obtained for other photometric systems such as 439–555 for the HST system and g′–r′, for the SDSS. Colour indices for the *uvby* intermediate band system are discussed below. It should be noted

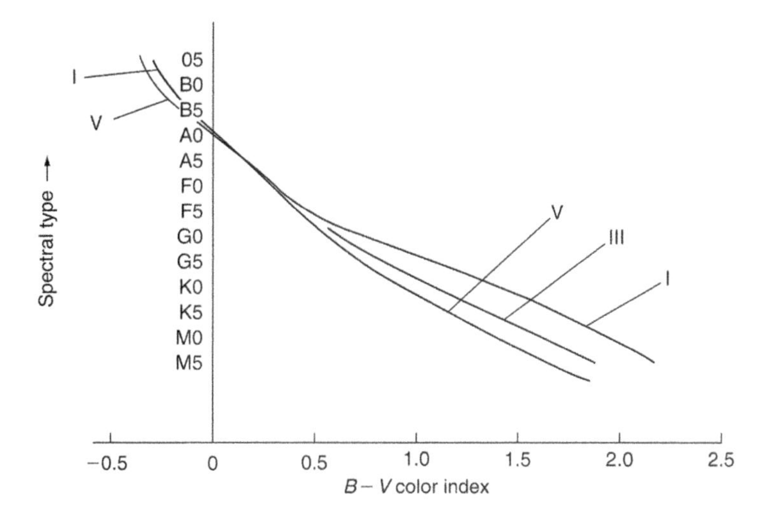

FIGURE 3.5 Relationship between spectral type and *B–V* colour index.

though that a fairly common alternative usage is for *C* to denote the so-called international colour index, which is based upon the photographic and photovisual magnitudes. The inter-relationship is

$$m_p - m_{pv} = C = B - V - 0.11 \tag{3.20}$$

The *B–V* colour index (for example) is then closely related to the spectral type (Figure 3.5) with an almost linear relationship for main sequence stars. This arises from the dependence of both spectral type and colour index upon temperature. For most stars, the *B* and *V* regions are located on the long wavelength side of the maximum spectral intensity. In this part of the spectrum, the intensity then varies approximately in a black body fashion for many stars and we get (semi-empirically):

$$T = \frac{8{,}540}{(B - V) + 0.865} \quad K \tag{3.21}$$

At higher temperatures, the relationship breaks down and the complete observed relationship is shown in Figure 3.6. Similar relationships may be found for other photometric systems which have filters near the wavelengths of the standard *B* and *V* filters. For example, the relationship between spectral type and the *b–y* colour index of the *uvby* system is shown in Figure 3.7.

For filters that differ from the *B* and *V* filters, the relationship of colour index with spectral type or temperature may be much more complex. In many cases the colour index is a measure of some other feature of the spectrum and the temperature relation is of little use or interest. The *U–B* index for the standard system is an example of such a case because the *U* and *B* responses bracket the Balmer discontinuity (Figure 3.8). The extent of the Balmer discontinuity is measured by a parameter, *D*, defined by

$$D = \log_{10}\left(\frac{I_{364+}}{I_{364-}}\right) \tag{3.22}$$

where I_{364+} is the spectral intensity at wavelengths just longer than the Balmer discontinuity (which is at or near 364 nm) and I_{364-} is the spectral intensity at wavelengths just shorter than the Balmer discontinuity.

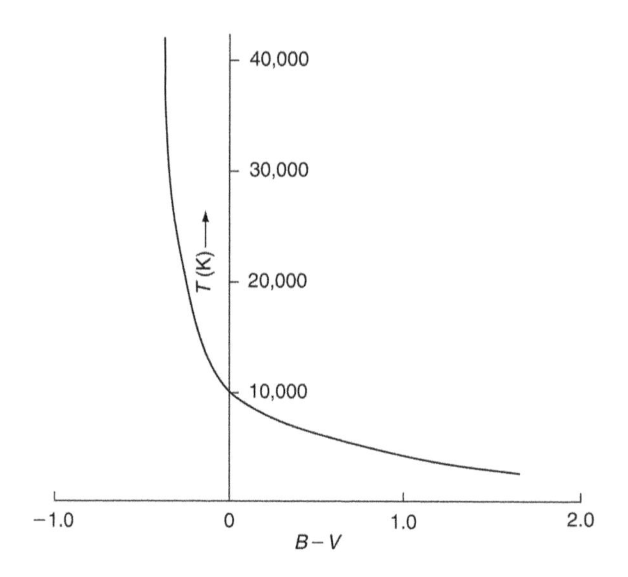

FIGURE 3.6 Observed $B-V$ versus T relationship for the whole of the main sequence.

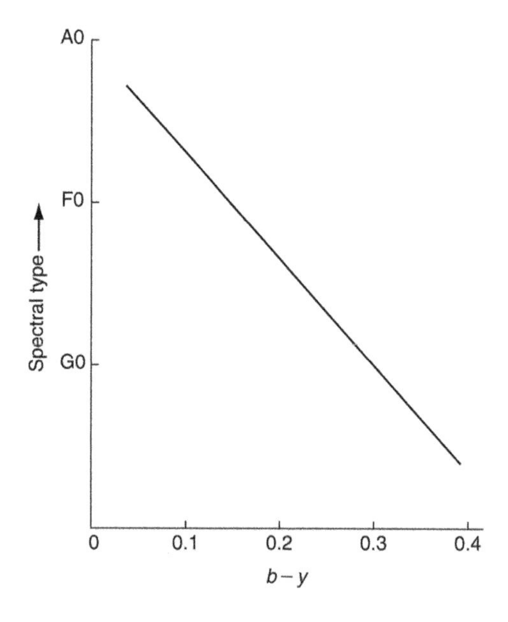

FIGURE 3.7 Relationship between spectral type and $b-y$ colour index.

The variation of D with the U–B colour index is shown in Figure 3.9. The relationship is complicated by the overlap of the B filter with the Balmer discontinuity and the variation of the effective position of that discontinuity with spectral and luminosity class. The discontinuity reaches a maximum at about $A0$ for main sequence stars and at about $F0$ for supergiants. It almost vanishes for hot stars and for stars cooler than about $G0$, corresponding to the cases of too high and too low temperatures for the existence of significant populations with electrons in the hydrogen $n = 2$ level. The colour index is also affected by the changing absorption coefficient of the negative hydrogen ion (Figure 3.10) and by line blanketing in the

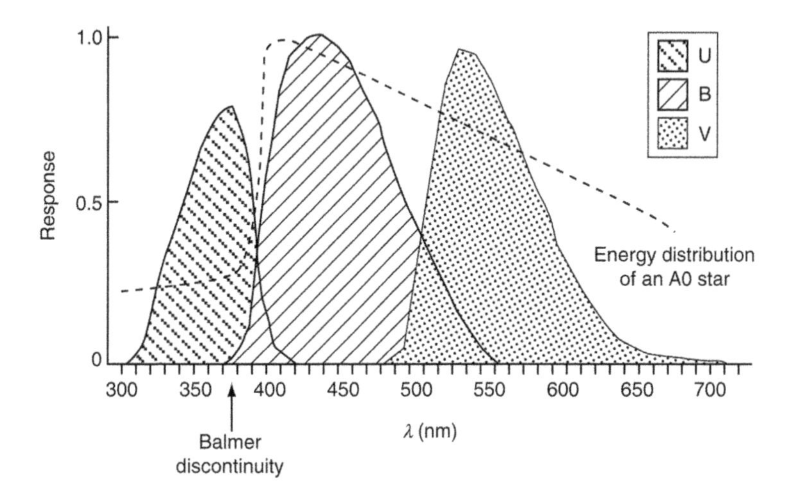

FIGURE 3.8 Position of the standard Ultraviolet, Blue, Visible (UBV) filters with respect to the Balmer discontinuity.

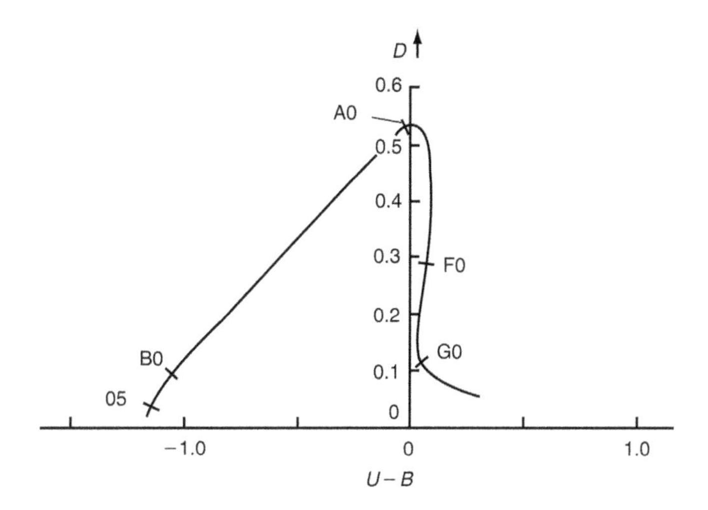

FIGURE 3.9 Variation of D with U–B for main sequence stars.

later spectral types. A similar relationship may be obtained for any pair of filters that bracket the Balmer discontinuity and another commonly used index is the c_1 index of the *uvby* system

$$c_1 = u + b - 2v \tag{3.23}$$

The filters are narrower and do not overlap the discontinuity, leading to a simpler relationship (Figure 3.11), but the effects of line blanketing must still be taken into account.

Thus, the B–V colour index is primarily a measure of stellar temperature, while the U–B index is a more complex function of both luminosity and temperature. A plot of one against the other provides a useful classification system analogous to the Hertzsprung-Russell diagram. It is commonly called the 'colour-colour diagram' and is shown in Figure 3.12. The deviations of the curves from that of a black body arise from the effects just mentioned: Balmer discontinuity, line blanketing, negative hydrogen ion absorption coefficient variation and also because the radiation originates over a region of the stellar atmosphere rather than in a single layer characterised by a single

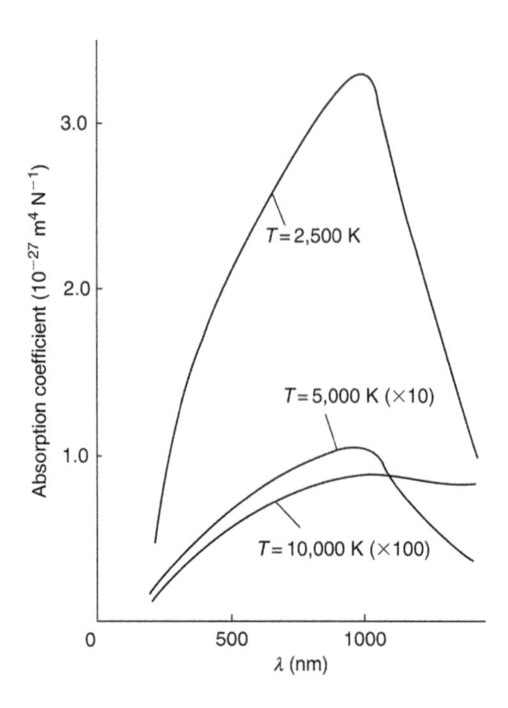

FIGURE 3.10 Change in the absorption coefficient of the negative hydrogen ion with temperature.

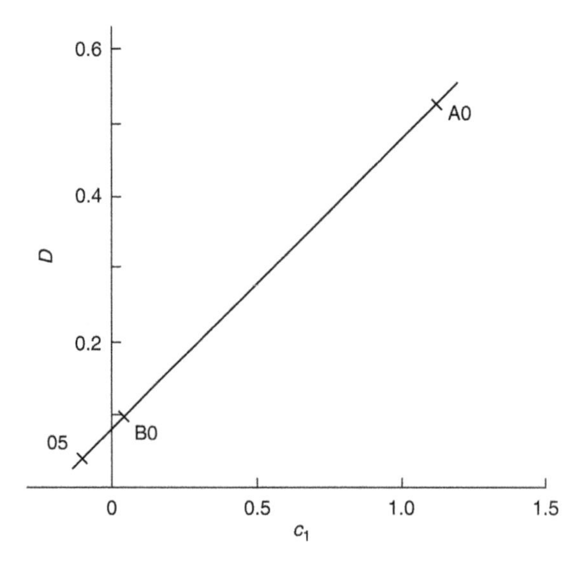

FIGURE 3.11 Variation of D with the c_1 index of the *uvby* system for main sequence stars.

temperature. The latter effect is due to limb darkening and because the thermalisation length (the distance between successive absorptions and re-emissions) normally corresponds to a scattering optical depth[9] many times unity. The final spectral distribution, therefore, contains contributions from regions ranging from perhaps many thousands of optical depths below the visible surface to well above the photosphere and so cannot be assigned a single temperature.

[9] A measure of the attenuation of a semi-transparent material. A radiation source at an optical depth of unity will have 63% (= $1-e^{-1}$) of its radiation absorbed before it arrives at the observer.

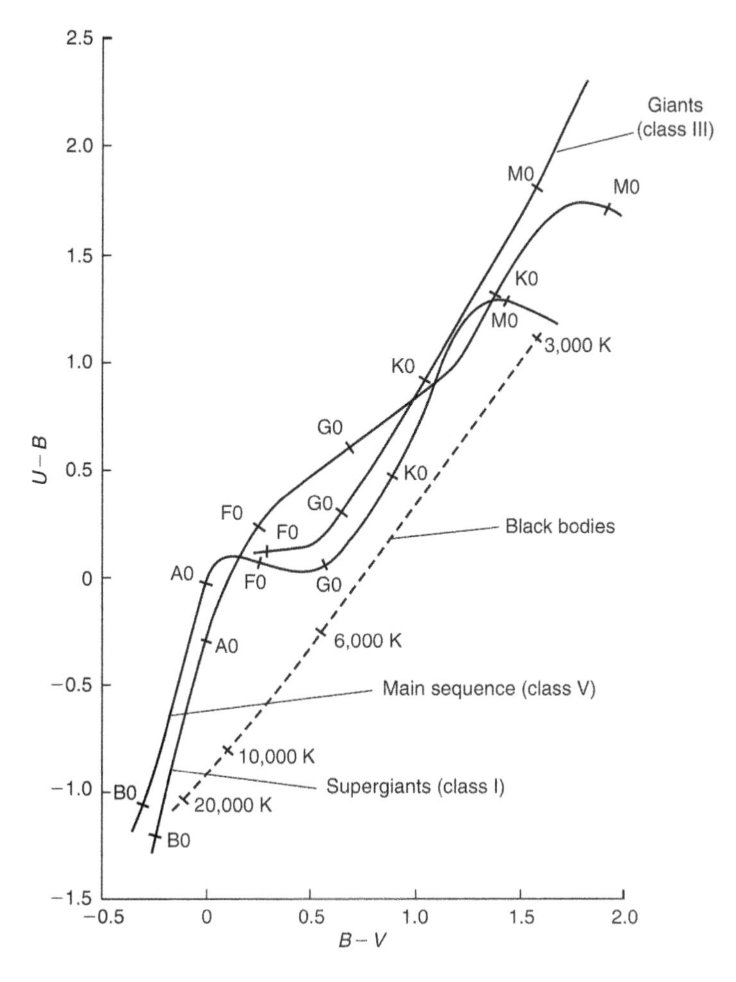

FIGURE 3.12 (*U–B*) versus (*B–V*) colour-colour diagram.

Figure 3.12 is based upon measurements of nearby stars. More distant stars are affected by interstellar absorption and because this is strongly inversely dependent upon wavelength (Figure 3.13), the *U–B* and the *B–V* colour indices are altered. The star's spectrum is progressively weakened at shorter wavelengths and so the process is often called 'interstellar reddening'.[10] In some cases, the degree of interstellar reddening can be used to determine the star's distance.

Similar or analogous relationships may be set up for other filter systems – Figures 3.14–3.16 for example.

By combining the photometry with model atmosphere work, much additional information such as stellar effective temperatures and surface gravities may be deduced. Thus, we have potentially a high return of information for a small amount of observational effort and the reader may see why the relatively crude methods of UBV and other wide band filter's photometry still remain popular.

[10] This is not the same as the redshift observed for distant galaxies. That results from the galaxies' velocities away from us and changes the observed wavelengths of the spectrum lines. With interstellar reddening the spectrum lines' wavelengths are unchanged.

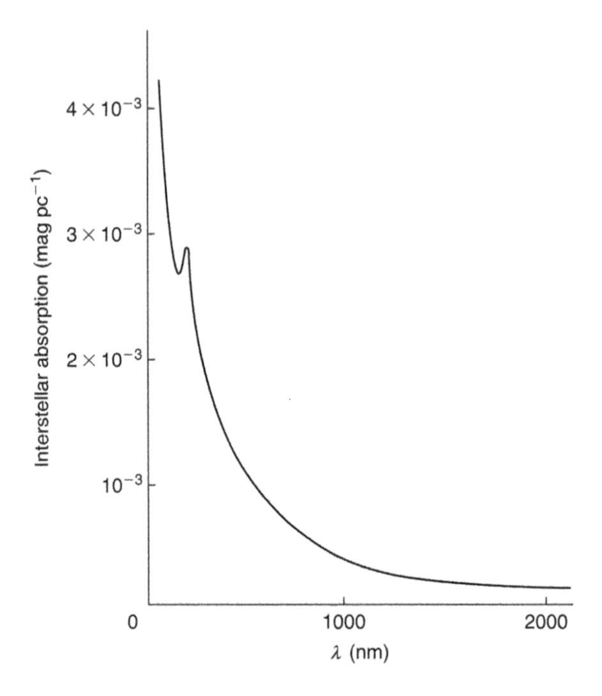

FIGURE 3.13 Average interstellar absorption.

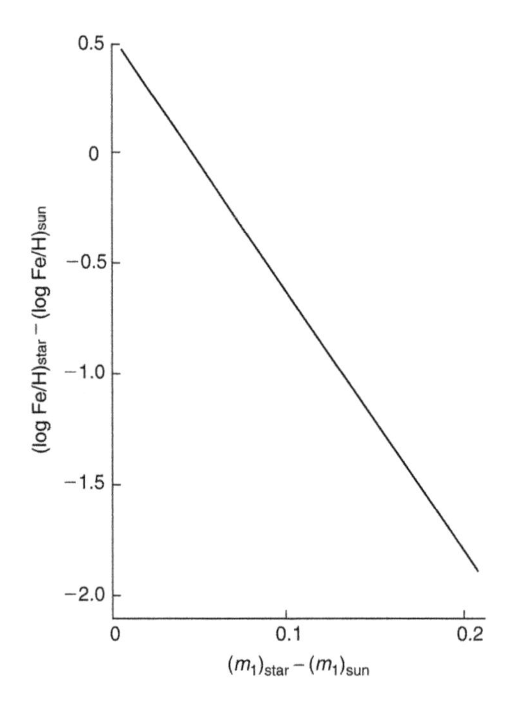

FIGURE 3.14 Relationship between metallicity index and iron abundance for solar type stars.

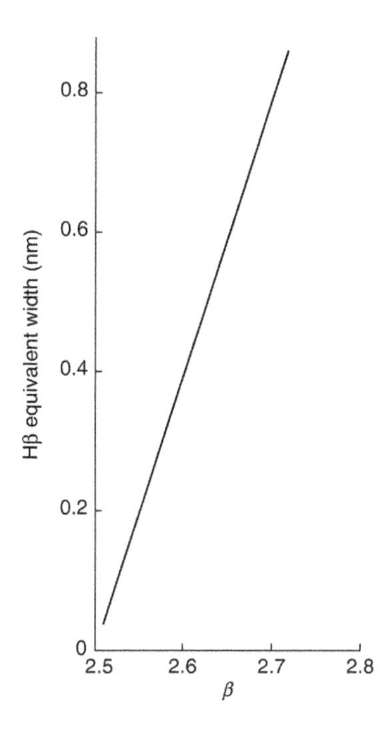

FIGURE 3.15 Relationship between β and the equivalent width of *H*β. Note the β is the β_n–β_w colour index of the *uvby* system.

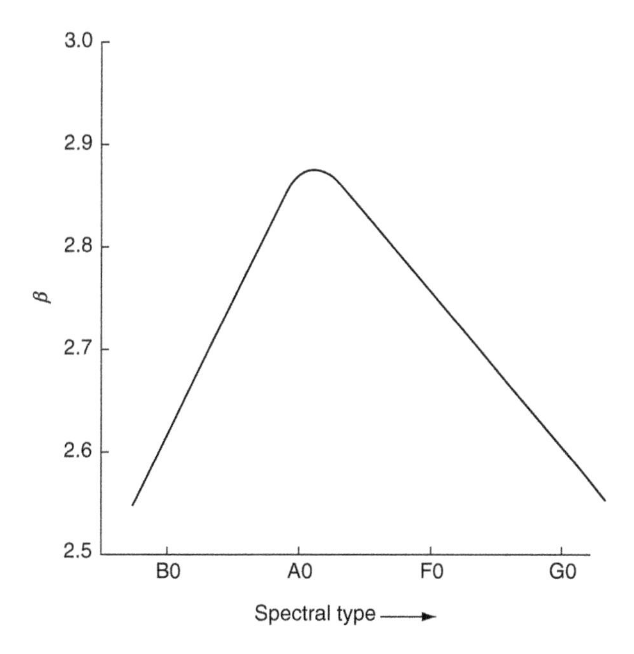

FIGURE 3.16 Relationship of β with spectral type.

3.2 PHOTOMETERS

3.2.1 INSTRUMENTS

3.2.1.1 Introduction

We may distinguish three basic methods of measuring stellar brightness, based upon the eye, the photographic plate or upon a variety of photoelectric devices. Only the latter are now of any significance, although a few amateur astronomers may still make eye estimates of brightness as part of long-term monitoring of variable stars. Forthcoming developments like the Vera C. Rubin Observatory (VRO; see Section 1.7), which will monitor tens of millions of stars every few days, will soon make even this usage obsolete. Photographic magnitudes may be encountered when working with archive material. The Hubble space telescope second Guide Star Catalogue (GSC 2) containing details of 500 million objects, for example, are based upon photographic plates obtained by the 1.2-m Schmidt cameras at Mount Palomar and the Anglo-Australian Observatory (AAO). Precision photometry, however, is now exclusively undertaken by means of photoelectric devices, with the CCD predominating in the optical region and solid-state devices in the infrared.

3.2.1.2 Photographic Photometry

As remarked in Section 2.2, this topic is now of little significance for a practising astronomer and interested readers are referred to historical sources for information.

3.2.1.3 CCD and Photoelectric Photometers

The most precise photometry relies upon photoelectric devices of various types now usually, in the form of arrays. These are primarily the CCD and other solid-state detectors (Section 1.1) in the visual and NIR regions and infrared array detectors (Section 1.1) at longer wavelengths. The PMTs continue to be used occasionally, as do p-i-n photodiodes in small photometers aimed at the amateur market. Photometry with a CCD differs little from ordinary imaging, except that several images of the same area usually need to be obtained through appropriate filters (Section 3.1). Anti-blooming CCDs should be avoided because their response is non-linear for objects nearing saturation. The normal data reduction techniques such as dark signal subtraction and flat fielding (Section 1.1) need to be applied to the images.

The brightness of a star is obtained directly by adding together the intensities of the pixels covered by the star image and subtracting the intensities in a similar number of pixels covering a nearby area of sky background. The HST, for example, undertakes aperture photometry by summing the intensities of all the pixels within a small circular area (aperture) that is slightly larger than the image and which is centred on it. The pixels in an annulus surrounding that circle (assuming no other images are contained within that annulus) are then also added together and then averaged to find the average background intensity per pixel. That average, multiplied by the number of pixels within the main image's circle, is then subtracted from the circle's summed intensity to give the background-subtracted intensity of the image. Contrary to most astronomical practice, the image may be slightly de-focussed for aperture photometry (or the beam spread by a diffusing screen[11]), so that more pixels are involved in the measurements and hence, the background noise level is (relatively) reduced. Small CCDs sold for the amateur market normally have software supplied by the manufacturer to enable stars' brightness to be found. This is accomplished by summing the intensities in a square of pixels that the user centres onto the stars' images one at a time. The size of the square is usually adjustable to take account of changes to the image sizes under different seeing conditions. Clearly, the same size of square is needed to obtain the background reading.

[11] These may now be fabricated with micro-structured surfaces which also shape the beam in some desired fashion.

Conversion to stellar magnitudes then requires the brightness of several standard stars also to be measured. With mosaics of several large area CCDs, there are likely to be several standard stars on any image. However, if this is not the case, or with detectors covering smaller areas of the sky, one or more separate calibration images will be needed. These are as identical to the main image as possible – as close in time as possible, as close in altitude and azimuth as possible, identical exposures, filters and data processing, etc.

Images from larger CCDs are usually processed automatically with general-purpose software (Section 2.9) or by using specialised programs written for the individual telescope or photometer or detector. Only when star's images overlap may the observer need to intervene in this process. Integrated magnitudes for extended objects such as galaxies are obtained in a similar fashion, though more input from the observer may be required to ensure that the object is delineated correctly before the pixel intensities are added together. From a good site under good observing conditions, photometry to an accuracy of $\pm 0.001^{m}$ is now achievable with many instruments.

At NIR and MIR wavelengths, the procedure is similar to that for CCDs. But in the long wavelength infrared regions, array detectors are still fairly small (though getting bigger – Section 1.1) and so only one or two objects are likely to be on each image – making separate calibration exposures essential. In many cases, the object being observed is much fainter than the background noise. For the MIR and FIR regions, special observing techniques such as chopping rapidly between the source and the background, subtracting the background from the signal and integrating the result over a long period need to be used. Some telescopes designed specifically for infrared work have secondary mirrors that can be oscillated to achieve this switching.

The PMTs (Section 1.1) continue to be used when individual photons need to be detected as in the neutrino and cosmic ray Čerenkov detectors (Sections 1.3 and 1.4) or when rapid responses are required as in the observation of occultations (Section 2.7 and below). They may also be used on board spacecraft for UV measurements in the 10- to 300-nm region where CCDs are insensitive.

3.2.2 Observing Techniques

Probably, the single most important 'technique' for successful ground-based photometry lies in the selection of the observing site (Section 1.1.24). Only the clearest and most consistent of skies are suitable for precision photometry. Haze, dust, clouds, excessive scintillation, light pollution, and so on and the variations in these, all render a site unsuitable for photometry. Furthermore, for infrared work, the amount of water vapour above the site should be as low as possible. For these reasons good photometric observing sites are rare. Generally, they are at high altitudes and are located where the weather is particularly stable. Oceanic islands and mountain ranges with a prevailing wind from an ocean, with the site above the inversion layer, are fairly typical of the best choices. Sites that are less than ideal can still be used for photometry, but their number of good nights will be reduced. Restricting the observations to near the zenith is likely to improve the results obtained at a mediocre observing site.

The second most vital part of photometry lies in the selection of the comparison star(s). This/these must be non-variable, close to the star of interest, of similar apparent magnitude[12] and spectral class and have its/their own magnitude(s) known reliably on the photometric system that is in use. Amongst the brighter stars there is usually some difficulty in finding a suitable standard that is close enough. With fainter stars, the likelihood that a suitable comparison star exists is higher, but the chances of its details being known are much lower. Thus, ideal comparison stars are found only rarely.

[12] An old technique, when the star of interest and its comparison *do* differ significantly in brightness and which may still be used occasionally today, is to place a mesh or grating over the telescope's aperture. This converts each stellar image into a short interference pattern. The first-order image (say) of the brighter star may then be compared with the zero-order image of the fainter star, so reducing the contrast.

The best known variables already have lists of good comparison stars, which may generally be found from the literature. But for less well-studied stars and those being investigated for the first time, there is no such useful information to hand. It may even be necessary to undertake extensive prior investigations, such as checking back through archive records to find non-variable stars in the region and then measuring these to obtain their magnitudes before studying the star itself. An additional star (or stars) may also need to be observed several times throughout an observing session to supply the data for the correction of atmospheric absorption (see below).

A single observatory can clearly only observe objects for a fraction of the time, even when there are no clouds. Rapidly changing objects however may need to be monitored continuously over 24 hours. A number of observatories distributed around the globe have, therefore, instituted various cooperative programmes so that such continuous photometric observations can be made. The Whole Earth Blazar Telescope (WEBT), for example, is a consortium of observatories studying compact quasars from visible light to radio waves with 24-hours coverage (see also Birmingham Solar Oscillations Network [BiSON] and Global Oscillation Network Group [GONG], Section 5.3). Similarly, the Liverpool telescope on La Palma and the two Faulkes telescopes in Hawaii and Australia are 2-m robotic telescopes that can respond to gamma-ray burst (GRB) alerts in less than 5 minutes.

3.2.3 Data Reduction and Analysis

The reduction of the data is performed in three (at least) stages – correction for the effects of the Earth's atmosphere, correction to a standard photometric system and correction to heliocentric time.

The atmosphere absorbs and reddens the star's light and its effects are expressed by Bouguer's law

$$m_{\lambda,0} = m_{\lambda,z} - a_\lambda \sec z \tag{3.24}$$

where $m_{\lambda,z}$ is the magnitude at wavelength λ and at zenith distance z. a_λ is a constant that depends on λ. The law is accurate for zenith distances up to about 60°, which is usually sufficient – because photometry is rarely carried out on stars whose zenith distances are greater than 45°. Correction for atmospheric extinction may be simply carried out once the value of the extinction coefficient a_λ is known. Unfortunately, a_λ varies from one observing site to another, with the time of year, from day to day and even throughout the night. Thus, its value must be found on every observing occasion. This is done by observing a standard star at several different zenith distances and by plotting its observed brightness against sec z. Now, for zenith angles less than 60° or 70°, sec z, to a good approximation, is just a measure of the air mass along the line of sight, so that we may reduce the observations to above the atmosphere by extrapolating them back to an air mass of zero (ignoring the question of the meaning of a value of sec z of zero – Figure 3.17). For the same standard star, this brightness should always be the same on all nights, so that an additional point, which is the average of all the previous determinations of the above-atmosphere magnitude of the star, is available to add to the observations on any given night. The extinction coefficient is simply obtained from the slope of the line in Figure 3.17. The coefficient is strongly wavelength dependent (Figure 3.18) and so it must be determined separately for every filter that is being used. Once the extinction coefficient has been found, the above-atmosphere magnitude of the star, m_λ, is given by

$$m_\lambda = m_{\lambda,z} - a_\lambda (1 + \sec z) \tag{3.25}$$

Thus, the observations of the star of interest and its standards must all be multiplied by a factor k_λ,

$$k_\lambda = 10^{0.4 a_\lambda (1 + \sec z)} \tag{3.26}$$

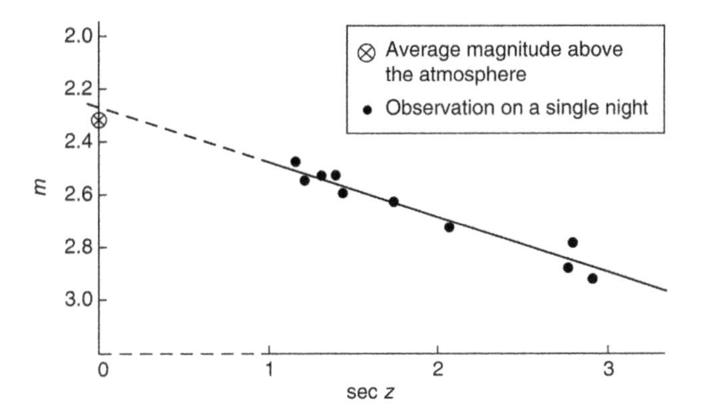

FIGURE 3.17 Schematic variation in magnitude of a standard star with zenith distance.

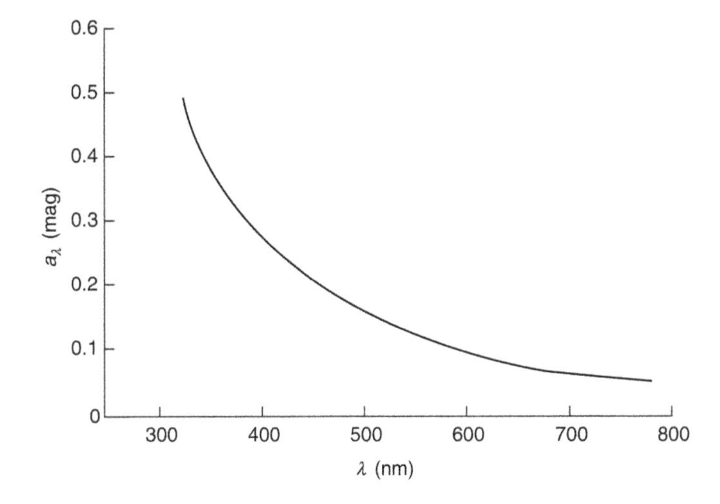

FIGURE 3.18 A typical dependence of the extinction coefficient with wavelength for a good observing site.

to correct them to their unabsorbed values. When the unknown star and its comparisons are close together in the sky, the differential extinction will be negligible and this correction need not be applied. But the separation must be small; for example, at a zenith distance of 45°, the star and its comparisons must be within 10′ of each other if the differential extinction is to be less than a thousandth of a magnitude. If E_λ and E'_λ are the original average signals for the star and its comparison, respectively, through the filter centred on λ then the corrected magnitude is given by

$$m_\lambda = m'_\lambda - 2.5\log_{10}\left(\frac{E_\lambda 10^{0.4a_\lambda(1+\sec z)}}{E'_\lambda 10^{0.4a_\lambda(1+\sec z')}}\right) \tag{3.27}$$

$$= m'_\lambda + a_\lambda\left(\sec z' - \sec z\right) - 2.5\log_{10}\left(\frac{E_\lambda}{E'_\lambda}\right) \tag{3.28}$$

where m_λ and m'_λ are the magnitudes of the unknown star and its comparison respectively and z and z' are similarly their zenith distances. The zenith angle is given by

$$\cos z = \sin\phi \sin\delta + \cos\phi \cos\delta \cos(\text{LST} - \alpha) \tag{3.29}$$

where ϕ is the latitude of the observatory, α and δ are the right ascension and declination of the star and LST is the local sidereal time of the observation.

If the photometer is working in one of the standard photometric systems, then the magnitudes obtained through Equation 3.28 may be used directly and colour indices, colour excesses and so on obtained as discussed in Section 3.1. Often, however, the filters or the detector may not be of the standard type. Then the magnitudes must be corrected to a standard system. This can only be attempted if the difference is small because absorption lines, ionisation edges and so on will make any large correction exceedingly complex. The required correction is best determined empirically by observing a wide range of the standard stars of the photometric system. Suitable correction curves may then be plotted from the known magnitudes of these stars.

Finally, and especially for variable stars, the time of the observation should be expressed in heliocentric Julian days. The geocentric Julian date may be found in the *Astronomical Almanac*[13] for the year and if the time of the observation is expressed on this scale, then the heliocentric time is obtained by correcting for the travel time of the light to the Sun,

$$T_{\text{Sun}} = T + 5.757 \times 10^{-3} \left[\sin\delta_* \sin\delta_{\text{Sun}} - \cos\delta_* \cos\delta_{\text{Sun}} \cos(\alpha_{\text{Sun}} - \alpha_*)\right] \tag{3.30}$$

where T is the actual time of observation in geocentric Julian days, T_{Sun} is the time referred to the Sun, α_{Sun} and δ_{Sun} are the right ascension and declination of the Sun at the time of the observation and α_* and δ_* are the right ascension and declination of the star. For precise work, the varying distance of the Earth from the Sun may need to be taken into account as well.

Further items in the reduction and analysis of the data such as the corrections for interstellar reddening, calculation of colour index, temperature and so on were covered in Section 3.1.

3.2.4 High-Speed Photometry

Measurements of objects' brightness at intervals of a millisecond or so are necessary for observing occultations (Section 2.7) and for real-time atmospheric compensation (Section 1.1). There are also many astrophysical processes, such as phenomena within pulsars and X-ray binary systems, surface oscillations on white dwarfs, solar and stellar flares and so on where observations at millisecond to microsecond intervals may be used profitably.

Almost any photometer based upon a detector with a short response time can be used to observe phenomena needing time resolutions in the 0.1–10 seconds range. As the required time resolution decreases below 0.1 seconds, however, the read-out times of array-type detectors start to limit photometers' responses. Also, the number of photons available to the detector decreases directly with the time resolution, and so shot and other noise sources become more significant. Thus, for time resolutions in the region of 1 μs to 1 ms, special adaptations and/or specially built instruments are needed. In particular only the few pixels containing the image of interest within an array detector may be read-out and/or large telescopes are needed to gather sufficient light. A recent proposal for high speed photometry incorporates both these requirements. It is to use the dishes of γ-ray air Čerenkov detectors, such as Major Atmospheric Gamma Imaging Čerenkov (MAGIC) and

[13] Published annually by HMNAO and the USNO (Her Majesty's Nautical Almanac Office and the US Naval Observatory), jointly.

Very Energetic Radiation Imaging Telescope Array System (VERITAS; see Sections 1.3 and 1.4) for the detection of optical transients. With VERITAS, each of the (four) dishes has between 2 and 16 PMTs, which can operate at kHz speeds and which have already been used to monitor Fast Radio Bursters' optical emissions.

Mercator Advanced Imager for Asteroseismology (MAIA), on the semi-robotic 1.2-m Mercator telescope located on La Palma, is a specially designed high-speed photometer primarily intended for astroseismology. It operates over three spectral regions (UV, green, and red) and uses high-speed frame-transfer CCDs cooled to 165 K as its detectors. The CCD arrays are 2 × 6 k pixels in size, with a 2 × 3-k imaging area and a similar-sized storage area. The frame transfer process takes about 300 ms, with a full read-out time of about 30 seconds. The HiPERCAM on the Gran Telescopio Canarias (GTC) has a cadence of more than 1,000 frames per second and also uses cooled frame-transfer CCDs. The CCDs have a 1,028 × 2,048-pixel imaging area and, for increased speed, two 512 × 2,048-pixel storage areas on either side of the imaging area. It uses four reflective dichroic beam splitters and covers five wavebands over the 300-nm to 1-μm region.

Flashes on the lunar surface arising from meteorite impacts were monitored by the purpose-built Near Earth Object Lunar Impacts and Optical Transits (NELIOTA) instrument (2017–2018). This used two cameras with Complementary Metal Oxide Superconductor (CMOS) detectors, observing in the R and I wavebands. It operated on the Greek 1.2-m Kryoneri telescope, obtaining images at 30 frames per second. Its field of view was 14.4 × 17′ (about one-third of the Full Moon's area), and it detected meteorite impacts about once a fortnight.

3.2.5 Exoplanets

The first planet orbiting a star other than the Sun (i.e., an exoplanet) was discovered in 1995.[14] The discovery was made by detecting changes in the radial velocity of the star as the planet orbited around it (see Chapter 4). Since then several thousand exoplanets have been found, the majority (~80%) being discovered via brightness changes (photometry) of the star/exoplanet system. Exoplanets may be detected via brightness variations arising from several different physical processes: transits (76.4%), transit timings (0.5%), eclipse timings (0.4%), orbital modulations (0.1%), and gravitational microlensing (2.1%).

When an exoplanet transits its host star as seen from the Earth, the brightness of that star decreases slightly because a small portion of its (bright) surface is blocked by the (dark) silhouette of the planet. When a star is already known to host an exoplanet, timings of the transits of that planet may sometimes be early and sometimes late due to another planet's gravitational field pulling the observed exoplanet forwards or backwards. The second exoplanet is thus discovered through timings of the transits of the first exoplanet. In a similar manner, an exoplanet in an eclipsing binary star system may affect the stellar eclipse times. The gravitational tides from a massive close-in exoplanet (of which there are many) may distort the shape of its host star, so that the projected area of the star, as seen from the Earth, changes as the two bodies orbit each other (orbital modulation).

Gravitational microlensing arises because gravity affects the paths of light beams passing near to the surface of a star. For a light (or radio, microwave, IR, UV, X-ray or γ-ray) beam just skimming the surface of the Sun, the deflection is about 1.7″, and this deflection decreases as the distance of the light beam from the Sun's surface increases. Thus, when we have a star hosting an exoplanet that, as seen from the Earth, is in front of a more distant star, the light from the latter is bent by the gravitational field of the former. Because both stars will be in relative motion with respect to the Earth, the gravitational lensing effect changes on a timescale of a few days to a few tens of days.

[14] Or 1992, if you accept that the incinerated remnants of 'something' (probably a small star) orbiting an un-vapourized/re-condensed supernova remnant constitute a normal planetary system.

The observed effect is that the brightness of the more distant star increases (sometimes by a factor of ten or more), as the nearer star passes between it and the Earth. If the nearer star has an exoplanet, then the gravitational field of the exoplanet will also distort the light variations of the distant star; adding brief, sharp brightening or dimming to the 'normal' light curve.[15] Microlensing is the only (current) technique capable of detecting exoplanets outside the Milky Way galaxy. In this way, in 2018, several free-floating exoplanets were found some 1–2 Gpc away in the quasar system RX J1131-1231, which is gravitationally lensed by a closer elliptical galaxy.

To detect an exoplanet through the brightness variations arising from any of these processes, stars' brightness must be measured to an accuracy of around one part in a thousand ($\pm 0.001^m$). Furthermore, because the chances of two stars lining up, as seen from the Earth, sufficiently closely for gravitational lensing to be significant are small, thousands of stars must be monitored for the brightness changes for months or years at a time. However, more than 3,000 exoplanets *have* been discovered though photometric measurements, so clearly these highly stringent observational requirements *are* being met.

Recently, amongst the large number of exoplanet photometric programmes, there has been NASA's spacecraft; Kepler (Figure 3.19). This was launched in 2009 into a Sun-centred orbit and it monitored the brightness of around 150,000 stars in a patch of the sky between Deneb (α Cyg) and Vega (α Lyr) to an accuracy of $\pm 0.002^m$ using a 0.95-m Schmidt camera (primary mirror; 1.4-m) and an array detector employing nearly 10^8 pixels. It added around 2,600 exoplanets to the list of discoveries and was deactivated in 2018, when its fuel finally ran out.

Transiting Exoplanet Survey Satellite (TESS) was launched just 6 months after Kepler ceased operating and is designed to follow on from Kepler's work. It is hoped to discover around another 20,000 exoplanets via transit observations. TESS observes, rather in contrast to Kepler's large Schmidt camera, with four 0.1-m aperture cameras (Section 1.1.19.2). The cameras, however, monitor a 24° × 96° area of the sky; an area some 400 times larger than that available to Kepler. Furthermore, unlike Kepler's observation of a single fixed area of the sky, TESS will cover most of the sky in 26 sets of observing campaigns, each lasting for 27.4 days (two orbits). The detectors (for each camera) are 4-million-pixel CCD arrays. The spacecraft is in a trans-lunar orbit which

FIGURE 3.19 The photometric exoplanet discoverer spacecraft, Kepler – An artist's impression of its appearance in space. N.B. The background exoplanetary system is completely imaginary – Kepler remained in orbit around the Sun; it did not travel to other planetary systems. (Courtesy of NASA, the Kepler mission and Wendy Stenzel.)

[15] For further details of these processes see the author's *Exoplanets: Finding, Exploring and Understanding Alien Worlds* (2012).

keeps it well above the Van-Allen belts, away from the Moon and which should be stable (i.e., TESS should use little manoeuvring fuel) for 20 or more years.

At the time of writing, Characterizing Extrasolar Planets by Opto-infrared Polarimetry and Spectroscopy (CHEOPS) has just been launched successfully. It will use a 0.3-m Ritchey-Chretien telescope with the primary mission of studying already known exoplanets in more detail. The Planetary Transits and Oscillations of the Stars (PLATO) mission is planned for a 2026 launch with the aim of monitoring a million stars for small (rocky) exoplanets. It is expected to use some twenty-four 0.12-m diameter cameras, each with 1,100 deg^2 fields of view and with 80 megapixel arrays as detectors.

There are many ongoing ground-based programmes aimed at discovering and studying exoplanets. Here, there is only room to choose (probably unfairly!) a couple to mention to give an idea of what is being done.

The WASP is based upon 0.11-m cameras. There are two robotic instruments, each with eight cameras and located in the Northern and Southern Hemispheres, respectively. The cameras use commercial telephoto lenses feeding CCD detectors and image up to a million stars every minute. Nearly, 200 exoplanets have been discovered from their transits to date.

Optical Gravitational Lensing Experiment (OGLE) uses the 1.3-m Warsaw telescope feeding thirty-two 2×4 k pixel CCDs giving it a 1.4 deg^2 total field of view. It monitors the Milky Way's galactic bulge and the Magellanic clouds for gravitational lensing events. Nearly, 60 exoplanets have been found by the programme at the time of writing, 50 planets were found by micro-lensing (including two possibly free-floating objects), and 8 planets were found from their transits.

4 Spectroscopy

4.1 SPECTROSCOPY

4.1.1 Introduction

Practical spectroscopes are usually based upon one or other of two quite separate optical principles – interference and differential refraction. The former produces instruments based upon diffraction gratings or interferometers, while the latter results in prism-based spectroscopes. There are also some hybrid designs. The details of the spectroscopes themselves are considered in the next section; here we discuss the basic optical principles which underlie their designs.

4.1.2 Diffraction Gratings

The operating principle of diffraction gratings relies upon the effects of diffraction and interference of light waves. Logically, therefore, it might seem that they should be included within the section on interference-based spectroscopes. Diffraction gratings, however, are in such widespread and common use in astronomy, that they merit a section to themselves.

We have already seen in Section 2.5 the structure of the image of a single source viewed through two apertures (Figure 2.10). The angular distance of the first fringe from the central maximum is λ/d, where d is the separation of the apertures. Hence, the position of the first fringe and, of course, the positions of all the other fringes, is a function of wavelength. If such a pair of apertures were to be illuminated with white light, all the fringes apart from the central maximum would thus be short spectra with the longer wavelengths furthest from the central maximum. In an image such as that of Figure 2.10, the spectra would be of little use because the fringes are so broad that they would overlap each other long before a useful dispersion could be obtained. However, if we add a third aperture in line with the first two and separated from the nearer of the original apertures by a distance, d, again, then we find that the fringes remain stationary but become narrower and more intense. Weak secondary maxima also appear between the main fringes (Figure 4.1). The peak intensities are, of course, modulated by the pattern from a single slit when looked at on a larger scale, in the manner of Figure 2.10. If further apertures are added in line with the first three and with the same separations, then the principal fringes continue to narrow and intensify and further weak maxima appear between them (Figure 4.2). The intensity of the pattern at some angle, θ, to the optical axis is given by (cf. Equation 1.12):

$$I(\theta) = I(0)\left[\frac{\sin^2\left(\frac{\pi D \sin\theta}{\lambda}\right)}{\left(\frac{\pi D \sin\theta}{\lambda}\right)^2}\right]\left[\frac{\sin^2\left(\frac{N\pi d \sin\theta}{\lambda}\right)}{\sin^2\left(\frac{\pi d \sin\theta}{\lambda}\right)}\right] \tag{4.1}$$

where D is the width of the aperture and N is the number of apertures. The term

$$\left[\frac{\sin^2\left(\frac{\pi D \sin\theta}{\lambda}\right)}{\left(\frac{\pi D \sin\theta}{\lambda}\right)^2}\right] \tag{4.2}$$

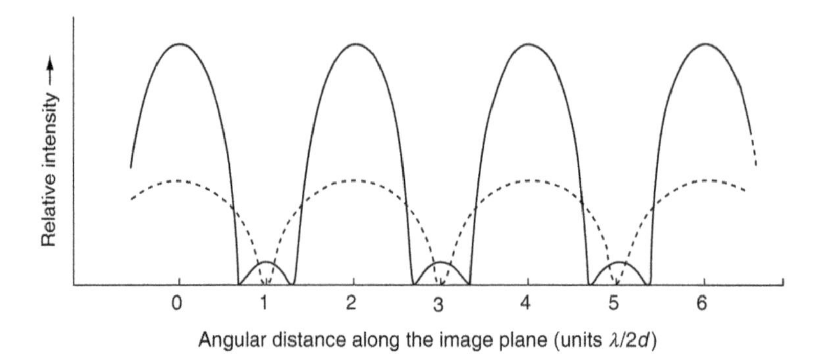

FIGURE 4.1 Small portion of the image structure for a single-point source viewed through two apertures (broken curve) and three apertures (full curve).

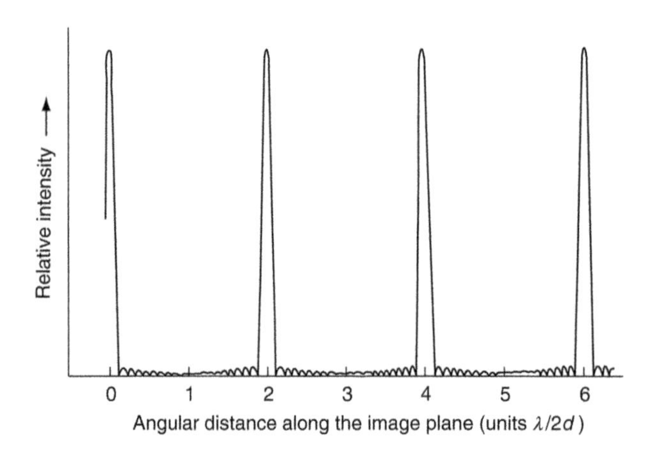

FIGURE 4.2 Small portion of the image structure for a single-point source viewed through 20 apertures.

represents the modulation of the image by the intensity structure for a single aperture, while the term

$$\left[\frac{\sin^2\left(\dfrac{N\pi d\sin\theta}{\lambda}\right)}{\sin^2\left(\dfrac{\pi d\sin\theta}{\lambda}\right)}\right] \tag{4.3}$$

represents the result of the interference between N apertures. We may write

$$\Delta = \left(\frac{\pi D\sin\theta}{\lambda}\right) \tag{4.4}$$

and

$$\delta = \left(\frac{\pi d\sin\theta}{\lambda}\right) \tag{4.5}$$

and Equation 4.1 then becomes

$$I(\theta) = I(0)\frac{\sin^2 \Delta}{\Delta^2}\frac{\sin^2(N\delta)}{\sin^2 \delta} \tag{4.6}$$

Now consider the interference component as δ tends to $m\pi$, where m is an integer. Substituting

$$P = \delta - m\pi, \tag{4.7}$$

we have

$$\lim_{\delta \to m\pi}\left(\frac{\sin(N\delta)}{\sin\delta}\right) = \lim_{P\to 0}\left(\frac{\sin[N(P+m\pi)]}{\sin(P+m\pi)}\right) \tag{4.8}$$

$$= \lim_{P\to 0}\left(\frac{\sin(NP)\cos(Nm\pi)+\cos(NP)\sin(Nm\pi)}{\sin P\cos(m\pi)+\cos P\sin(m\pi)}\right) \tag{4.9}$$

$$= \lim_{P\to 0}\left(\pm\frac{\sin(NP)}{\sin P}\right) \tag{4.10}$$

$$= \pm N \lim_{P\to 0}\left(\frac{\sin(NP)}{NP}\,\frac{P}{\sin P}\right) \tag{4.11}$$

$$= \pm N \tag{4.12}$$

Hence, integer multiples of π give the values of δ for which we have a principal fringe maximum. The angular positions of the principal maxima are given by

$$\theta = \sin^{-1}\left(\frac{m\lambda}{d}\right) \tag{4.13}$$

and m is usually called the 'order of the fringe'. The zero intensities in the fringe pattern will be given by

$$N\delta = m'\pi \tag{4.14}$$

where m' is an integer, but excluding the cases where $m' = mN$ that are the principal fringe maxima. Their positions are given by

$$\theta = \sin^{-1}\left(\frac{m'\lambda}{Nd}\right) \tag{4.15}$$

The angular width of a principal maximum, W, between the first zeros on either side of it is thus given by

$$W = \frac{2\lambda}{Nd\cos\theta} \tag{4.16}$$

The width of a fringe is, therefore, inversely proportional to the number of apertures, while its peak intensity, from Equations 4.6 and 4.12, is proportional to the square of the number of apertures. Thus, for a bichromatic source observed through a number of apertures we obtain the type

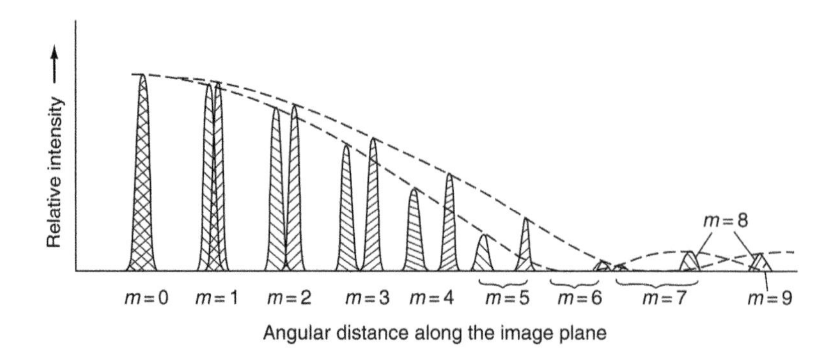

FIGURE 4.3 A portion of the image structure for a single bichromatic point source viewed through several apertures.

of image structure shown in Figure 4.3. The angular separation of fringes of the same order for the two wavelengths, for small values of θ, can be seen from Equation 4.13 to be proportional to both the wavelength and to the order of the fringe, while the fringe width is independent of the order (Equation 4.16). For a white light source, by a simple extension of Figure 4.3, we may see that the image will consist of a series of spectra on either side of a white central image. The Rayleigh resolution within this image is obtained from Equation 4.16

$$W' = \frac{\lambda}{Nd\cos\theta} \tag{4.17}$$

and is independent of the fringe order. The ability of a spectroscope to separate two wavelengths is called the 'spectral resolution' and is denoted by W_λ, and it may now be found from Equation 4.17

$$W_\lambda = W'\frac{d\lambda}{d\theta} \tag{4.18}$$

but from Equation 4.13

$$\frac{d\lambda}{d\theta} = \frac{d}{m}\cos\theta \tag{4.19}$$

so that

$$W_\lambda = \frac{\lambda}{Nm} \tag{4.20}$$

The spectral resolution thus improves directly with the fringe order because of the increasing dispersion of the spectra.

More commonly, the resolution is expressed as the ratio of the operating wavelength to the spectral resolution and denoted by R (often and confusingly also called the 'spectral resolution'), and it is given by

$$R = \frac{\lambda}{W_\lambda} = Nm \tag{4.21}$$

The resolution for a series of apertures is thus just the product of the number of apertures and the order of the spectrum. It is independent of the width and spacing of the apertures.

From Figure 4.3 we may see that at higher orders the spectra are overlapping. This occurs at all orders when white light is used. The difference in wavelength between two superimposed wavelengths from adjacent spectral orders is called the 'free spectral range', Σ. From Equation 4.13 we may see that if λ_1 and λ_2 are two such superimposed wavelengths, then

$$\sin^{-1}\left(\frac{m\lambda_1}{d}\right) = \sin^{-1}\left(\frac{(m+1)\lambda_2}{d}\right) \tag{4.22}$$

that is, for small angles

$$\Sigma = \lambda_1 - \lambda_2 \approx \frac{\lambda_2}{m} \tag{4.23}$$

For small values of m, Σ is, therefore, large and the unwanted wavelengths in a practical spectroscope may be rejected by the use of filters. Some spectroscopes, such as those based on Fabry-Perot etalons and echelle gratings, however, operate at high spectral orders, and *both* of the overlapping wavebands may be needed. Then, it is necessary to use a cross disperser, so that the final spectrum consists of a two-dimensional (2D) array of short sections of the spectrum (see Figure 4.24 and the discussion later in this section).

A practical device for producing spectra by diffraction uses a large number of closely spaced, parallel, narrow slits or grooves and is called a 'diffraction grating'. Typical gratings for astronomical use have between 100 and 1,000 grooves per millimetre and 1,000–50,000 grooves in total. They are used at orders ranging from one up to 200 or so. Thus, the spectral resolutions range from 10^2 to 10^5. Although the previous discussion was based upon the use of clear apertures, a narrow plane mirror can replace each aperture without altering the results. So, diffraction gratings can be used either in transmission or reflection modes. Most astronomical spectroscopes are in fact based upon reflection gratings. Often the grating is inclined to the incoming beam of light, but this changes the discussion only marginally. There is a constant term, $d \sin i$, added to the path differences, where i is the angle made by the incoming beam with the normal to the grating. The whole image (Figure 4.3) is shifted at an angular distance i along the image plane. Equation 4.13 then becomes

$$\theta = \sin^{-1}\left[\left(\frac{m\lambda}{d}\right) - \sin i\right] \tag{4.24}$$

and in this form is often called the 'grating equation'.

Volume phase holographic gratings (VPHGs) are now used within many astronomical spectroscopes. These have a grating in the form of a layer of gelatine within which the refractive index changes (Section 4.2), with the lines of the grating produced by regions of differing refractive indices. The VPHGs operate through Bragg diffraction (Section 1.3, Figure 1.97 and Equation 1.85). Their efficiencies can thus be up to 95% in the first order. They can be used either as transmission or reflection gratings and replace conventional gratings in spectroscopes at the appropriate Bragg angle for the operating wavelength.

To form a part of a spectroscope, a grating must be combined with other optical elements. The basic layout is shown in Figure 4.4; practical designs are discussed in Section 4.2. The grating is illuminated by parallel light that is usually obtained by placing a slit at the focus of a collimating lens but sometimes may be obtained simply by allowing the light from a distant object to fall directly onto the grating. After reflection from the grating, the light is focused by the imaging lens to form the required spectrum, and this may then be recorded, observed through an eyepiece, projected onto a screen and so on as desired. The collimator and imaging lenses may be simple lenses as shown, in which case the spectrum will be tilted with respect to the optical axis because of chromatic aberration, or they may be achromats or mirrors (see the next section).

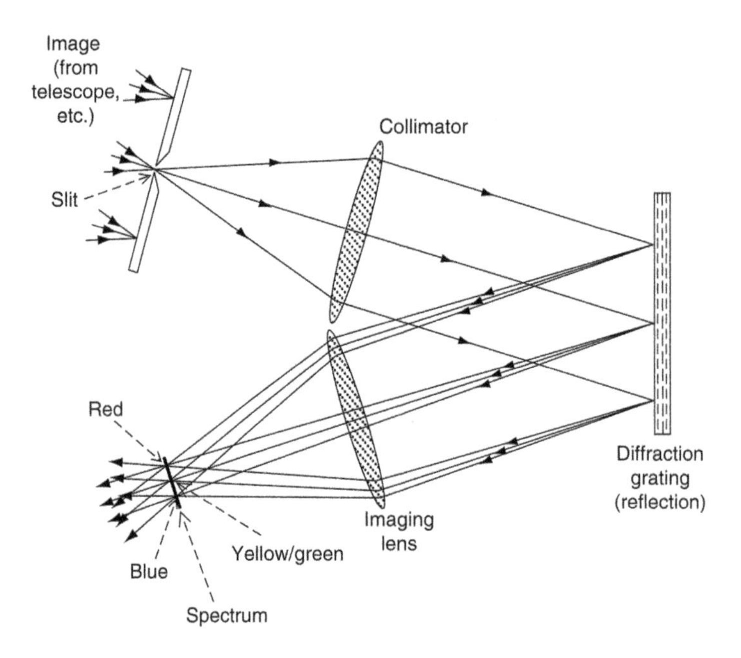

FIGURE 4.4 Basic optical arrangement of a reflection grating spectroscope.

The angular dispersion of a grating is not normally used as a parameter of a spectroscopic system. Instead it is combined with the focal length of the imaging element to give either the linear dispersion or the reciprocal linear dispersion. If x is the linear distance along the spectrum from some reference point, then we have for an achromatic imaging element of focal length, f_2,

$$\frac{dx}{d\lambda} = f_2 \frac{d\theta}{d\lambda} \tag{4.25}$$

where θ is small and is measured in radians. From Equation 4.24, the linear dispersion within each spectrum is thus given by

$$\frac{dx}{d\lambda} = \pm \frac{mf_2}{d\cos\theta} \tag{4.26}$$

Now, because θ varies little over an individual spectrum, we may write

$$\frac{dx}{d\lambda} \approx \text{constant} \tag{4.27}$$

The dispersion of a grating spectroscope is thus roughly constant compared with the strong wavelength dependence of a prism spectroscope (see below). More commonly, the reciprocal linear dispersion, $\frac{d\lambda}{dx}$, is quoted and used. For practical astronomical spectrometers, this usually has values in the range

$$10^{-7} < \frac{d\lambda}{dx} < 5 \times 10^{-5} \tag{4.28}$$

The commonly used units are nanometres change of wavelength per millimetre along the spectrum so that the above range is from 0.1 to 50 nm mm^{-1}. The use of Å mm^{-1} is still fairly common practice amongst astronomers; the magnitude of the dispersion is then a factor of ten larger than the standard measure. The advent of electronic detectors has also made the use of nm or Å per pixel a measure of dispersion.

The resolving power of a spectroscope is limited by the spectral resolution of the grating, the resolving power of its optics (Section 1.1) and by the projected slit width. The spectrum is formed from an infinite number of monochromatic images of the entrance slit. It is easy to see that the width of one of these images, S, is given by

$$S = s\frac{f_2}{f_1} \tag{4.29}$$

where s is the slit width, f_1 is the collimator's focal length, and f_2 is the imaging element's focal length. In wavelength terms, the slit width, $S\frac{d\lambda}{dx}$, is sometimes called the 'spectral purity of the spectroscope'. The entrance slit must have a physical width of s_{max} or less, if it is not to degrade the spectral resolution, where

$$s_{max} = \frac{\lambda f_1}{Nd\cos\theta} \tag{4.30}$$

If the optics of the spectroscope are well corrected then we may ignore their aberrations and consider only the diffraction limit of the system. When the grating is fully illuminated, the imaging element will intercept a rectangular beam of light. The height of the beam is just the height of the grating and has no effect upon the spectral resolution. The width of the beam, D, is given by

$$D = L\cos\theta \tag{4.31}$$

where L is the length of the grating and θ the angle of the exit beam to the normal to the plane of the grating. The diffraction limit is just that of a rectangular slit of width D. So that from Figure 1.33, the linear Rayleigh limit of resolution, W'', is given by

$$W'' = \frac{f_2\lambda}{D} \tag{4.32}$$

$$= \frac{f_2\lambda}{L\cos\theta} \tag{4.33}$$

If the beam is limited by some other element of the optical system and/or is of circular cross section, then D must be evaluated as may be appropriate or the Rayleigh criterion for the resolution through a circular aperture (Equation 1.43) used in place of that for a rectangular aperture. Optimum resolution occurs when

$$S = W'' \tag{4.34}$$

i.e.,

$$s = \frac{f_1\lambda}{D} \tag{4.35}$$

$$= \frac{f_1\lambda}{L\cos\theta} \tag{4.36}$$

The major disadvantage of a grating as a dispersing element is immediately obvious from Figure 4.3; the light from the original source is spread over a large number of spectra. The grating's efficiency

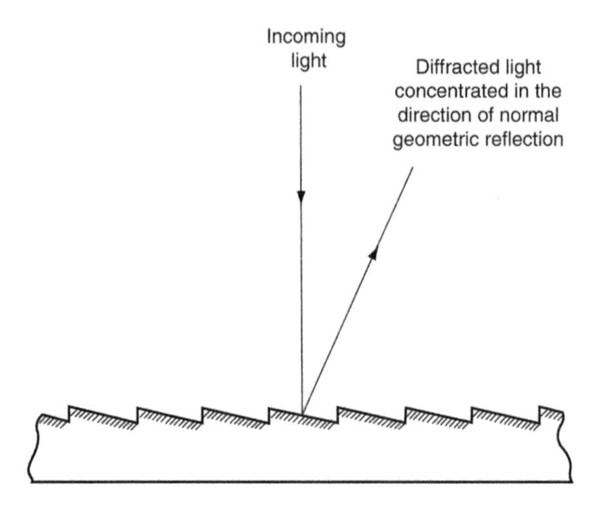

FIGURE 4.5 Enlarged section through a blazed reflection grating.

in terms of the fraction of light concentrated into the spectrum of interest is, therefore, low. This disadvantage, however, may be largely overcome with reflection gratings through the use of the technique of blazing the grating. In this technique, the individual mirrors that comprise the grating are angled so that they concentrate the light into a single narrow solid angle (Figure 4.5). For instruments based upon the use of gratings at low orders, the angle of the mirrors is arranged so that the light is concentrated into the spectrum to be used and by this means up to 90% efficiency can be achieved. In terms of the interference patterns, the grating is designed so that central peak as a result of an individual aperture just extends over the width of the desired spectrum. The blaze angle then shifts that peak along the array of spectra until it coincides with the desired order.

For those spectroscopes that use gratings at high orders, the grating can still be blazed, but then the light is concentrated into short segments of many different orders of spectra. By a careful choice of parameters, these short segments can be arranged so that they overlap slightly at their ends and so continuous coverage of a much wider spectral region may be obtained by producing a montage of the segments.

Transmission gratings can also be blazed although this is less common. Each of the grooves then has the cross section of a small prism, the apex angle of which defines the blaze angle. Blazed transmission gratings for use at infrared wavelengths can be produced by etching the surface of a block of silicon in a similar manner to the way in which integrated circuits are produced.

Another problem that is intrinsically less serious but which is harder to counteract is that of shadowing. If the incident and/or reflected light makes a large angle to the normal to the grating, then the step-like nature of the surface (Figure 4.5) will cause a significant fraction of the light to be intercepted by the vertical portions of the grooves and so lost to the final spectrum. There is little that can be done to eliminate this problem except either to accept the light loss or to design the system so that large angles of incidence or reflection are not needed.

Curved reflection gratings are frequently produced. By making the curve that of an optical surface, the grating itself can be made to fulfil the function of the collimator and/or the imaging element of the spectroscope, thus reducing light losses and making for greater simplicity of design and reduced costs. The grooves should be ruled so that they appear parallel and equally spaced when viewed from infinity. The simplest optical principle employing a curved grating is that due to Henry Rowland. The slit, grating, and spectrum all lie on a single circle which is called the 'Rowland circle' (Figure 4.6). This has a diameter equal to the radius of curvature of the grating. The use of a curved grating at large angle to its optical axis introduces astigmatism and spectral lines may also be curved due to the varying angles of incidence for rays from the centre and ends of the slit.

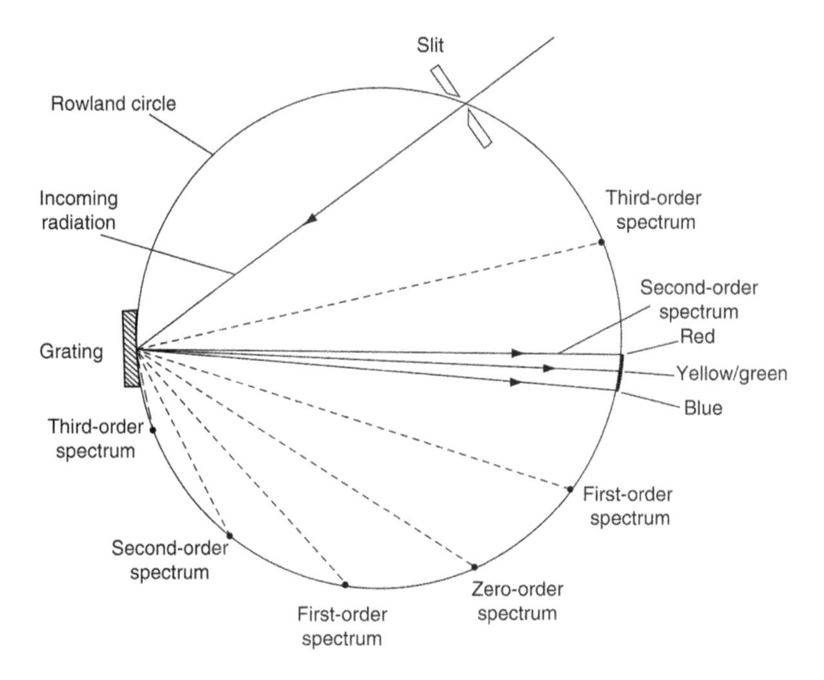

FIGURE 4.6 Schematic diagram of a spectroscope based upon a Rowland circle, using a curved grating blazed for the second order.

Careful design, however, can reduce or eliminate these defects, and there are several practical designs for spectroscopes based upon the Rowland circle (Section 4.2). Aspherical curved gratings are also possible and can be used to provide highly corrected designs with few optical components.

Higher dispersion can be obtained by using immersed reflection gratings. The light in these interacts with the grating within a medium other than air (or a vacuum). The refractive index of the medium shortens the wavelength so that the groove spacing is effectively increased relative to the wavelength and so the dispersion is augmented (Equation 4.19). One approach to producing an immersion grating is simply to illuminate the grating from the back (i.e., through the transparent substrate). A second approach is to flood the grating with a thin layer of oil kept in place by a cover sheet. For example, the Giant Magellan Telescope (GMT) is expected to use silicon immersion gratings in its first-generation near infrared (NIR) spectrograph and to achieve spectral resolutions from 65,000 to 85,000 over several wavebands covering the 1.0- to 5.3-μm region.

A grating spectrum generally suffers from unwanted additional features superimposed upon the desired spectrum. Such features are usually much fainter than the main spectrum and are called 'ghosts'. They arise from a variety of causes. They may be due to overlapping spectra from higher or lower orders or to the secondary maxima associated with each principal maximum (Figure 4.2). The first of these is usually simple to eliminate by the use of filters because the overlapping ghosts are of different wavelengths from the overlapped main spectrum. The second source is usually unimportant since the secondary maxima are faint when more than a few tens of apertures are used, though they still contribute to the wings of the point spread function (PSF).

Of more general importance are the ghosts that arise through errors in the grating. Such errors most commonly take the form of a periodic variation in the groove spacing. A variation with a single period gives rise to Rowland ghosts which appear as faint lines close to and on either side of strong spectrum lines. Their intensity is proportional to the square of the order of the spectrum. Echelle gratings (see below), therefore, must be of high quality because they may use spectral orders of several hundred. If the error is multi-periodic, then Lyman ghosts of strong lines may appear. These are similar to the Rowland ghosts, except that they can be formed at large distances from the line that is producing them. Some compensation for these errors can be obtained through deconvolution of the PSF (Section 2.1), but for critical work, the only real solution is to use a grating without periodic errors, such as a holographically produced grating.

Wood's anomalies may also sometimes occur. These do not arise through grating faults but are the result of light that should go into spectral orders behind the grating (were that to be possible) reappearing within lower order spectra. The anomalies have a sudden onset and a slower decline towards longer wavelengths and are almost 100% plane polarized. They are rarely important in efficiently blazed gratings.

By increasing the angle of a blazed grating, we obtain an echelle grating (Figure 4.7). This is illuminated more or less normally to the groove surfaces and, therefore, at a large angle to the normal to the grating. It is usually a coarse grating – ten lines per millimetre is not uncommon – so that the separation of the apertures, d, is large. The reciprocal linear dispersion

$$\frac{d\lambda}{dx} = \pm \frac{d \cos^3 \theta}{m f_2} \tag{4.37}$$

is, therefore, also large. Such gratings concentrate the light into many overlapping high-order spectra and so from Equation 4.21, the resolution is high. A spectroscope that is based upon an echelle grating requires a second low dispersion grating or prism, called a 'cross-disperser', whose dispersion is perpendicular to that of the echelle to separate out each of the orders (Section 4.2).

A quantity known variously as 'throughput', 'etendu', or 'light gathering power', is useful as a measure of the efficiency of the optical system. It is the amount of energy passed by the system when its entrance aperture is illuminated by unit intensity per unit area per unit solid angle and it is denoted by u

$$u = \tau A \Omega \tag{4.38}$$

where Ω is the solid angle accepted by the instrument, A is the area of its aperture and τ is the fractional transmission of its optics, incorporating losses due to scattering, absorption, imperfect reflection, etc. For a spectroscope, Ω is the solid angle subtended by the entrance slit at the collimator or, for slitless spectroscopes, the solid angle accepted by the telescope-spectroscope combination. A is the area of the collimator or the effective area of the dispersing element, whichever is the smaller, τ will depend critically upon the design of the system, but as a reasonably general rule it may be taken to be the product of the transmissions for all the surfaces. These will usually be in the region of 0.8–0.9 for each surface, so that τ for the design illustrated in Figure 4.4 will have a value of about 0.4. Older spectroscope designs often had low throughputs – less than 10% was not uncommon. Even today, a throughput of 40% such as has been achieved by the bench-mounted High Resolution Optical Spectrometer (bHROS; now decommissioned) instrument for the Gemini South telescope and NN-EXPLORE Exoplanet Investigations with Doppler spectroscopy (NEID) for the Wisconsin-Indiana-Yale-NOAO (WIYN) telescope is considered excellent. Thus, much light is lost

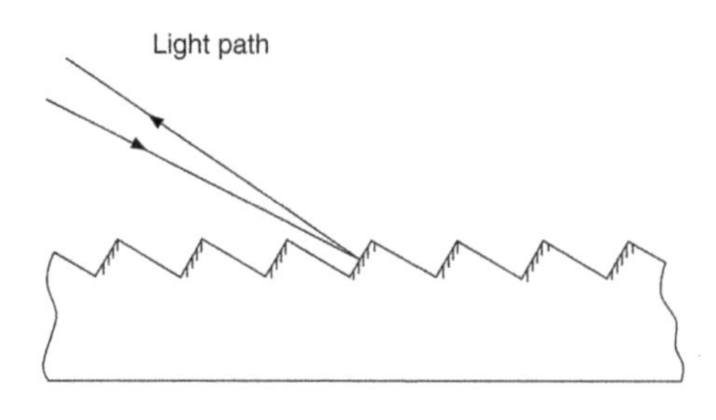

FIGURE 4.7 Enlarged view of an echelle grating.

in spectroscopy compared with direct imaging and because the remaining light is then spread out over the spectrum, the exposures needed for spectroscopy are typically a hundred or more times those required for imaging.

The product of resolution and throughput, P, is a useful figure for comparing the performances of different spectroscope systems

$$P = Ru \tag{4.39}$$

Normally, it will be found that, other things being equal, P will be largest for Fabry-Perot spectroscopes (see below), of intermediate values for grating-based spectroscopes and lowest for prism-based spectroscopes.

4.1.3 Prisms

Pure prism-based spectroscopes will rarely be encountered today, except within instruments constructed some time ago. An exception to that, though, is an option of the James Webb Space Telescope's (JWST's) Near InfraRed Spectrograph (NIRSpec) instrument, which will be a low-resolution double-pass prism-based NIR spectrograph. The prism is fabricated from CaF_2, will have a spectral resolution of 30–300 and cover the 600-nm to 5-μm wavelength range.

However, prisms *are* commonly used in conjunction with gratings in some modern instruments. The combination is known as a 'grism', and the deviation of the light beam by the prism is often used to counteract that of the grating, so that the light passes straight through the instrument. The spectroscope can then be used for direct imaging just by removing the grism and without having to move the camera. Prisms are also often used as cross-dispersers for high spectral order spectroscopes based upon echelle gratings or etalons and may be used non-spectroscopically for folding light beams.

In view of the now fairly restricted astronomical applications of prisms, only a brief outline of their optics is included here – more details may be found in previous editions of this book and within most sources on optics.

The basic physics underlying the production of a spectrum by a prism is differential refraction[1] (i.e., the refractive index of the prism's material changes as the radiation's wavelength changes). Thus, for a typical flint glass (often used for visible-region prisms), the refractive index (μ_λ) has values of about;

1.684 at 400 nm wavelength
1.661 at 500 nm wavelength
1.649 at 600 nm wavelength and
1.640 at 700 nm wavelength

If we now pass light beams symmetrically through a 60° prism made from such a glass, then the deviations (the angle between the incoming and outgoing beams) will be

36.5° at 400 nm wavelength
35.2° at 500 nm wavelength
34.5° at 600 nm wavelength and
34.0° at 700 nm wavelength

Thus, the blue beam (400 nm) will be deviated by the prism by 2.5° more than the red (700 nm) beam.

[1] This phenomenon was previously encountered in Section 1.1 as chromatic aberration, and there we were concerned with eliminating or minimising its effects; for spectroscopy, by contrast, we are interested, here, in maximising the dispersion.

Putting a parallel light beam which ranges continuously in colour from 400 to 700 nm through such a prism, followed by a 1-m focal length lens to image the outgoing radiation, we will then see a continuous blue to red spectrum on the screen, whose length is about 44 mm. Replace the screen with an array detector, and you have a prism-based spectrograph! In fact, for astronomical purposes, the prism will require slits, collimators and so on just like the grating (c.f. Figure 4.4 and see Section 4.2) to produce a useable instrument. There are many detailed designs for prism-based spectrographs, including ones similar to Figure 4.4, where the reflective diffraction grating would be replaced a by a 30°/60°/90° prism which has a reflective coating on its back (i.e., the beam goes twice through the prism making it the equivalent of a 60° prism; see NIRSpec above). Also, several prisms may be combined to increase the dispersion of the system.

Unlike the diffraction grating in normal use, where the dispersion is more or less constant with wavelength (Equation 4.27), the dispersion produced by a prism increases rapidly towards shorter wavelengths.

$$\frac{dx}{d\lambda} \propto \sim\lambda^{-2} \tag{4.40}$$

Thus, using the figures for flint glass above (and the 1-m focal length imaging lens), we have an average reciprocal linear dispersion of ~7 nm mm^{-1}, but this ranges from ~11.5 nm mm^{-1} at 650 to ~4.4 nm mm^{-1} at 450 nm.[2]

In a similar way to the grating spectroscope, the resolving power of a prism spectroscope is limited by the spectral resolution of the prism, the resolving power of its optics (Section 1.1) and by the projected slit width. The spectrum is formed from an infinite number of monochromatic images of the entrance slit. The width of one of these images, *S*, is again given by

$$S = s\frac{f_2}{f_1} \tag{4.41}$$

where s is the slit width, f_1 is the collimator's focal length and f_2 is the imaging element's focal length. If the optics of the spectroscope are well corrected, then we may ignore their aberrations and consider only the diffraction limit of the system. When the prism is fully illuminated, the imaging element will intercept a rectangular beam of light. The height of the beam is just the height of the prism and has no effect upon the spectral resolution. The width of the beam, *D*, is given by

$$D = L\left[1-\mu_\lambda^2 \sin^2\left(\frac{\alpha}{2}\right)\right]^{1/2} \tag{4.42}$$

where L is the length of a prism face and α its apex angle. The diffraction limit is then just that of a rectangular slit of width *D*. So that from Figure 1.33, the linear Rayleigh limit of resolution, W'', is given by

$$W'' = \frac{f_2\lambda}{D} \tag{4.43}$$

$$= \frac{f_2\lambda}{L\left[1-\mu_\lambda^2 \sin^2\left(\frac{\alpha}{2}\right)\right]^{1/2}} \tag{4.44}$$

[2] Remember this is *reciprocal* linear dispersion, so the *smaller* the figure, the *better* (higher), the dispersion.

If the beam is limited by some other element of the optical system and/or is of circular cross section, then D must be evaluated as may be appropriate, or the Rayleigh criterion for the resolution through a circular aperture (Equation 1.43) used in place of that for a rectangular aperture. Optimum resolution occurs (c.f. Equation 4.34 and previous derivation) when

$$S = W'' \tag{4.45}$$

i.e.,

$$s = \frac{f_1 \lambda}{D} \tag{4.46}$$

and the prism's spectral resolution is

$$R = \frac{\lambda}{W_\lambda} \tag{4.47}$$

For a dense flint prism with an apex angle of 60° and a side length of 0.1 m, we then obtain in the visible

$$R \approx 1.5 \times 10^4 \tag{4.48}$$

and this is a fairly typical value for the resolution of a prism-based spectroscope.

The resolution varies slightly across the width of the spectrum, unless cylindrical lenses or mirrors are used for the collimator, and these have disadvantages of their own. The variation arises because the light rays from the ends of the slit impinge on the first surface of the prism at an angle to the optical axis of the collimator. The peripheral rays, therefore, encounter a prism whose effective apex angle is larger than that for the paraxial rays. The deviation is also increased for such rays and so the ends of the spectrum lines are curved towards shorter wavelengths. Fortunately, astronomical spectra are mostly so narrow that both these effects can be neglected; the wider beams encountered in integral field and multi-object spectroscopes (Section 4.2), though, may be affected in this way.

The material used to form prisms depends upon the spectral region that is to be studied. In the visual region, the normal types of optical glass may be used, but these mostly start to absorb in the near ultraviolet (UV). Fused silica and crystalline quartz can be formed into prisms to extend the limit down to 200 nm. Crystalline quartz, however, is optically active (Section 5.2) and, therefore, must be used in the form of a Cornu prism. This has the optical axis parallel to the base of the prism so that the ordinary and extraordinary rays coincide and is made in two halves cemented together. The first half is formed from a right-handed crystal and the other half from a left-handed crystal; the deviations of left- and right-hand circularly polarized beams are then similar. If required, calcium fluoride or lithium fluoride can extend the limit down to 140 nm or so, but astronomical spectroscopes working at such short wavelengths are normally based upon gratings. In the infrared, quartz can again be used for wavelengths out to about 3.5 μm. Rock salt can be employed at even longer wavelengths, but it is extremely hygroscopic which makes it difficult to use. More commonly Fourier spectroscopy (see below) is adopted when high-resolution spectroscopy is required in the far infrared (FIR).

4.1.4 Interferometers

We only consider here in any detail two main types of spectroscopic interferometry: the Fabry-Perot interferometer or etalon and the Michelson interferometer or Fourier-transform spectrometer. Other systems exist but at present are of little importance for astronomy.

4.1.4.1 Fabry-Perot Interferometer

Two parallel, flat, partially reflecting surfaces are illuminated at an angle θ (Figure 4.8). The light undergoes a series of transmissions and reflections as shown, and pairs of adjoining emergent rays differ in their path lengths by ΔP, where

$$\Delta P = 2t \cos. \tag{4.49}$$

Constructive interference between the emerging rays will then occur at those wavelengths for which

$$\mu \, \Delta P = m\lambda \tag{4.50}$$

where m is an integer, i.e.,

$$\lambda = \frac{2t\mu\cos\theta}{m} \tag{4.51}$$

If such an interferometer is used in a spectroscope in place of the prism or grating (Figure 4.9), then the image of a point source is still a point. However, the image is formed from *only* those

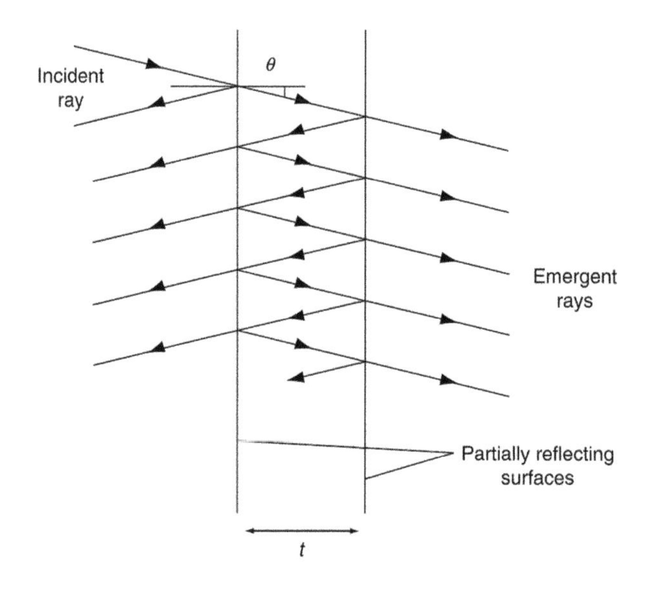

FIGURE 4.8 Optical paths in a Fabry-Perot interferometer.

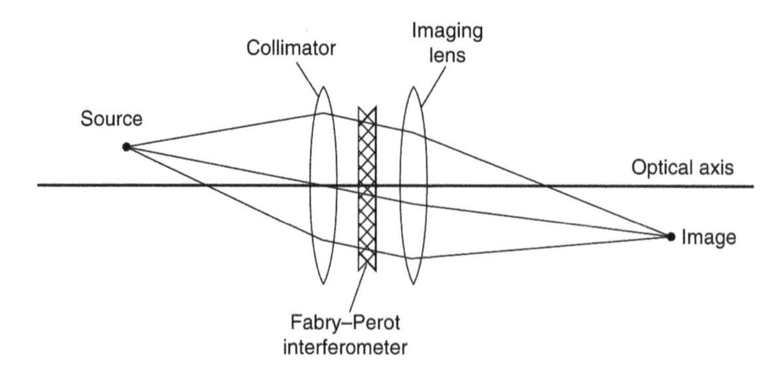

FIGURE 4.9 Optical paths in a Fabry-Perot spectroscope.

wavelengths for which Equation 4.51 holds. If the image is then fed into another spectroscope it will be broken up into a series of monochromatic images. If a slit is now used as the source, the rays from the different points along the slit will meet the etalon at differing angles and the image will then consist of a series of superimposed short spectra. If the second spectroscope is then set so that its dispersion is perpendicular to that of the etalon (a cross-disperser), then the final image will be a rectangular array of short parallel spectra. The widths of these spectra depend upon the reflectivity of the surfaces. For a high reflectivity, we get many multiple reflections, while with a low reflectivity the intensity becomes negligible after only a few reflections. The monochromatic images of a point source are, therefore, not truly monochromatic but are spread over a small wavelength range in a similar but not identical manner to the intensity distributions for several collinear apertures (Figures 4.1 and 4.2). The intensity distribution varies from that of the multiple apertures because the intensities of the emerging beams decrease as the number of reflections required for their production increases. Examples of the intensity distribution with wavelength are shown in Figure 4.10. In practice, reflectivities of about 90% are usually used.

In the absence of absorption, the emergent intensity at a fringe peak is equal to the incident intensity at that wavelength. This often seems a puzzle to readers, who on inspection of Figure 4.8, might expect there to be radiation emerging from the left as well as on the right, so that at the fringe peak, the total emergent intensity appears to be greater than the incident intensity. If we examine the situation more closely, however, we find that when at a fringe peak for the light emerging on the right, there is zero intensity in the beam emerging on the left. If the incident beam has intensity I and amplitude a ($I = a^2$), then the amplitudes of the successive beams on the left in Figure 4.8 are

$$-aR^{1/2},\ aR^{1/2}T,\ aR^{3/2}T,\ aR^{5/2}T,\ aR^{7/2}T, \ldots$$

where the first amplitude is negative because it results from an internal reflection (it has therefore an additional phase delay of 180° compared with the other reflected beams). T is the fractional intensity transmitted and R the fractional intensity reflected by the reflecting surfaces (note that $T + R = 1$ in the absence of absorption). Summing these terms (assumed to go on to infinity) gives zero amplitude and therefore, zero intensity for the left-hand emergent beam. Similarly, the beams emerging on the right have amplitudes

$$aT, aTR, aTR^2, aTR^3, aTR^4, \ldots \tag{4.52}$$

Summing these terms to infinity gives, at a fringe peak, the amplitude on the right as 'a'. This is the amplitude of the incident beam, and so at a fringe maximum the emergent intensity on the right equals the incident intensity (see also Equation 4.60).

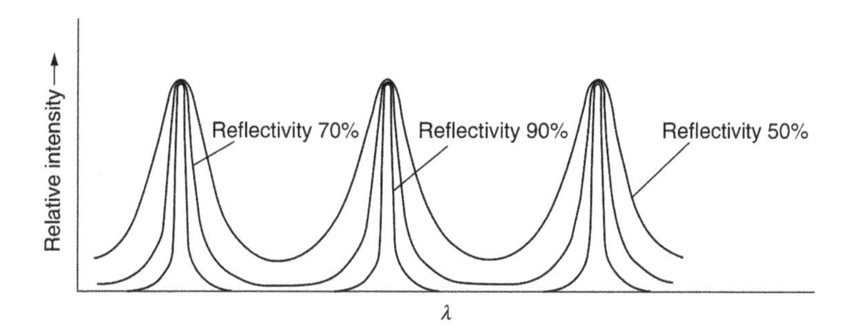

FIGURE 4.10 Intensity versus wavelength in the image of a white light point source in a Fabry-Perot spectroscope, assuming negligible absorption.

The dispersion of an etalon may easily be found by differentiating Equation 4.51

$$\frac{d\lambda}{d\theta} = -\frac{2t\mu}{m}\sin\theta \tag{4.53}$$

Because the material between the reflecting surfaces is usually air and the etalon is used at small angles of inclination, we have

$$\mu \approx 1 \tag{4.54}$$

$$\sin\theta \approx \theta \tag{4.55}$$

and from Equation 4.51

$$\frac{2t}{m} \approx \lambda \tag{4.56}$$

so that

$$\frac{d\lambda}{d\theta} \approx \lambda\theta \tag{4.57}$$

Thus, the reciprocal linear dispersion for a typical system, with $\theta = 0.1°$, $f_2 = 1$ m and used in the visible, is 0.001 nm mm^{-1}, which is a factor of 100 or so better than that achievable with more common dispersing elements.

The resolution of an etalon is rather more of a problem to estimate. Our usual measure – the Rayleigh criterion – is inapplicable because the minimum intensities between the fringe maxima do not reach zero, except for a reflectivity of 100%. However, if we consider the image of two equally bright point sources viewed through a telescope at its Rayleigh limit (Figure 2.7) then the central intensity is 81% of that of either of the peak intensities. We may, therefore, replace the Rayleigh criterion by the more general requirement that the central intensity of the envelope of the images of two equal sources falls to ~81% of the peak intensities. Consider, therefore, an etalon illuminated by a monochromatic slit source perpendicular to the optical axis (Figure 4.9). The image will be a strip, also perpendicular to the optical axis, and the intensity will vary along the strip accordingly as the emerging rays are in or out of phase with each other. The intensity variation is given by

$$I(\theta) = \frac{T^2 I_\lambda}{(1-R)^2 + 4R\sin^2\left(\dfrac{2\pi t\mu\cos\theta}{\lambda}\right)} \tag{4.58}$$

where I_λ is the incident intensity at wavelength, λ. The image structure will resemble that shown in Figure 4.11. If the source is now replaced with a bichromatic one, then the image structure will be of the type shown in Figure 4.12. Consider just one of these fringes, its angular distance, θ_{max}, from the optical axis is, from Equation 4.51,

$$\theta_{max} = \cos^{-1}\left(\frac{m\lambda}{2t\mu}\right) \tag{4.59}$$

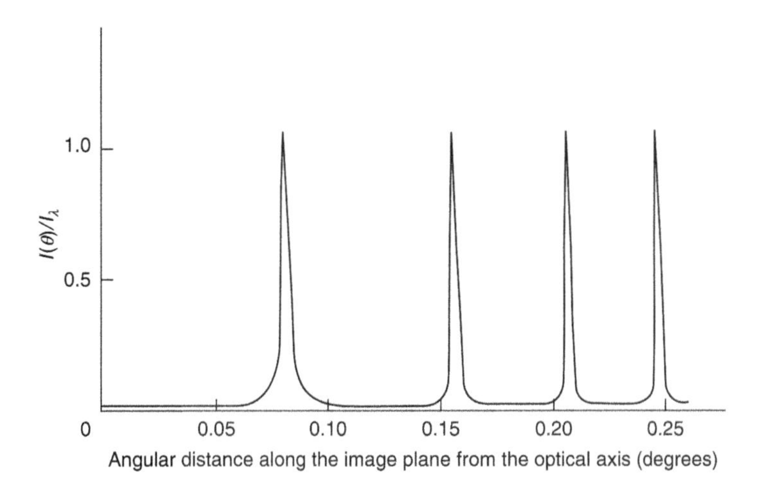

FIGURE 4.11 Image structure in a Fabry-Perot spectroscope viewing a monochromatic slit source, with $T = 0.1$, $R = 0.9$, $t = 0.1$ m, $\mu = 1$ and $\lambda = 550$ nm.

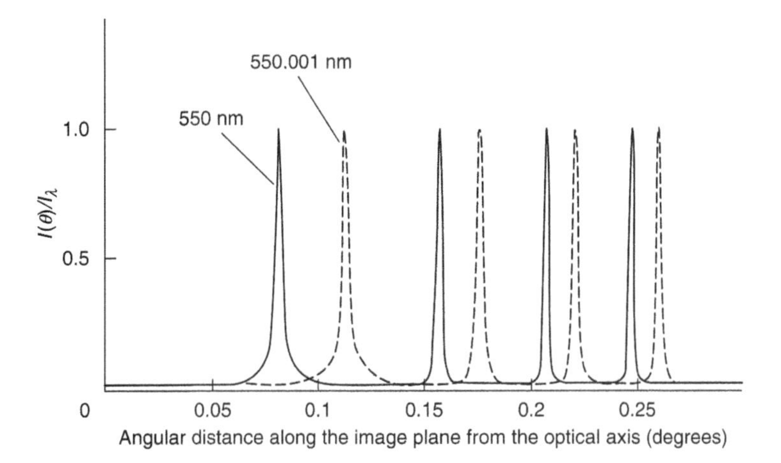

FIGURE 4.12 Image structure in a Fabry-Perot spectroscope viewing a bichromatic slit source, with $T = 0.1$, $R = 0.9$, $t = 0.1$ m, $\mu = 1$ and $\lambda = 550$ nm (full curve) and $\lambda = 550.001$ nm (broken curve).

and so the peak intensity from Equation 4.58 is

$$I(\theta_{\max}) = \frac{T^2 I_\lambda}{(1-R)^2} \tag{4.60}$$

$$= I_\lambda \text{ (when there is no absorption)} \tag{4.61}$$

Let the angular half width of a fringe at half intensity be $\Delta\theta$, then a separation of twice the half-half width of the fringes gives a central intensity of 83% of either of the peak intensities. So that if α is the resolution by the extended Rayleigh criterion, we may write

$$\alpha \approx 2\Delta\theta = \frac{\lambda(1-R)}{2\pi\mu t R^{0.5}\theta_{\max}\cos\theta_{\max}} \tag{4.62}$$

Hence, from Equation 4.57, we obtain the spectral resolution

$$W_\lambda = \alpha \frac{d\lambda}{d\theta} \tag{4.63}$$

$$= \frac{\lambda^2 (1-R)}{2\pi\mu t R^{0.5} \cos\theta_{max}} \tag{4.64}$$

and so the resolution of the system (previously given the symbol *R*) is

$$\frac{\lambda}{\Delta\lambda} = \frac{2\pi\mu t R^{0.5} \cos\theta_{max}}{\lambda(1-R)} \tag{4.65}$$

or, since θ_{max} is small and μ is usually close to unity,

$$\frac{\lambda}{\Delta\lambda} \approx \frac{2\pi t R^{0.5}}{\lambda(1-R)} \tag{4.66}$$

Thus for typical values of $t = 0.1$ m, $R = 0.9$ and for visible wavelengths, we have

$$\frac{\lambda}{\Delta\lambda} \approx 10^7 \tag{4.67}$$

which is almost two orders of magnitude higher than typical values for prisms and gratings. It is comparable with the resolution for a large echelle grating, while physically the device is much less bulky.

An alternative measure of the resolution that may be encountered is the finesse. This is the reciprocal of the half-width of a fringe measured in units of the separation of the fringes from two adjacent orders. It is given by

$$\text{Finesse} = \frac{\pi R^{0.5}}{1-R} = \frac{\lambda}{2t} \times \text{resolution} \tag{4.68}$$

For a value of *R* of 0.9, the finesse is therefore about 30.

The free spectral range of an etalon is small because it is operating at high spectral orders. From Equation 4.59, we have

$$\Sigma = \lambda_1 - \lambda_2 = \frac{\lambda_2}{m} \tag{4.69}$$

where λ_1 and λ_2 are superimposed wavelengths from adjacent orders (cf. Equation 4.23). Thus, the device must be used with a cross disperser (as already mentioned) and/or the free spectral range increased. The latter may be achieved by combining two or more different etalons, then, only the maxima that coincide, will be transmitted through the whole system and the intermediate maxima will be suppressed.

Practical etalons are made from two plates of glass or quartz whose surfaces are flat to 1%–2% of their operating wavelength. They are held accurately parallel to each other by low thermal expansion spacers, with a spacing in the region of 10–200 mm. The inner faces are mirrors, which are usually produced by a metallic or a multi-layer dielectric coating. The outer faces are inclined by a small angle to the inner faces so that the plates are the basal segments of low angle prisms.

Any multiple reflections (other than the desired ones) are then well angularly displaced from the required image. The limit to the resolution of the instrument is generally imposed by departures of the two reflecting surfaces from absolute flatness. This limits the main use of the instrument to the visible and infrared regions. The absorption in metallic coatings also limits the short-wave use, so that 200 nm represents the shortest practicable wavelength even for laboratory usage. Etalons are commonly used as scanning instruments. By changing the air pressure by a few per cent, the refractive index of the material between the plates is changed, and so the wavelength of a fringe at a given place within the image is altered (Equation 4.58). The astronomical applications of Fabry-Perot spectroscopes are comparatively few for direct observations. However, the instruments are used extensively in determining oscillator strengths and transition probabilities upon which much of the more conventional astronomical spectroscopy is based.

Another important application of etalons and one that *does* have many direct applications for astronomy is in the production of narrow band filters. These are usually known as 'interference filters' and are etalons in which the separation of the two reflecting surfaces is small. For materials with refractive indices near 1.5 and for near normal incidence, we see from Equation 4.51 that if t is 167 nm (say) then the maxima will occur at wavelengths of 500, 250, 167 nm and so on, accordingly as m is 1, 2, 3, etc. While from Equation 4.62, the widths of the transmitted regions will be 8.4, 2.1, 0.9 nm, and so on for 90% reflectivity of the surfaces. Thus, a filter centred upon 500 nm with a bandwidth of 8.4 nm can be made by combining such an etalon with a simple dye filter to eliminate the shorter wavelength transmission regions (or in this example, just by relying on the absorption within the glass substrates). Other wavelengths and bandwidths can easily be chosen by changing t, μ, and R. Such a filter would be constructed by evaporating a partially reflective layer onto a sheet of glass. A second layer of an appropriate dielectric material such as magnesium fluoride or cryolite would then be evaporated on top of this to the desired thickness, followed by a second partially reflecting layer. A second sheet of glass is then added for protection. The reflecting layers may be silver or aluminium, or they may be formed from a double layer of two materials with different refractive indices to improve the overall filter transmission. In the FIR, pairs of inductive meshes can be used in a similar way for infrared filters. The band-passes of interference filters can be made squarer by using several superimposed Fabry-Perot layers.

Tunable filters have also been developed (see also the Lyot birefringent filter, Section 5.3), that are especially suited to observing the emission lines of gaseous nebulae. The reflecting surfaces are mounted on stacks of piezoelectric crystals so that their separations can be altered. Thus, the Maryland-Magellan Tunable Filter (MMTF, in use 2006–2018) for the 6.5-m Magellan-Baade telescope covered the wavelength range from 500 to 920 nm with a bandwidth that could be varied between 0.5 and 2.5 nm. The Gran Telescopio Canarias' (GTC's) Optical System for Imaging and Low-Resolution Integrated Spectroscopy (OSIRIS)[3] imaging spectrograph has been operating for several years now and observes over the 365-nm to 1-μm region with bandwidths between 2 and 0.9 nm. Two optional tunable filters can be included within it, covering from 650 to 935 nm and 450 to 670 nm with bandwidths of ~1 nm and tuning accuracies of ~0.1 nm. Most other large optical telescopes have either recently added tunable filters to their range of ancillary equipment or are in the process of doing so. By making the layers forming the etalon shaped like very-low angle wedges, a tapered or linear etalon is produced. The transmitted wavelength of the filter then varies along the physical length of the filter and a continuous spectrum may be produced. The Lisa Hardaway Infrared Mapping Spectrometer (LEISA) spectral imager which forms a part of the Ralph instrument on board the New Horizons Pluto probe uses a linear etalon covering the 1.25- to 2.5-μm region.

4.1.4.2 Michelson Interferometer

This Michelson interferometer should not be confused with the Michelson stellar interferometer which was discussed in Section 2.5. The instrument discussed here is similar to the device used by

[3] Not to be confused with the Keck telescopes' instrument.

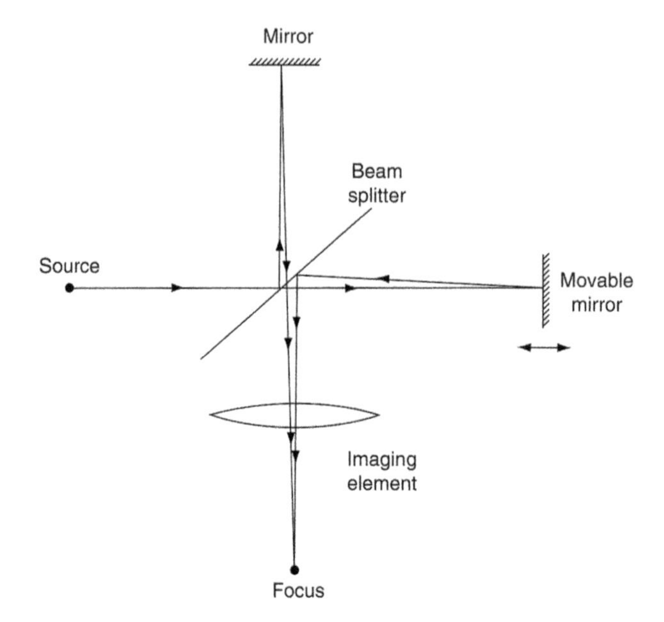

FIGURE 4.13 Optical pathways in a Michelson interferometer.

Michelson and Morley to try and detect the Earth's motion through the aether. Its optical principles are shown in Figure 4.13 (see also Section 1.6). The light from the source is split into two beams by the beam splitter and then recombined as shown. For a particular position of the movable mirror and with a monochromatic source, there will be a path difference, ΔP, between the two beams at their focus. The intensity at the focus is then

$$I_{\Delta P} = I_m \left[1 + \cos\left(\frac{2\pi\Delta P}{\lambda} \right) \right] \tag{4.70}$$

where I_m is a maximum intensity. If the mirror is moved, then the path difference will change and the final intensity will pass through a series of maxima and minima (Figure 4.14). If the source is bichromatic, then two such variations will be superimposed with slightly differing periods and the final output will have a beat frequency (Figure 4.15). The difference between the two

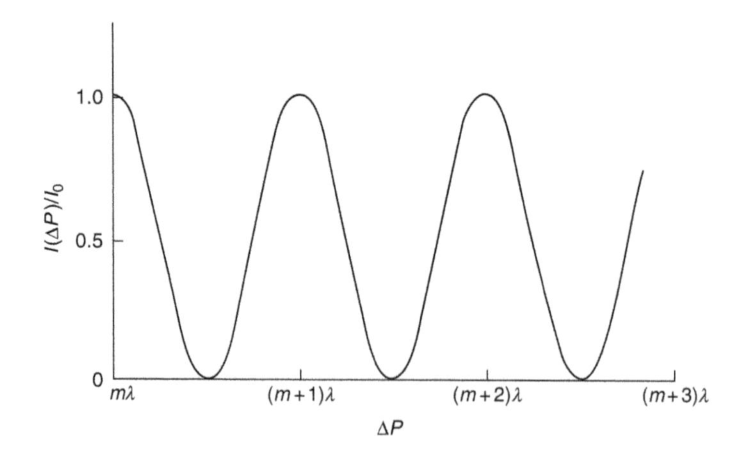

FIGURE 4.14 Variation of fringe intensity with mirror position in a Michelson interferometer.

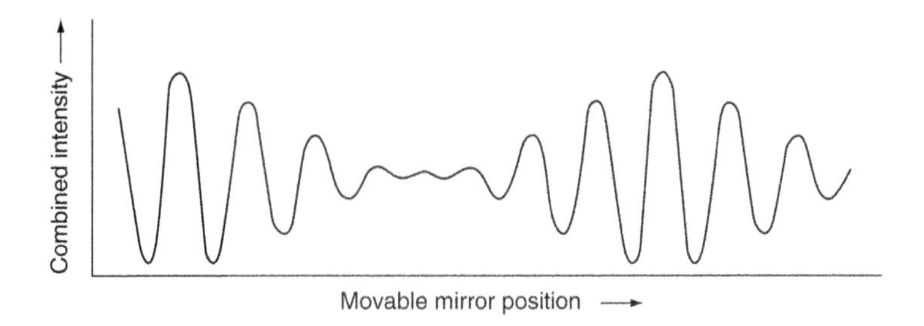

FIGURE 4.15 Output of a Michelson interferometer observing a bichromatic source.

outputs (Figures 4.20 and 4.22) gives the essential principle of the Michelson interferometer when it is used as a spectroscope. Neither output in any way resembles an ordinary spectrum, yet it would be simple to recognise the first as due to a monochromatic source and the second as due to a bichromatic source. Furthermore, the spacing of the fringes could be related to the original wavelength(s) through Equation 4.70. More generally of course, sources emit a broad band of wavelengths, and the final output will vary in a complex manner. To find the spectrum of an unknown source from such an output, therefore, requires a rather different approach than this simple visual inspection.

Let us consider, therefore, a Michelson interferometer in which the path difference is ΔP observing a source whose intensity at wavelength λ is $I(\lambda)$. The intensity in the final image resulting from the light of a particular wavelength, $I'_{\Delta P}(\lambda)$ is then

$$I'_{\Delta P}(\lambda) = K\,I(\lambda)\left[1+\cos\left(\frac{2\pi\Delta P}{\lambda}\right)\right] \tag{4.71}$$

where K is a constant that takes account of the losses at the various reflections, transmissions, etc. Thus, the total intensity in the image for a given path difference is just

$$I'_{\Delta P} = \int_0^\infty I'_{\Delta P}(\lambda)\,d\lambda \tag{4.72}$$

$$= \int_0^\infty K\,I(\lambda)\,d\lambda + \int_0^\infty K\,I(\lambda)\cos\left(\frac{2\pi\Delta P}{\lambda}\right)d\lambda \tag{4.73}$$

Now the first term on the right-hand side of Equation 4.73 is independent of the path difference and is simply the mean intensity of the image. We may, therefore, disregard it and concentrate instead on the deviations from this average level, $I(\Delta P)$. Thus,

$$I(\Delta P) = K\int_0^\infty I(\lambda)\cos\left(\frac{2\pi\Delta P}{\lambda}\right)d\lambda \tag{4.74}$$

or in frequency terms

$$I(\Delta P) = K'\int_0^\infty I(\nu)\cos\left(\frac{2\pi\Delta P\nu}{c}\right)d\nu \tag{4.75}$$

Now the Fourier transform, $F(u)$, of a function, $f(t)$, (see also Section 2.1) is defined by

$$F(f(t)) = F(u) = \int_{-\infty}^{\infty} f(t) e^{-2\pi i u t} \, dt \tag{4.76}$$

$$= \int_{-\infty}^{\infty} f(t) \cos(2\pi u t) dt - i \int_{-\infty}^{\infty} f(t) \sin(2\pi u t) dt \tag{4.77}$$

Thus, we see that the output of the Michelson interferometer is akin to the real part of the Fourier transform of the spectral intensity function of the source. Furthermore, by defining

$$I(-\nu) = I(\nu) \tag{4.78}$$

we have

$$I(\Delta P) = \frac{1}{2} K^{I} \int_{-\infty}^{\infty} I(\nu) \cos\left(\frac{2\pi \Delta P \nu}{c}\right) d\nu \tag{4.79}$$

$$= K^{II} \operatorname{Re}\left\{ \int_{-\infty}^{\infty} I(\nu) \exp\left[-i \left(\frac{2\pi \Delta P}{c}\right) \nu \right] d\nu \right\} \tag{4.80}$$

where K^{II} is an amalgam of all the constants. Now by inverting the transformation

$$F^{-1}(F(u)) = f(t) = \int_{-\infty}^{\infty} F(u) e^{2\pi i u t} \, du \tag{4.81}$$

and taking the real part of the inversion, we may obtain the function that we require – the spectral energy distribution, or as it is more commonly known, the 'spectrum'. Thus,

$$I(\nu) = K^{III} \operatorname{Re}\left\{ \int_{-\infty}^{\infty} I\left(\frac{2\pi \Delta P}{c}\right) \exp\left[i \left(\frac{2\pi \Delta P}{c}\right) \nu \right] d\left(\frac{2\pi \Delta P}{c}\right) \right\} \tag{4.82}$$

or

$$I(\nu) = K^{IV} \int_{-\infty}^{\infty} I\left(\frac{2\pi \Delta P}{c}\right) \cos\left(\frac{2\pi \Delta P \nu}{c}\right) d(\Delta P) \tag{4.83}$$

where K^{III} and K^{IV} are again amalgamated constants. Finally, by defining

$$I\left(\frac{-2\pi \Delta P}{c}\right) = I\left(\frac{2\pi \Delta P}{c}\right) \tag{4.84}$$

we have

$$I(\nu) = 2K^{IV} \int_{0}^{\infty} I\left(\frac{2\pi \Delta P}{c}\right) \cos\left(\frac{2\pi \Delta P \nu}{c}\right) d(\Delta P) \tag{4.85}$$

and so, the spectrum is obtainable from the observed output of the interferometer as the movable mirror scans through various path differences. We may now see why a Michelson interferometer when used as a scanning spectroscope is often called a 'Fourier transform spectroscope' (FTS). The inversion of the Fourier transform is carried out on computers using the fast Fourier transform algorithm.

In practice of course, it is not possible to scan over path differences from zero to infinity and also measurements are usually made at discrete intervals rather than continuously, requiring the use of the discrete Fourier transform equations (Section 2.1). These limitations are reflected in a reduction in the resolving power of the instrument. To obtain an expression for the resolving power, we may consider the Michelson interferometer as equivalent to a two-aperture interferometer (Figure 4.1) because its image is the result of two interfering beams of light. We may, therefore, write Equation 4.20 for the resolution of two wavelengths by the Rayleigh criterion as

$$W_\lambda = \frac{\lambda^2}{2\Delta P} \tag{4.86}$$

However, if the movable mirror in the Michelson interferometer moves a distance x, then ΔP ranges from 0 to $2x$ and we must take the average value of ΔP rather than the extreme value for substitution into Equation 4.86. Thus, we obtain the spectral resolution of a Michelson interferometer as

$$W_\lambda = \frac{\lambda^2}{2x} \tag{4.87}$$

so that the system's resolution is

$$\frac{\lambda}{W_\lambda} = \frac{2x}{\lambda} \tag{4.88}$$

Since x can be as much as 2 m, we obtain a spectral resolution of up to 4×10^6 for such an instrument used in the visible region.

The measurements must be obtained sufficiently frequently to preserve the resolution, but not more often than this, or time and effort will be wasted. If the final spectrum extends from λ_1 to λ_2, then the number of useful intervals, n, into which it may be divided, is given by

$$n = \frac{\lambda_1 - \lambda_2}{W_\lambda} \tag{4.89}$$

so that if λ_1 and λ_2 are not too different, then

$$n \approx \frac{8x(\lambda_1 - \lambda_2)}{(\lambda_1 + \lambda_2)^2} \tag{4.90}$$

However, the inverse Fourier transform gives both $I(\nu)$ and $I(-\nu)$, so that the total number of separate intervals in the final inverse transform is $2n$. Hence, we must have at least $2n$ samples in the original transformation and, therefore, the spectroscope's output must be sampled $2n$ times. Thus, the interval between successive positions of the movable mirrors, Δx, at which the image intensity is measured, is given by

$$\Delta x = \frac{(\lambda_1 + \lambda_2)^2}{16(\lambda_1 - \lambda_2)} \tag{4.91}$$

A spectrum between 500 and 550 nm, therefore, requires step lengths of 1 μm, while between 2,000 and 2,050 nm a spectrum would require step lengths of 20 μm. This relaxation in the physical constraints required on the positional accuracy of the movable mirror for longer wavelength spectra, combined with the availability of other methods of obtaining visible spectra, has led to the majority of applications of Fourier transform spectroscopy, to date, being in the infrared and FIR. Even there though, the increasing size of infrared arrays is now leading to the use of more conventional diffraction grating spectroscopes.

The basic PSF of the FTS is of the form $\frac{\sin\Delta\lambda}{\Delta\lambda}$ (Figure 4.16), where $\Delta\lambda$ is the distance from the central wavelength, λ, of a monochromatic source. This is not a particularly convenient form for the profile, and the secondary maxima may be large enough to be significant, especially where the spectral energy undergoes abrupt changes such as at ionisation edges or over molecular bands. The effect of the PSF may be reduced at some cost to the theoretical resolution by a technique known as 'apodisation' (see also Sections 1.1, 2.1, 2.5, and 5.1). The transform is weighted by some function, ω, known as the 'apodisation function'. Many functions can be used, but the commonest is probably a triangular weighting function, i.e.,

$$\omega(\Delta P) = 1 - \frac{\Delta P}{2x} \tag{4.92}$$

The resolving power is halved, but the PSF becomes that for a single rectangular aperture (Figure 1.33, Equation 1.12) and has much reduced secondary maxima.

A major advantage of the Michelson interferometer over the etalon when the latter is used as a scanning instrument lies in its comparative rapidity of use when the instrument is detector-noise limited. Not only is the total amount of light gathered by the Michelson interferometer higher (the Jacquinot advantage), but even for equivalent image intensities, the total time required to obtain a spectrum is much reduced. This arises because all wavelengths are contributing to every reading in the Michelson interferometer, whereas a reading from the etalon gives information for just a single wavelength. The gain of the Michelson interferometer is called the 'multiplex advantage' or 'Fellget advantage' and is similar to the gain of the Hadamard masking technique over simple scanning (Section 2.4). If t is the integration time required to record a single spectral element, then the etalon requires a total observing time of nt. The Michelson interferometer, however, requires a time of only $t/\sqrt{n/2}$ to record each sample because it has contributions from n spectral elements. It must obtain $2n$ samples so the total observing time for the same spectrum is, therefore, $2nt/\sqrt{n/2}$, and it thus has an advantage over the etalon by a factor of $\sqrt{n/8}$ in observing time.

Michelson interferometers have another advantage in that they require no entrance slit to preserve their spectral resolution. This is of considerable significance for large telescopes where the stellar image may have a physical size of a millimetre or more due to atmospheric turbulence, while the spectroscope's slit may be only a few tenths of a millimetre wide. Thus, either much of the light is wasted or complex image dissectors (Section 4.2) must be used.

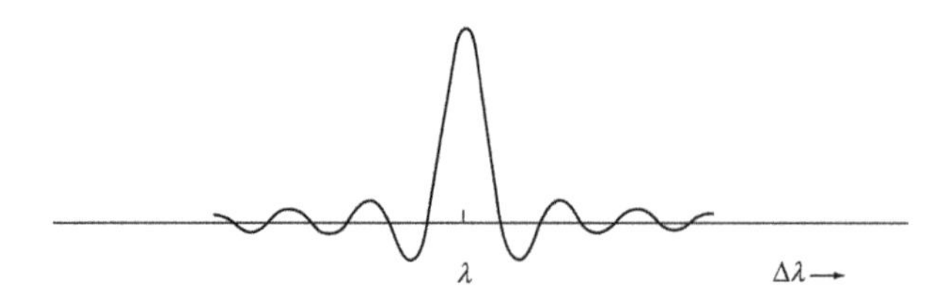

FIGURE 4.16 Basic instrumental profile (point spread function [PSF]) of a Fourier transform spectroscope.

4.1.5 Fibre-Optics

Fibre-optic cables are now widely used to connect spectroscopes to telescopes, enabling the spectroscope to be mounted separately from the telescope. This reduces the problems of flexure that occur within telescope-mounted spectroscopes because the gravitational loads no longer change with the different telescope positions. It also enables the spectroscope to be kept in a temperature-controlled room and, in some cases, cooled and inside a vacuum chamber. Fibre-optics can also be used to reformat stellar images so that all the light enters the spectroscope and so to enable extended objects or multiple objects to be observed efficiently. This application is similar to some of the other applications of photonics – specific examples are discussed previously (Sections 1.1.17, 2.5.2 and 2.5.4) and in the next section. Here we are concerned with the optics of fibre-optic cables.

Fibre-optic cables usually consist of a thin (10–500 μm) filament of glass encased in a cladding of another glass with a lower refractive index. One or more (sometimes thousands) of these strands make up the cable as a whole. Light entering the core is transmitted through the core by multiple internal reflections off the interface between the two glasses, provided that its angle of incidence exceeds the critical angle for total internal reflection. The sine of the critical angle q_c is given by the ratio of the refractive index of the cladding ($\mu_{cladding}$) to the refractive index of the core (μ_{core}) and this determines the maximum angle to the axis of the fibre that may be transmitted along it without loss $(90 - q_c)$. However, for fibres immersed in air or a vacuum, a wider-angle beam may be fed into the end of the fibre, because, when refracted into the core material, its angle to the fibre axis will change to $(90 - q_c)$ or less. This larger angle, q_{max} is given by,

$$\theta_{max} = \sin^{-1}\sqrt{(\mu_{core}^2 - \mu_{cladding}^2)} \tag{4.93}$$

Fibre optic cables are usually characterised by their numerical aperture (NA) and this is simply equal to sin (q_{max}). The minimum focal ratio, f_{min} that may be fed into the fibre and which will be transmitted by the core is then

$$f_{min} = \frac{\sqrt{1 - NA_2}}{2NA} \tag{4.94}$$

Commercially produced fibre optics have numerical apertures ranging from about 0.15–0.35, giving minimum focal ratios ranging from f3.3 to f1.4.

Silica glass is widely used as the core of the fibre for the spectral region 400 nm–2 μm and is sufficiently transparent that cables can be tens of metres long without significant absorption. Further into the infrared, specialist glasses such as zirconium fluoride need to be used. Imperfections in the walls of the fibres and internal stresses lead to focal ratio degradation. This is a decrease in the focal ratio that can lead to light loss if the angle of incidence exceeds the critical angle. It may also cause problems in matching the output from the cable to the focal ratio of the instrument being used. Focal ratio degradation affects long focal ratio light beams worst, so short (faster) focal ratios are better for transmitting light into the fibres.

Cables comprising many individual strands will not transmit all the light that they receive because of the area occupied by the cladding. This is typically 40% of the cross-sectional area of the cable. Many astronomical applications, therefore, use single-strand cables with a core diameter sufficient to accept a complete stellar image. Multi-strand cables can be coherent or non-coherent. In the former, the individual strands have the same relative positions at the input end of the cable as at the output. In non-coherent cables, the relationship between strand positions at the input and output ends of the cable is random. For some special applications, such as reformatting stellar images to match the shape of the spectroscope's entrance slit the input and output faces of the fibre-optic cable are of different shapes. Such cables need to be coherent in the sense that the positions of the strands at the output face are related in a simple and logical manner to those at the input face.

Optical fibres also have other potential astronomical applications which use them in rather more sophisticated ways than just simply getting the light to where it is needed. For some purposes – such as improving the uniformity and stability of images – the modes in multi-mode fibres may best be scrambled, and this can be accomplished by the use of non-circular cross-section fibres. The High Accuracy Radial velocity Planet Searcher (HARPS) spectrographs, for example, use octagonal-shaped fibres for this reason. We have already seen (Section 1.1), that by varying the refractive index within the fibre, a Bragg grating can be formed which cuts out the atmospheric OH lines forming the main source of background noise in the NIR. Another possibility may soon result in a new type of spectroscope. The integrated photonic spectroscope uses a number of fibres of different lengths to form a phased array (see Section 1.2 for an account of a radio analogue). The fibres are fed by a 2D waveguide that acts as a multiplexor and their outputs are recombined to form the spectrum within a second waveguide. Integrated photonic spectroscopes, including echelle and FTS designs, hold out the possibility of replacing the massive and cumbersome spectroscopes often used today, with an instrument that is just a few centimetres in size. Photonic Arrayed Waveguides (AWGs) can replace gratings; the radiation is sent through several single-mode optical fibres/waveguides of differing lengths but with *identical* path length *differences* and then recombined. The path differences within the recombining beams can then result in interference effects and spectra in a similar fashion to diffraction gratings.

4.2 SPECTROSCOPES

4.2.1 Basic Design Considerations

The specification of a spectroscope[4] usually begins from just three parameters. One is the focal ratio of the telescope upon which the spectroscope is to operate (though this can be easily changed), the second is the required spectral resolution and the third is the required spectral range. Thus, in terms of the notation used in Section 4.1, we have f', W_λ and λ specified (where f' is the effective focal ratio of the telescope at the entrance aperture of the spectroscope), and we require the values of f_1, f_2, s, R, $\frac{d\lambda}{d\theta}$, L and D.

We may immediately write down the resolution required of the dispersion element

$$R = \frac{\lambda}{W_\lambda} \tag{4.95}$$

For a grating, the resolution depends upon the number of lines and the order of the spectrum (Equation 4.21). A reasonably typical astronomical grating (ignoring echelle gratings, etc.) will operate in its third order and have some 500 lines mm^{-1}. Thus, from Equation 4.21, we may write the resolution of a grating as

$$R \approx 1.5 \times 10^6 L \tag{4.96}$$

where L is the width of the ruled area of the grating in metres. Thus, Equations 4.95 and 4.99 provide a first estimate of the required size of the grating. It may be calculated more accurately in second and subsequent iterations through this design process.

The diameter of the exit beam from the dispersing element, assuming that it is fully illuminated, can then be found from

$$D = L \cos \phi \tag{4.97}$$

[4] We now only consider the design of grating-based spectrographs in this edition; the design process for prism-based instruments has been given in previous editions and, anyway, is similar.

where ϕ is the angular deviation of the exit beam from the perpendicular to the exit face of the dispersing element. For a grating used in the third order with equal angles of incidence and reflection, this would be around 25°. Thus, we get

$$D = 0.9\,L \tag{4.98}$$

The dispersion can now be determined by setting the angular resolution of the imaging element equal to the angle between two just resolved wavelengths from the dispersing element

$$\frac{\lambda}{D} = W_\lambda \frac{d\theta}{d\lambda} \tag{4.99}$$

giving

$$\frac{d\theta}{d\lambda} = \frac{R}{D} \tag{4.100}$$

Because the exit beam is rectangular in cross section, we must use the resolution for a rectangular aperture of the size and shape of the beam (Equation 1.13) for the resolution of the imaging element and not its actual resolution, assuming that the beam is wholly intercepted by the imaging element and that its optical quality is sufficient not to degrade the resolution below the diffraction limit.

The final parameters now follow easily. The physical separation of two just-resolved wavelengths on the charge-coupled device (CCD) or other imaging detector must be greater than or equal to the separation of two pixels. The CCD pixels are typically around 15–20 μm in size, so the focal length in metres of the imaging element must be at least

$$f_2 \geq \frac{\text{Pixel size}}{W_\lambda} \frac{d\theta}{d\lambda} \tag{4.101}$$

The diameter of the imaging element, D_2, must be sufficient to contain the whole of the exit beam. So, for a square cross-section exit beam

$$D_2 = \sqrt{2}\,D \tag{4.102}$$

The diameter of the collimator, D_1, must be similar to that of the imaging element in general if the dispersing element is to be fully illuminated. Thus again

$$D_1 = \sqrt{2}\,D \tag{4.103}$$

Now for the collimator to be fully illuminated in its turn, its focal ratio must equal the effective focal ratio of the telescope. Hence, the focal length of the collimator, f_1, is given by

$$f_1 = \sqrt{2}\,Df' \tag{4.104}$$

Finally, from Equation 4.35, we have the slit width

$$s = \frac{f_1 \lambda}{D} \tag{4.105}$$

and a first approximation has been obtained to the design of the spectroscope.

The low light levels involved in astronomy usually require the focal ratios of the imaging elements to be small, so that it is fast in imaging terms. Satisfying this requirement usually means a compromise in some other part of the design, so that an optimally designed system is rarely achievable in practice. The slit may also need to be wider than specified to use a reasonable fraction of the star's light and, of course, costs are a major constraint under most circumstances.

The limiting magnitude of a telescope-spectroscope combination is the magnitude of the faintest star for which a *useful* spectrum may be obtained. This is an imprecise quantity because it depends upon the type of spectrum and the purpose for which it is required, as well as the properties of the instrument and the detector. For example, if strong emission lines in a spectrum are the features of interest, then fainter stars may be studied than if weak absorption lines are desired. Similarly, spectra of sufficient quality to determine radial velocities may be obtained for fainter stars than if line profiles are wanted. A guide, however, to the limiting magnitude may be gained through the use of Bowen's formula

$$m = 12 + 2.5\log_{10}\left(\frac{s\,D_1 T_D g\,q\,t\left(\dfrac{d\lambda}{d\theta}\right)}{f_1 f_2 \alpha H}\right) \tag{4.106}$$

where m is the faintest B magnitude that will give a usable spectrum in t seconds of exposure, T_D is the telescope objective's diameter, g is the optical efficiency of the system (i.e., the ratio of the usable light at the focus of the spectroscope to that incident upon the telescope – typically it has a value of ~0.2 – note, however, that g does *not* include the effect of any curtailment of the image by the slit), q is the quantum efficiency of the detector (typical values are 0.4–0.9 for CCDs – Section 1.1), α is the angular size of the stellar image at the telescope's focus (typically 5×10^{-6} to 2×10^{-5} radians) and H is the height of the spectrum.

This formula gives quite good approximations for spectroscopes in which the star's image is larger than the slit and it is trailed along the length of the slit to broaden the spectrum. At one time this was the commonest mode of use for astronomical spectroscopes, but now many spectroscopes are fed by optical fibres that may or may not intercept the whole of the star's light. Other situations such as an untrailed image, or an image smaller than the slit, require the formula to be modified. Thus, when the slit is wide enough for the whole stellar image to pass through it, the exposure varies inversely with the square of the telescope's diameter, while for extended sources, it varies inversely with the square of the telescope's focal ratio (cf. Equation 1.55).

The slit is quite an important part of the spectroscope because in many astronomical spectroscopes it fulfils two functions. First, it acts as the entrance aperture of the spectroscope. For this purpose, its sides must be accurately parallel to each other and perpendicular to the direction of the dispersion. It is also usual for the slit width to be adjustable or for slits of different widths to be provided, so that alternative detectors may be used and/or changing observing conditions catered for. Although we have seen how to calculate the optimum slit width, it is usually better in practice to find the best slit width empirically. The slit width is optimised by taking a series of images of a sharp emission line in the comparison spectrum through slits of different widths. As the slit width decreases, the image of the line's width should also decrease at first, but should eventually become constant. The change-over point occurs when some part of the spectroscope system, other than the slit, starts to limit the resolution. The slit width at the changeover is then the required optimum value. As well as allowing the desired light through, the slit must additionally reject unwanted radiation. The jaws of the slit are, therefore, usually of a knife-edge construction with the chamfering on the inside of the slit so that light is not scattered or reflected into the spectroscope from the edges of the jaws.

On some instruments, a secondary purpose of the slit is to assist in the guiding of the telescope on the object. When the stellar image is larger than the slit width, it will overlap onto the slit jaws. By polishing the front of the jaws to an optically flat mirror finish, these overlaps can be observed via an auxiliary detector, and the telescope driven and guided to keep the image bisected by the slit. However, guiding in this manner is wasteful of the expensively collected light from the star of interest. Most modern instruments, therefore, use another star within the field of view upon which to guide. If the stellar image is then still larger than the slit, it can be reformatted so that all its light enters the spectroscope.

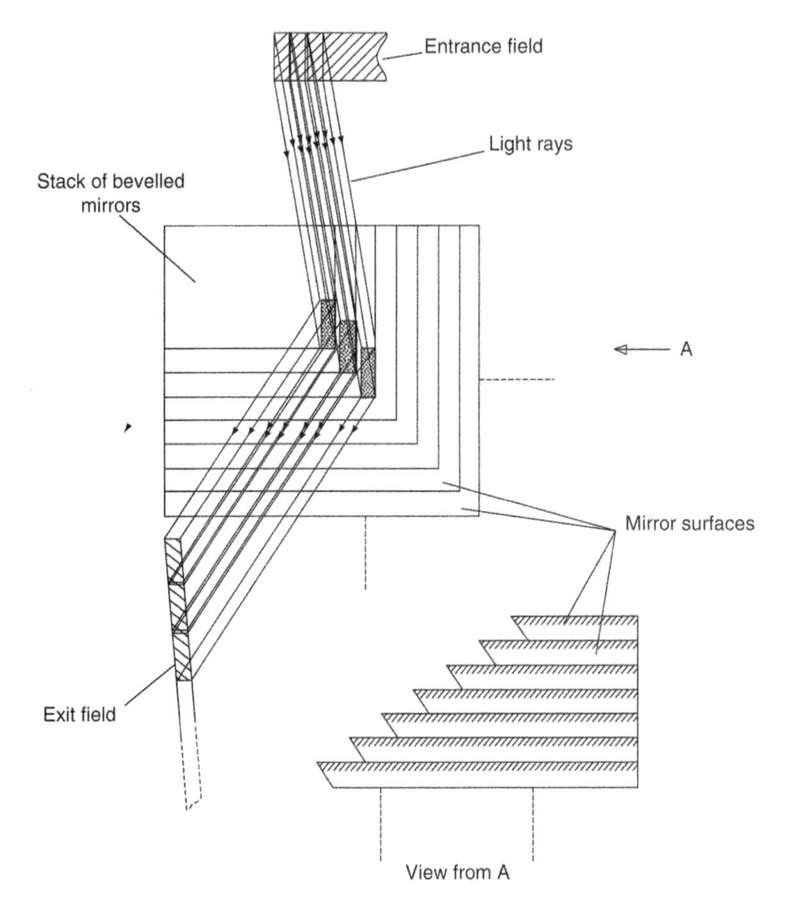

FIGURE 4.17 Bowen image slicer.

There are several ways of reformatting the image. Early approaches, such as that due to Bowen, are still in use.[5] The Bowen image slicer consists of a stack of overlapped mirrors (Figure 4.17) which section the image and then rearrange the sections to be end to end and so to form a linear image suitable for matching to a spectroscope slit. The Bowen-Walraven image slicer uses multiple internal reflections. A prism with a chamfered side is used and has a thin plate attached to it. Because of the chamfered side, the plate only touches the prism along one edge. A light beam entering the plate is repeatedly internally reflected wherever the plate is not in contact with the prism but is transmitted into the prism along the contact edge (Figure 4.18).

The simplest concept, though, for reformatting an image is a bundle of optical fibres whose cross section matches the slit at one end and the star's image at the other. So that, apart from the reflection and absorption losses within the fibres, all the star's light is conducted into the spectroscope. The disadvantages of fibre-optics are mainly the degradation of the focal ratio of the beam due to imperfections in the walls of the fibre so that not all the light is intercepted by the collimator and the multi-layered structure of the normal commercially available units which leads to other light losses since only the central core of fibre transmits the light. Hence, specially designed fibre-optic cables are usually required (see also references elsewhere to optical fibres and photonics), and these are often made 'in house' at the observatory needing them. They are usually much thicker than normal

[5] For example, on the William Herschel Telescope's (WHT's) Intermediate-Dispersion Spectrograph and Imaging System (ISIS).

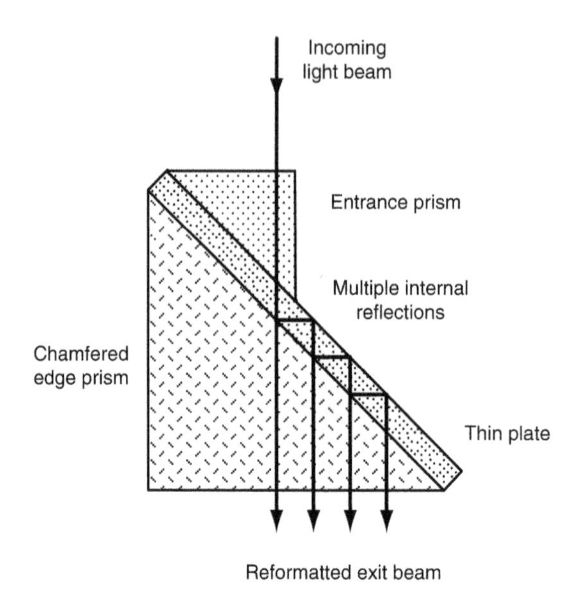

FIGURE 4.18 The Bowen-Walraven image slicer.

fibre-optics and are formed from plastic or fused quartz. On telescopes with adaptive optics, the size of the star's image is much reduced and this allows not only the slit width to be smaller but also a more compact design to be used for the whole spectroscope.

For extended sources, it is common practice to use long slits. Provided that the image does not move with respect to the slit (and no spectrum widener is used), then each point in that portion of the image falling onto the slit has its individual spectrum recorded at the appropriate height within the final spectrum. Several slits can be used as the entrance aperture of the spectroscope provided that the spectra do not overlap. Then, all the information that is derivable from a single-slit spectrogram is available, but the whole source can be covered in a fraction of the time. The ultimate development of this procedure, known as 'integral field spectroscopy' or 'three-dimensional (3D) spectroscopy', is to obtain a spectrum for every resolved point within an extended source, and this is discussed further below.

For several purposes the slit may be dispensed with and some specific designs are considered later in this section. Apart from the FTS (Section 4.1), they fall into two main categories. In the first, the projected image size on the spectrum is smaller than some other constraint on the system's resolution. The slit and the collimator may then be discarded, and parallel light from the source allowed to impinge directly onto the dispersing element. In the second type of slitless spectroscope, the source is producing a nebular type of spectrum (i.e., a spectrum consisting almost entirely of emission lines, with little or no continuum). If the slit alone is then eliminated from the telescope-spectroscope combination, the whole of the image of the source passes into the spectroscope. The spectrum then consists of a series of monochromatic images of the source in the light of each of the emission lines. Slitless spectroscopes are difficult to calibrate so that radial velocities can be found from their spectra, but they may be much more optically efficient than a slit spectroscope. In the latter, perhaps 1%–10% of the incident light is eventually used in the image, but some types of slitless spectroscope can use as much as 75% of the light. Furthermore, some designs, such as the objective prism (see below), can image as many as 10^5 stellar spectra in one exposure.

A system that is closely related to the objective prism, places the disperser shortly before the focal point of the telescope. Although the light is no longer in a parallel beam, the additional aberrations that are produced may be tolerable if the focal ratio is long. Using a zero-deviation grism (Sections 4.1.3 and 4.2.3) in combination with correcting optics enables a relatively wide field to be covered, without needing the large sizes required for objective prisms. With suitable blazing for the grating part of the grism, the zero order images provide wavelength reference points for the spectra. For example, the Visible Multi-Object Spectrograph (VIMOS; decommissioned in 2018) instrument on the Very Large Telescope (VLT) used a range of grisms (VPH transmission gratings in combination with two prisms) to produce a zero-deviation disperser covering a $7' \times 8'$ field of view and various parts of the 370-nm to 1-μm spectral region at spectral resolutions ranging from 180 to 2,500.

Spectroscopes, as we have seen, contain many optical elements which may be separated by large distances and arranged at large angles to each other. For the spectroscope to perform as expected, the relative positions of these various components must be kept correct and stable to within tight limits. The two major problems in achieving such stability arise through flexure and thermal expansion. Flexure primarily affects the smaller spectroscopes that are attached to and move around with the telescope. Their changing attitudes as the telescope moves causes the stresses within them to alter, so that if in correct adjustment for one telescope position, they may be out of adjustment in other positions. The light beam from the telescope is often folded so that it is perpendicular to the telescope's optical axis. The spectroscope is then laid out in a plane that is parallel to the back of the main mirror. Hence, the spectroscope components can be rigidly mounted onto a stout metal plate which in turn is bolted flat onto the back of the telescope. Such a design can be made rigid, and the flexure reduced to acceptable levels. In some modern spectroscopes, active supports are used to compensate for flexure along the lines of those used for telescope mirrors (Section 1.1).

Temperature changes affect all spectroscopes, but the relatively short light paths in the small instruments that are attached directly to telescopes mean that generally the effects are unimportant.

The large fixed spectroscopes, which operate at Coudé and Nasmyth foci or which have the light brought to them through fibre-optic cables, are obviously unaffected by changing flexure, and usually there is little difficulty other than that of cost in making them as rigid as desired. They experience much greater problems, however, from thermal expansion. The size of the spectrographs may be large, with optical path lengths measured in tens of metres. Hence, the temperature control must be correspondingly strict. A major problem is that the thermal inertia of the system may be so large that it may be impossible to stabilise the spectroscope at ambient temperature before the night has ended. Hence, many such spectroscopes are housed in temperature-controlled sealed rooms and/or have some or all of their components cooled and enclosed in a vacuum chamber.

Any spectroscope except the Michelson interferometer can be used as a monochromator. That is; a device to observe the object in a restricted range of wavelengths. Most scanning spectroscopes are in effect monochromators whose waveband may be varied. The most important use of the devices in astronomy, however, is in the spectrohelioscope. This builds up a picture of the Sun in the light of a single wavelength, and this is usually chosen to be coincident with a strong absorption line. Further details are given in Section 5.3. A related instrument for visual use on small telescopes is called a 'prominence spectroscope'. This has the spectroscope offset from the telescope's optical axis so that the (quite wide) entrance slit covers the solar limb. A small direct-vision prism or transmission grating then produces a spectrum and a second slit isolates an image in Hα light, allowing prominences and other solar features to be discerned.

Spectroscopy is undertaken throughout the entire spectrum. In the infrared and UV regions, techniques, designs and so on are almost identical to those for visual work except that different materials may need to be used. Some indication of these has already been given in Section 4.1. Generally, diffraction gratings and reflection optics are to be preferred because there is then no worry over absorption within the optical components.

The technique of Fourier transform spectroscopy, as previously mentioned, has so far had its main applications in the infrared region. However, Spectromètre Imageur à Transformée de Fourier pour l'Etude en Long et en Large de raies d'Emission (SITELLE) on the Canada-France-Hawaii Telescope (CFHT), operates from 350 to 900 nm. It is an imaging FTS. It obtains spectra of every object within an $11'$ field of view, at spectral resolutions ranging from 2 to >10,000 and it uses two 2,048 × 2,048-pixel CCD arrays as its detectors.

A recent development is the externally dispersed interferometer (EDI) which has the potential to measure radial velocities to high accuracies at low cost and can also be retrospectively added to existing spectrographs. It consists simply of sending the incoming radiation though a Michelson interferometer with fixed, but slightly different, path lengths for the two light beams before the recombined radiation enters the 'ordinary' spectrograph.

If we imagine the EDI interferometer observing a monochromatic point source then the output will be a series of point images along the image plane wherever there is constructive interference (see the dotted line curve in Figure 4.1). Because of the path difference between the light beams within the interferometer, the fringes near the optical axis will have orders larger than zero. Now imagine that the source is bichromatic. The new wavelength will also produce fringes and the zero orders of both sets of fringes will coincide, but higher orders will be displaced from each other in a line along the image plane. Adding a third wavelength to the source will add a third set of displaced fringes still along the same line along the image plane. Making the source a white light emitter will make the fringes merge continuously into each other along the line in the image plane and will also cause many different orders of fringes to overlay each other. However, if that line of radiation is now directed through the slit of the 'normal' spectrograph, the overlapping orders will be separated out (the spectrograph simply acting as a cross disperser).

The appearance of the image resulting from the combined effects of the interferometer and the spectrograph is shown in Figure 4.19. The continuum portions of the spectrum result in slanting fringes and the spectrum lines cut across these fringes.[6] Because the slant of the fringes is shallow, the darkest portions of two lines are separated in the vertical direction by *more* than the horizontal separation of the lines. In effect, any change in the wavelength of the line due to its Doppler shift is *amplified* in the vertical movement of the dark portions of the lines and, hence, is more easily and more accurately measurable.

At short UV and at X-ray wavelengths, glancing optics (Section 1.3) and diffraction gratings can be used to produce spectra using the same designs as for visual spectroscopes, the appearance and layout, however, will usually look different because of the off-axis optical elements, even though the optical principles may be the same. Also, many of the detectors at short wavelengths have some intrinsic spectral resolution. Radio spectroscopes were described in Section 1.2.

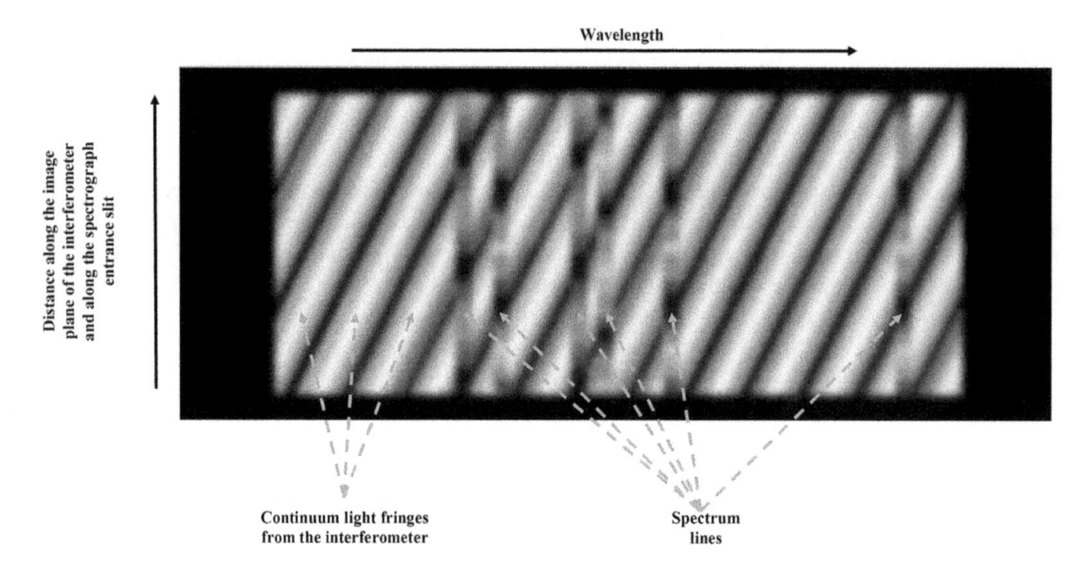

FIGURE 4.19 A schematic image from an Externally Dispersed Interferometer (EDI).

[6] If the spectrum lines formed a regular grid the result would be a Moiré-type pattern.

4.2.2 Prism-Based Spectroscopes

As noted previously, spectroscopes using prisms as the sole dispersing element will now rarely be encountered except for some mid-infrared (MIR) instruments such as that planned for the JWST. Many of the specific designs for prism-based spectroscopes are closely related to those for diffraction gratings, especially if a 30° prism with a reflecting back is used in place of the grating. Thus, the basic grating-based spectroscope (Figure 4.4) and the Littrow-type of prism-based spectroscope (Figure 4.20) do not have any fundamental differences.[7]

Also as mentioned, some prism-based spectroscopes may use a chain of several prisms. This is to increase the dispersion of the instrument. The dispersion of a prism is more or less fixed because most have apex angles close to 60° (or 30° plus a reflective back), and the choice of suitable materials to form the prism is quite limited. By contrast, the dispersion of a grating is easily changed (Equation 4.26) by changing the line spacing and/or the spectral order. However, if the spectrum coming from one prism is fed into a second prism, then the dispersion will be doubled – and be tripled, if three prisms are used. The spectral resolution is unchanged and remains that for a single prism. Thus, such an arrangement is of use when the resolution of the system is limited by some element of the spectroscope other than the prism. Because prisms, especially towards their bases, are physically thick, absorption in the prism's material increases rapidly with the number of prisms being used, so instruments using more than three prisms were/are rare. The recently commissioned NIR Planet Searcher (NIRPS) instrument, though, uses five low-apex-angle ZnSe prisms for its cross dispersion. It operates simultaneously with HARPS on European Southern Observatory's (ESO's) 3.6-m telescope and has a 1 m s^{-1} velocity resolution for exoplanet detection over the 970-nm to 1.81-μm waveband, with spectral resolutions up to 100,000.

The deviation of the optical axis caused by the prism can be a disadvantage for some purposes. Direct-vision spectroscopes overcome this problem and have zero deviation for some selected wavelength. There are several designs but most consist of combinations of prisms made in two different types of glass with the deviations arranged so that they cancel out while some remnant of the dispersion remains. This is the inverse of the achromatic lens discussed in Section 1.1 and, therefore, usually uses crown and flint glasses for its prisms. The condition for zero deviation, assuming that the light passes through the prisms at minimum deviation is

$$\sin^{-1}\left[\mu_1 \sin\left(\frac{\alpha_1}{2}\right)\right] - \frac{\alpha_1}{2} = \sin^{-1}\left[\mu_2 \sin\left(\frac{\alpha_2}{2}\right)\right] - \frac{\alpha_2}{2} \tag{4.107}$$

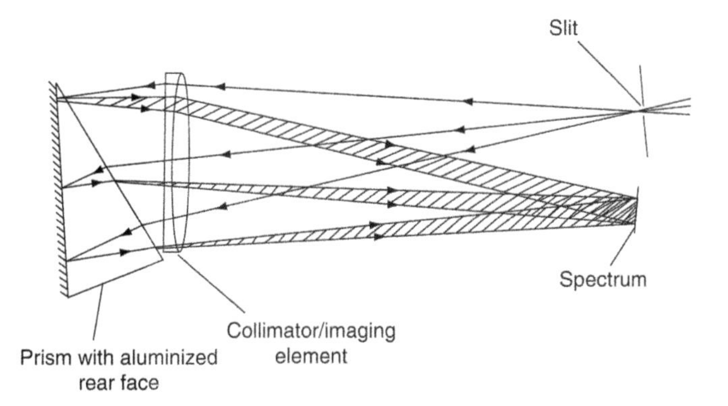

FIGURE 4.20 Light paths in a Littrow spectroscope.

[7] The combination of the collimator and imaging lens into a single component within the Littrow design makes no difference to the optical paths within the two instruments – and is frequently used for grating-based spectroscopes, anyway.

where α_1 is the apex angle of prism number 1, α_2 is the apex angle of prism number 2, μ_1 is the refractive index of prism number 1 and μ_2 is the refractive index of prism number 2. In practical designs, the prisms are cemented together so that the light does not pass through all of them at minimum deviation. Nonetheless, Equation 4.107 still gives the conditions for direct vision to a good degree of approximation. Direct vision spectroscopes can also be based upon grisms, where the deviations of the prism and grating counteract each other.

Applications of most direct-vision spectroscopes are non-astronomical because they are best suited to visual work. They are, though, being increasingly used within instruments which obtain direct images as well as spectra – such as many integral field spectroscopes (see also grisms, below). Spectra are obtained when the direct-vision spectroscope is placed into the optical path and direct images are obtained when it is removed – without the need to adjust the instrument in other respects between the two operating modes.

Another application of the prism, that is now largely of historical interest, having been supplanted by integral field spectroscopy (see below), is the simplest spectroscope of all, the objective prism. This is just a thin prism that is large enough to cover completely the telescope's objective and which is positioned immediately before the telescope's entrance aperture. The starlight is already parallel, so that a collimator is unnecessary, while the scintillation disc of the star replaces the slit. The telescope acts as the imaging element (Figure 4.21). This has the enormous advantage that a spectrum is obtained for *every* star in the normal field of view. Thus, if the telescope is a Schmidt camera, up to 10^5 spectra may be obtainable in a single exposure. The system has three main disadvantages. First, the dispersion is low; second, the observed star field is at an angle to the telescope axis (although direct vision objective prisms can be made from two prisms of different glasses oriented in opposite directions to each other); and finally there is no reference point for wavelength measurements (though a number of ingenious adaptations of the device have been tried to overcome this latter difficulty).

4.2.3 Grating Spectroscopes

Most of the gratings used in astronomical spectroscopes are of the reflection type. This is because the light can be concentrated into the desired order by blazing quite easily, whereas for transmission gratings, blazing is much more difficult and expensive. Transmission gratings are, however, often used in grisms, and these are finding increasing use in integral field spectroscopes, etc.

Plane gratings are most commonly used in astronomical spectroscopes and are almost invariably incorporated into the compact basic spectroscope (Figure 4.4 – sometimes called a 'Czerny-Turner system' when it is based upon a grating) and the Littrow spectroscope (Figure 4.20) which is called an 'Ebert spectroscope' when based upon a grating and reflection optics.

Most of the designs of spectroscopes that use curved gratings are based upon the Rowland circle (Figure 4.6). The Paschen-Runge mounting in fact is identical with that shown in Figure 4.6. It is a common design for laboratory spectroscopes because wide spectral ranges can be accommodated, but its size and awkward shape make it less useful for astronomical purposes. A more compact design based upon the Rowland circle is called the 'Eagle spectroscope' (Figure 4.22). However, the vertical displacement of the slit and the spectrum (see the side view in Figure 4.22) introduces some

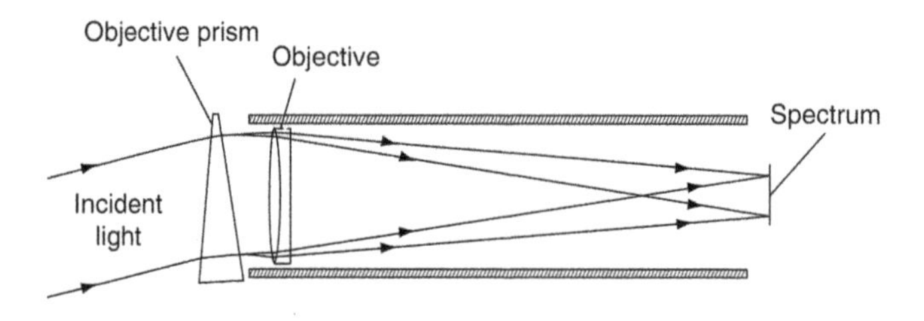

FIGURE 4.21 An objective prism spectroscope.

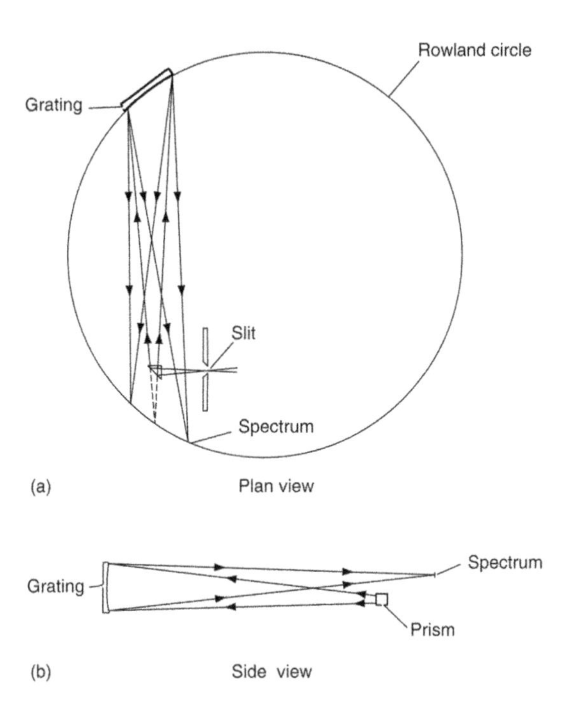

FIGURE 4.22 Optical arrangement of an Eagle spectroscope: (a) Plan view and (b) side view.

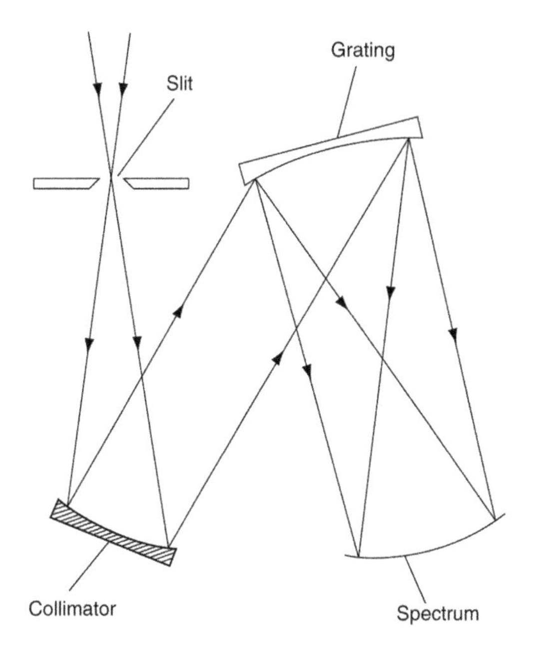

FIGURE 4.23 The Wadsworth spectroscope mounting.

astigmatism. The Wadsworth design abandons the Rowland circle but still produces a stigmatic image through its use of a collimator (Figure 4.23). The focal surface, however, becomes paraboloidal, and some spherical aberration and coma are introduced. Furthermore, the dispersion for a given grating, if it is mounted into a Wadsworth system, is only half what it could be if the same grating were mounted into an Eagle system because the spectrum is produced at the prime focus of the grating and not at its radius of curvature.

With designs such as the Wadsworth and its variants, the imaging element is likely to be a Schmidt camera system (Section 1.1) to obtain high-quality images with a fast system. Some recent spectroscopes though, have used dioptric cameras to avoid the light loss involved with the detector holder in a Schmidt system, Exotic optical materials such as calcium fluoride often need to be used in these designs to achieve the required imaging speed, flat field and elimination of other aberrations. Gratings can also be used, as discussed previously, in various specialised applications such as nebular and prominence spectroscopes.

Spectroscopes are usually optimised for one spectral region. If spectra are needed that extend over a wider range than is covered by the spectroscope, then it may need adjusting to operate in another region, or a different instrument entirely may be required. To overcome this problem to some extent, several spectroscopes have two or three channels optimised for different wavelength ranges. The incoming light from the telescope is split into the channels by dichroic mirrors, so that (with some designs) the spectra can be obtained simultaneously. In other designs, the spectra are obtained in quick succession with little down time needed to adjust the spectroscope. The ESO's Ultraviolet/Visual Echelle Spectroscope (UVES), for example, has a blue channel covering 300–500 nm and a red channel covering 420–1100 nm. VIMOS similarly could select from six different wavebands within the 370-nm to 1-μm spectral range by using six different grisms. Other examples of multi-channel instruments are mentioned below.

The design of a spectroscope is generally limited by the available size and quality of the grating and these factors are, of course, governed by cost. The cost of a grating, in turn, is dependent upon its method of production.

The best gratings are originals, which are produced in the following manner. A glass or other low expansion substrate is over-coated with a thin layer of aluminium. The grooves are then scored into the surface of the aluminium by lightly drawing a diamond across it. The diamond's point is precisely machined and shaped so that the required blaze is imparted to the rulings. An extremely high-quality machine is required for controlling the diamond's movement because not only must the grooves be straight and parallel, but also their spacings must be uniform if Rowland and Lyman ghosts are to be avoided. The position of the diamond is, therefore, controlled by a precision screw thread and is monitored interferometrically. We have seen that the resolution of a grating is dependent upon the number of grooves, while the dispersion is a function of the groove spacing (Section 4.1). So, ideally a grating should be as large as possible and the grooves should be as close together as possible (at least until their separation is less than their operating wavelength). Unfortunately, both of these parameters are limited by the wear on the diamond point. Thus, it is possible to have large coarse gratings and small fine gratings, but not large fine gratings. The upper limits on size are about 0.5 m^2 and on groove spacing, about 1,500 lines mm^{-1}. A typical grating for an astronomical spectroscope might be 0.1-m across and have 500 lines mm^{-1}.

Often, though, a replica grating will be adequate, especially for lower resolution instruments. Because many replicas can be produced from a single original, their cost is a small fraction of that of an original grating. Some loss of quality occurs during replication, but this is acceptable for many purposes. A replica improves on an original in one way however and that is in its reflection efficiency. This is better than that of the original because the most highly burnished portions of the grooves are at the bottoms of the grooves on the original but are transferred to the tops of the grooves on the replicas. The original has of course to be the inverse of the finally desired grating. Covering the original with a thin coat of liquid plastic or epoxy resin, which is stripped off after it has set, produces the replicas. The replica is then mounted onto a substrate, appropriately curved if necessary and then aluminised.

More recently, large high-quality gratings have been produced holographically. An intense monochromatic laser beam is collimated and used to illuminate a photoresist-covered surface. The reflections from the back of the blank interfere with the incoming radiation and the photoresist is exposed along the nodal planes of the interference field. Etching away the surface then leaves a grating that can be aluminised and used directly, or it can have replicas formed from it as described. The wavelength of the illuminating radiation and its inclination can be altered to give almost any required groove spacing and blaze angle.

A related technique to the last produces VPHGs (see also Section 4.1). A layer of gelatine around 10-μm thick and containing a small proportion of ammonia or potassium dichromate is coated onto a glass substrate. It is then illuminated with an interference pattern from two laser beams. Instead of being etched, however, the layer is then treated in water and alcohol baths so that the refractive index within the layer varies according to the exposure it has received. The layer thus forms a grating with the lines of the grating produced by regions of differing refractive indices. Because the gelatine is hygroscopic, it must be protected after production of the grating by a cover sheet. The VPHGs can have up to 95% efficiency in their blaze region and currently can be produced up to 300 mm in diameter and with from 100 to 6,000 lines per millimetre. They can be used in transmission or reflection, and they seem likely to find increasing use within astronomical spectroscopes. It is also possible to produce two different gratings within a single element. Such gratings can be tilted with respect to each other so that the spectra are separated. Different wavelength regions can then be observed using a single spectroscope. Five VPHGs are used in the WHT Enhanced Area Velocity Explorer (WEAVE)[8] instrument of the William Herschel Telescope (WHT) – two for the lower spectral resolution ($R \leq 7{,}500$) mode and three for the higher resolution ($R \leq 25{,}000$) mode, and it has red and blue spectral channels.

Echelle gratings are used for many recently built spectroscopes. The rectangular format (Figure 4.24) of the group of spectral segments after the cross-disperser matches well to the shape of large CCD arrays, so that the latter may be used efficiently. The UVES for ESO's VLT, for example, operates in the blue, red and NIR regions with two 0.2 × 0.8-m echelle gratings. The gratings have 41 and 31 lines mm^{-1} and spectral resolutions of 80,000 and 115,000, respectively (Figure 4.24). Also, for the VLT, the X-Shooter instrument has three independent spectrographs each of which uses an echelle grating with a cross disperser. The three components cover the spectral regions 300–559.5 nm, 559.5 nm–1.024 μm and 1.024–2.489 μm and receive 'their' portion of the incoming radiation via dichroic mirrors. The maximum spectral resolution is 18,200. Similarly, the Magellan Inamori Kyocera Echelle (MIKE) instrument for the Magellan telescope, which started science operations in 2003 and is still in use at the time of writing, uses two echelle gratings with prism cross dispersers and a dichroic mirror to cover the spectral regions 335–500 nm and 490–950 nm at resolutions up to 83,000.

On a quite different scale there is the commercially produced Basic Echelle Spektrograph (BACHES)[9] instrument that is designed for use on small ($\geq$0.2 m) telescopes and, thus, potentially available to amateur astronomers. This instrument can obtain spectra over the 390- to 750-nm region for stars brighter than 5^{m} at a spectral resolution of 19,000 with a 15-minutes exposure on a 0.35-m telescope (Figure 4.25). It uses a 79 lines mm^{-1} echelle grating with a diffraction grating as the cross-disperser and is designed to monitor spectrum variables such as Be stars.

Most spectroscopic observations require exposures ranging from tens of seconds to hours or more. The read-out times from their detectors are, therefore, negligible in comparison. However, for some applications, such as observing rapidly varying or exploding stars, a series of short exposures may be needed, and then the read-out and processing times can become significant. Recently, therefore, instruments optimised for high-speed spectroscopy have started to be developed. The ULTRASPEC, for example, a visitor instrument on ESO's 3.6-m telescope, uses a

---

[8] Being commissioned at the time of writing.

[9] Marketed by the Baader Planetarium; the cost is about twice that of a 0.3-m Schmidt-Cassegrain Telescope (SCT) commercially produced for the amateur astronomy market.

(a)

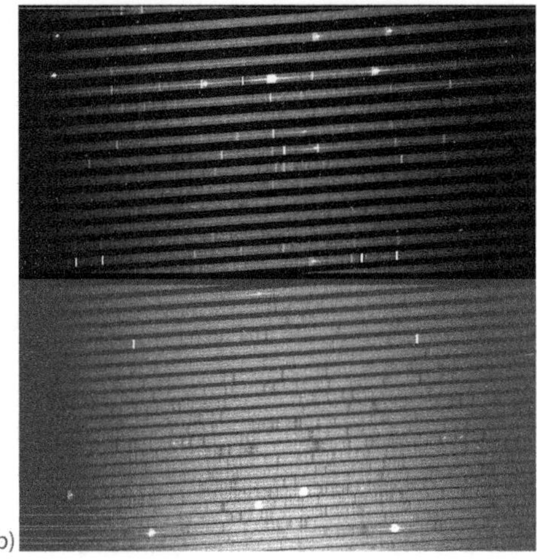

(b)

FIGURE 4.24 FIGURE 4.24 (a) The Ultraviolet/Visual Echelle Spectroscope (UVES) spectrograph shown on one of the Nasmyth platforms of the Kueyen VLT telescope and with its cover removed. (Courtesy of ESO.) (b) An echelle spectrum of supernova SN1987A in the Large Magellanic Cloud (LMC) obtained with UVES on European Southern Observatory's (ESO's) Very Large Telescope (VLT). The individual segments of the spectrum can clearly be seen as almost horizontal bands. The whole coverage is from 479 nm (bottom) to 682 nm (top). Each segment covers about 10 nm with about a 2.5-nm overlap on the left-hand side with the segment above and about a 2.5-nm overlap on the right-hand side with the segment below. The supernova's spectrum is the narrow bright lines running along the centre of each segment and several strong emission lines can be seen, such as Hα near the centre of the fourth segment from the top. The supernova's spectrum is superimposed on a faint solar absorption spectrum arising from scattered light from the Full Moon. Also, numerous emission lines from the Earth's atmosphere can be seen which are narrower than the supernova emission lines and have the same height as the solar spectrum. (Courtesy of ESO.) .

frame-transfer electron multiplying charge-coupled device (EMCCD) which enables hundreds of spectra per second to be obtained (c.f. Mercator Advanced Imager for Asteroseismology [MAIA], see Section 3.2.4). Clearly, such short exposures can only be used for objects with bright apparent magnitudes; however as the 20- to 40-m class telescopes, that are currently being planned, come on stream, high-speed spectroscopy is likely to be extended to fainter objects and become more widely used.

FIGURE 4.25 A solar spectrum obtained using the Basic Ecehlle Spektrograph (BACHES). The $H\alpha$ line is near the centre of the second segment from the top and the sodium D lines near the centre of the sixth segment from the top. (Courtesy of Burwitz/Club of Aficionados in Optical Spectroscopy [CAOS]. For more information on astronomical spectroscopy using small telescopes and by amateur astronomers see http://spectroscopy.wordpress.com/)

4.2.4 Integral Field Spectroscopy

Where the spectra of several individual parts of an extended object such as a gaseous nebula or galaxy are needed, then, as already discussed, a long slit may be used so that a linear segment of the object is covered by a single exposure. Where spectra of every resolution element of an object are required then repeated adjacent exposures using a long-slit spectroscope are one possible, but time-consuming, approach. Several techniques that come under the heading of integral field spectroscopy or 3D spectroscopy have thus been developed recently to obtain spectra for every resolved point within an extended source more efficiently.

Most simply several long slits can be used as the entrance aperture of the spectroscope provided that the spectra do not overlap. Then, all the information that is derivable from a single-slit spectrogram is available, but the source can be covered in a fraction of the time. Another approach is to use a scanning narrow band filter and obtain numerous images at slightly differing wavelengths. Examples of this are the OSIRIS and MMTF tunable filters discussed in Section 4.1. Alternatively, imaging detectors, such as superconducting tunnel junctions (STJs; see Section 1.1), that are also intrinsically sensitive to wavelength can be used to obtain the whole data set in a single exposure. A colour photograph or image is, of course, essentially a low-resolution 3D spectrogram. A possible extension to existing colour imaging techniques which may have a high enough spectral resolution to be useful, is through the use of dye-doped polymers. These are thin films of polymers such as polyvinyl butyral, containing a dye such as chlorin that has a narrow absorption band. Changing conditions within the substrate cause the wavelength of the absorption band to scan through a few nm and potentially provide high-efficiency direct imaging combined with spectral resolutions of perhaps 10^5 or more.

Most integral field spectroscopy however relies on three other approaches. The first is to use an image slicer, though this will need to cover a larger area than that of the stellar image slicers discussed previously. A stack of long plane mirrors whose widths match that of the slit and that are slightly twisted with respect to each other around their short axes is used. The stack is placed at the focal pane of the telescope, and the segments of the sliced image then re-imaged into a line along the length of the slit. SPectrometer for Infrared Faint Field Imaging (SPIFFI), for example, which formed the heart of the Spectrograph for Integral Field Observations in the Near Infrared (SINFONI; decommissioned in 2019) adaptive optics integral field spectroscope on the VLT, used two sets of 32 mirrors to split fields of view of up to 8″ × 8″ into narrow rectangular arrays matching the shape of spectroscope's entrance aperture.

The second approach is to use a large fibre-optic bundle with a square or circular distribution of the fibres at the input and a linear arrangement at the output. The arrangement of the fibres at the output must correspond in a simple and logical fashion to their arrangement at the input so that the position within the final spectrum of the individual spectrum from a point in the image can be found. As with the stellar image slicers, the cladding of the fibres would mean that significant amounts of light would be lost if the fibre-optics were to be placed directly at the telescope's focus. Instead an array of small lenses is used to feed the image into the fibres. The array of square or hexagonal lenses is placed some distance behind the telescope focus and each lens images the telescope pupil onto the end of a fibre. Providing that these sub-images are contained within the transmitting portions of the fibres, no light is lost due to the cladding.

The Kilo-Fibre Optical AAT Lenslet Array (KOALA) Integral Field Unit (IFU), for example offers a 1,000-element lenslet fibre-optic feed to the AAOmega spectroscope on the Anglo-Australian Telescope (AAT). It can provide 0.7″ or 1.25″ angular resolution over 18 × 24″ or 32 × 43″ fields of view, respectively. Spectral resolutions range from 1,300 to 10,000, and the 370- to 950-nm spectral region is coverable. A data-processing pipeline written in python is used.

The third approach also uses an array of small lenses but dispenses with the spectroscope slit. The lenses produce a grid of images that is fed directly into the spectroscope and results in a grid of spectra. By orienting the lens array at a small angle to the dispersion, the spectra can be arranged to lie side by side and not to overlap.

A fairly recent development is the photonic lantern. The radiation is input into a multi-mode optical fibre. The large size[10] of multi-mode fibres enables the whole of the stellar image to be accepted. The various modes are then separated from each other within a flared section of the fibre and each is then fed into a single-mode optical fibre. The single-mode fibres are much smaller in diameter than the multi-mode fibre and feed the radiation directly into the spectrograph. A prototype photonic spectrograph was successfully tested in 2012 using the 3.9-m AAT and a NIR spectrum of π^1 Gru obtained at a spectral resolution of 2,500. Gemini Near infrared OH Suppression IFU System (GNOSIS; see Section 1.1.17) followed from that first trial and PRAXIS (Figure 1.25) is the current instrument based around a photonic lantern.

The ideal output from an integral field spectroscope is a spectral cube (hence, the alternative name of 3D spectroscopy). The spectral cube has two sides that are the x and y positional coordinates of a conventional 2D image, with the third side being wavelength or frequency. Thus, each pixel in the 2D image of the object or area of sky has its spectrum recorded – alternatively we may regard the cube as a series of images at different wavelengths each separated by the spectral resolution of the system. The term 'Voxel'[11] is sometimes used for a single datapoint within a 3D grid of data. Spectral cubes can in principle be obtained for any wavelength but currently are only available in the visible, NIR and the microwave regions (where, for example, Heterodyne Array Receiver Program (HARP) on the James Clerk Maxwell Telescope (JCMT) observes a 16-pixel image over the 325- to 375-GHz region – Section 1.2). Multi Unit Spectroscopic Explorer (MUSE), a second-generation instrument for the VLT's Yepun telescope, is based upon 24 IFU modules each with a 4,096 × 4,096-pixel CCD array as its detector and fed by a reflective image slicer (Section 4.2.1). MUSE observes over the 480- to 930-nm region with a 1 arcminute^2 field of view. It has a spectral resolution of up to 3,600 and its angular resolution about 200 milliarcseconds using Yepun's adaptive optics system (Section 1.1.22.1). The resulting spectral cube contains some 324 million voxels.

---

[10] Multi-mode optical fibres can be up to 200 μm in diameter, single-mode fibres are around 10-μm across. The physical size of the image of a star from a large telescope with good adaptive optics atmospheric correction is around 100 μm.

[11] Volume Element - c.f. 'pixel' for a 2D image.

One of the first light instruments for the Extremely Large Telescope (ELT) will be High Angular Resolution Monolithic Optical and Near-Infrared Spectrograph (HARMONI). Its details are still under discussion, but it is clear that it will be one of the ELT's prime workhorse instruments. It is expected to have its own adaptive optics system that will use one natural and six artificial guide stars and operate via the main telescope's fourth mirror. HARMONI will be an IFU, possibly based upon image slicers, with four spectrographs, spectral resolutions between 3,500 and 20,000 and covering the 470-nm to 2.45-μm spectral region. Its field of view will range from 600 × 800 milliarcseconds to 6 × 9″ using array detectors with up to 32,000 pixels and it will be cooled to 140 K.

4.2.5 Multi-Object Spectroscopy

Some versions of integral field spectroscopes, especially the earlier designs, use multiple slits to admit several parts of the image of the extended object into the spectroscope simultaneously. Such an instrument can easily be extended to obtain spectra of several different objects simultaneously by arranging for the slits to be positioned over the images of those objects within the telescope's image plane. Practical devices that use this approach employ a mask with narrow linear apertures cut at the appropriate points to act as the entrance slits. However, a new mask is needed for every field to be observed, or even for the same field if different objects are selected, and cutting the masks is generally a precision job requiring several hours work for a skilled technician. This approach has therefore rather fallen out of favour, though the two Gemini telescopes' Multi Object Spectroscopes (GMOSs) continue to use masks. Their masks are generated automatically using a laser cutter, with the slit positions determined from a direct image of the required area of the sky obtained by GMOS operating in an imaging mode. The Large Binocular Telescope's (LBT's) LBT Utility Camera in the Infrared (LUCI)[12] can also operate with masks to provide multi-object spectroscopy over the 900-nm to 2.5-μm region as well as being available in other operating modes. One mode of operation of the Keck II telescope's Deep Imaging Multi-Object Spectrograph (DEIMOS) instrument uses up to 130 slits cut into a mask for spectroscopy over the 410- to 1,100-nm region covering an 80 arcminute^2 area of the sky. Up to 11 pre-cut masks can be stored in the instrument to facilitate rapid exchanges. Big Baryon Oscillation Spectroscopic Survey (BOSS; see Section 1.7.2.4) on 2.5-m Sloan Digital Sky Survey (SDSS) telescope uses an aluminium mask with 1,000 holes drilled through it feeding two CCD-based spectroscopes (red and blue channels). It has a 3° × 3° field of view and a best spectral resolution of 2,650.

A related technique has been developed for NIRSpec on the JWST. The source mask is made from four arrays, each with about 62,400 microscopic windows, which can be opened to allow the light from the object through to the spectrograph or otherwise be closed. Individual windows (often called 'shutters' and the array; a micro-shutter device) are 100 × 200 μm in size (a human hair is typically about 50-μm thick). The micro-shutter arrays will allow NIRSpec to select up to a hundred objects for viewing simultaneously within a single exposure. Several adjacent shutters may be opened to form a slit. The shutters are individually addressed by row and column electrodes and the open shutters held in position by an electrostatic charge during an exposure. Perhaps not surprisingly, given the complexity and microscopic size of the arrays, about 15% of the shutters are stuck – either open or closed. This, however, normally will only cause a problem with planning the observations; it will not affect any observations that *are* obtained. NIRSpec will cover the spectral range from 600 nm to 5.3 μm, its field of view will be 3′ across, its spectral resolution be optional at 100, 1,000, and 2,700 and it will be cooled to 38 K. It will also be able to operate as an imager to align the masks' windows accurately with the desired sources. Micro-shutter arrays up to 420 × 840 elements in size (more than 350,000 shutters) have now been manufactured and a smaller prototype was tested successfully during a Black Brant rocket-based mission to study M33 in 2019.

[12] Previously named LUCIFER.

The Large Ultraviolet Optical Infrared Survey (LUVOIR) astro2020 proposal has already been mentioned (Section 1.1.23). Its LUVOIR Ultraviolet Multi-Object Spectrograph (LUMOS) spectroscope is expected to be a multi-object and a point-source instrument covering from 100 nm to 1 μm at spectral resolutions ranging from 500 to 65,000 using holographic gratings. It would have a 1.6 × 3′ field of view and use micro-shutter arrays to select its targets (c.f. NIRSpec above). There would be six micro-shutter arrays, each 480 × 840 shutters in size, with the individual shutters at 100 × 200 μm centres and each of them accepting an area of the sky 68 × 136 milliarcseconds in size.

Most multi-object spectroscopy however is now undertaken using fibre-optics to transfer the light from the telescope to the spectroscope. The individual fibre-optic strands are made large enough to be able to contain the whole of the seeing disk of a stellar image or in some cases the nucleus of a distant galaxy or even the entire galaxy. Each strand then has one of its ends positioned in the image plane of the telescope so that it intercepts one of the required images. The other ends are aligned along the length of the spectroscope slit, so that hundreds of spectra may be obtained in a single exposure. Initially, the fibre-optics were positioned by being plugged into holes drilled through a metal plate. But this has the same drawbacks as cutting masks for integral field spectroscopes and has been superseded. There are now two main methods of positioning the fibres, both of which are computer-controlled and allow hundreds of fibres to be repositioned in a matter of minutes.

The first approach is to attach the input ends of the fibre-optic strands to small magnetic buttons that cling to a steel back plate. The fibres are reconfigured, one at a time, by a robot arm that picks up a button and moves it to its new position as required. A positional accuracy for the buttons of 10–20 μm is needed.

The Dark Energy Spectroscopic Instrument (DESI), on the Mayall telescope and its 5,000 robotically positioned fibre-optic cables has already been mentioned (Section 1.7.2.4). Its fibre-optic cables' positions can be changed in three minutes and the cables, in groups of 500, feed ten spectroscopes some 50 m away from the telescope. As another example, the AAT's AAOmega, may be combined with the fibre-positioner from the two degree field (2dF)[13] facility as well as with KOALA (see above). This provides up to 400 optical fibre pick-offs (392 for science, 8 for guiding) and it can be reconfigured in about 45 minutes.

The OzPoz[14] system for ESO's VLT originally could position up to 560 fibres in the same way. It is now used with the Fibre Large Array Multi Element Spectrograph (FLAMES) instrument. FLAMES has several modes of operation enabling it to observe between 8 and 130 areas of the sky simultaneously with spectral resolutions up to 47,000. In some modes fibre-optic bundles replace the single fibres each covering a 2″ × 3″ area of the sky. Thus, each fibre-optic bundle represents a mini integral field feeding one of two spectroscopes, so enabling the whole of the visible spectrum to be observed. The system uses two backing plates so that one can be in use while the other is being reset. Swapping the plates takes about 5 minutes, thus minimising the dead time between observations.

The second approach mounts the fibres on the ends of computer-controlled arms that can be moved radially and/or from side to side. The arms may be moved by small motors, or, as in the Subaru telescope's (decommissioned) Echidna system, by an electromagnetic system. Subaru's current prime focus spectrograph observes over a 1.3° field of view and employs 2,394 optical fibres. The fibres can each be positioned within a 9.5-mm diameter circle (its patrol area) with a precision of 5 μm by a system of actuators (called 'Cobra'), wherein each 7.5 mm-wide actuator houses two rotary motors to give the required two-degrees of freedom fibre positioning. The fibre's movement within its patrol area arises from its being mounted on the end of a rotatable arm and with that arm mounted on the edge of a rotatable disk (rather like the Moon's motion around the Earth and

[13] Now decommissioned, although its fibre-positioner continues in use.

[14] Oz for Australia and Poz for positioner.

Sun, though with rather different radii for the 'orbits'). The patrol areas overlap so that the whole of the field of view is accessible (although collisions between the fibres do need to be avoided). The 4-m Multi-Object Spectroscopic Telescope (4MOST) instrument, planned for first light in 2022 on ESO's Visible and Infrared Survey Telescope for Astronomy (VISTA) telescope, will use Echidna-type fibre-optic positioners. The Echidna system is based upon fibre-optic cables mounted on arms that can be extended radially and rotated slightly from side to side to cover the field of view. The low-resolution ($R \leq 7{,}800$) spectrographs will employ 1,624 fibres, whilst the high-resolution ones ($R > 18{,}000$) will use 812 fibres. Wavelengths from 390 to 680 nm will be observable in three separate wavebands.

Schmidt cameras can also be used with advantage for multi-object spectroscopy because their wide fields of view provide more objects for observation and their short focal ratios are well matched to transmission through optical fibres. Thus, for example, the (long decommissioned) six degree field (6dF) project on the United Kingdom (UK) Schmidt camera could obtain up to 150 spectra simultaneously over a 40 deg^2 field of view with automatic positioning of the fibre-optics using the magnetic button system.

Once the light is travelling through the fibre-optic cables, it can then be led anywhere. There is thus no requirement for the spectroscope to be mounted on the telescope, although this remains the case in a few systems. More frequently, the fibre-optics take the light to a fixed spectroscope that can be several tens of metres away (see DESI, for example). This has the advantage that the spectroscope is not subject to changing gravitational loads and so does not flex and it can be in a temperature-controlled room. ESO's HARPS[15] spectroscope on the 3.6-m telescope is intended for extra-solar planet finding and so needs to determine stellar velocities to within ±1 m/s. Thus, a stable instrument is required, and HARPS is not only mounted 38 m away from the telescope but is enclosed in a temperature-controlled vacuum chamber. The just-commissioned NEID spectrograph is expected to better HARPS' velocity resolution by about a factor of three (i.e., to ± ~300 mm/s). The NEID is mounted on the WIYN telescope and uses an echelle plus prism-cross-disperser design to provide continuous cover over the 380- to 930-nm spectral region. It uses CCD array detectors and a laser frequency comb (LFC) calibration spectrum and at some wavelengths, achieves a throughput of 40% or more.

4.2.6 Techniques of Spectroscopy

There are several problems and techniques which are peculiar to astronomical spectroscopy and that are essential knowledge for the intending astrophysicist.

One of the greatest problems and one which has been mentioned several times already, is that the image of the star may be broadened by atmospheric turbulence until its size is several times the slit width. Only a small percentage of the already pitifully small amount of light from a star, therefore, enters the spectroscope. The design of the spectroscope and in particular the use of a large focal length for the collimator in comparison with that of the imaging element, can enable wider slits to be used. Even then, the slit is generally still too small to accept the whole stellar image, and the size of the spectroscope may have to become large if reasonable resolutions and dispersions are to be obtained. The alternative approach to the use of a large collimator focal length lies in the use of an image slicer and/or adaptive optics as discussed previously.

Another problem in astronomical spectroscopy concerns the width of the spectrum. If the stellar image is held motionless at one point of the slit during the exposure, then the final spectrum may only be a few microns high. Not only is such a narrow spectrum difficult to measure, but also individual spectral features are recorded by only a few pixels and so the noise level is high. Both of these difficulties can be overcome, though at the expense of longer exposure times, by widening

[15] This instrument is sometimes called HARPS-S because it is in the Southern Hemisphere. An almost identical instrument, HARPS-N is located on La Palma and uses the 3.6-m Telescopio Nazionale Galileo (TNG).

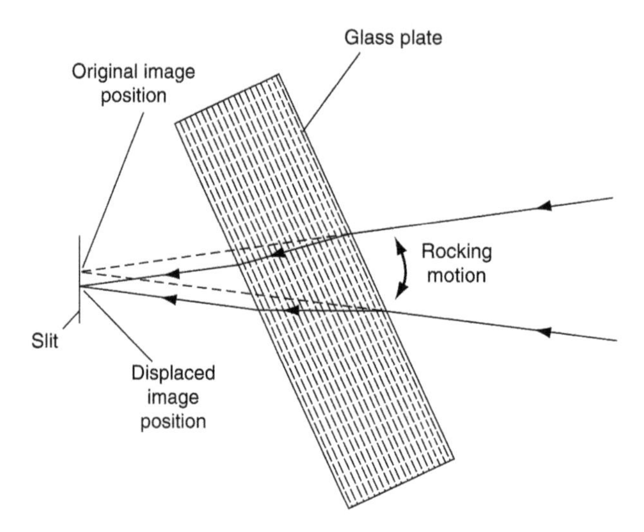

FIGURE 4.26 Image displacement by a plane-parallel glass plate.

the spectrum artificially. There are several ways of doing this, such as introducing a cylindrical lens to make the images astigmatic or by trailing the telescope during the exposure so that the image moves along the slit. For single-object spectroscopes the spectrum can be widened by using a rocker block.[16] This is just a thick piece of glass (etc.) with polished plane-parallel sides through which the light passes. The block is oscillated about an axis parallel to the spectrum and the displacement of the light beam (Figure 4.26) moves the beam up and down, so broadening the spectrum. In multi-object spectroscopes using fibre-optic links the fibre size is generally sufficient to provide a useable spectrum. Astronomical spectra are typically widened to about 0.1–1 mm. Clearly, though, the spectra in integral field spectroscopes cannot be widened without destroying the imaging information

Spectrographs operating in the NIR (1–5 μm) are generally of conventional designs, but need to be cooled to reduce the background noise level. Typically, instruments such as ESO's Cryogenic Infrared Echelle Spectrograph (CRIRES), which operates from 950 nm to 5.2 μm, have their main components cooled to 70 K, while the detectors are held at 25–4 K or less and the whole instrument is enclosed in a vacuum chamber. The third-generation instrument, Multi-Object Spectrometer For Infrared Exploration (MOSFIRE), on the Keck I telescope, uses a cooled mask of 46 slits to select objects over a 6.1′ × 6.1′ area in the NIR and its detector is cooled to 77 K.

At MIR wavelengths, the entire spectroscope may need to be cooled to very low temperatures. The VLT's Imager and Spectrometer for the Mid-Infrared (VISIR),[17] which started operations in 2004 but is still in use at the time of writing, for example, is housed in a vacuum vessel and has most of its structure and optics cooled to 29 K, with the parts near the detectors cooled to <15 K and the rs to 10 K. The Stratospheric Observatory for Infrared Astronomy's (SOFIA's) High Resolution MIR Spectrometer (HIRMES) MIR spectroscope is currently under development for a possible operational start in 2020–2021. It is planned to operate with TES detectors cooled to 0.1 K and in a vacuum, over the 25- to 122-μm spectral region, with spectral resolutions up to 50,000, tunable filters and with a 100″ field of view.

The UV spectroscopy is not possible from ground-based telescopes (see also Section 1.3), except in the most limited sense from the eye's cut-off wavelength of about 380 nm[18] to the point where the Earth's atmosphere becomes opaque (300–340 nm depending upon altitude, state of the ozone

[16] Also called a dekker or decker.

[17] VLT's Imager and Spectrometer for the Mid-Infrared.

[18] Persons having had cataract operations may be able to see further in to the UV than this because the eye's lens absorbs short wave radiation strongly.

layer, and other atmospheric conditions), i.e., the UVA and UVB regions as they are popularly known. All except the longest wavelength UV spectroscopy, therefore, needs spectroscopes flown onboard balloons, rockets or spacecraft to lift them above most or all of the Earth's atmosphere. The still operating Cosmic Origins Spectrograph (COS) instrument, for example, was installed on board the HST in 2009 and uses windowless microchannel plates (MCPs) as its detectors. It covers the 115–320 nm region at a spectral resolution of up to 24,000. Similarly, India's Astrosat carries the UVIT instrument (Section 1.1.16.1) for UV imaging and slitless spectroscopy with a spectral resolution of ~100. It is based upon a 0.4-m telescope, has an angular resolution of 1.8″ in the UV and its detectors are CsI-based image intensifiers. The Russian led World Space Observatory-Ultraviolet (WSO-UV) is scheduled for launch in 2023 and will carry a 1.7-m telescope for imaging and spectroscopy over the 115–315 nm spectral region. The spectrographs will use echelle gratings with designs based upon the Rowland circle. The highest available spectral resolution will be ~50,000.

Atmospheric dispersion is another difficulty that needs to be considered. Refraction in the Earth's atmosphere changes the observed position of a star from its true position. But the refractive index varies with wavelength. For example, at standard temperature and pressure, we have Cauchy's formula for the refractive index of the atmosphere:

$$\mu = 1.000287566 + \frac{1.3412 \times 10^{-18}}{\lambda^2} + \frac{3.777 \times 10^{-32}}{\lambda^4} \tag{4.108}$$

so that the angle of refraction changes from one wavelength to another and the star's image is drawn out into a short vertical spectrum. In a normal basic telescope system (i.e., without the extra mirrors and so on required, for example, by the Coudé system), the long wavelength end of the spectrum will be uppermost. To avoid spurious results, particularly when undertaking spectrophotometry, the atmospheric dispersion of the image must be arranged to lie along the slit, otherwise certain parts of the spectrum may be preferentially selected from the image. Alternatively, an atmospheric dispersion corrector may be used. This is a low but variable dispersion direct vision spectroscope that is placed before the entrance slit of the main spectroscope and whose dispersion is equal and opposite to that of the atmosphere. The VLT's X-Shooter, for example, has pairs of atmospheric dispersion corrector prisms in its UV and visible component spectrographs. The problem, however, is usually only significant at large zenith distances, so that it is normal practice to limit spectroscopic observations to zenith angles of less than 45°. The increasing atmospheric absorption and the tendency for telescope tracking to deteriorate at large zenith angles also contributes to the wisdom of this practice.

In many cases, it will be necessary to try and remove the degradation introduced into the observed spectrum because the spectroscope is not perfect. The many techniques and their ramifications for this process of deconvolution are discussed in Section 2.1.

To determine radial velocities, it is necessary to be able to measure the actual (i.e., observed) wavelengths of the lines in the spectrum and to compare these with their laboratory wavelengths. The difference, $\Delta\lambda$, then provides the radial velocity via the Doppler shift formula

$$\mathrm{v} = \frac{\Delta\lambda}{\lambda} c = \frac{\lambda_{\mathrm{Observed}} - \lambda_{\mathrm{Laboratory}}}{\lambda_{\mathrm{Laboratory}}} c \tag{4.109}$$

where c is the velocity of light. The observed wavelengths of spectrum lines are most usually determined by comparison with the positions of emission lines in an artificial spectrum. This comparison spectrum is normally that of an iron or copper arc or comes from a low-pressure gas emission lamp such as sodium or neon or combinations such as thorium and argon (or uranium and neon for the NIR). The light from the comparison source is fed into the spectroscope and appears as one or two spectra on one or both sides of the main spectrum (Figure 4.27). The emission lines in the comparison spectra are at their rest wavelengths and are known precisely. The observed wavelengths

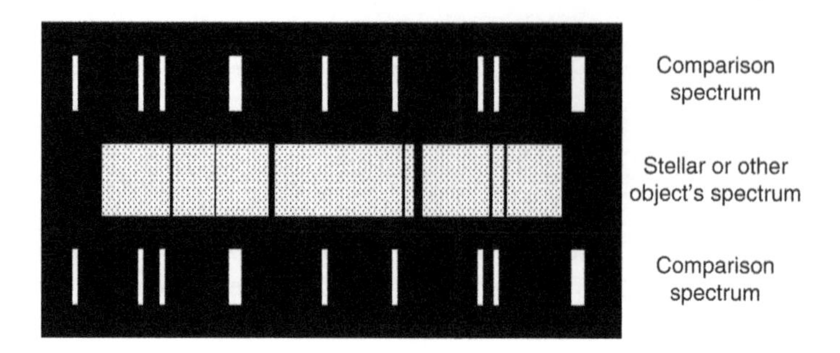

FIGURE 4.27 The wavelength comparison spectrum.

of the stellar (or other object's) spectrum lines are found by comparison with the positions of the artificial emission lines. For high precision radial velocities needed to detect exoplanets (see below), an absorption cell may be used to produce the comparison lines.

However, the emission lines produced by real atoms have drawbacks when used to produce comparison spectra. This problem is becoming particularly apparent with the requirement to be able to measure radial velocities to a precision of ± 1 m s^{-1} or even $\pm$ 0.1 m s^{-1} and for this precision to be stable over several years – which is needed for the detection of terrestrial-sized exoplanets. Thus, the lines from real atoms are not distributed uniformly, and there may be long gaps between usable lines, the intensities of the lines vary widely and also many lines are blends whose exact median wavelength will depend upon the relative strengths of the two or more individual lines contributing to the blend – and these can vary with the physical conditions within the emission lamp, etc. Finally the wavelengths of the lines are not known a priori, but have to be measured individually in the laboratory.

At the time of writing, therefore, the production of a 'comb' of close, uniformly intense and regularly spaced emission features using femto-second lasers is coming into astronomical use (see also the next subsection). The LFC is based upon the repetitive emissions from a mode-locked laser. Mode-locked lasers emit pulses of radiation separated by the round-trip time of the laser cavity. Stability is ensured by synchronising the repetitions with an atomic clock (perhaps from the GPS satellites). The Fourier transform of the output of such a laser is the frequency comb, and its emission features are spaced at frequencies of $1/T$ and their spectral widths are $1/D$, where T is the laser repetition interval and D is the laser pulse duration.

For rapid determinations of radial velocity, the cross-correlation spectroscope, originally devised by Roger Griffin, can be used, although this device will not now be much encountered. This instrument places a mask over the image of the spectrum and re-images the radiation passing through the mask onto a point source detector. The mask is a negative version (either a photograph or artificially generated) of the spectrum. The mask is moved across the spectrum and when the two coincide exactly, there is a sharp drop in the output from the detector. The position of the mask at the correlation point can then be used to determine the radial velocity with a precision, in the best cases, of a few metres per second.

4.2.7 Exoplanets

About 20% of the 4,000 or so exoplanets that have currently been discovered have been found through the radial velocity variations of their host stars as the exoplanet and star orbit their common centre of mass. The first conventional exoplanet to be found was 51 Peg b by Michel Mayor and Didier Queloz in 1995. The host star, 51 Peg, was changing its radial velocity by 120 m s^{-1} every 4.2 days. They used the Elodie spectrograph for the 1.93-m telescope at the Observatoire de Haute Provence to obtain the spectra. This spectroscope is of a relatively convention optical design but is housed separately from the telescope in a temperature-controlled room and fed by fibre-optic cable. It was able to measure velocities to an accuracy of about 10 m s^{-1}.

The host star's velocity changes for 51 Peg b, however, are unusually large. More typical is the recently discovered[19] exoplanet, HD 213885c, orbiting around a host star that is almost a solar-twin. The planet has a mass 20 times that of the Earth and it orbits ~0.06 astronomical units (AU) out from its star every 4.8 days at about 130 km s^{-1} – yet the radial velocity[20] of the *host star* resulting from the effect of this exoplanet is just 7 m s^{-1}. Spectrographs searching for exoplanets must thus be able to measure radial velocities to a precision of ±1.0 to ±0.1 m s^{-1} or so. Most such instruments now in use are, therefore, purpose-designed and in particular have high levels of stability and special arrangements for the comparison spectra. HARPS, for example, is housed in a vacuum chamber and has its temperature controlled to ±0.01 K. The star's light and that of the comparison spectrum (originally a thorium-argon emission lamp, now an LFC – see below) are fed to the spectroscope via identical fibre-optic cables. It can measure radial velocities to ±1 m s^{-1} and has discovered some 230 exoplanets to date.

An alternative means of producing the comparison spectrum is via an absorption cell. These have the advantage that the light paths for the star and comparison spectra are identical. An absorption cell is just a container for a suitable gas that has optically flat windows and through which the light from the star is passed. The gas in the cell then absorbs at its characteristic wavelengths, and these lines are superimposed upon the stellar spectrum. Most absorption cells use molecular iodine at a pressure of about 10 milli-bars and a temperature around 50°C. The majority of iodine's spectrum lines are in the green. A cell length around 100 mm suffices to produce measurable absorption lines.

Confirmation of the discovery of 51 Peg B b was made by Geoff Marcy and Paul Butler using an iodine absorption cell spectrograph on the Hamilton spectrograph of the 3.05-m Shane telescope. Many large telescopes now have spectrographs equipped with absorption cells as optional facilities, such as the High Resolution Echelle Spectrometer (HIRES) on the Keck I telescope which achieves ±1 m s^{-1} accuracy with an iodine absorption cell and has enabled the discovery of nearly 200 exoplanets.

As mentioned, a set of closely and regularly spaced emission lines, a laser comb, can be produced by a mode-locked laser (Figure 4.28). Laser combs have been used for some time as calibration sources for other types of wavelength comparison spectra and are now starting to be used as the primary calibrators. Thus, the Habitable Zone Planet Finder (HPF) on the Hobby-Eberly telescope uses a laser comb based upon a 1.064-μm Nd:YAG laser. The initial comb spectrum only covers about a 10-nm waveband and the HPF spectrum is some 500-nm long. The comb spectrum is, therefore, expanded to cover the full spectrum by multiplying it up within a non-linear SiN waveguide. The initial laser power output also has to be amplified to about 2 W (Section 1.6.3.2.1.1.4) and filtered to ensure that the comb's lines have roughly equal intensities across the whole waveband. The comb is stabilised by comparison with

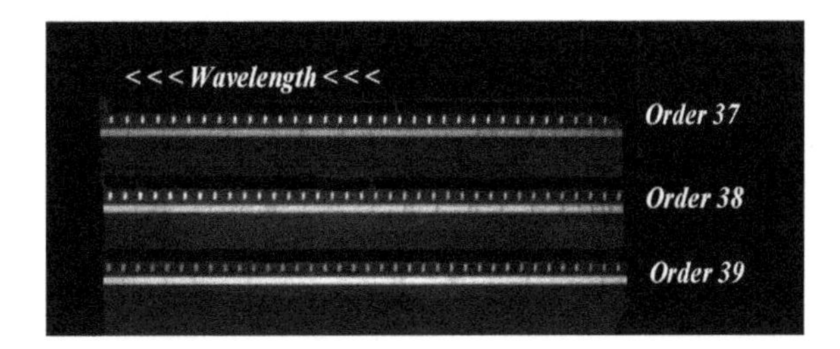

FIGURE 4.28 The spectrum of HD 168723 together with a laser comb comparison spectrum. The image shows three orders from the stellar spectrum running from 1.54 to 1.63 μm with the laser comb spectrum above the stellar spectrum in each case. (University of Colorado/National Institute of Science and Technology (NIST) H band Laser Comb Spectra obtained with the Penn State Pathfinder Spectrograph on the Hobby Eberly Telescope. Courtesy of NIST/Tech Beat).

[19] December 2019.

[20] That is, along the Earth-star line and what must be measured to detect the exoplanet.

a dedicated atomic clock, and its current accuracy is about ±1.5 m s^{-1}. The HARPS instruments are now also operating with laser comb calibrators. The combs have emission lines at 16-GHz (0.016 nm at 550 nm) intervals over the visible spectrum. The measurement accuracy is around ±1 m s^{-1}, with an improvement to around ±25 mm s^{-1} perhaps eventually being possible.

Obtaining the spectrum of an exoplanet itself is a technique that is still in its infancy but has been accomplished. The spectrum may be obtained directly (either via modifications to the reflected host star's spectrum or from the exoplanet's own emissions, if it is hot enough) or the spectrum of the exoplanet's atmosphere, if any, obtained whilst the exoplanet is transiting the host star from the changes in the host star's spectrum. The former is likely to become much more common as more and better planetary coronagraphs are produced (Section 5.3). Thus, HR 8799 e's spectrum was recently studied directly by the Gravity (Section 2.5.4) instrument on the Very Large Telescope Interferometer (VLTI) and showed the presence of carbon monoxide, silicates and iron. The VLTI was able to separate the host star and planet, which are about 390 milliarcseconds apart, using single-mode optical fibres with fields of view of 60 milliarcseconds and so obtain separate spectra at spectral resolutions of 500. Some years previously, the VLT's NaCo instrument obtained a direct spectrum of another exoplanet in the system; HR 8799 c and that exoplanet's spectrum has also been observed by several other spectroscopes, including the Hale telescope's Project 1640 and the Keck's OH Suppressing Infrared Integral Field Spectrograph (OSIRIS)[21] and NIRSpec Adaptive Optics (NIRSPAO) instruments.

The transit approach is exemplified by HARPS-S' recent observation of HD189733b in and out of transit, which showed the presence of sodium in that exoplanet's atmosphere and by the detection of water and TiO in WASP-19b's atmosphere by the HST's WFC3 and the VLT's Focal Reducer and Low Dispersion Spectrograph (FORS2) spectroscopes, respectively.

4.2.8 Future Developments

The major foreseeable developments in spectroscopy seem likely to lie in the direction of improving the efficiency of existing systems rather than in any radically new systems or methods. The lack of efficiency of a typical spectroscope arises mainly from the loss of light within the overall system. To improve on this requires gratings of greater efficiency, reduced scattering and surface reflection and so on from the optical components and so on. Because these factors are already quite good and it is mostly the total number of components in the telescope-spectroscope combination that reduces the efficiency, improvements are thus likely to be slow and gradual.

A possible alternative to laser frequency combs could be based upon the etalon (Section 4.1). If an etalon is illuminated by a white light point source, then only those wavelengths given by Equation 4.51 will emerge from it. By operating with high orders (large values of m), numerous sharp emission lines will be produced whose wavelengths can be accurately and easily calculated. The practical realisation of this idea lies in the Fibre Optic Fabry-Perot (FFP) filter. This comprises a fibre-optic waveguide with multi-layer mirrors at each end. However, the FFP is more difficult to keep stable than the laser comb, and so the latter may continue to be the preferred development.

Adaptive optics is already used on most large telescopes to reduce the seeing disk size of stellar images and so allow more compact spectroscopes to be used. This approach is likely to spread to smaller instruments in the near future. Direct energy detectors such as STJs are likely to be used more extensively, although at present, they have relatively low spectral resolutions and require extremely low operating temperatures. The use of integral field and multi-object spectroscopes is likely to become more common, with wider fields of view and more objects being studied for individual instruments. The extension of high-resolution spectroscopy to longer infrared wavelengths is likely to be developed, even though this may involve cooling the fibre-optic connections and large parts of the telescope as well as most of the spectroscope. Plus, of course, the continuing increase in the power and speed of computers will make real-time processing of the data from spectroscopes much more commonplace.

[21] Not to be confused with the GTC's instrument.

5 Other Techniques

5.1 ASTROMETRY

5.1.1 Introduction

Astrometry is probably the most ancient branch of astronomy, dating back to at least several centuries BCE and possibly to a couple of millennia BCE. Indeed, until the late eighteenth century, astrometry was the whole of astronomy. Yet although it is such an ancient sector of astronomy, it is still alive and well today and employing the most modern techniques, plus some that William Herschel would recognise. Astrometry is the science of measuring the positions in the sky of galaxies, stars, planets, comets, asteroids and recently, of spacecraft. From these positional measurements come determinations of distance via parallax, motions in space via proper motion, orbits and hence, sizes and masses within binary systems and a reference framework that is used by the whole of the rest of astronomy and astrophysics as well as by space scientists to direct and navigate their spacecraft. Astrometry also leads to the production of catalogues of the positions and sometimes the natures, of objects which are then used for other astronomical purposes.

From the invention of the telescope until the 1970s, absolute positional accuracies of about 0.1″ (100 milliarcseconds) were the best that astrometry could deliver. That has now improved to better than 1 milliarcsecond and space missions planned for the next couple of decades should improve that by another factor ×100 at least. Positional accuracies of a few microarcseconds, potentially allow proper motions of galaxies to be determined, though the diffuse nature of galaxy images may render this difficult. However, such accuracies *will* enable the direct measurement of stellar distances throughout the whole of the Milky Way galaxy and out to the Magellanic clouds and the Andromeda galaxy (M31).

Astrometry may be Absolute (sometimes called 'fundamental') when the position of a star is determined without knowing the positions of other stars, or Relative when the star's position is found with respect to the positions of its neighbours. Relative astrometry may be used to give the absolute positions of objects in the sky, provided that some of the reference stars have their absolute positions known. It may also be used for determinations of parallax, proper motion, binary star orbital motion, without needing to convert to absolute positions. An important modern application of relative astrometry is to enable the optical fibres of multi-object spectroscopes (Section 4.2) to be positioned correctly in the focal plane of the telescope to intercept the light from the objects of interest.

The current[1] absolute reference frame is called the International Celestial Reference Frame (ICRF3) and contains 4,536 extragalactic sources; 303 of these sources (called the 'defining sources') form the primary basis of the system, with their positions determined by radio interferometry to an accuracy (at best) of about ±30 microarcseconds. Until 1998, the ICRF1 was based upon optical astrometric measurements and, when space-based interferometric systems produce their results, the definition may well revert to being based upon optical measurements.

[1] Since 2019.

5.1.2 Background

5.1.2.1 Coordinate Systems

The measurement of a star's position in the sky must be with respect to some coordinate system. There are several such systems in use by astronomers, but the commonest is that of right ascension and declination. This system, along with most of the others, is based upon the concept of the celestial sphere; a hypothetical sphere, centred upon the Earth and enclosing all objects observed by astronomers. The space position of an object is related to a position on the celestial sphere by a radial projection from the centre of the Earth.

Henceforth in this section, we talk about the position of an object as its position on the celestial sphere and we ignore the differing radial distances that may be involved. We also extend the polar axis, the equatorial and orbital planes of the Earth until these too meet the celestial sphere (Figure 5.1). These intersections are called the 'celestial North Pole', the celestial equator, etc. Usually, there is no ambiguity if the 'celestial' qualification is omitted, so that they are normally referred to as the North Pole, the equator and so on. The intersections (or nodes) of the ecliptic and the equator are the vernal and autumnal equinoxes. The former is additionally known as the 'first point of Aries' (from its position in the sky some two thousand years ago). The ecliptic is also the apparent path of the Sun across the sky during a year and the vernal equinox is defined as the node at which the Sun passes from the Southern to the Northern Hemisphere. This passage occurs within a day of 21 March each year. The position of a star or other object is thus given with respect to these reference points and planes.

The declination of an object, δ, is its angular distance north or south of the equator. The right ascension, α, is its angular distance around from the meridian (or great circle) that passes through the vernal equinox and the poles, measured in the same direction as the solar motion (Figure 5.2). By convention, declination is measured from $-90°$ to $+90°$ in degrees, arc-minutes and arcseconds

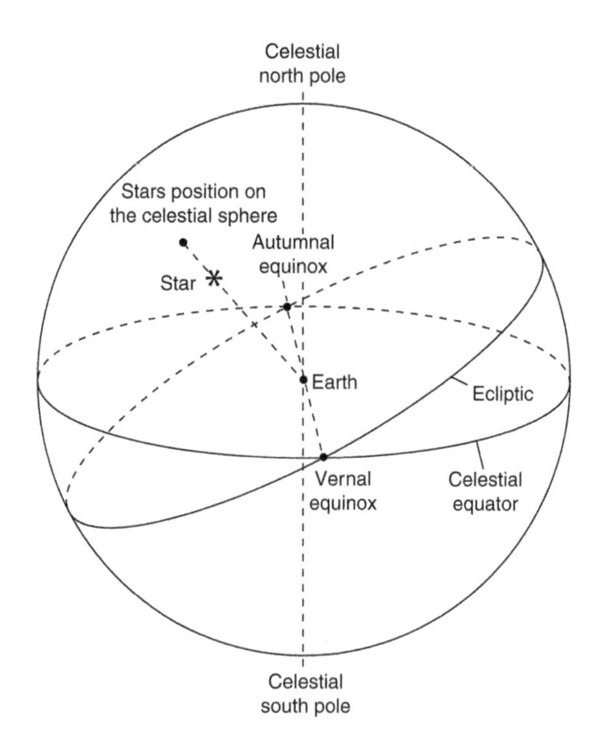

FIGURE 5.1 The celestial sphere.

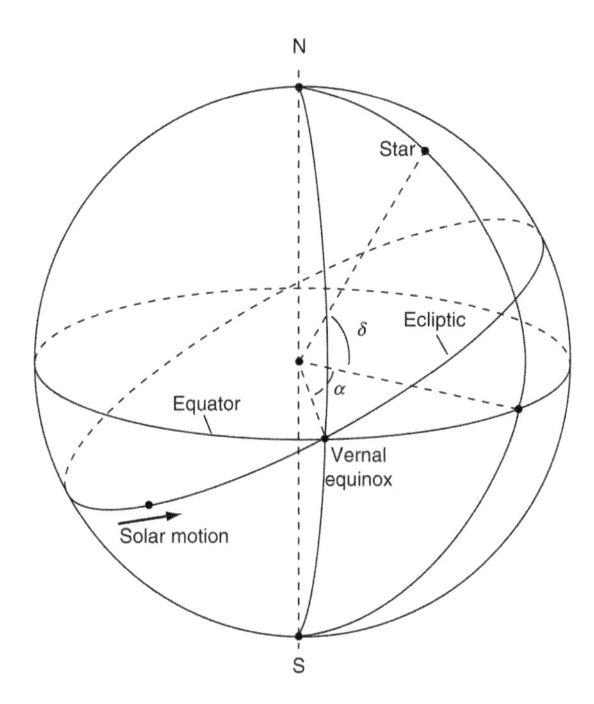

FIGURE 5.2 Right ascension and declination.

and is positive to the north of the equator and negative to the south. Right ascension is measured from 0° to 360°, in units of hours, minutes and seconds of time where, at the equator,

$$1 \text{ hour} = 15° \tag{5.1}$$

$$1 \text{ minute} = 15' \tag{5.2}$$

$$1 \text{ second} = 15'' \tag{5.3}$$

The direction of the Earth's axis moves in space with a period of about 25,750 years – a phenomenon known as 'precession'. Hence, the celestial equator and poles also move. The positions of the stars, therefore, slowly change with time.[2] Catalogues of stars thus customarily give the date, or epoch, for which the stellar positions that they list are valid. To obtain the position at some other date, the effects of precession must be added to the catalogue positions

$$\delta_T = \delta_E + (\theta \sin \varepsilon \cos \alpha_E)\, T \tag{5.4}$$

$$\alpha_T = \alpha_E + [\theta(\cos \varepsilon + \sin \varepsilon \sin \alpha_E \tan \delta_E)]\, T \tag{5.5}$$

where α_T and δ_T are the right ascension and declination of the object at an interval, T, years after the epoch E, α_E, δ_E are the coordinates at the epoch and θ is the precession constant

$$\theta = 50.40'' \text{ yr}^{-1} = 3.36s \text{ yr}^{-1} \tag{5.6}$$

[2] N.B. – This is an *apparent* positional change for the object only arising from the *actual* change in the coordinate system; a *true* change in the position of an object in the sky with respect to the other stars is called its proper motion.

ε is the angle between the equator and the ecliptic – more commonly known as the 'obliquity of the ecliptic'

$$\varepsilon = 23°27'\ 8'' \tag{5.7}$$

Frequently used epochs are the beginnings of the years 1900, 1950, 2000, 2050, and so on with 1975 and 2025 also being encountered. Other effects upon the position, such as nutation, proper motion, and so on may also need to be taken into account in determining an up-to-date position for an object.

Two alternative coordinate systems that are in use are, first, celestial latitude (β) and longitude (λ), which are respectively the angular distances up or down from the ecliptic and around the ecliptic from the vernal equinox and second, galactic latitude (b) and longitude (l), which are respectively the angular distances above or below the galactic plane and around the plane of the galaxy measured from the direction to the centre of the galaxy.[3]

5.1.2.2 Position Angle and Separation

The separation of a double star is just the angular distance between its components. The position angle is the angle from the north, measured in the sense, North → East → South → West → North, from 0° to 360° (Figure 5.3) of the fainter star with respect to the brighter star. The separation and position angle are related to the coordinates of the star by

$$\text{separation} = \left\{\left[(\alpha_F - \alpha_B)\cos\delta_B\right]^2 + (\delta_F - \delta_B)^2\right\}^{1/2} \tag{5.8}$$

$$\text{position angle} = \tan^{-1}\left(\frac{\delta_F - \delta_B}{(\alpha_F - \alpha_B)\cos\delta}\right) \tag{5.9}$$

where α_B and δ_B are the right ascension and declination of the brighter star and α_F and δ_F are the right ascension and declination of the fainter star with the right ascensions of both stars being converted to normal angular measures.

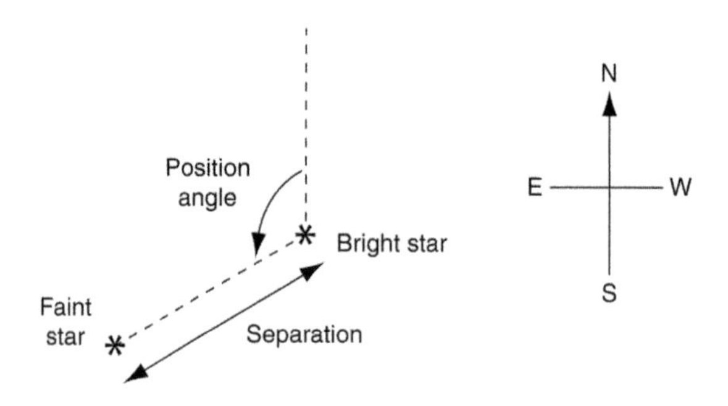

FIGURE 5.3 Position angle and separation of a visual double star (as seen on the sky directly).

[3] The system is based upon a position for the galactic centre of RA_{2000} 17 h 45 m 36 s, Dec_{2000} -28° 56′ 18″. It is now known that this is in error by about 4′; however, the incorrect position continues to be used. Prior to about 1960, the intersection of the equator and the galactic plane was used as the zero point and this is about 30° away from the galactic centre. Coordinates based upon this old system are sometimes indicated by a superscript 'I' and those using the current system by a superscript 'II', i.e., b^{I} or b^{II} and l^{I} or l^{II}.

5.1.3 Transit Telescopes

Transit telescopes (Figure 5.4), which are also sometimes called 'meridian circles', historically provided the basic absolute measurements of stellar positions for the calibration of other astrometric methods, though they have now largely been superseded by the results from Hipparcos (see below) and radio interferometers. They developed from the mural quadrants that were used before the invention of the telescope. These were just sighting bars, pivoted at one end, restricted to moving in just one plane and with a divided arc to measure the altitude at the other end; nonetheless, they provided surprisingly good measurements when in capable hands. The one used by Tycho Brahe (1546–1601), for example, was mounted on a north-south aligned wall and had a 2-m radius. He was able to measure stellar positions using it to a precision of about ±30″.

The principle of the transit telescope is simple, but great care was required in practice if it were to produce reliable results. The instrument was almost always a refractor because of the greater stability of the optics, and it was pivoted only about a horizontal east-west axis. The telescope was thereby constrained to look just at points on the prime meridian. A star was observed as it crossed the centre of the field of view (i.e., when it transited the prime meridian) and the precise time of the passage was noted. The declination of the star is obtainable from the altitude of the telescope, while the right ascension is given by the local sidereal time at the observatory at the instant of transit,

$$\delta = A + \phi - 90^\circ \tag{5.10}$$

$$\alpha = LST \tag{5.11}$$

where A is the altitude of the telescope, ϕ is the latitude of the observatory and LST is the local sidereal time at the instant of transit. To achieve an accuracy of a tenth of an arcsecond in the absolute position of the star, a large number of precautions and corrections are needed. A few of the more

FIGURE 5.4 A transit telescope – the Carlsberg Meridian Telescope on La Palma. Decommissioned 2013. (Courtesy of D. W. Evans.)

important ones are: temperature control; use of multiple or driven cross wires in the micrometer eyepiece; reversal of the telescope on its bearings; corrections for flexure, non-parallel objective and eyepiece focal planes; rotational axis not precisely horizontal; rotational axis not precisely east-west; errors in the setting circles; incorrect position of the micrometer cross wire, personal setting errors, etc. Then of course all the normal corrections for refraction, aberration and so on have also to be added.

Modern versions of the instrument, such as the Carlsberg Meridian Telescope (CMT; see Figure 5.4), use charge-coupled devices (CCDs) for detecting transits. The CMT was a refractor with a 17.8 cm diameter objective and a 2.66-m focal length sited on La Palma in the Canary Islands. It was operated remotely, until 2013, via the internet and measured between 100,000 and 200,000 star positions in a single night. The CCD detector on the CMT used charge transfer to move the accumulating electrons from pixel to pixel at the same rate as the images drift over the detector (also known as time delayed integration [TDI], drift scanning or image tracking – see Section 1.1 [liquid mirror telescopes] and Section 2.7). This enabled the instrument to detect stars down to 17^m and also to track several stars simultaneously. The main programme of the CMT was to link the positions of the bright stars measured by Hipparcos to those of fainter stars.

5.1.4 Photographic Zenith Tube and the Impersonal Astrolabe

Two more modern instruments that until recently have performed the same function as the transit telescope are the photographic zenith tube (PZT) and the astrolabe.[4] Both of these use a bath of mercury to determine precisely the zenith position. The PZT obtained photographs of stars that transit close to the zenith and provided an accurate determination of time and the latitude of an observatory, but it was restricted in its observations of stars to those that culminated within a few tens of arc-minutes of the zenith. The PZT has been superseded by long baseline radio interferometers (below and Section 2.5) and measurements from spacecraft which are able to provide positional measurements for far more stars and with significantly higher accuracies.

The astrolabe observes near an altitude of 60° and so can give precise positions for a wider range of objects than the PZT. In its most developed form (Figure 5.5), it is known as the 'Danjon astrolabe' or 'impersonal astrolabe' because its measurements are independent of focusing errors. Two separate beams of light are fed into the objective by the 60° prism, one of the beams having

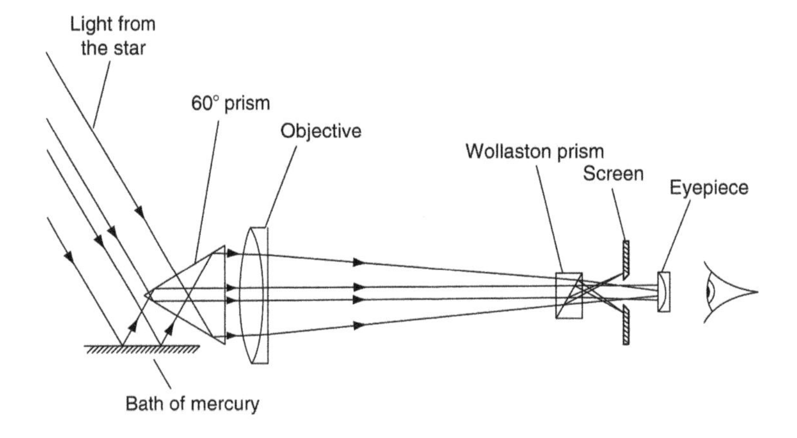

FIGURE 5.5 Optical paths in the impersonal astrolabe.

[4] The astrolabe used centuries ago for measuring stellar altitudes and time-keeping had a similar function to the impersonal astrolabe, but the devices are *very* different in their practical manifestations.

been reflected from a bath of mercury. A Wollaston prism (Section 5.2) is used to produce two focused beams parallel to the optical axis with the other two non-parallel emergent beams being blocked off. Two images of the star are then visible in the eyepiece. The Wollaston prism is moved along the optical axis to compensate for the sidereal motion of the star. The measurement of the star's position is accomplished by moving the 60° prism along the optical axis until the two images merge into one. The position of the prism at a given time for this coincidence to occur can then be converted into the position of the star in the sky. The astrolabe has also largely fallen out of use; the CCD-based Danjon astrolabe of the Observatorio Nacional, Rio de Janeiro, used for measuring the solar angular diameter, for example, was replaced in 2009 by a split-mirror heliometer (Section 5.3).

5.1.5 Micrometers

The earliest device used on telescopes for measuring double stars and the angular sizes of extended objects was the bifilar micrometer. This comprises a fixed cross wire at the focus of an eyepiece, with a third thread parallel to one of the fixed ones. The third thread is displaced slightly from the others, although still within the eyepiece's field of sharp focus, and it may be moved perpendicularly to its length by a precision screw thread. The screw has a calibrated scale so that the position of the third thread can be determined to within 1 μm. The whole assembly is mounted on a rotating turret whose angular position may also be measured. The bifilar micrometer and other variants such as double image micrometer[5] will hardly ever be encountered now outside museums, so are not considered further here.

5.1.6 Astrographs and Other Telescopes

Almost any telescope can be used to obtain images from which the positions of objects in the sky can be determined, but some telescopes are optimised for the work. The astrograph has a wide (a degree or more) field of view so that many faint stars' positions can be determined from those of a few brighter reference stars. The astrographs designed around the beginning of the twentieth century, were refractors with highly corrected multi-element objectives. Their apertures were in the region of 0.1–0.25 m and their focal lengths between 2 and 4 m. Measurements using these instruments formed the basis of the ICRF. More recently, Schmidt cameras and conventional reflectors with correcting lenses have also been used.

Smaller fields of view are adequate for parallax determination. Long focus refractors are favoured for this work because of their greater stability when compared with reflectors, because of their closed tubes, because their optics may be left undisturbed for many years and because flexure is generally less deleterious in them. However, a few specially designed reflectors, especially the US Naval Observatory's 1.55-m Strand reflector, are used successfully for astrometry. The Extremely Large Telescope's (ELT's) proposed Multi-AO Imaging Camera for Deep Observations (MICADO) instrument is likely to achieve near infrared (NIR) positional measurement accuracies into the sub-milliarcsecond region and so may be usable for astrometry (if the scientific priority is sufficiently important).

Against the advantages of the refractor, their disadvantages must be set. These are primarily their residual chromatic aberration and coma, temperature effects and slow changes in the optics such as the rotation of the elements of the objective in their mounting cells. Refractors in use for astrometry usually have focal ratios of *f*15 to *f*20 and range in diameter from about 0.3 to 1.0 m. Their objectives are achromatic (Section 1.1) and filters are usually used so that the bandwidth is limited to a few tens of nanometres either side of their optimally corrected wavelength. Coma may

[5] Though this is related to the split-mirror heliometer (Section 5.3).

need to be limited by stopping down the aperture, by introducing correcting lenses near the focal plane, or by compensating the measurements later, in the analysis stages, if the magnitudes and spectral types of the stars are well known. Observations are usually obtained within a few degrees of the meridian to minimise the effects of atmospheric refraction and dispersion and to reduce changes in flexure. The position of the focus may change with temperature and needs to be checked frequently throughout the night. Other corrections for such things as plate tilt collimation errors, astigmatism, distortion, emulsion movement and so on need to be known and their effects allowed for when reducing the raw data.

5.1.7 Interferometers

As we have seen (Section 2.5), interferometers have the capability of providing much higher resolutions than single dish-type instruments because their baselines can be extended easily and cheaply. High resolution can be translated into high positional accuracy providing that the mechanical construction of the interferometer is adequate. Interferometers also provide absolute positions for objects. The disadvantage of interferometers for astrometry is that only one object can be observed at a time (although the Very Large Telescope's [VLT's] Phase-Referenced Imaging and Microarcsecond Astrometry [PRIMA][6] and Galaxy observe a reference star as well as the object of interest simultaneously – see below), and each observation requires several hours. Both radio and optical interferometers are used for the purpose.

Optical interferometers such as the US Naval Observatory's Navy Precision Optical Interferometer (NPOI; see Section 2.5.4) are essentially Michelson stellar interferometers (Section 2.5) incorporating a delay line into one or both light beams. The delay line is simply an arrangement of several mirrors that reflects the light around an additional route. By moving one or more of the mirrors in the delay line, the delay can be varied. The observational procedure is to point both telescopes (or the flat mirrors feeding the telescopes) at the same star and adjust the delay until the fringe visibility (Section 2.5) is maximum. The physical length of the delay line then gives the path difference between the light beams to the two telescopes. Combining this with the baseline length gives the star's altitude. Several such observations over a night will then enable the star's position in the sky to be calculated.

Currently, absolute positions may be obtained in this manner to accuracies of submilliarcsecond accuracy, with the relative positions of close stars being determined to perhaps an order of magnitude better precision. Relative measurements for widely separated objects may be obtained from the variation of the delay over several hours. The delay varies sinusoidally over a 24-hours period and the time interval between the maxima for two stars equals the difference in their right ascensions. As with conventional interferometers, the use of more than two telescopes enables several delays to be measured simultaneously, speeding up the process and enabling instrumental effects to be corrected. The Center for High Angular Resolution Astronomy (CHARA) optical interferometer on Mount Wilson, for example, uses six 1-m telescopes in a two-dimensional (2D) array and has a maximum baseline of 350 m. Gravity (Sections 2.5.4 and 4.2.7), a second-generation instrument for the Very Large Telescope Interferometer (VLTI), achieved first light in 2016 and combines the light from four of the VLT instruments and observes both the object of interest and a close reference star. Operating at 2.2 μm, it can reach 10 microarcsecond precision. Gravity built on the capabilities of PRIMA, a similar instrument which used two VLT telescopes and operated from 2008 to 2015.

Radio interferometers are operated in a similar manner for astrometry and Very Long Baseline Interferometry (VLBI) systems (Section 2.5) generally provide the highest accuracies, at the time of writing. For some of the radio sources this accuracy may better 100 microarcseconds.

[6] Decommissioned in 2015.

The measurements by VLBI of some 303 compact radio sources have, thus, been used to define the ICRF3 (Section 5.1.1) since 2019. Potentially, radio interferometry could be used to monitor star positions with sufficient accuracy to discover exoplanets (Section 5.1.12) via their host stars' movements, but this technique has yet to be used in practice.

A recent ingenious adaptation of VLBI has enabled details of the pulsar PSR 0834+06 to be studied with less than microarcsecond resolution. The approach uses the speckles introduced into the radio image by interstellar material (cf. atmospheric speckles – Figure 1.61), giving in effect an interferometer with 5 astronomical units (AU) baseline. For this pulsar, whose distance is 640 pc, the interstellar material is mostly around 415 pc away from us. For observations at 300 MHz (1 m), the measurement precision is about 0.3 microarcseconds. With the interstellar scattering material about 220 pc from the pulsar, this corresponds to a distance of around 10,000 km at the pulsar.

5.1.8 Space-Based Systems

By operating instruments in space and so removing the effects of the atmosphere and of gravitational loading, absolute astrometry is expected to reach accuracies of a few microarcseconds in the next decade or so. Space astrometry missions, actual and planned, divide into two types: scanning (or survey) and point and stare. Scanning means that exposures are short and so only the brighter stars can be observed and with relatively low accuracy; the accuracy also depends upon the star's brightness. However, large numbers of star positions may be measured quickly. Point and stare missions, as the name suggests, look at individual stars (or a few stars close together) for long periods of time. Such missions can observe relatively few star positions but do so with high accuracy and to faint magnitudes. Hipparcos and Gaia (see below) are examples of scanning systems, while the Hubble Space Telescope (HST) is point and stare.

The first astrometric spacecraft, the European Space Agency's (ESA's) Hipparcos, used a telescope fed by two flat mirrors at an angle of 29° to each other enabling two areas of the sky 58° apart to be observed simultaneously. It determined the relative phase difference between two stars viewed via the two mirrors to obtain their angular separation. A precision grid at the focal plane modulated the light from the stars, and the modulated light was then detected from behind the grid to determine the transit times. The satellite rotated once every 2 hours and measured all the stars down to 9.0 m about 120 times, observing each star for about 4 seconds on each occasion. Each measurement was from a slightly different angle, and the final positional catalogue was obtained by the processing of some 10^{12} bits of such information. The data from Hipparcos are to be found in the Hipparcos, Tycho, and Tycho-2[7] catalogues. The Hipparcos catalogue contains the positions, distances, proper motions and magnitudes of some 118,000 stars to an accuracy of ±700 microarcsecond. The Tycho catalogue contains the positions and magnitudes for about 1 million stars and has positional accuracies ranging from ±7 milliarcsecond to ±25 milliarcsecond. The Tycho-2 catalogue uses ground-based data as well as that from Hipparcos and has positional accuracies ranging from ±10 milliarcsecond to ±70 milliarcsecond, depending upon the star's brightness. It contains positions, magnitudes and proper motions for some 2 ½ million stars.

The HST can use the Wide Field Camera 3 (WFC3) (and previously the WFPC2) and Fine Guidance Sensor (FGS) instruments for astrometry. The planetary camera provides measurements to an accuracy of ±1 milliarcsecond for objects down to magnitude 26 m but only has a 160″ field of view. The telescope has three FGSs but only needs two for guidance purposes. The third is, therefore, available for astrometry. The FGSs are interferometers based upon Köster prisms (Section 2.5) and provide positional accuracies of ±5 milliarcsecond.

[7] A separate instrument, the star mapper, observed many more stars, but for only 0.25 second and its data produced the Tycho and Tycho-2 catalogues.

FIGURE 5.6 An artist's impression of the Gaia spacecraft in orbit. (Courtesy of European Space Agency.)

Several other spacecraft missions have been proposed for astrometry, but most have been cancelled. However, ESA's Gaia mission (Figure 5.6) was launched in 2013 with a planned lifetime of nine years. It is positioned at the Sun-Earth inner Lagrange point, some 1.5×10^6 km from the Earth, to avoid eclipses and occultations and so to provide a stable thermal environment. It operates in a similar fashion to Hipparcos but using two separate telescopes set at a fixed angle and with elongated apertures 1.7×0.7 m. The long axis of the aperture in each case is aligned along the scanning direction to provide the highest resolution. The read-out is by CCDs using TDI. The aim is to determine positions for a billion stars to ± 10 microarcsecond accuracy down to a visual magnitude 15 m. Gaia also determines radial velocities for stars down to 17 m using a separate spectroscopic telescope. The objective for the Gaia mission is to establish a precise three-dimensional (3D) map of the galaxy and provide a massive data base for other investigations of the Milky Way.

A few years ago, an ESA medium-class mission concept was for Near Earth Astrometric Telescope (NEAT), which might have involved two spacecraft flying in formation about 40 m apart and have aimed to reach a few times 0.1-microarcsecond levels of precision. However, this mission was not funded. Another medium-class mission, Euclid (Sections 1.1.19.2 and 1.7.2.4), though, is going ahead for a 2022 launch. Whilst not quite an astrometric mission in the sense of Hipparcos and Gaia, its aim is to map the 3D geometry of the Universe by determining the shapes, red-shifts and positions of some 10^{10} galaxies and other extragalactic sources, covering a third of the entire sky to a depth of about 3 Gpc over a six-year period. It will be placed into the Earth-Sun Lagrangian L2 point and operate in a staring mode, examining about half a square degree at a time.

5.1.9 Detectors

The CCDs and related solid-state detectors (Section 1.1) are now widely used as detectors for astrometry (Figure 1.12). They have the enormous advantage that the stellar or other images can be identified with individual pixels whose positions are automatically defined by the structure of the detector. The measurement of the relative positions of objects within the image is, therefore, vastly simplified. Software to fit a suitable point spread function (PSF; see Section 2.1) to the image enables the position of its centroid to be defined to a fraction of the size of the individual pixels.

The disadvantage of electronic images is their small scale. Most CCD chips are only a few centimetres in size (though mosaics are larger) compared with tens of centimetres for astrographic photographic plates. Therefore, they cover only a small region of the sky, and so it can become difficult to find suitable reference stars for the measurements. However, the greater dynamic range of CCDs compared with photographic emulsion and the use of anti-blooming techniques (Section 1.1) enables the positions of many more stars to be usefully measured. With transit telescopes (see above), TDI may be used to reach fainter magnitudes. The TDI is also used for the Gaia spacecraft detection system (see above). Otherwise the operation of a CCD for positional work is conventional (Section 1.1). In other respects, the processing and reduction of electronic images for astrometric purposes is the same as that for a photographic image.

The CCD imaging may usefully be used to observe double stars when the separation is greater than about 3″. Many exposures are made with a slight shift of the telescope between each exposure. Averaging the measurements can give the separation to better than a hundredth of an arcsecond and the position angle to within a few arc-minutes. For double stars with large magnitude differences between their components, an objective grating can be used (see below), or the shape of the aperture can be changed so that the fainter star lies between two diffraction spikes and has an improved noise level. The latter technique is a variation of the technique of apodisation mentioned in Sections 1.1, 1.2, 2.5, 4.1, and 5.3.

5.1.10 Measurement and Reduction

Transit telescopes and interferometers give absolute positions directly as discussed. In most other cases, the position of an unknown star is obtained by comparison with reference stars that also appear on the image and whose positions are known. If absolute positions are to be determined, then the reference stars' positions must have been found via a transit instrument, interferometer or astrolabe, etc. For relative positional work, such as that involved in the determination of parallax and proper motion, any distant star may be used as a comparison star.

It is advantageous to have the star of interest and its reference stars of similar brightness, although the much greater dynamic ranges of CCDs and other electronic detectors compared with that of photographic emulsions render this requirement less important than it was in the past. If still needed, the brightness of one or more stars may be altered through the use of variable density filters, by the use of a rotating chopper in front of the detector or by the use of an objective grating (Section 4.2). The latter device is arranged so that pairs of images, which in fact, are the first-order or higher-order spectra of the star, appear on either side of it. By adjusting the spacing of the grating, the brightness of these secondary images for the brighter stars may be arranged to be comparable with the brightness of the primary images of the fainter stars. The position of the brighter star is then taken as the average of the two positions of its secondary images. The apparent magnitude, m, of such a secondary image resulting from the nth order spectrum is given by

$$m = m_0 + 5 \log_{10}(N\eta \operatorname{cosec} \eta) \tag{5.12}$$

where m_0 is the apparent magnitude of the star, η is given by

$$\eta = \pi N \, d / D \tag{5.13}$$

where D is the separation of the grating bars, d is the width of the gap between the grating bars and N is the total number of slits across the objective.

Once an image of a star field has been obtained, the positions of the stars' images must be measured to a high degree of accuracy and the measurements converted into right ascension and declination or separation and position angle. For most astrometric photographic archive material, this process has already been undertaken. The CCDs and similar arrays give positions directly (see above) once the physical structure of the device has been calibrated. Usually, the position will be found by fitting a suitable PSF (Section 2.1) for the instrument used to obtain the overall image to the stellar images when they spread across several pixels. The stars' positions may then be determined to sub-pixel accuracy.

Howsoever, the raw data may have been obtained, the process of converting the measurements into the required information is known as 'reduction'. For accurate astrometry, it is a lengthy process (and ready-made pipeline data-processing packages will usually be available). A number of corrections have already been mentioned and in addition to those, the distortion caused by the projection of the curved sky onto the flat photographic plate or CCD (known as the 'tangential plane'), or in the case of Schmidt cameras its projection onto the curved focal plane must be corrected. For the simpler case of the astrograph, the projection is shown in Figure 5.7 and the relationship between the coordinates on the flat image and in the curved sky is given by

$$\rho = \tan^{-1}\left(\frac{y}{F}\right) \tag{5.14}$$

$$\tau = \tan^{-1}\left(\frac{x}{F}\right) \tag{5.15}$$

where x and y are the coordinates on the image, τ and ρ are the equivalent coordinates on the image of the celestial sphere and F is the focal length of the telescope.

One instrument, which, although it is not strictly of itself a part of astrometry, is frequently used to identify the objects that are to be measured and is called the 'blink comparator'. In this machine, the worker looks alternately at two aligned images of the same star field obtained some time apart from each other, the frequency of the interchange being about 2 Hz. Stars, asteroids or comets and so on which have moved in the interval between the images then call attention to themselves by appearing to jump backwards and forwards, whilst the remainder of the stars are stationary. The device can also pick out variable stars for further study by photometry (Chapter 3). Although these remain stationary, they appear to blink on and off (hence, the name of the instrument). Early blink comparators were basically binocular microscopes which had a mechanical arrangement for swapping the view of two photographic plates. Now, the images are usually viewed on a computer

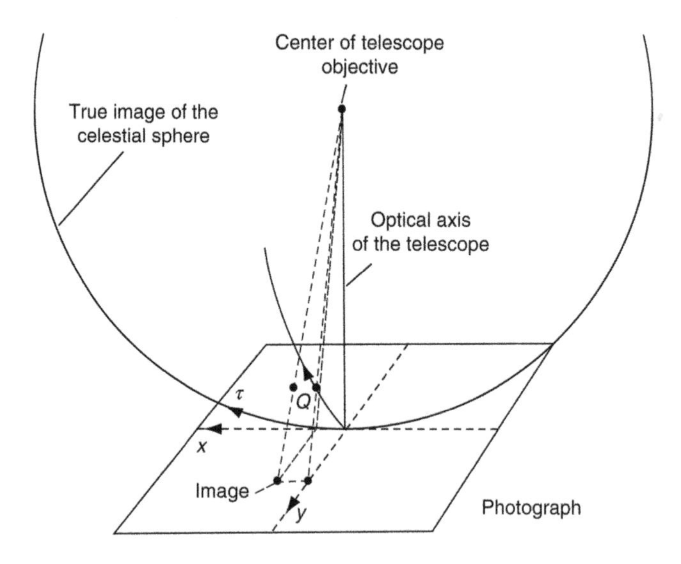

FIGURE 5.7 Projection onto a photographic plate or flat array detector, etc.

screen and software aligns the images and provides the alternating views. Software can also be used to identify the moving or changing objects directly. But this is still one area where the human eye-brain functions as efficiently as the computer. Some non-astronomical image-processing packages have blink functions which may be used for this purpose.

5.1.11 Sky Surveys and Catalogues

The end result of most astrometry is a catalogue of positions and other properties of (usually) a large number of objects in the sky. The Hipparcos, Tycho and Tycho-2 (see above) catalogues are recent examples of the process (see also Section 5.5). Other recent astrometric catalogues include the Fourth USNO CCD Astrograph Catalog (UCAC4) which has the positions and proper motions of some 113 million stars in the 8^m to 16^m range with ±15-milliarcsecond accuracy and the USNO-B1.0 catalogue containing data on a billion stars with 200-milliarcsecond positional accuracy and which is based upon Schmidt camera photographic images from several instruments. The 2012 Naval Observatory Merged Astrometric Dataset (NOMAD) catalogue is an aggregation of the data from the Hipparcos, Tycho-2, UCAC2, and USNO-B1.0 catalogues. The USNO Robotic Astrometric Telescope (URAT) catalogues are the successors to the UCAC with URAT1, published in 2015, containing data on 220 million stars.

Older astrometric catalogues include the Fundamental Katalog series which culminated in FK5 in 1998, containing 1,500 stars with positional accuracies of better than ±100 milliarcsecond, the AC Astrographic Catalogue and Astronomische Gesellschaft Katalog (AGK) series, plus arguably, Argelander's, Flamsteed's, Brahe's and even the original Hipparchus' catalogues because these gave state-of-the-art stellar positions for their day.

By 2021, the Vera C. Rubin Observatory (VRO) should be swamping all these surveys and catalogues by obtaining more than 1,000 images per night, with each image covering a 9.6 deg^2 area of the sky down to a limiting magnitude of $+24.5^m$, or better. Its astrometric accuracy should be around ±10 to ±50 milliarcsecond and its photometric accuracy around $\pm0.02^m$. The survey is planned to last for 10 years so that every area of the available sky will be imaged more than 2,000 times.

There have been many other sky surveys and catalogues (and more are being produced all the time) that are non-astrometric. That is to say; their main purpose is other than that of providing accurate positions and the position (if determined at all for the catalogue) is well below the current levels of astrometric accuracy. Indeed, most such catalogues just use the already known astrometric positions. Non-astrometric catalogues are also produced for all regions of the spectrum, not just from optical and radio sources. There are tens of thousands of non-astrometric catalogues produced for almost as many different reasons and ranging in content from a few tens to half a billion objects. Examples of such catalogues include the HST's second Guide Star Catalogue (GSC2) containing 500 million objects, the Two-Micron All Sky Survey (2MASS) and Deep Near Infrared Survey of the southern sky (DENIS) surveys in the infrared and the Sloan Digital Sky Survey (SDSS) with its million red shifts of galaxies and quasars.

5.1.12 Exoplanets

Before any exoplanets had been definitively discovered, most astronomers expected that astrometry would be the method which did eventually find them. That detection would be through the cyclic change in the position of the host star in the sky as it and its exoplanet orbited around their common centre of mass – the same way in which Freidrich Bessel discovered the white dwarf, Sirius B, in 1844.

In fact, at the time of writing – more than two decades since the first exoplanets were found – only one out of over 4,000 exoplanets, currently catalogued in NASA's Exoplanet Archive, is listed as having been discovered astrometrically. That is DENIS-P J082303.1-491201, which is 54 million km out from a 0.08 $M_\odot$ brown dwarf in Vela. Furthermore, that object's mass is estimated at

~29 M_J (~0.027 $M_{\odot}$) which makes it more likely to be a brown dwarf than an exoplanet. The system is 21 pc away from us and the exoplanet/brown dwarf has an eight-month orbital period. It was detected in 2013 using the VLT's FORS2 instrument (Section 4.2.7) in imaging mode.

Astrometry has, however, detected the host star's changing position in the case of the already known exoplanet Gliese 876 by using data from the HST. Exoplanet discoveries through astrometric measurements are thus not impossible, and they may start to be made more frequently as astrometric positional accuracies improve further. For exoplanet studies, astrometry does have the advantage of determining their masses unambiguously and of being likely to discover exoplanets at distances of tens of AU from their host stars (compared with transits and radial velocities which favour close-in exoplanets) because larger orbits lead to larger movements in the sky.

5.2 POLARIMETRY

5.2.1 Background

Although the discovery of polarized light from astronomical sources dates back to the beginning of the nineteenth century, when Dominique Arago detected its presence in moonlight, the extensive development of its study is a relatively recent phenomenon. This is largely due to the technical difficulties that are involved and initially, at least, to the lack of any expectation of polarized light from stars by astronomers. Many phenomena, however, can contribute to the polarisation of radiation and so, conversely, its observation can potentially provide information upon an equally wide range of basic causes.

5.2.1.1 Stokes' Parameters

Polarisation of radiation is simply the non-random angular distribution of the electric vectors of the photons in a beam of radiation. Customarily two types are distinguished – linear and circular polarisations. In the former, the electric vectors are all parallel and their direction is constant, while in the latter, the angle of the electric vector rotates with time at the frequency of the radiation. These are not really different types of phenomena, however, and all types of radiation may be considered to be different aspects of partially elliptically polarised radiation. This has two components, one of which is unpolarised, the other being elliptically polarised. Elliptically polarised light is similar to circularly polarised light in that the electric vector rotates at the frequency of the radiation, but in addition the magnitude varies at twice that frequency, so that plotted on a polar diagram the electric vector would trace out an ellipse (Figure 5.8). The properties of partially elliptically polarised light are completely described by four parameters that are called the 'Stokes' parameters'. These fix the intensity of the unpolarised light, the degree of ellipticity, the direction of the major axis of the ellipse and the sense (left- or right-handed rotation) of the elliptically polarised light. If the radiation is imagined to be propagating along the z axis of a three-dimensional rectangular coordinate system, then the elliptically polarised component at a point along the z-axis may have its electric vector resolved into components along the x- and y-axes (Figure 5.8), these being given by

$$E_x(t) = e_1 \cos\,(2\pi\,\nu\,t) \tag{5.16}$$

$$E_y(t) = e_2 \cos\,(2\pi\,\nu\,t + \delta) \tag{5.17}$$

where ν is the frequency of the radiation, δ is the phase difference between the x and y components and e_1 and e_2 are the amplitudes of the x and y components. It is then a tedious but straightforward matter to show that

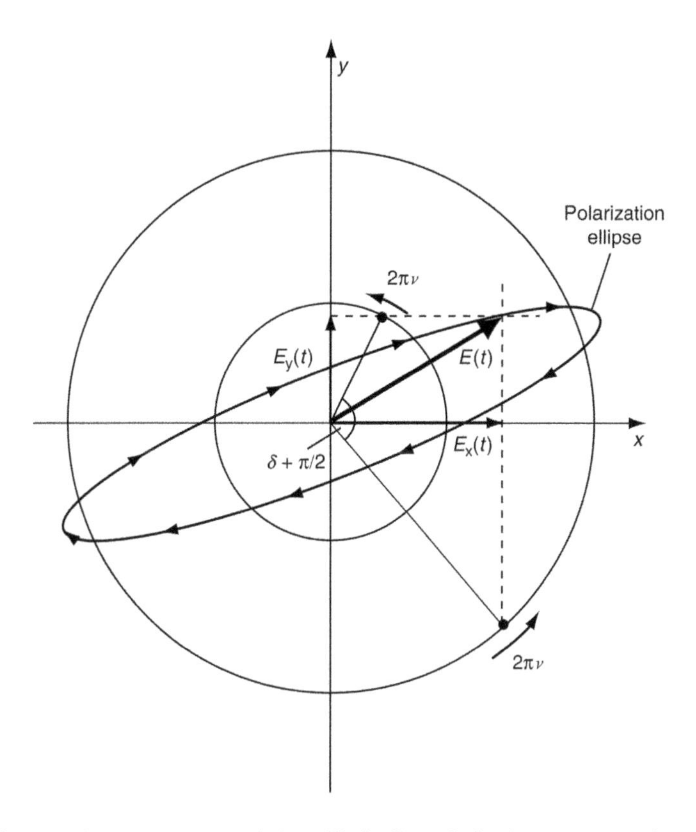

FIGURE 5.8 The x and y components of the elliptically polarised component of partially elliptically polarised light.

$$a = \left(\frac{\left(e_1^2 + e_2^2\right)}{1 + \tan^2 \left[\frac{1}{2} \sin^{-1} \left\{ \left[\frac{2e_1e_2}{\left(e_1^2 + e_2^2\right)} \right] \sin \delta \right\} \right]} \right)^{1/2} \tag{5.18}$$

$$b = a \tan \left[\frac{1}{2} \sin^{-1} \left\{ \left[\frac{2e_1e_2}{\left(e_1^2 + e_2^2\right)} \right] \sin \delta \right\} \right] \tag{5.19}$$

$$a^2 + b^2 = e_1^2 + e_2^2 \tag{5.20}$$

$$\psi = \frac{1}{2} \tan^{-1} \left\{ \left[\frac{2e_1e_2}{\left(e_1^2 - e_2^2\right)} \right] \cos \delta \right\} \tag{5.21}$$

where a and b are the semi-major and semi-minor axes of the polarisation ellipse, and ψ is the angle between the x-axis and the major axis of the polarisation ellipse. The Stokes' parameters are then defined by

$$Q = e_1^2 - e_2^2 = \frac{a^2 - b^2}{a^2 + b^2}\cos(2\psi)I_p \tag{5.22}$$

$$U = 2e_1e_2\cos\delta = \frac{a^2 - b^2}{a^2 + b^2}\sin(2\psi)I_p \tag{5.23}$$

$$V = 2e_1e_2\sin\delta = \frac{2ab}{a^2 + b^2}I_p \tag{5.24}$$

where I_p is the intensity of the polarised component of the light. From Equations 5.22–5.24, we have

$$I_p = \left(Q^2 + U^2 + V^2\right)^{1/2} \tag{5.25}$$

The fourth Stokes' parameter, I, is just the total intensity of the partially polarised light

$$I = I_u + I_p \tag{5.26}$$

where I_u is the intensity of the unpolarised component of the radiation. (*Note*: The notation and definitions of the Stokes' parameters can vary. Whilst that given here is probably the commonest usage, a check should always be carried out in individual cases to ensure that a different usage is not being employed.)

The degree of polarisation, π, of the radiation is given by

$$\pi = \frac{\left(Q^2 + U^2 + V^2\right)^{1/2}}{I} = \frac{I_p}{I} \tag{5.27}$$

while the degree of linear polarisation, π_L and the degree of ellipticity, π_e, are

$$\pi_L = \frac{\left(Q^2 + U^2\right)^{1/2}}{I} \tag{5.28}$$

$$\pi_e = \frac{V}{I} \tag{5.29}$$

When $V = 0$ (i.e., the phase difference, δ, is 0 or π radians), we have linearly polarised radiation. The degree of polarisation is then equal to the degree of linear polarisation, and this is the quantity that is commonly determined experimentally

$$\pi = \pi_L = \frac{I_{\max} - I_{\min}}{I_{\max} + I_{\min}} \tag{5.30}$$

where $I_{\max}$ and $I_{\min}$ are the maximum and minimum intensities that are observed through a polariser as it is rotated. The value of π_e is positive for right-handed and negative for left-handed radiation.

When several incoherent beams of radiation are mixed, their various Stokes' parameters combine individually by simple addition. A given monochromatic partially elliptically polarised beam of radiation may, therefore, have been formed in many different ways, and these are indistinguishable by the intensity and polarisation measurements alone. It is, therefore, often customary to regard partially elliptically polarised light as formed from two separate components, one of which is unpolarised and the other completely elliptically polarised. The Stokes' parameters of these components are then:

	I	*Q*	*U*	*V*
Unpolarised component	I_u	0	0	0
Elliptically polarised component	I_p	Q	U	V

and the normalised Stokes' parameters for more specific mixtures are:

Type of Radiation	Stokes' Parameters			
	I/I	*Q/I*	*U/I*	*V/I*
Right-hand circularly polarised (clockwise)	1	0	0	1
Left-hand circularly polarised (anticlockwise)	1	0	0	−1
Linearly polarised at an angle ψ to the *x* axis	1	cos 2ψ	sin 2ψ	0

The Stokes' parameters are related to more familiar astronomical quantities by

$$\theta = \frac{1}{2}\tan^{-1}\left(\frac{U}{Q}\right) = \psi \tag{5.31}$$

$$e = \left\{1 - \tan^2\left[\frac{1}{2}\sin^{-1}\left(\frac{V}{\left(Q^2 + U^2 + V^2\right)^{1/2}}\right)\right]\right\}^{1/2} \tag{5.32}$$

where θ is the position angle (Section 5.1) of the semi-major axis of the polarisation ellipse (when the *x*-axis is aligned North-South), *e* is the eccentricity of the polarisation ellipse – which is 1 for linearly polarised radiation and 0 for circularly polarised radiation.

5.2.2 Optical Components for Polarimetry

Polarimeters usually contain a number of components that are optically active in the sense that they alter the state of polarisation of the radiation. They may be grouped under three headings: polarisers, converters, and depolarisers. The first produces linearly polarised light, the second converts elliptically polarised light into linearly polarised light, or vice versa, while the last eliminates polarisation. Most of the devices rely upon birefringence for their effects, so we must initially discuss some of its properties before looking at the devices themselves.

5.2.2.1 Birefringence

The difference between a birefringent material and a more normal optical one may best be understood in terms of the behaviour of the Huygens' wavelets. In a normal material, the refracted ray can be constructed from the incident ray by taking the envelope of the Huygens' wavelets, which spread out from the incident surface with a uniform velocity. In a birefringent material, however, the velocities of the wavelets depend upon the polarisation of the radiation, and for at least one component, the velocity will also depend on the orientation of the ray with respect to the structure of the material.

In some materials, it is possible to find a direction for linearly polarised radiation for which the wavelets expand spherically from the point of incidence as in 'normal' optical materials. This ray is then termed the 'ordinary ray', and its behaviour may be described by normal geometrical optics.

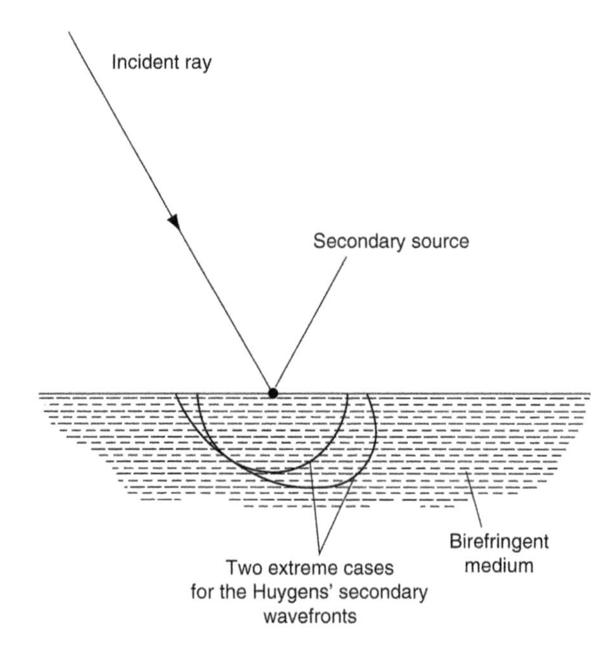

FIGURE 5.9 Huygens' secondary wavelets in a birefringent medium.

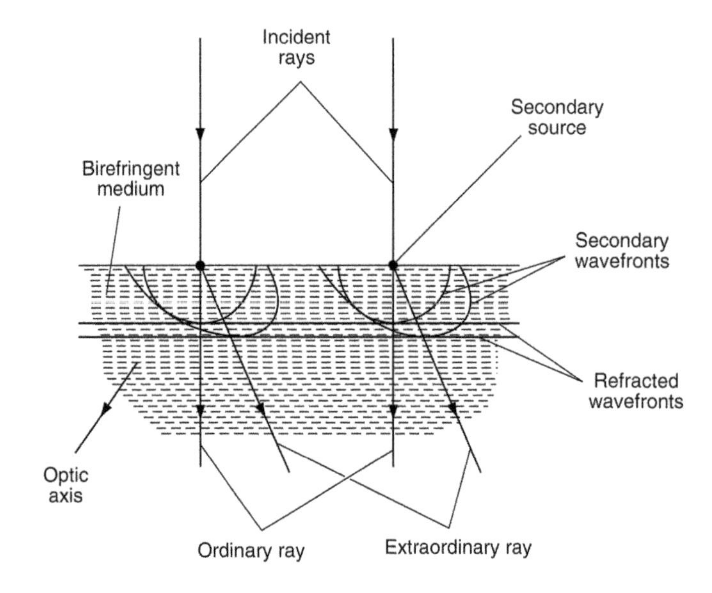

FIGURE 5.10 Formation of ordinary and extraordinary rays in a birefringent medium.

The ray that is polarised orthogonally to the ordinary ray is then termed the 'extraordinary ray', and wavelets from its point of incidence spread out elliptically (Figure 5.9). The velocity of the extraordinary ray thus varies with direction. We may construct the two refracted rays in a birefringent material by taking the envelopes of their wavelets as before (Figure 5.10). The direction along which the velocities of the ordinary and extraordinary rays are equal is called the 'optic axis of the material'. When the velocity of the extraordinary ray is in general larger than that of the ordinary ray (as illustrated in Figures 5.9 and 5.10), then the birefringence is negative. It is positive when the situation is reversed. The degree of the birefringence may be obtained from the principal extraordinary refractive index, μ_E. This is the refractive index corresponding to the maximum velocity of the extraordinary ray for negative materials and the minimum velocity for positive materials. It will be obtained from rays travelling perpendicularly to the optic axis of the material. The degree of birefringence, which is often denoted by the symbol *J*, is then simply the difference between the principal extraordinary refractive index and the refractive index for the ordinary ray, μ_O

$$J = \mu_E - \mu_O \tag{5.33}$$

Most crystals exhibit natural birefringence, and it can be introduced into many more and into amorphous substances such as glass by the presence of strain in the material. One of the most commonly encountered birefringent materials is calcite. The cleavage fragments form rhombohedrons and the optic axis then joins opposite blunt corners if all the edges are of equal length (Figure 5.11). The refractive index of the ordinary ray is 1.658, whilst the principal extraordinary refractive index is 1.486, giving calcite the high degree of birefringence of −0.172.

Crystals such as calcite are uniaxial and have only one optic axis. Uniaxial crystals belong to the tetragonal or hexagonal crystallographic systems. Crystals that belong to the cubic system are not usually birefringent, whilst crystals in the remaining systems – orthorhombic, monoclinic or triclinic – generally produce biaxial crystals. In the latter case, normal geometrical optics breaks down completely and all rays are extraordinary.

FIGURE 5.11 The beam from the laser of a surveying level being shone through a calcite cleavage rhomb. The normally single line is clearly seen split by the calcite's birefringence into the ordinary and extraordinary rays producing two lines – see magnified inset. (Courtesy of C. E. Danes. Copyright © C.E. Danes 2013.)

Some crystals such as quartz that are birefringent ($J = +0.009$) are in addition optically active. This is the property whereby the plane of polarisation of a beam of radiation is rotated as it passes through the material. Many substances other than crystals, including most solutions of organic chemicals, can be optically active. Looking down a beam of light, against the motion of the photons, a substance is called 'dextro-rotatory' or 'right-handed' if the rotation of the plane of vibration is clockwise. The other case is called 'laevo-rotatory' or 'left-handed'. Unfortunately and confusingly the opposite convention is also in occasional use.

The description of the behaviour of an optically active substance in terms of the Huygens' wavelets is rather more difficult than was the case for birefringence. Incident light is split into two components as previously, but the velocities of both components vary with angle and in no direction do they become equal. The optic axis, therefore, has to be taken as the direction along which the difference in velocities is minimised. Additionally, the nature of the components changes with angle as well. Along the optic axis they are circularly polarised in opposite senses, while perpendicular to the optic axis they are orthogonally linearly polarised. Between these two extremes the two components are elliptically polarised to varying degrees and in opposite senses (Figure 5.12).

5.2.2.2 Polarisers

Polarisers[8] are devices that only allow the passage of light that is linearly polarised in some specified direction. There are several varieties that are based upon birefringence, of which the Nicol prism is the best known. This consists of two calcite prisms cemented together by Canada balsam. Since the refractive index of Canada balsam is 1.55, it is possible for the extraordinary ray to be

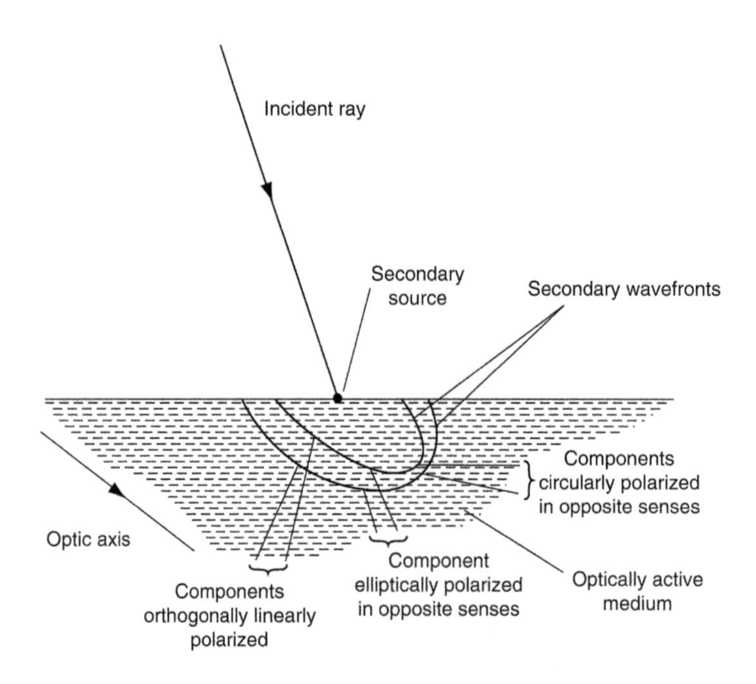

FIGURE 5.12 Huygens' secondary wavelets in an optically active medium.

[8] Also often known as analysers.

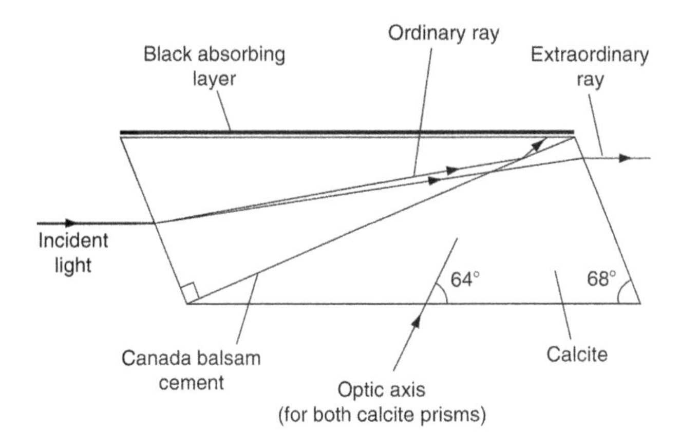

FIGURE 5.13 The Nicol prism.

transmitted, whilst the ordinary ray is totally internally reflected (Figure 5.13). The Nicol polariser has the drawbacks of displacing the light beam and of introducing some elliptical polarisation into the emergent beam. Its inclined faces also introduce additional light losses by reflection. Various other designs of polarisers have therefore been developed, some of which in fact produce mutually orthogonally polarised beams with an angular separation. Examples of several such designs are shown in Figure 5.14. If both the ordinary ray and the extraordinary ray are required *and* their path lengths must be identical (perhaps so that the focal position of both is in the same plane, or because mutual interference is needed at some point), then Savart plates may be used. These are made from two equally thick, parallel-sided plates of calcite or quartz that are cemented together. The plates are cut so that their optical axes are at 45° to their surfaces and the orientation of one plate is rotated by 90° with respect to the other. The first plate splits an incoming light beam into the ordinary and extraordinary rays as usual. Within the second plate, the ordinary ray emerging from the first plate becomes an extraordinary ray and the extraordinary ray emerging from the first plate becomes an ordinary ray. When both beams finally emerge, they have therefore travelled equal optical distances within the calcite or quartz. Magnesium fluoride and quartz can also be used to form polarisers for the visible region, while lithium niobate and sapphire are used in the infrared.

The ubiquitous polarising sunglasses are based upon another type of polariser. They employ dichroic crystals that have nearly 100% absorption for one plane of polarisation and less than 100% for the other. Generally, the dichroism varies with wavelength so that these polarisers are not achromatic. Usually, however, they are sufficiently uniform in their spectral behaviour to be usable over quite wide wavebands. The polarisers are produced commercially in sheet form and contain many small aligned crystals rather than one large one. The two commonly used compounds are polyvinyl alcohol impregnated with iodine and polyvinyl alcohol catalysed to polyvinylidene by hydrogen chloride. The alignment is achieved by stretching the film. The use of microscopic crystals and the existence of the large commercial market means that dichroic polarisers are far cheaper than birefringent polarisers, and so they may be used even when their performance is poorer than that of the birefringent polarisers.

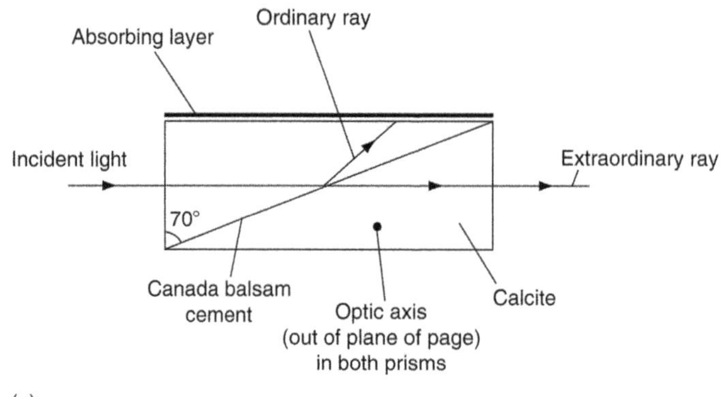

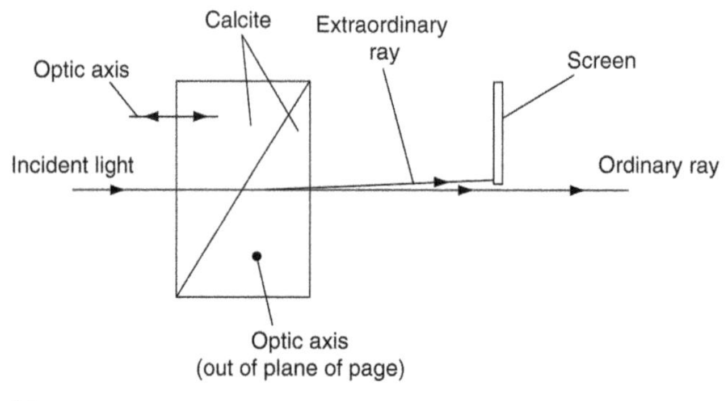

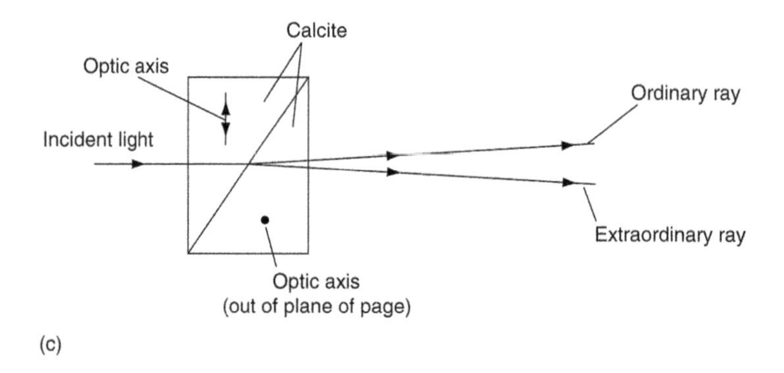

FIGURE 5.14 Examples of birefringence polarisers. (a) Glan-Thompson polarizer, (b) Rochon polarizer, and (c) Wollaston polarizer.

Polarisation by reflection can be used to produce a polariser. A glass plate inclined at the Brewster angle will reflect a totally polarised beam. This, of course, is why in bright sunlight, the glare (reflected light) from objects can be preferentially reduced by wearing polarising sunglasses. Although polarisers can be made by stacking several reflecting plates, they are rarely used today (although they may be needed for far ultraviolet [UV] and soft X-ray polarimeters – see the Large Ultraviolet Optical Infrared [LUVOIR] concept space mission – Section 5.2.3.1).

Outside the optical, NIR and near-UV regions, the polarisers tend to be somewhat different in nature. In the radio region, linear dipoles and related antennae are naturally only sensitive to radiation polarised along their lengths. Sometimes, though, dedicated polarimeters similar to those described later may be used. Thus, the B-Machine polarimeter, located on California's White Mountain and which observes the cosmic microwave background (CMB) at around 40 GHz, uses a half-wave plate (Section 5.2.2.3) to switch between polarisation states.

In the microwave region and the medium to far infrared, wire-grid polarisers are suitable. These, as their name suggests, are grids of electrically conducting wires. Their spacing is about five times their thickness, and they transmit the component of the radiation that is polarised so that its electric vector is orthogonal to the length of the wires.[9] They work efficiently for wavelengths longer than their spacing. In the microwave region, polarisation can also be studied using transition edge sensor (TES) detectors coupled with microstrip antennas. In the X-ray region, Bragg reflection is polarised (Section 1.3), and so a rotating Bragg spectrometer can also act as a linear polarisation detector.

The behaviour of a polariser may be described mathematically by its effect upon the Stokes' parameters of the radiation. This is most easily accomplished by writing the parameters as a column vector. A matrix multiplying the vector on the left may then represent the effect of the polariser and also of the other optical components discussed in this section. The technique is sometimes given the name 'Mueller calculus'. There is also an alternative and to some extent complementary formulation, termed 'Jones calculus'. In Mueller calculus, the effect of passing the beam through several optical components is simply found by successive matrix multiplications of the Stokes' vector. The first optical component's matrix is closest to the original Stokes' vector, and the matrices of subsequent components are positioned successively further to the left as the beam passes through the optical system. For a perfect polariser whose transmission axis is at an angle, θ, to the reference direction, we have

$$\begin{bmatrix} I' \\ Q' \\ U' \\ V' \end{bmatrix} = \begin{bmatrix} \frac{1}{2} & \frac{1}{2}\cos 2\theta & \frac{1}{2}\sin 2\theta & 0 \\ \frac{1}{2}\cos 2\theta & \frac{1}{2}\cos^2 2\theta & \frac{1}{2}\cos 2\theta \sin 2\theta & 0 \\ \frac{1}{2}\sin 2\theta & \frac{1}{2}\cos 2\theta \sin 2\theta & \frac{1}{2}\sin^2 2\theta & 0 \\ 0 & 0 & 0 & 0 \end{bmatrix} \begin{bmatrix} I \\ Q \\ U \\ V \end{bmatrix} \tag{5.34}$$

where the primed Stokes' parameters are for the beam *after* passage through the polariser and the unprimed ones are for it *before* such passage.

5.2.2.3 Converters

These are devices that alter the type of polarisation and/or its orientation. They are also known as 'retarders', 'modulators', 'wave plates', and 'phase plates'. We have seen previously (Equations 5.16 and 5.17 and Figure 5.8) that elliptically polarised light may be resolved into two orthogonal linear components with a phase difference. Altering that phase difference will alter the degree of ellipticity (Equations 5.18 and 5.19) and the inclination (Equation 5.21) of the ellipse. We have also seen that the velocities of mutually orthogonal linearly polarised beams of radiation will in general differ from each other when the beams pass through a birefringent material. From inspection of Figure 5.13 it will be seen that if the optic axis is rotated until it is perpendicular to the incident radiation, then the ordinary and extraordinary rays will travel in the same direction.

[9] If this seems counterintuitive, then it is because radiation with its electric vector parallel to the (conducting) wires is able to induce electric currents in them and so is reflected by the grid. The wires are too thin for significant currents to be induced across their widths and so the radiation component with its electric vector at right angles to the wires' lengths is able to pass through the grid.

Thus, they will pass together through a layer of a birefringent material that is oriented in this way and will recombine upon emergence, but with an altered phase delay due to their differing velocities. The phase delay, δ', is given to a first-order approximation by

$$\delta' = \frac{2\pi d}{\lambda} J \tag{5.35}$$

where d is the thickness of the material and J is the birefringence of the material. Now let us define the x-axis of Figure 5.8 to be the polarisation direction of the extraordinary ray. We then have from Equation 5.21, the intrinsic phase difference, δ, between the components of the incident radiation

$$\delta = \cos^{-1}\left\{\frac{\left(e_1^2 - e_2^2\right)}{2e_1e_2}\tan 2\psi\right\} \tag{5.36}$$

The ellipse for the emergent radiation then has a minor axis given by

$$b' = a'\tan\left[\frac{1}{2}\sin^{-1}\left\{\left[\frac{2e_1e_2}{\left(e_1^2 + e_2^2\right)}\right]\sin(\delta + \delta')\right\}\right] \tag{5.37}$$

where the primed quantities are for the emergent beam. So,

$$b' = 0 \qquad \text{for } \delta + \delta' = 0 \tag{5.38}$$

$$b' = a' \qquad \text{for } \delta + \delta' = \sin^{-1}\left[\frac{\left(e_1^2 + e_2^2\right)}{2e_1e_2}\right] \tag{5.39}$$

and also

$$\psi = \frac{1}{2}\tan^{-1}\left\{\left[\frac{2e_1e_2}{\left(e_1^2 - e_2^2\right)}\right]\cos(\delta + \delta')\right\} \tag{5.40}$$

Thus,

$$\psi' = -\psi \qquad \text{for } \delta' = 180° \tag{5.41}$$

and

$$a' = a \qquad \text{and} \qquad b' = b \tag{5.42}$$

Thus, we see that, in general, elliptically polarised radiation may have its degree of ellipticity altered and its inclination changed by passage through a converter. In particular cases, it may be converted into linearly polarised or circularly polarised radiation, or its orientation may be reflected about the fast axis of the converter.

In real devices, the value of δ' is chosen to be 90° or 180° and the resulting converters are called 'quarter-wave plates' or 'half-wave plates', respectively, because one beam is delayed with respect to the other by a quarter or a half of a wavelength. The quarter-wave plate is then used to convert elliptically or circularly polarised light into linearly polarised light or vice versa, while the half-wave plate is used to rotate the plane of linearly polarised light.

Many substances can be used to make converters, but mica is probably the commonest because of the ease with which it may be split along its cleavage planes. Plates of the right thickness (about 40 μm) are, therefore, simple to obtain. Quartz cut parallel to its optic axis can also be used, while for UV work, magnesium fluoride is suitable. Amorphous substances may be stretched or compressed to introduce stress birefringence. It is then possible to change the phase delay in the material by changing the amount of stress. Extremely high acceptance angles and chopping rates can be achieved with low power consumption if a small acoustic transducer at one of its natural frequencies drives the material. This is then commonly called a 'photoelastic modulator'. An electric field can also induce birefringence. Along the direction of the field, the phenomenon is called the 'Pockels effect' and perpendicular to the field it is called the 'Kerr effect'. Suitable materials abound and may be used to produce quarter- or half-wave plates. Generally, these devices are used when rapid switching of the state of birefringence is needed, as for example in Babcock's solar magnetometer (Section 5.3). In glass, the effects take up to a minute to appear and disappear, but in other substances the delay can be far shorter. The Kerr effect in nitrobenzene, for example, permits switching at up to 1 MHz and the Pockels effect can be used similarly with ammonium dihydrogen phosphate or potassium dihydrogen phosphate.

All these converters will normally be chromatic and usable over only a restricted wavelength range. Converters that are more nearly achromatic can be produced which are based upon the phase changes that occur during total internal reflection. The phase difference between components with electric vectors parallel and perpendicular to the plane of incidence is shown in Figure 5.15. For two of the angles of incidence (approximately 45° and 60° in the example shown) the phase delay is 135°. Two such reflections, therefore, produce a total delay of 270° or, as one may also view the situation, an advance of the second component with respect to the first by 90°. The minimum value of the phase difference shown in Figure 5.15 is equal to 135° when the refractive index is 1.497. There is then only one suitable angle of incidence and this is 51° 47′. For optimum results, the optical design should approach as close to this ideal as possible. A quarter-wave retarder that is nearly achromatic can thus be formed using two total internal reflections at appropriate angles. The precise angles usually have to be determined by trial and error because additional phase changes are produced when the beam interacts with the entrance and exit faces of the component. Two practical designs – the

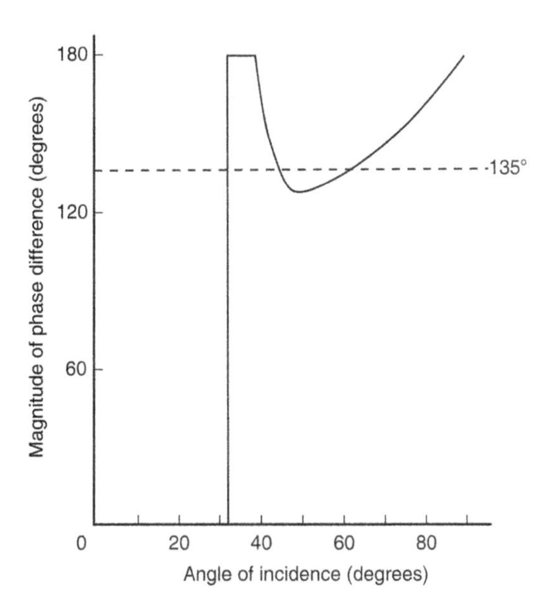

FIGURE 5.15 Differential phase delay upon internal reflection in a medium with a refractive index of 1.6.

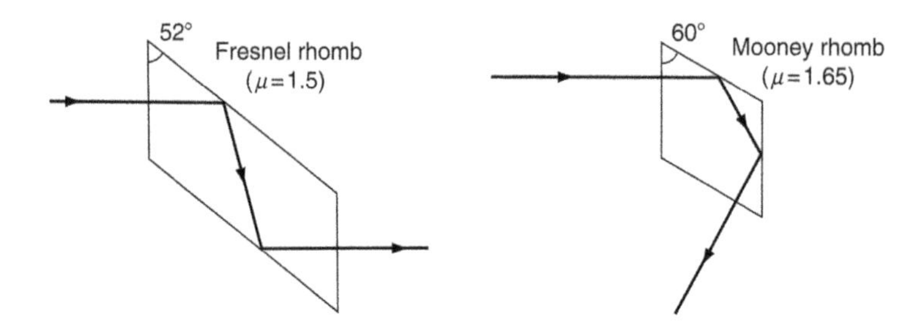

FIGURE 5.16 Quarter-wave retarders using total internal reflection.

Fresnel rhomb and the Mooney rhomb – are illustrated in Figure 5.16. Half-wave retarders can be formed by using two such quarter-wave retarders in succession.

Combining three retarders can make pseudo quarter- and half-wave plates that are usable over several hundred nanometre-wide wavebands in the visual. The Pancharatnam design employs two retarders with the same delay and orientation, together with a third sandwiched between the first two. The inner retarder is rotated with respect to the other two and possibly has a differing delay. A composite 'half-wave plate' with actual delays in the range 180° ± 2° over wavelengths from 300 to 1200 nm can, for example, be formed from three 180° retarders with the centre one oriented at about 60° to the two outer ones. Quartz and magnesium fluoride are commonly used materials for the production of such super-achromatic waveplates.

The Mueller matrices for converters are

$$\begin{bmatrix} 1 & 0 & 0 & 0 \\ 0 & \cos^2 2\psi & \cos 2\psi \sin 2\psi & -\sin 2\psi \\ 0 & \cos 2\psi \sin 2\psi & \sin^2 2\psi & \cos 2\psi \\ 0 & \sin 2\psi & -\cos 2\psi & 0 \end{bmatrix} \tag{5.43}$$

for a quarter-wave plate and

$$\begin{bmatrix} 1 & 0 & 0 & 0 \\ 0 & \cos^2 2\psi - \sin^2 2\psi & 2\cos 2\psi \sin 2\psi & 0 \\ 0 & 2\cos 2\psi \sin 2\psi & \sin^2 2\psi - \cos^2 2\psi & 0 \\ 0 & 0 & 0 & -1 \end{bmatrix} \tag{5.44}$$

for a half-wave plate, while in the general case we have

$$\begin{bmatrix} 1 & 0 & 0 & 0 \\ 0 & \cos^2 2\psi + \sin^2 2\psi \cos\delta & (1-\cos\delta)\cos 2\psi \sin 2\psi & -\sin 2\psi \sin\delta \\ 0 & (1-\cos\delta)\cos 2\psi \sin 2\psi & \sin^2 2\psi + \cos^2 2\psi \cos\delta & \cos 2\psi \sin\delta \\ 0 & \sin 2\psi \sin\delta & -\cos 2\psi \sin\delta & \cos\delta \end{bmatrix} \tag{5.45}$$

5.2.2.4 Depolarisers

The ideal depolariser would accept any form of polarised radiation and produce unpolarised radiation. No such device exists, but pseudo-depolarisers can be made. These convert the polarised radiation into radiation that is unpolarised when it is averaged over wavelength, time or area.

A monochromatic depolariser can be formed from a rotating quarter-wave plate that is in line with a half-wave plate rotating at twice its rate. The emerging beam at any given instant will have some form of elliptical polarisation, but this will change rapidly with time and the output will average to zero polarisation over several rotations of the plates.

The Lyot depolariser averages over wavelength. It consists of two retarders with phase differences very much greater than 360°. The second plate has twice the thickness of the first and its optic axis is rotated by 45° with respect to that of the first. The emergent beam will be polarised at any given wavelength, but the polarisation will vary rapidly with wavelength. In the optical region, averaging over a waveband of a few tens of nanometres is then sufficient to reduce the net polarisation to 1% of its initial value.

If a retarder is left with a rough surface and immersed in a liquid whose refractive index is the average refractive index of the retarder, then a beam of light will be undisturbed by the roughness of the surface because the hollows will be filled in by the liquid. However, the retarder will vary in its effect on the scale of its roughness. Thus, the polarisation of the emerging beam will vary on the same scale and a suitable choice for the parameters of the system can lead to an average polarisation of zero over the whole beam. The Cornu depolariser is also based upon a variable retarder and averaging over an area. It comprises two wedge-shaped blocks of quartz (or other optically active material) cemented together to form a box shape. The wedges have perpendicular fast axes which are oriented equally, but in opposite directions, to the plane of the wedge faces. The block is thus a retarder whose effect varies across the output face and averaging the output beam produces the depolarising effect.

The Mueller matrix of an ideal depolariser is simply

$$\begin{bmatrix} 1 & 0 & 0 & 0 \\ 0 & 0 & 0 & 0 \\ 0 & 0 & 0 & 0 \\ 0 & 0 & 0 & 0 \end{bmatrix} \tag{5.46}$$

5.2.3 Polarimeters

A polarimeter is an instrument that measures the state of polarisation, or some aspect of the state of polarisation, of a beam of radiation. Ideally, the values of all four Stokes' parameters should be determinable, together with their variations with time, space and wavelength. In practice, this is rarely possible, at least for astronomical sources, when only the degree of linear polarisation and its direction are found most of the time. Astronomical optical polarimeters now normally use CCDs, photomultipliers (PMTs) or avalanche photodiodes as their detectors but photographic plates have been used in the past.

For a few astronomical sources, the degree of polarisation can be several tens of percent (up to 60% for the Crab nebula, for example). Mostly, however, the degree of polarisation is less than 1% and for those observers hoping to detect exoplanets through polarimetric measurements, the instruments need to detect levels of polarisation of 0.001%–0.0001%. Polarimeters, therefore, vary in their designs depending upon the purpose for which they are intended. At the simplest, a sheet of polaroid film can be inserted before the CCD detector on a small telescope and images obtained with the polaroid rotated through 45° or 90° between successive exposures. This will certainly show up the Crab nebula's polarisation. At a slightly more sophisticated level, a higher quality polariser may be placed before the entrance aperture to a photometer or spectrograph (Chapters 3 and 4) and estimates made of the source's degree of linear polarisation and its direction and in the latter case, of its variation along the spectrum as well. However, most polarimeters and especially those aimed at achieving the highest levels of sensitivity have to be designed for the purpose.

5.2.3.1 Photoelectric Polarimeters

Most of these devices bear a distinct resemblance to photoelectric photometers (Section 3.2) and indeed, many polarimeters can also function as photometers. The major differences, apart from the components that are needed to measure the polarisation, arise from the necessity of reducing or eliminating the instrumentally induced polarisation. The instrumental polarisation originates primarily in inclined planar reflections or in the detector. Thus, Newtonian and Coudé telescopes cannot be used for polarimetry because of their inclined subsidiary mirrors. Inclined mirrors must also be avoided within the polarimeter. Cassegrain and other similar designs should not, in principle, introduce any net polarisation because of their symmetry about the optical axis. However, problems may arise, even with these telescope designs, if mechanical flexure or poor adjustments move the mirrors and/or the polarimeter away from their line of symmetry.

Many detectors, especially PMTs (Section 1.1), are sensitive to polarisation, and furthermore, this sensitivity may vary over the surface of the detector. If the detector is not shielded from the Earth's magnetic field, the sensitivity may vary with the position of the telescope as well. In addition, although an ideal Cassegrain system will produce no net polarisation, this sensitivity to polarisation arises because rays on opposite sides of the optical axis are polarised in opposite but equal ways. The detector non-uniformity may then result in incomplete cancellation of the telescope polarisation. Thus, a Fabry lens and a depolariser immediately before the detector are almost always essential components of a polarimeter.

Alternatively, the instrumental polarisation can be reversed using a Bowen compensator. This comprises two retarders with delays of about $\lambda/8$ that may be rotated with respect to the light beam and to each other until their polarisation effects are equal and opposite to those of the instrument. Small amounts of instrumental polarisation can also be corrected by the insertion of a tilted glass plate into the light beam to introduce opposing polarisation. In practice, an unpolarised source is observed and the plate tilted and rotated until the observed polarisation is minimised or eliminated.

There are many detailed designs of polarimeter, but we may group them into single- and double-channel devices, with the former further subdivided into discrete and continuous systems.[10] The single-channel discrete polarimeter is the basis of most of the other designs and its main components are shown in Figure 5.17, although the order is not critical for some of them. Since stellar polarisations are usually 1% or less, it is also often necessary to incorporate a chopper and use a phase-sensitive detector (also known as a 'lock-in amplifier') plus integration to obtain sufficiently accurate measurements. Switching between the sky and background is also vital because the background radiation is likely to be polarised, especially if the Moon is in the sky. The polariser is rotated either in steps or continuously, but at a rate which is slow compared with the chopping and integration times. The output for a linearly polarised source is then modulated with the position of the polariser.

The continuous single-beam systems use a rapidly rotating polariser. This gives an oscillating output for a linearly polarised source, and this may be fed directly into a computer or into a phase-sensitive detector that is driven at the same frequency. Alternatively, a quarter-wave plate and a photoelastic modulator can be used. The quarter-wave plate produces circularly polarised light, and this is then converted back into linearly polarised light by the photoelastic modulator but with its plane of polarisation rotated by 90° between alternate peaks of its cycle. A polariser that is stationary will then introduce an intensity variation with time that may be detected and analysed as before. The output of the phase-sensitive detector in systems such as these is directly proportional to the degree of polarisation, with unpolarised sources giving zero output. They may also be used as photometers to determine the total intensity of the source, by inserting an additional fixed polariser

[10] They may also be divided into imaging versus point source designs and/or monochromatic polarimeters versus spectropolarimeters. But with most modern instruments, these options are available just as different settings for the instrument.

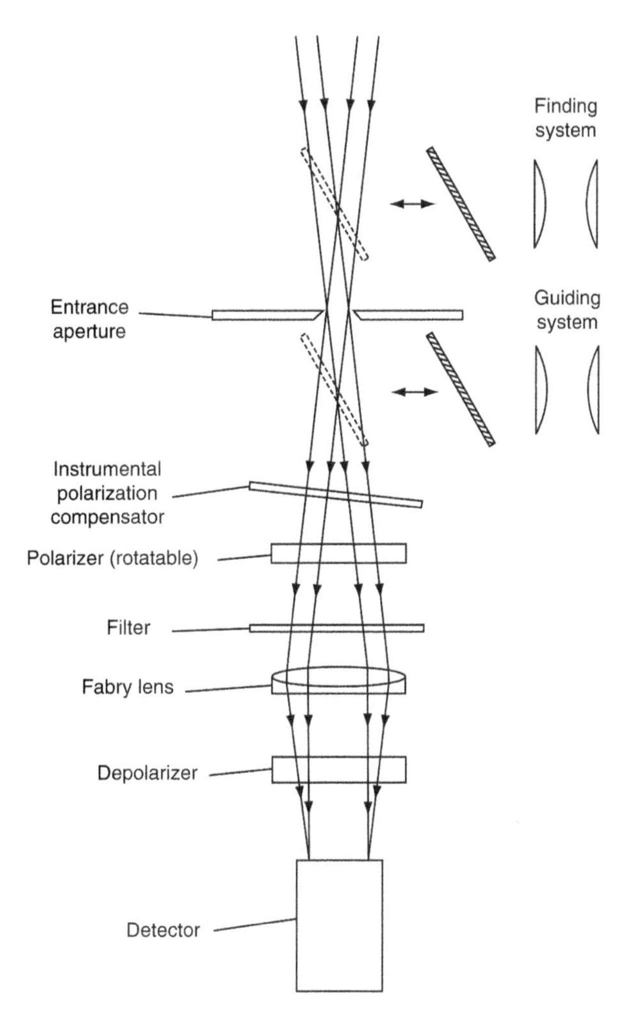

FIGURE 5.17 The basic polarimeter: schematic design and optical paths.

before the entrance aperture, to give in effect a source with 100% polarisation and half the intensity of the actual source (provided that the intrinsic polarisation is small). An early polarimeter system that is related to these is due to Dolfus. It uses a quarter-wave plate to convert the linearly polarised component of the radiation into circularly polarised light. A rotating disc that contains alternate sectors made from zero- and half-wave plates then chops the beam. The circular polarisation alternates in its direction but remains circularly polarised. Another quarter-wave plate converts the circularly polarised light back to linearly polarised light, but the plane of its polarisation switches through 90° at the chopping frequency. Finally, a polariser is used to eliminate one of the linearly polarised components, so that the detector produces a square wave output.

The continuous single-beam polarimeter determines both linear polarisation components quasi-simultaneously, but it still requires a separate observation to acquire the total intensity of the source. In double-beam devices, the total intensity is found with each observation, but several observations are required to determine the polarisation. These instruments are essentially similar to that shown in Figure 5.17, but a double-beam polariser such as a Wollaston prism or a Savart plate is used in place of the single-beam polariser shown there. Two detectors then observe both beams simultaneously. The polariser is again rotated either in steps or continuously but slowly to measure the polarisation. Alternatively, a half-wave plate can be rotated above a fixed polariser. The effects of scintillation and other atmospheric changes are much reduced by the use of dual beam instruments.

The VLT's Gravity instrument (Sections 2.5.4, 4.2.7 and 5.1.7) can act as a basic spectropolarimeter by inserting MgF_2 Wollaston prisms into the light paths of its two spectroscope cameras. The two linear polarisation spectra can then be simultaneously and separately recorded. Also on the VLT, as already mentioned (Section 2.7.4), the Spectro-Polarimetric High-contrast Exoplanet Research instrument (SPHERE) has the Zurich Imaging Polarimeter (ZIMPOL) as an optional polariser. The ZIMPOL undertakes differential polarimetric imaging over a 3.5′ field of view using two 2 × 4 k frame transfer CCD detectors. It has a polarisation sensitivity better than 0.1% over the 500- to 900-nm waveband. It uses dichroic beam splitters to allow other VLT instruments to make simultaneous observations and a polarising beam splitter to separate its 'own' linear polarised components. A liquid crystal half-wave plate,[11] driven at up to 1 kHz, rotates the polarisations of the incoming light beam through 90° so that the two CCDs rapidly interchange their images. The CCDs frame-shift read-out is then synchronised with the modulations to produce the output (c.f. phase sensitive detection above). Various filters and a field de-rotator are optionally available.

The two High Precision Polarimetric Instrument (HIPPI2) polarimeters have been used on several telescopes, including Gemini North and the Anglo-Australian Telescope (AAT), and they are physically compact and so could be used on smaller instruments. Their optical layout resembles that of ZIMPOL; a 500-Hz liquid crystal half-wave plate precedes a Wollaston prism and then the two beams are directed to PMT detectors through achromatic Fabry lenses. The whole instrument can be rotated around the main optical axis and typically four measurements will be obtained for each observation; at position angles of 0°, 45°, 90°, and 135°. Under optimum conditions, the polarisation measurements can reach 0.001% accuracies.

The Large Binocular Telescope's (LBT's) fibre-fed high-resolution spectrograph, Potsdam Echelle Polarimetric and Spectroscopic Instrument (PEPSI), has a twin polarimeter option. These instruments are based upon super-achromatic retarders, Foster prism polarisers[12] and CCD detectors and measure all four Stokes' parameters for spectra with spectral resolutions up to 130,000. The PEPSI can also be used (using a *much* smaller telescope) for solar work and can be fibre-optic fed from the 1.8-m Vatican Advanced Technology Telescope (ATT).

Smaller observatories may be inspired by the RoboPol instrument on the 1.3-m Ritchey-Chrétien telescope at the Skinakas Observatory on Crete. This is based upon a pair of quartz Wollaston prisms and a pair non-rotating MgF_2/quartz half-wave plates with a single CCD detector. The orientations of the half-wave plates and the Wollaston prisms are arranged so that four images of the source are produced (the ordinary and extraordinary rays plus rotations through 45° with respect to each other). It thus obtains, to an accuracy of about 0.1%, both the normalised Stokes' parameters (Q/I and U/I) and the intensity (I) of the source, in a single observation.

The Astro 2020 concept instrument, LUVOIR (Sections 1.1.23 and 4.2.5), has POLLUX proposed as its spectropolarimeter. This might cover the UV from 90 to 400 nm at a spectral resolution up to 120,000. The possibilities for the polarimeter part of the instrument are currently suggested as: MgF_2 retarders for the near UV and reflectors for the mid and far UV combined with MgF_2 Wollaston prisms (near and mid UV) or a Brewster angle SiC plate (far UV).

[11] Often called a 'modulator' when used for rapid switching, as here.

[12] A variety of the Glan-Thompson polariser (Figure 5.14) which transmits both polarised beams.

5.2.4 Data Reduction and Analysis

The output of a polarimeter is usually in the form of a series of intensity measurements for varying angles of the polariser. These must be corrected for instrumental polarisation, if this has not already been removed or has only been incompletely removed by the use of an inclined plate. The atmospheric contribution to the polarisation must then be removed by comparison of the observations of the star and its background. Other photometric corrections, such as for atmospheric extinction (Section 3.2) must also be made, especially if the observations at differing orientations of the polariser are made at significantly different times, so that the position of the source in the sky has changed.

The normalised Stokes' parameters are then obtained from

$$\frac{Q}{I} = \frac{I(0) - I(90)}{I(0) + I(90)} \tag{5.47}$$

and

$$\frac{U}{I} = \frac{I(45) - I(135)}{I(45) + I(135)} \tag{5.48}$$

where $I(\theta)$ is the intensity at a position angle θ for the polariser. Since for most astronomical sources the elliptically polarised component of the radiation is 1% or less of the linearly polarised component, the degree of linear polarisation, π_L, can then be found via Equation 5.28 or more simply from Equation 5.30 if sufficient observations exist to determine I_{max} and I_{min} with adequate precision. The position angle of the polarisation may be found from Equation 5.31 or again directly from the observations if these are sufficiently closely spaced. The fourth Stokes' parameter, V, is rarely determined in astronomical polarimetry. However, the addition of a quarter-wave plate to the optical train of the polarimeter is sufficient to enable it to be found if required (Section 5.4 for example). The reason for the lack of interest is, as already remarked, the low levels of circular polarisation that are usually found. However, recently, the European Southern Observatory's (ESO's) 3.6-m telescope has had interchangeable linear and circular polarisation detectors installed (ESO's Faint Object Spectrograph and Camera [EFOSC2]), the latter being similar to the former except for the addition of a quarter-wave plate. The first Stokes' parameter, I, is simply the intensity of the source, and so if it is not measured directly as a part of the experimental procedure, can simply be obtained from

$$I = I(\theta) + I(\theta + 90°) \tag{5.49}$$

A flowchart to illustrate the steps in an observation and analysis sequence to determine all four Stokes' parameters is shown in Figure 5.18.

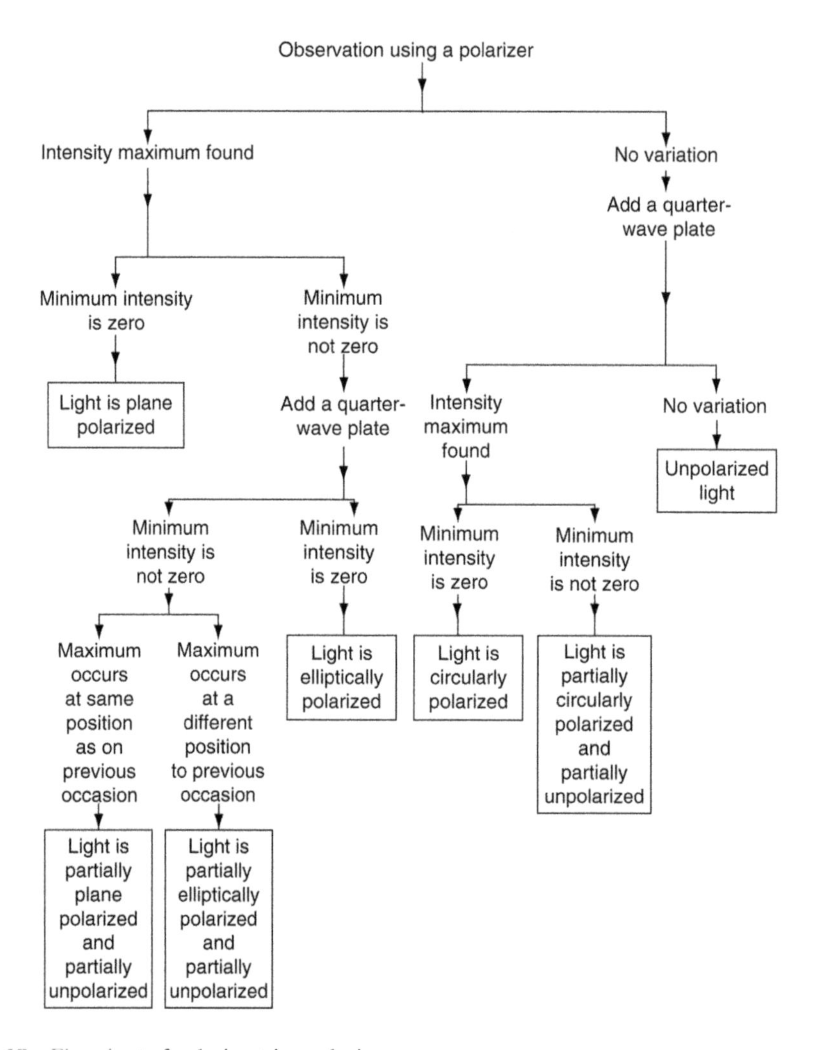

FIGURE 5.18 Flowchart of polarimetric analysis.

5.3 SOLAR STUDIES

WARNING

OBSERVING THE SUN CAN BE DANGEROUS: PERMANENT EYE DAMAGE OR BLINDNESS CAN EASILY OCCUR FROM LOOKING AT THE SUN

This warning applies

BOTH to unaided observations (i.e., with the naked eye)

AND to observations made through telescopes, binoculars, monoculars or any other optical aid or instrument.

For unaided observations; ONLY filters made and sold specifically for viewing the Sun and which meet the current ISO 12312-2 standard should ever be used.

You should NEVER look at the Sun without using such a filter which SAFELY covers BOTH eyes completely.

(Continued)

WARNING (Continued)

N.B. Filters may deteriorate over time and/or become scratched or otherwise damaged and so should be replaced frequently.

Furthermore, for recent solar eclipses, there have been cases of unsafe, counterfeit, or at least untested, eclipse viewing glasses being sold, although they were marked as meeting the International Organisation for Standardization (ISO) standard.

Your National Astronomical Society or Organisation (easily found via an internet search at such times), should be able to provide a list of reputable suppliers of eclipse viewing glasses and related products.

Typically, filters for the safe unaided viewing of the Sun need to have optical densities of 4.5 or more (transmissions of 0.003% or less) across the UVB, UVA and visible parts of the spectrum, and optical densities of 3.5 or more (transmissions of 0.03% or less) across the infrared.

For observing the Sun using a telescope, binoculars, monocular or any other optical aid, full aperture filters (see main text) are widely available. Your own telescope provider, in many cases, should be able to supply you with a suitable full aperture filter to fit your instrument.

Where your original telescope supplier is not able to provide you with a suitable solar filter, there are other independent sources (see the popular *Astronomy* magazines or contact your National Astronomical Society for reputable suppliers).

Although the *surface brightness* of the Sun is *not* increased by the use of a telescope (Equations 1.55 to 1.57), the *total brightness* is increased. Full aperture filters, generally, therefore should have optical densities of 5 or 6 or more (transmissions of 0.001%–0.0001% or less) for comfortable viewing. You may also find it useful to reduce the collecting area of your telescope by fitting a sub-aperture stop (i.e., a sheet of cardboard with a hole smaller than your telescope) over the top end of its tube.

The widely promoted eyepiece projection method of solar viewing is safe from the point of view of personal injury but can damage the eyepiece and/or the telescope, especially with the shorter focal ratio instruments. Its use may also invalidate the guarantee on your telescope.

Older practical astronomy books/sources may recommend the Herschel Wedge (aka the Solar Diagonal) and/or the use of small optically dense filters fitted into/onto/near/after the eyepiece.

These should NOT be used – the heat from the Sun may cause them to become hot enough to cause burns, or to stress them until they shatter, or at least, cause them to become so distorted that the resulting image quality is much degraded.

PUBLIC OUTREACH

These warnings and advice should be carefully noted by all intending solar observers and passed onto inexperienced persons with whom the observer may be working.

In particular, the dangers of solar observing should be made quite clear to any groups of lay people touring the observatory because the temptation to try looking at the Sun for themselves is great, *especially* when there is a solar eclipse.

Despite the following dire warning, intending solar observers should not be put off their planned observing programmes. Observing the Sun *can* be quite safe and is certainly enjoyable; you just have to take the right precautions.

Inexperienced observers should research what is needed for solar work *before* actually undertaking it. There are plenty of available sources, both hard copy and on the internet – and your National Astronomical Society/Organisation should be able to give you help and guidance. There may also be a local Amateur Astronomy Society near to you; however, you should be somewhat wary of advice from such a source – this author, at least, has encountered highly inadvisable solar observing procedures being suggested by well-meaning but ill-informed members of such societies.

5.3.1 Introduction

The Sun is a reasonably typical star, and as such, it can be studied by many of the techniques discussed elsewhere in this book. In one major way, however, the Sun is quite different from all other stars and that is that it may easily be studied using an angular resolution smaller than its diameter. This, combined with the vastly greater amount of energy that is available compared with that from any other star, has led to the development of numerous specialised instruments and observing techniques. Many of these have already been mentioned or discussed in some detail in other sections. The main such techniques include: solar cosmic ray and neutron detection (Section 1.4); neutrino detection (Section 1.5); radio observations (Sections 1.2 and 2.5); X-ray and γ-ray observations (Section 1.3); adaptive optics (Section 2.5), radar observations (Section 2.8); magnetometry (Section 5.4); spectroscopy; and prominence spectroscopes (Section 4.2). The other remaining important specialist instrumentation and techniques for solar work are covered in this section.

5.3.2 Solar Telescopes – Part 1

With appropriate adaptations (see WARNING above and further discussions below) almost any small(ish) telescope may be used to image the Sun, but some are more suited to the purpose than others. The specialist solar telescopes discussed later in this section will clearly have been designed for their job, but even with these, there will be safe observing procedures to be followed. By the time, though, that observers arrive at (say) the Big Bear Solar Observatory bearing their precious approved allocation of telescope time, they are likely to be fully experienced in all aspects of solar observing.

This first part of this section, therefore, is aimed at less experienced observers, perhaps using their own 0.1- or 0.2-m telescopes in their backyards to attempt solar observing for the first time. Problems, dangers and injuries are more likely to arise when the owners of telescopes, binoculars and the like, which they purchased for viewing the night sky, for bird-watching, for use at sea, and so on decide to use them to observe the Sun. This situation is especially liable to arise when a solar eclipse is due.

Until, therefore, you have fully read and understood the requirements (below) for SAFE solar telescopic observations

DO NOT LOOK AT THE SUN BY ANY METHOD.

Even looking at the Sun without any optical aids should never be done except by looking through ISO 12312-2 compliant eclipse-viewer-type spectacles (see WARNING).

Eclipse watchers should, in any case, note that *nothing* more will be seen of any solar eclipse by using a telescope, binoculars, and so on than can be seen with the unaided eye through proper eclipse-viewer spectacles – except during the few minutes of totality during a total eclipse, when quite different observing requirements apply anyway.

5.3.2.1 Binoculars

In practice, binoculars and bird-watching-type monoculars present the most widespread dangers for opportunistic or spur-of-the-moment solar eclipse (or solar) viewers. Probably, more than 99% of binoculars are purchased for non-astronomical reasons. When it comes to solar observations, therefore, the full aperture solar filters as discussed and generally sold for astronomical telescopes, will not be available (and both halves of the binoculars would need filters anyway).

The eyepiece projection method also has its drawbacks (discussed below), even for simple telescopes. Binoculars, however, are *not* simple optical devices. Most designs incorporate internal prisms and lenses and the zooming and/or image-stabilising variants are even more complex inside.

Using binoculars for eyepiece projection of the Sun will thus be highly likely to damage their optical components and/or other internal mechanisms. So, binocular owners can only be counselled:

DO NOT **EVER** OBSERVE THE SUN USING YOUR BINOCULARS
BY ANY METHOD **WHAT-SO-EVER – EVER**!

5.3.2.2 Telescopes

Telescopes, despite usually having much larger apertures than binoculars, may though, be used for solar work. This is partially because in optical terms most small telescopes are simpler than binoculars, but more because many telescopes will have been purchased for astronomical reasons and so their owners will be more knowledgeable about their use.

The problem with observing the Sun is just that its intensity is high enough to cause injury to people and animals (especially to eyes) and/or physical damage to objects. The solution, therefore, is to reduce its intensity.[13]

Several ways of reducing solar brightness have already been mentioned above and we may subdivide them into:

1. methods which reduce the solar intensity *before* its radiation enters the telescope and
2. methods which reduce the solar intensity *before* its radiation enters the *eye*, but *after* it has passed through some, or all, of the telescope.

The best approach is '(i)' and today, this means the use of the telescope with a full-aperture solar filter. '(ii)' includes eye-piece projection which, with care, may still be usable, but also Herschel Wedges and eyepiece filters – which are *not* now recommended. These various approaches are now considered in more detail individually.

Whilst properly fabricated and undamaged full aperture solar filters will protect the telescope as well as the viewer, you may find that your telescope manufacturer's guarantee will be invalidated for *any* type of solar observation – so check on this *before* attempting solar observations of whatever type.

5.3.2.2.1 Full-Aperture Solar Filters

Full aperture filters are produced commercially for use on small telescopes. They cover the whole aperture of the telescope, as their name suggests and may be made sufficiently opaque for use on telescopes up to 0.5 m in diameter. They are much the preferred approach to converting a small (-ish) telescope to solar observation. As noted previously, the filters need to have an optical density of five or six and larger telescopes may need to be stopped-down to smaller apertures. With the filter in place, the telescope may be used for observing the Sun with all the normal eyepieces, cameras, and so on available for night-time observations. It may, however, be necessary to shade the telescope and/or observer and/or instruments from the much higher-than-night-time ambient light.

One of the earliest types of full-aperture solar filter to be produced and which is still available today, consists of a very thin Mylar film that has been coated with a reflective layer of aluminium, stainless steel, etc. The film is so thin that little degradation of the image occurs; however, this also makes the film liable to damage and its viable life, when in regular use, can be quite short. Such filters are produced with single or double layers of reflective coating. The double layer is to be preferred because any pinholes in one layer are likely to be covered by the second layer and vice versa.

[13] Rather ironically, when, on 2 July 2019, a total solar eclipse occurred directly over the ESO's La Silla observatory (see cover image), most of the observations were made using small portable telescopes brought in by visitors. Of the numerous permanent instruments at the observatory, only the 3.6-m New Technology Telescope (NTT) was used for observing the eclipse – and then only for spectroscopy of the corona.

Reflective coatings on optically flat glass substrates have become available recently, though they are usually somewhat more expensive than the coated plastic filters. Finally, the filter may simply be a thin sheet of absorbing plastic (do not be tempted to use any old piece of black plastic that may be lying around – it may not be sufficiently opaque in the UV or infrared – always purchase a purpose-made filter from a reliable supplier).

An additional note of warning – do not forget to put a full aperture filter on the guide/finder telescope (or blank it off with an opaque screen). Without such a filter, the guide/finder telescope must *not* be used to align the telescope – instead the telescope's shadow should be circularised – this is usually sufficient to bring the solar image into the field of view of the (filtered) main telescope. Alternatively, if a small portable telescope's mounting is correctly aligned and has accurate setting circles, or for larger permanently mounted instruments, the Sun may be found in the telescope by setting onto its current position (listed in the *Astronomical Almanac* each year). Most, recently produced, telescopes for the amateur astronomy market will now have computer systems and mountings that allow the telescope to set directly onto the Sun – although these systems may need to be set up/aligned at night, using stars.

5.3.2.2.2 *Eyepiece Projection*

First stop-down your telescope until its aperture is 0.05-m (50 mm, 2 inches) or less. Then, in this approach, the solar image is projected onto a sheet of white card held behind the eyepiece. The telescope is pointed towards the Sun (see Section 5.3.2.2.1), the sheet of card mounted behind the eyepiece, perpendicular to the optical axis and the eyepiece is then adjusted to bring the image of the Sun on the card into focus (DO NOT LOOK THROUGH THE TELESCOPE WHILST DOING ANY OF THIS!). The solar image is then observed on the screen. Using different eyepieces or different projection distances will enable the image size to be changed.

This method may be preferred when demonstrating to a group of people because everyone can see the image at the same time and they are looking away from the Sun itself to see its image.

The principle of eyepiece projection is shown in Figure 5.19. The eyepiece is moved outwards from its normal position for visual use and a real image of the Sun is obtained that may easily be projected onto a sheet of white paper or cardboard and so on for viewing purposes. We have the effective focal length (EFL) of the telescope

$$\text{EFL} = \frac{Fd}{e} \tag{5.50}$$

where F is the focal length of the objective, e is the focal length of the eyepiece and d is the projection distance (see Figure 5.19). The final image size, D, for the Sun is then simply

$$D = \frac{0.0093Fd}{e} \tag{5.51}$$

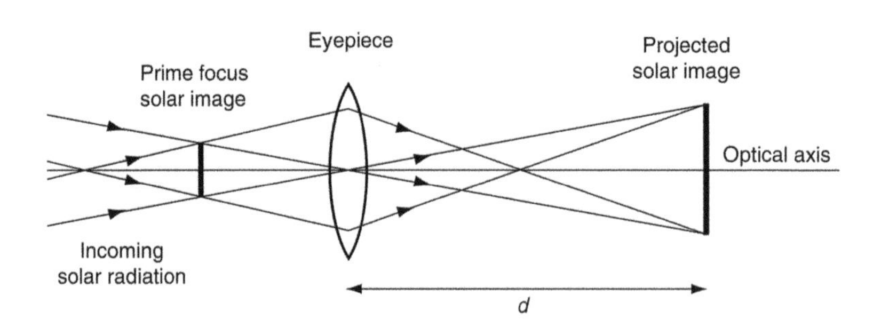

FIGURE 5.19 Projection of the solar image.

The main drawback to solar observing via eyepiece projection arises with the widely available Schmidt-Cassegrain, Maksutov and related designs of telescopes which usually have a short focal length (fast *f*-ratio) primary mirror. This will produce a small, but intense image of the Sun inside the telescope, which is quite capable of damaging parts of the telescope. Even setting fire to them should it remain stationary on a single point for any length of time. Thus, although eyepiece projection is safe for the *viewer*, it is still possible to damage the *telescope*. This is especially, but not exclusively, the case whilst setting the telescope onto the Sun, because the primary mirror's solar image is quite likely to impinge onto parts of the internal structure of the telescope during that process. Even for other telescope designs, the heat passing through the eyepiece may be sufficient to damage it, especially with the more expensive multi-component types.

5.3.2.2.3 No-Longer Recommended Solar Observing Methods

The Herschel wedge or solar diagonal (Figure 5.20) was, in the past, a common telescope accessory, but its use is now strongly to be discouraged. When the typically available commercial telescopes had apertures of only 25–50 mm, then the Herschel wedge did reduce the solar brightness to not-very-dangerous levels provided that high (>×300 or so) magnifications were also used. But with many amateur astronomers now using 200- to 300-mm aperture telescopes, the Herschel wedge's effective optical density of about 1.3 (transmission ~5%) is quite inadequate and observers still using them are at serious risk of incurring significant eye damage. The device is a thin prism with unsilvered faces. The first face is inclined at 45° to the optical axis and, thus, reflects about 5% of the solar radiation into an eyepiece in the normal star diagonal position. The second face also reflects about 5% of the radiation, but its inclination to the optical axis is not 45° and so it is easy to arrange for this radiation to be intercepted by a baffle before it reaches the eyepiece. The remaining radiation passes through the prism and can be absorbed in a heat trap, or more commonly (and dangerously) just allowed to emerge as an intense beam of radiation at the rear end of the telescope. The device should also incorporate a separate infrared filter (but often does not).

A second approach which is also strongly to be discouraged with today's larger amateur telescopes, was simply a dense filter placed at the eyepiece end of the telescope, sometimes being incorporated into the eyepiece itself. By now, it should be obvious to the reader that such a filter is likely to become hot and highly stressed. The heat (if you are lucky) will be conducted into the main body of the telescope and/or the eyepiece, which will then become uncomfortably or even dangerously hot.

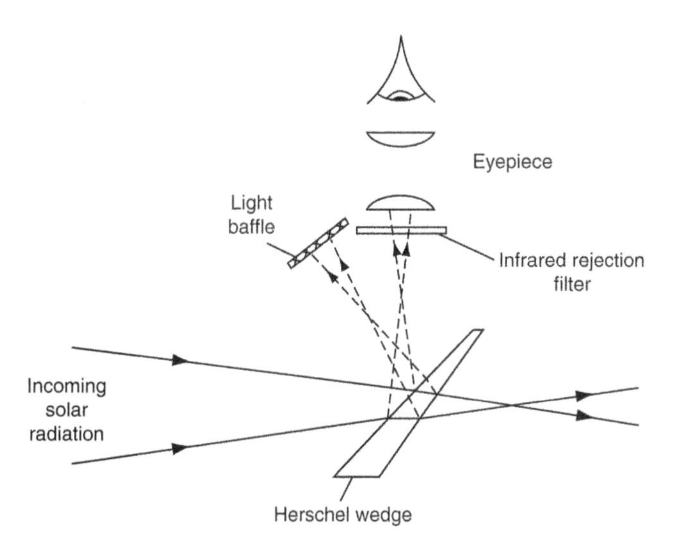

FIGURE 5.20 Optical paths in a Herschel wedge.

The resulting convection currents inside and outside the telescope will also lead to poor-quality images of the Sun. If you are not so lucky, then the eyepiece components, especially any lenses that are cemented together, may be permanently damaged. At the worst, the filter and/or eyepiece lenses may shatter under their thermal stresses, with possible physical injuries to the eye from the glass fragments and the sudden exposure of the eye to the full intensity of the solar image.

5.3.3 Solar Telescopes – Part 2

Larger instruments built specifically for solar observing come in several varieties. They are designed to try and overcome the major problem of the solar observer, which is the turbulence within and outside the instrument, whilst taking advantage of the observational opportunity afforded by the plentiful supply of radiation. The most intractable of the problems is that of external turbulence. This arises from the differential heating of the observatory, its surroundings and the atmosphere which leads to the generation of convection currents. The effect reduces with height above the ground and so one solution is to use a tower telescope. This involves a long focal length telescope that is fixed vertically with its objective supported on a tower at a height of several tens of metres. The solar radiation is fed into it using a coelostat (Section 1.1) or heliostat. The latter uses a single plane mirror rather than the double mirror of the coelostat (cf. siderostat – Section 1.1) but has the disadvantage that the field of view rotates as the instrument tracks the Sun. The 150-foot (46-m) tower telescope on Mount Wilson for example was constructed by George Ellery Hale in 1911 with a 0.3-m lens. It has been observing the Sun every clear day since 1912 and continues to do so to this day.

The other main approach to the turbulence problem (the two approaches are not necessarily mutually exclusive) is to try and reduce the extent of the turbulence. The planting of low growing tree or bush cover of the ground is one method of doing this; another is to surround the observatory by water. The Big Bear Solar Observatory is thus built on an artificial island in Big Bear Lake, California, and houses several solar telescopes, including the 1.6-m Goode Solar Telescope (see below). Painting the telescope and observatory with titanium dioxide-based white paint will help to reduce heating, because this reflects the solar incoming radiation, but allows the long wave infrared radiation from the building to be radiated away. A gentle breeze also helps to clear away local turbulence and some tower telescopes such as the 0.45-m Dutch Open Solar Telescope on La Palma (mothballed in 2008) are completely exposed to the elements to facilitate the effect of the wind.

Within the telescope, the solar energy can create convection currents, especially from heating the objective. Sometimes these can be minimised by careful design of the instrument, or the telescope may be sealed and filled with helium whose refractive index is only 1.000036 compared with 1.000293 for air and whose high conductivity helps to keep the components cool and so to minimise convection. But the ultimate solution to them is to evacuate the optical system to a pressure of a few milli-bars, so that all the light paths after the objective, or a window in front of the objective, are effectively in a vacuum. Several modern solar telescopes are therefore of vacuum designs, such as the 0.7-m German Vacuum Tower Telescope (VTT) and the 0.9-m French Télescope Héliographique pour l'Etude du Magnétismeet des Instabilités Solaires (THEMIS), both on Tenerife. While the 0.76-m Dunn solar telescope in New Mexico was the first vacuum telescope to be constructed (in 1969) and is still in active use. The largest vacuum solar telescope is currently the 1-metre New Vacuum Solar Telescope (NVST) Gregorian instrument at the Fuxian solar observatory in southwest China.

Given the problems that solar telescopes have with excessive energy, it might seem that at least they would never suffer from too little light for their observations. However, the use of narrow band filters and short exposures means that some types of solar observations *are* limited by too few photons. Solar telescopes, therefore, are starting to follow the trend of night-time instruments towards larger and larger sizes.

The Goode Solar telescope at the Big Bear Observatory thus has a 1.6-m aperture. It is of an off-axis Gregorian design on an equatorial mounting. It is too large to be a vacuum telescope, so the mirrors have thermal control systems that keep them to within one degree of the ambient air temperature and air knives[14] ensure streamlined air flows over the mirrors. Gregor is a 1.5-m Gregorian design solar telescope, commissioned in 2009. It operates without any enclosure, even in quite strong winds to minimise turbulence. It is sited on Tenerife and uses a high-order adaptive optics system.

However, the venerable McMath-Pierce solar telescope, constructed in 1961 with a 1.61-m primary mirror remains, at the time of writing, as the world's largest solar telescope. It uses a 2.1-m heliostat mounted at the top of a 30-m tower to feed its main mirror which is at the bottom of an 80-m-long slanting tunnel lying partly underground.

The McMath-Pierce telescope, though, will not hold its world record for much longer because work started in Hawaii on the Daniel K. Inouye Solar Telescope (DKIST) in 2013, and it is now nearing completion. The DKIST is planned to have a 4.24-m primary mirror (although the clear aperture will be 4 m) and to be of an off-axis Gregorian design. It is expected to be able to resolve 30-km (50 milliarcsecond) features on the Sun using an adaptive optics system to overcome turbulence. First light for the DKIST is currently planned for 2020. Other possible planned projects include the 1.8-m Chinese Large Solar Telescope (CLST; currently nearing completion), the 4-m European Solar Telescope (EST; currently advertising for tenders for its construction) and India's 2-m National Large Solar Telescope (NLST) tower telescope, China's 4- to 8-m concept Chinese Giant Solar Telescope (CGST) and the US's 1.5-m Coronal Solar Magnetism Observatory (COSMO), the latter three instruments all being still in their early planning stages.

Optically, solar telescopes are fairly conventional (Section 1.1) except that long focal lengths can be used to give large image scales because there is usually plenty of available light. It is generally undesirable to fold the light paths since the extra reflections will introduce additional scattering and distortion into the image. Thus, the instruments are often cumbersome and many are fixed in position and use a coelostat to acquire the solar radiation.

Conventional large optical telescopes can be used for solar infrared observations if they are fitted with a suitable mask. The mask covers the whole of the aperture of the telescope and protects the instrument from the visible solar radiation. It also contains one or more apertures filled with infrared transmission filters. The resulting combination provides high-resolution infrared images without running the normal risk of thermal damage when a large telescope is pointed towards the Sun.

Smaller instruments may conveniently be mounted onto a solar spar. This is an equatorial mounting that is driven so that it tracks the Sun and which has a general-purpose mounting bracket in the place normally reserved for a telescope. Special equipment may then be attached as desired. Often several separate instruments will be mounted on the spar together and may be in use simultaneously.

Numerous spacecraft, including many manned missions, have carried instruments for solar observing and many missions have had solar studies as their primary objective. At the time of writing, the Parker Solar Probe is the most recently launched (August 2018) solar mission and is also the human artefact which is closest to the Sun (by far). Its highly elliptical and evolving orbit will take it to within 6.2×10^9 km of the surface of the Sun by 2025 – well into the lower part of the solar corona. It makes its solar observations only for intervals of about 10 days around its perihelion points, when it will be moving at up to 190 km/s and be protected from the 700,000 Wm^{-2} heat blast from the Sun behind a 0.11-m-thick carbon-composite heat shield. Its primary science aims are to study the energy flows, structure and dynamics of the corona, solar wind, high-energy solar particles and solar magnetic fields. The Parker Solar Probe carries, amongst other things, fluxgate and search-coil magnetometers (Section 5.4) and two wide-field (58° × 95° FoV) refractive telescopes

[14] Essentially a pipe containing air at a high pressure that has a line of small holes along one side producing a series of air jets.

for directly imaging the corona (but not the Sun itself) employing 2,000 × 2,000-pixel radiation-hardened Complementary Metal Oxide Semiconductor-Active Pixel Sensor (CMOS-APS) detectors (Section 1.1.16). The 0.051-m, f/0.4 and 0.042-m, f/0.66 telescopes (c.f. Transiting Exoplanet Survey Satellite [TESS] Section 1.1.19.2) observe over a spectral region from about 480 to 740 nm with angular resolutions of about 6′. Solar and Heliospheric Observatory (SOHO; 1995 to date) monitors solar oscillations (see below) and the solar wind. Its instrumentation includes an Extreme Ultraviolet (EUV) telescope (Sections 1.1 and 1.3), a coronagraph (see below) a spectroscope and Doppler imager (Section 4.2). Hinode (Sunrise in Japanese) was launched in 2006 carrying a 0.5-m Gregorian optical telescope – the largest solar telescope in space. Additionally, Hinode has an X-ray telescope and an EUV spectroscope for studying the solar corona and is still operating today.

The two Solar Terrestrial Relations Observatory (STEREO 2006–2016) spacecraft were in orbits just inside and just outside that of the Earth around the Sun. They, therefore, gained and lost on the position of the Earth in its orbit by 22° per year (60 million km per year). Their spatial separation of tens of millions of kilometres enabled them to obtain 3D images of the Sun and in particular of coronal mass ejections. The spacecraft each carried three telescopes and two coronagraphs. Other notable solar missions of the past include: Solar Maximum Mission (SMM; 1980 to 1989 – a mission to study solar flares and the first unmanned spacecraft to be repaired in space by astronauts from the space shuttle, Challenger), Ulysses (1990 to 2009 – in a high inclination orbit to see the Solar poles), and Genesis (2001 to 2004 – a mission to sample the solar wind and return the samples to Earth).

ESO's Solar Orbiter mission is scheduled to be launched at the time of writing. Its orbit should take to within about 40 million km of the Sun every five months. It will be carrying magnetometers, EUV telescopes and spectrographs, a polarimeter and a coronagraph. India's Aditya-L1 is planned for a 2020–2021 launch to the Earth-Sun Lagrange L-1 point. It is expected to have a payload which includes a coronagraph, a UV imaging telescope and a magnetometer.

However, spacecraft are not the only option. The Sunrise (not to be confused with Hinode – above) telescope was a balloon-borne 1-m Gregorian instrument flown for several days above northern Canada in 2009 and 2013. It imaged the Sun at five wavelengths ranging from 214 to 397 nm. While the EUV-imaging Hi-C Imager (Section 1.3.8) which was launched on a 10-min suborbital flight by sounding rocket in 2012 has already been mentioned in Section 1.3. Its telescope had a 0.24-m aperture and it obtained solar images at wavelengths down to 19 nm.

5.3.4 Spectrohelioscope[15]

This is a monochromator (Section 4.2) which is adapted to provide an image of the whole or of a substantial part of the Sun. It operates as a normal spectroscope, but with a second slit at the focal plane of the spectrum that is aligned with the position of the desired wavelength in the spectrum (Figure 5.21). The spectroscope's entrance slit is then oscillated so that it scans across the solar image. As the entrance slit moves, so also will the position of the spectrum. Hence, the second slit must move in sympathy with the first if it is to continue to isolate the required wavelength. As an alternative to moving the slits, rotating prisms can be used to scan the beam over the grating and the image plane. If the frequency of the oscillation is higher than 15–20 Hz, then the monochromatic image may be viewed directly by eye. Alternatively, an image may be taken and a spectroheliogram is produced.

Usually, the wavelength that is isolated in a spectrohelioscope is chosen to be that of a strong absorption line, so that particular chromospheric features such as prominences, plages, mottling and so on are highlighted. In the visual region, birefringent filters (see below) can be used in place of spectrohelioscopes, but for UV imaging and scanning, satellite-based spectrohelioscopes are used exclusively.

[15] Not to be confused with the helioscope in Section 1.7.2.1.6,

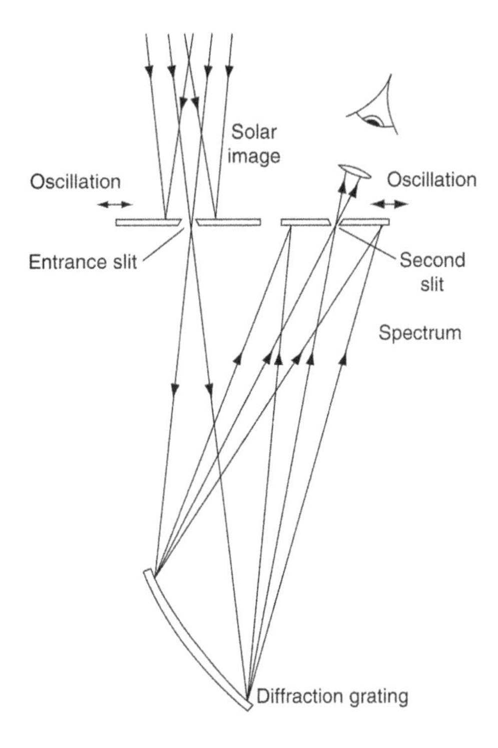

FIGURE 5.21 Principle of the spectroheliosocope.

5.3.5 Narrow Band Filters

The spectrohelioscope produces an image of the Sun over a narrow range of wavelengths. However, a similar result can be obtained by the use of a narrow band filter in the optical path of a normal telescope. Since the bandwidth of the filter must lie in the region of 0.01–0.5 nm if the desired solar features are to be visible, normal dye filters or interference filters (Section 4.1) are not suitable. Instead a filter based upon a birefringent material (Section 5.2) has been developed by Bernard Lyot. It has various names – quartz monochromator, Lyot monochromator, or birefringent filter being amongst the commonest (Figure 5.22).

The filter's operational principle is based upon a slab of quartz or other birefringent material that has been cut parallel to its optical axis. As we saw in Section 5.2, the extraordinary ray will then travel more slowly than the ordinary ray and so the two rays will emerge from the slab with a phase difference (Figure 5.23). The rays then pass through a sheet of Polaroid film whose axis is midway between the directions of polarisation of the ordinary and extraordinary rays. Only the components of each ray which lie along the Polaroid's axis are transmitted by it. Thus, the rays emerging from the Polaroid film have parallel polarisation directions and a constant phase difference and so they can mutually interfere. If the original electric vectors of the radiation, E_o and E_e, along the directions of the ordinary and extraordinary rays' polarisations respectively, are given at some point by

$$E_o = E_e, = a \cos(2\pi \nu t) \tag{5.52}$$

then after passage through the quartz, we will have at some point

$$E_o = a \cos(2\pi \nu t) \tag{5.53}$$

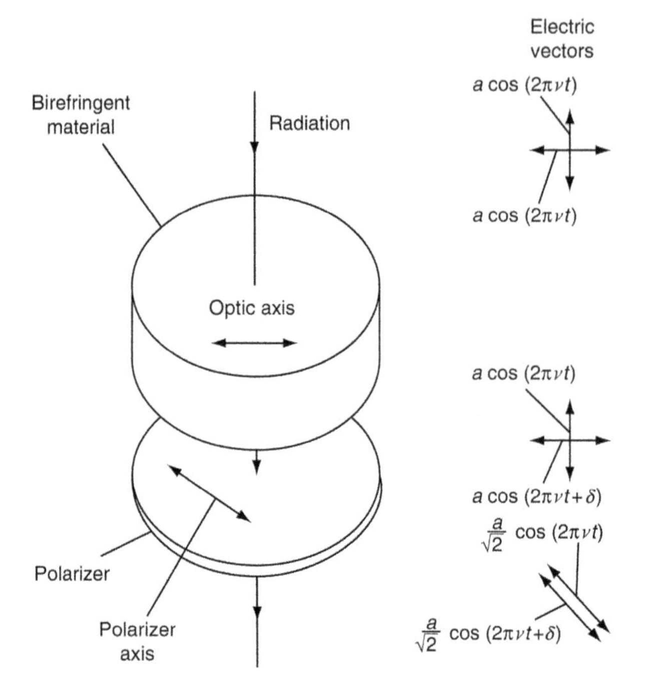

FIGURE 5.22 Basic unit of a birefringent monochromator.

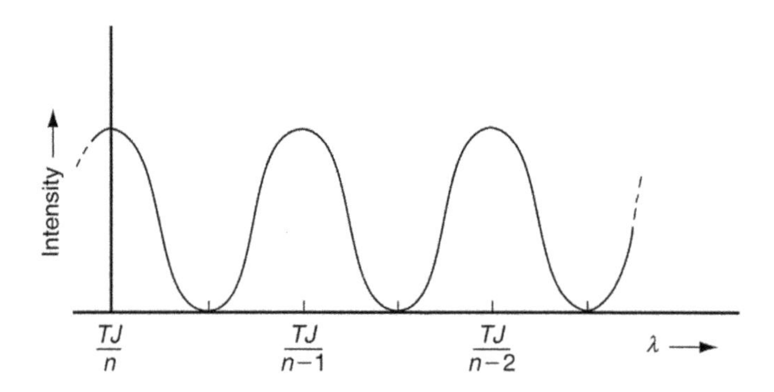

FIGURE 5.23 Spectrum of the emerging radiation from a basic unit of a birefringent monochromator.

and

$$E_{e}, = a \cos(2\pi\, \nu\, t + \delta) \tag{5.54}$$

where δ is the phase difference between the two rays. After passage through the polaroid, we then have

$$E_{45,o} = \frac{a}{\sqrt{2}} \cos(2\pi \nu t) \tag{5.55}$$

$$E_{45,e} = \frac{a}{\sqrt{2}} \cos(2\pi \nu t + \delta) \tag{5.56}$$

where $E_{45,x}$ is the component of the electric vector along the Polaroid's axis for the xth component. Thus, the total electric vector of the emerging radiation, E_{45}, will be given by

$$E_{45} = \frac{a}{\sqrt{2}}\left[\cos(2\pi \nu t) + \cos(2\pi \nu t + \delta)\right] \tag{5.57}$$

and so the emergent intensity of the radiation, I_{45}, is

$$I_{45} = 2a^2 \quad \text{for} \quad \delta = 2n\pi \tag{5.58}$$

$$I_{45} = 0 \quad \text{for} \quad \delta = (2n+1)\pi \tag{5.59}$$

where n is an integer. Now

$$\delta = \frac{2\pi c \Delta t}{\lambda} \tag{5.60}$$

where Δt is the time delay introduced between the ordinary and extraordinary rays by the material. If v_o and v_e are the velocities of the ordinary and extraordinary rays in the material then

$$\Delta t = \frac{T}{v_o} - \frac{T}{v_e} \tag{5.61}$$

$$= T\left(\frac{v_e - v_o}{v_e v_o}\right) \tag{5.62}$$

$$= \frac{TJ}{c} \tag{5.63}$$

where J is the birefringence of the material (Section 5.2) and T is the thickness of the material. Thus,

$$\delta = \frac{\left[2\pi c\left(\frac{TJ}{c}\right)\right]}{\lambda} \tag{5.64}$$

$$= \frac{2\pi TJ}{\lambda} \tag{5.65}$$

The emergent ray therefore reaches a maximum intensity at wavelengths, λ_{max}, given by

$$\lambda_{max} = \frac{TJ}{n} \tag{5.66}$$

and is zero at wavelengths, λ_{min}, given by

$$\lambda_{min} = \frac{TJ}{(2n+1)} \tag{5.67}$$

(see Figure 5.23).

Now if we require the eventual filter to have a whole bandwidth of $\Delta\lambda$ centred on λ_c, then the parameters of the basic unit must be such that one of the maxima in Figure 5.23 coincides with λ_c and the width of one of the fringes is $\Delta\lambda$; that is

$$\lambda_c = \frac{TJ}{n_c} \tag{5.68}$$

$$\Delta\lambda = TJ\left[\frac{1}{n_c - 1/2} - \frac{1}{n_c + 1/2}\right] \tag{5.69}$$

Since selection of the material fixes J and n_c is deviously related to T, the only truly free parameter of the system is T and, thus for a given filter, we have

$$T = \frac{\lambda_c}{2J}\left[\frac{\lambda_c}{\Delta\lambda} + \left(\frac{\lambda_c^2}{\Delta\lambda^2} + 1\right)^{1/2}\right] \tag{5.70}$$

and

$$n_c = \frac{TJ}{\lambda_c} \tag{5.71}$$

Normally, however,

$$\lambda_c >> \Delta\lambda \tag{5.72}$$

and quartz is the birefringent material, for which in the visible

$$J = +\,0.0092 \tag{5.73}$$

so that

$$T \approx \frac{109\lambda_c^2}{\Delta\lambda} \tag{5.74}$$

Thus, for a quartz filter to isolate the $H\alpha$ line at 656.2 nm with a bandwidth of 0.1 nm, the thickness of the quartz plate should be about 470 mm.

Now from just one such basic unit, the emergent spectrum will contain many closely spaced fringes as shown in Figure 5.23. But if we now combine it with a second basic unit oriented at 45° to the first and whose central frequency is still λ_c, but whose bandwidth is $2\Delta\lambda$, then the final output will be suppressed at alternate maxima (Figure 5.24). From Equation 5.74, we see that the thickness required for the second unit, for it to behave in this way, is just half the thickness of the first unit. Further basic units may be added, with thicknesses 1/4, 1/8, 1/16 and so on that of the first, whose transmissions continue to be centred upon λ_c, but whose bandwidths are 4, 8, 16, and so on times that of the original unit. These continue to suppress additional unwanted maxima. With six such units, the final spectrum has only a few narrow maxima that are separated from λ by multiples of $32\Delta\lambda$ (Figure 5.25). At this stage, the last remaining unwanted maxima are sufficiently well

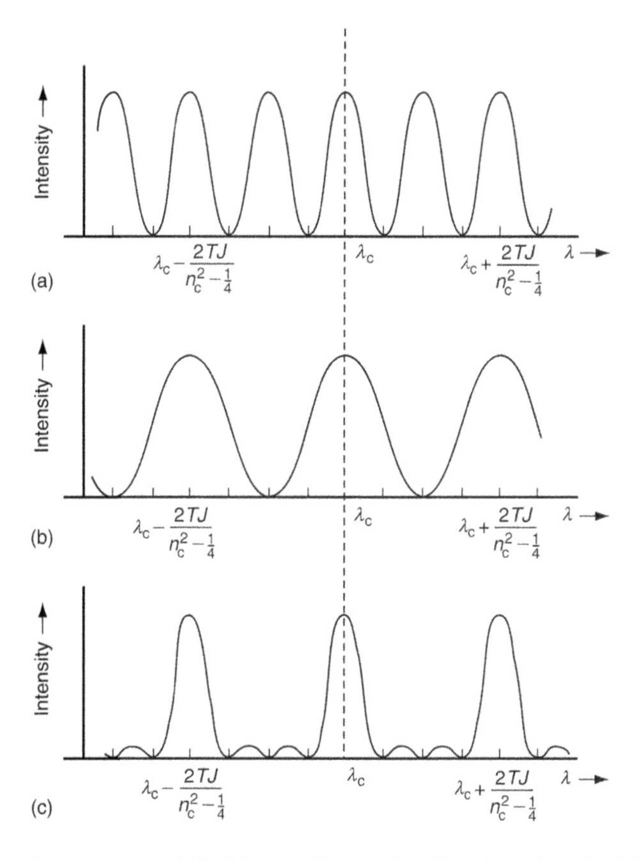

FIGURE 5.24 Transmission curves of birefringent filter basic units. (a) Basic unit of thickness, *T*; (b) basic unit of thickness, *T*/2; and (c) combination of the units in (a) and (b).

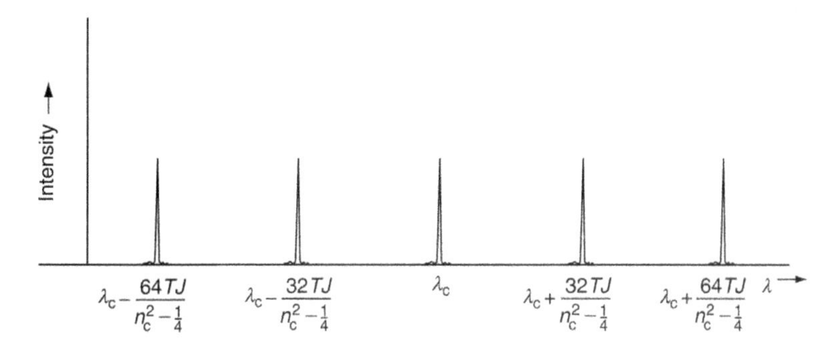

FIGURE 5.25 Transmission curve of a birefringent filter comprising six basic units with a maximum thickness *T*.

separated from λ_c for them to be eliminated by conventional dye or interference filters, so that just the desired transmission curve remains.

One further refinement to the filter is required and that is to ensure that the initial intensities of the ordinary and extraordinary rays are equal because this was assumed in obtaining Equation 5.52. This is easily accomplished, however, by placing an additional sheet of polaroid before the first unit that has its transmission axis at 45° to the optical axis of that first unit. The complete unit is shown in Figure 5.26. Neglecting absorption and other losses, the peak transmitted intensity is half

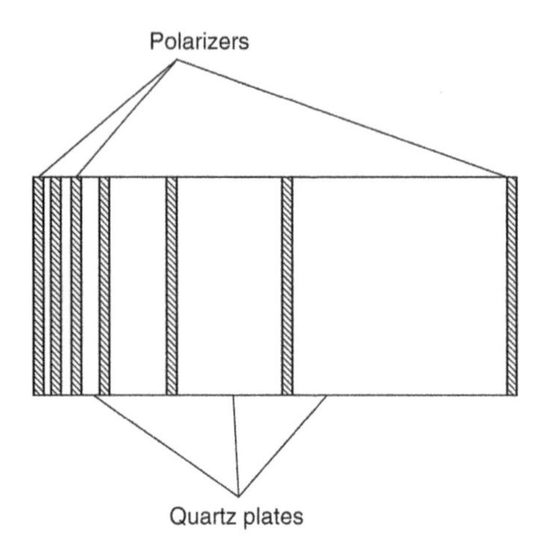

FIGURE 5.26 Six-element quartz birefringent filter.

the incident intensity because of the use of the first polaroid. The whole filter, though, uses such a depth of quartz that its actual transmission is a few percent. The properties of quartz are temperature dependent and so the whole unit must be enclosed and its temperature controlled to a fraction of a degree when it is in use. The temperature dependence, however, allows the filter's central wavelength to be tuned slightly by changing the operating temperature. Much larger changes in wavelength may be encompassed by splitting each retarder into two opposed wedges. Their combined thicknesses can then be varied by displacing one set of wedges with respect to the other in a direction perpendicular to the axis of the filter. The wavelength may also be decreased by slightly tilting the filter to the direction of the incoming radiation, so increasing the effective thicknesses of the quartz plates. For the same reason, the beam of radiation must be collimated for its passage through the filter, or the bandwidth will be increased.

A closely related filter is due to Solč.[16] It uses only two polarisers placed before and after the retarders. All the retarders are of the same thickness, and they have their optic axes alternately oriented with respect to the axis of the first polariser in clockwise and anticlockwise directions at some specified angle. The output consists of a series of transmission fringes whose spacing in wavelength terms increases with wavelength. Two such units of differing thicknesses can then be combined so that only at the desired wavelength do the transmission peaks coincide.

Another device for observing the Sun over a narrow wavelength range is the magneto-optical filter (MOF). This, however, can only operate at the wavelengths of strong absorption lines due to gases. It comprises two polarisers on either side of a gas cell. The polarisers are oriented orthogonally to each other if they are linear or are left- and right-handed if circular in nature. In either case, no light is transmitted through the system. A magnetic field is then applied to the gas cell, inducing the Zeeman effect. The Zeeman components of the lines produced by the gas are then linearly and/or circularly polarised (Section 5.4) and so permit the partial transmission of light at their wavelengths through the whole system. The gases currently used in MOFs are vapours of sodium or potassium.

Relatively inexpensive Hα filters can be made as solid Fabry-Perot etalons (Section 4.1). A thin fused-silica spacer between two optically flat dielectric mirrors is used, together with a blocking filter. Peak transmissions of several tens of percent and bandwidths of better than a tenth of a nanometre can be achieved in this way.

[16] Pronounced 'Sholts'.

5.3.6 Coronagraph

This instrument enables observations of the corona to be made at times other than during solar eclipses (see also stellar coronagraphs, Section 2.7). It does this by producing an artificial eclipse. The principle is simple; an occulting disc at the prime focus of a telescope obscures the photospheric image while allowing that of the corona to pass by. The practice, however, is considerably more complex because scattered and/or diffracted light and so on in the instrument and the atmosphere can still be several orders of magnitude brighter than the corona. Extreme precautions have, therefore, to be taken in the design and operation of the instrument to minimise this extraneous light.

The most critical of these precautions lies in the structure of the objective. A single simple lens objective is used to minimise the number of surfaces involved, and it is formed from a glass blank that is as free from bubbles, striae and other imperfections as possible. The surfaces of the lens are polished with extreme care to eliminate all scratches and other surface markings. In use, they are kept dust-free by being tightly sealed when not in operation and by the addition of a long tube lined with grease, in front of the objective to act as a dust trap.

The occulting disc is a polished metal cone or an inclined mirror so that the photospheric radiation may be safely reflected away to a separate light and heat trap. Diffraction from the edges of the objective is eliminated by imaging the objective with a Fabry lens after the occulting disc and by using a Lyot[17] stop that is slightly smaller than the image of the objective to remove the edge effects. Alternatively, the objective can be apodised (Section 2.5); its transparency decreases from its centre to its edge in a Gaussian fashion. This leads to full suppression of the diffraction halo although with some loss in resolution. Recently, the use of aspherical mirrors for this same purpose has been investigated (see Phase-Induced Amplitude Apodization [PIAA], Section 2.7.4). A second occulting disc (the Lyot stop) before the final objective removes the effects of multiple reflections within the first objective. The final image of the corona is produced by this second objective and this is placed after the diffraction stop. The full system is shown in Figure 5.27.

A Lyot coronagraph with an additional occulting disk placed before the objective so that the whole instrument is shielded from direct photospheric light is called the 'Newkirk design'. It has been used for many of the solar coronagraphs flown on spacecraft including the solar maximum mission (1980–1989). A Lyot type coronagraph based upon mirrors and two Newkirk coronagraphs form the Large Angle Spectroscopic Coronagraph (LASCO) instrument on board the still operating SOHO spacecraft. Whilst the STEREO spacecraft carried classical Lyot coronagraphs as well as externally occulted coronagraphs.

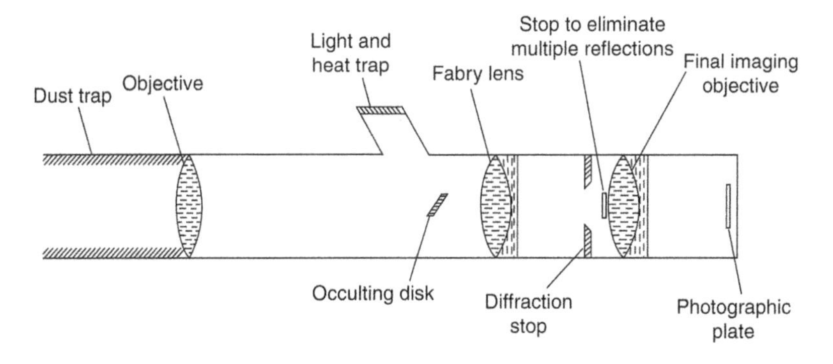

FIGURE 5.27 Schematic optical arrangement of a coronagraph.

[17] After Bernard Lyot who obtained the first photograph of the solar corona outside a total eclipse with his newly constructed coronagraph in 1931.

For ground-based instruments, the atmospheric scattering can only be reduced by a suitable choice of observing site. Thus, the early coronagraphs are to be found at high altitude observatories.

The use of a simple lens results in a chromatic prime focus image. A filter must, therefore, normally be added to the system. This is desirable in any case because the coronal spectrum is largely composed of emission lines superimposed upon a diluted solar photospheric spectrum. Selection of a narrow band filter that is centred upon a strong coronal emission line, therefore, considerably improves the contrast of the final image. White light or wideband imaging of the corona is only possible using Earth-based instruments on rare occasions, and it can only be attempted under absolutely optimum observing conditions. Spacecraft-borne instruments may be so used more routinely.

Improvements to the basic coronagraph may be justified for balloon or spacecraft-borne instruments because the sky background is then less than the scattered light in a more normal device and they have taken two different forms. A reflecting objective can be used. This is formed from an uncoated off-axis parabola. Most of the light passes through the objective and is absorbed. Bubbles and striations in the glass are of much less importance because they cause scattering primarily in the forward direction. The mirror is uncoated since metallic films are sufficiently irregular to cause a considerable amount of scattering. In other respects, the coronagraph then follows the layout of Figure 5.27. A current outline proposal envisages using titanium spheres with surface coatings of super-absorbent carbon nanotubes as the occulters in its coronagraphs. Solar Polar Observing Constellation (SPOC) might comprise two spacecraft in polar orbits around the Sun and would probably need one or more proof-of-concept cube-sat-based missions in advance of making any further progress.

The second approach to the improvement of coronagraphs is quite different and consists simply of producing the artificial eclipse outside the instrument rather than at its prime focus (see also the Newkirk design, above). An occulting disc is placed well in front of the first objective of an otherwise fairly conventional telescope. The disc must be large enough to ensure that the first objective lies entirely within its umbral shadow. The inner parts of the corona will, therefore, be badly affected by vignetting. However, this is of little importance because these are the brightest portions of the corona, and it may even be advantageous because it will reduce the dynamic range that must be covered by the detector. A simple disc produces an image with a bright central spot due to diffraction, but this can be eliminated by using a disc whose edge is formed into a zigzag pattern of sharp teeth, or by the use of multiple occulting discs. By such means the instrumentally scattered light can be reduced to 10^{-4} of that of a basic coronagraph.

The final image in a coronagraph may be imaged directly but more commonly is fed to a spectroscope, photometer or other ancillary instrument. From the Earth, the corona may normally only be detected out to about one solar radius, but satellite-based coronagraphs have been used successfully out to six solar radii or more.

Devices similar to the coronagraph are also sometimes carried on spacecraft so that they may observe the atmospheres of solar system planets, whilst shielding the detector from the radiation from the planet's surface. The contrast is generally far smaller than in the solar case so that such planetary 'coronagraphs' can be far simpler in design.

5.3.7 Pyrheliometer/Radiometer

This is an instrument intended to measure the total flux of solar radiation at all wavelengths. In practice, current devices measure the energy from the microwave to the soft X-ray region. Modern pyrheliometers are active cavity radiometers. The radiation is absorbed within a conical cavity. The cavity is held in contact with a heat sink and maintained at a temperature about 1 K higher than the heat sink by a small heater. The difference between the power used by the heater when the cavity is exposed to the Sun and that when a shutter closes it off, provides the measure of the solar energy. The ESA's PROBA-2 spacecraft, launched in 2009 and still functioning, for example, carries an ultraviolet to X-ray radiometer monitoring the Sun over four wavebands. While Centre National d'ÉtudesSpatiales' (CNES') Picard spacecraft (2010–2014) carried a radiometer with a bolometric detector.

5.3.8 Solar Oscillations

Whole-body vibrations of the Sun reveal much about its inner structure. They are small-scale effects and their study requires precise measurements of the velocities of parts of the solar surface. The resonance scattering spectrometer originated by the solar group at the University of Birmingham is capable of detecting motions of a few tens of millimetres per second. It operates by passing the 770-nm potassium line from the Sun through a container of heated potassium vapour. A magnetic field is directed through the vapour and the solar radiation is circularly polarised. Depending upon the direction of the circular polarisation either the light from the long-wave wing (I_L) or the short-wave wing (I_S) of the line is resonantly scattered. By switching the direction of the circular polarisation rapidly, the two intensities may be measured nearly simultaneously. The line of sight velocity is then given by

$$\mathrm{v} = k\frac{I_S - I_L}{I_S + I_L} \tag{5.75}$$

where k is a constant with a value around 3 km s^{-1}.

Birmingham Solar Oscillations Network (BiSON) has six automated observatories around the world so that almost continuous observations of the Sun can be maintained. It measures the average radial velocity over the whole solar surface using resonance scattering spectrometers. The Global Oscillation Network Group (GONG) project likewise has six observatories and measures radial velocities on angular scales down to 8″ using Michelson interferometers to monitor the position of the nickel 676.8-nm spectrum line. Live feeds from BiSON can be seen at http://bison.ph.bham.ac.uk/ and from GONG at https://gong.nso.edu/.

The SOHO spacecraft carries three helioseismology instruments – Global Oscillations at Low Frequencies (GOLF), Michelson Doppler Interferometer (MDI), and Variability of Solar Irradiance and Gravity Oscillations (VIRGO).[18] The GOLF is a resonance scattering device based upon sodium D line that detects motions to better than 1 mm s^{-1} over the whole solar surface. The MDI is a Michelson Doppler imager that measures magnetic fields as well as velocities, and VIRGO measures the solar constant and detects oscillations via variations in the Sun's brightness.

5.3.9 Other Solar Observing Methods

Slitless spectroscopes (Section 4.2) are of considerable importance in observing those regions of the Sun whose spectra consist primarily of emission lines. They are simply spectroscopes in which the entrance aperture is large enough to accept a significant proportion of the solar image. The resulting spectrum is, therefore, a series of monochromatic images of this part of the Sun in the light of each of the emission lines. They have the advantage of greatly improved speed over a slit spectroscope combined with the obvious ability to obtain spectra simultaneously from different parts of the Sun. Their most well-known application is during solar eclipses, when the 'flash spectrum' of the chromosphere may be obtained in the few seconds immediately after the start of totality or just before its end. More recently, they have found widespread use as spacecraft-borne instrumentation for observing solar flares in the UV part of the spectrum.

One specialised satellite-borne instrument based upon slitless spectroscopes is the Ly–α camera. The solar Lyman-α line is a strong emission line. If the image of the whole disc of the Sun from an UV telescope enters a slitless spectroscope, then the resulting Ly–α image may be isolated from the rest of the spectrum by a simple diaphragm, with little contamination from other wavelengths. A second identical slitless spectroscope whose entrance aperture is this diaphragm and whose

[18] Not to be confused with Virgo (Section 1.6).

dispersion is perpendicular to that of the first spectroscope will then provide a stigmatic spectroheliogram at a wavelength of 121.6 nm.

The heliometer is a modern specialist instrument designed to monitor the angular diameter of the Sun (c.f. the Danjon astrolabe, Section 5.1.4). Its origins, though, date back to Römer in 1675, when it was known as a double-image micrometer, and its recipe is easily described:

- take a refracting telescope
- saw its objective lens in half along a diameter
- move the two halves of the lens sideways with respect to each other until the western limb, say, of one of the (now) two images of the Sun aligns with the eastern limb of the second image
- read off the angular diameter of the Sun using the scale attached to the lens-halves which you have (thoughtfully) calibrated some time before hand.

The 0.11-m heliometer at the Observatorio Nacional in Rio de Janeiro replaced the existing Danjon astrolabe in 2009 (Section 5.1.4) to continue its programme of solar observations. It uses a split parabolic mirror and makes observations to a precision of a few tens of milliarcseconds. The heliometer's major advantage over the astrolabe is that it can measure the diameter of the Sun over all position angles – not just the vertical diameter – and at any time of day.

A solar 'instrument' that has recently become available is the virtual solar observatory (see also Section 5.5). This is a software system linking various solar data archives and which provides tools for searching and analysing the data. It may be accessed by anyone who is interested at http://vso.nascom.nasa.gov/cgi-bin/search.

In the radio region, solar observations tend to be undertaken by fairly conventional equipment, although there are now a few dedicated solar radio telescopes.

The Siberian Solar Radio Telescope, for example, comprises two hundred and fifty-six 2.5-m dishes laid out in the shape of a cross (cf. Mills Cross, Section 1.2) and monitors solar activity at 5.7 GHz with a 15″ resolution. In 2017, the instrument was joined by the Siberian Radio Heliograph with 48 antennae in a T-shaped array observing at 4–8 GHz. In India, the Gauribidanur Radio Heliograph (GRAPH) instrument obtains 2D images of the corona over the 40- to 150-MHz region using 384 log-periodic dipole antennae, which are also arranged in a T-shape. The GRAPH is supplemented by a radio spectrograph, Gauribidanur Low-Frequency Solar Spectrograph (GLOSS), with eight antennae. Perhaps surprisingly, Atacama Large Millimeter Arrary (ALMA), with its 66 solid-surface reflectors *can* be used for solar observations. This is because it was designed from the beginning with the possibility of solar radio observations being undertaken. The reflecting surfaces thus have a small-scale roughness which scatters the visible and infrared radiation and so no focussed images are produced at those wavelengths. Nonetheless ALMA's radio receivers have to be rated for temperatures over 500°C. Solar observations at 23 and 100 GHz were started from 2015.

There is also a proposal, which is currently at the concept stage, for an array of many antennas forming a radio aperture synthesis system either on the surface of the Moon or carried on board a number of micro-spacecraft for observing the inner parts of the solar wind (c.f. Dark Ages Radio Explorer [DARE] and Farside Array for Radio Science Investigations of the Dark ages and Exoplanets [FARSIDE] – Section 1.2.3).

One exception to the usual conventionality of solar radio instruments though, is the use of multiplexed receivers to provide a quasi-instantaneous radio spectrum of the Sun. This is of particular value for the study of solar radio bursts because their emitted bandwidths may be quite narrow and their wavelengths drift rapidly with time. These radio spectroscopes and the acousto-optical radio spectroscope were discussed more fully in Section 1.2. The data from them is usually presented as a frequency/time plot, from whence the characteristic behaviour patterns of different types of solar bursts may easily be recognised.

This account of highly specialised instrumentation and techniques for solar observing could be extended almost indefinitely and might encompass all the equipment designed for parallax observations, solar radius determinations, oblateness determinations, eclipse work, tree rings and C^{14} determination and so on. However, at least in the author's opinion, these are becoming too specialised for inclusion in a general book like this, and so readers referred to more specialised texts and the scientific journals for further information upon them.

5.4 MAGNETOMETRY

5.4.1 Background

The measurement of astronomical magnetic fields is accomplished in two quite separate ways. The first is direct measurement by means of apparatus carried by spacecraft, whilst the second is indirect and is based upon the Zeeman effect of a magnetic field upon spectrum lines (or more strictly upon the inverse Zeeman effect because it is usually applied to absorption lines). A third approach suggested recently is to observe the X-rays arising from the solar wind interactions with the Earth's magnetosheath and, thereby, infer the shape of the terrestrial magnetic fields involved but this has yet to be tried in practice.

5.4.1.1 Zeeman Effect

The Zeeman effect describes the change in the structure of the emission lines in a spectrum when the emitting object is in a magnetic field. The simplest change arises for singlet lines – that is lines arising from transitions between levels with a multiplicity of one, or a total spin quantum number, M_s of zero, for each level. For these lines, the effect is called the 'normal Zeeman effect'. The line splits into two or three components depending on whether the line of sight is along, or perpendicular, to the magnetic field lines. An appreciation of the basis of the normal Zeeman effect may be obtained from a classical approach to the problem. If we imagine an electron in an orbit around an atom, then its motion may be resolved into three simple harmonic motions along the three coordinate axes. Each of these in turn we may imagine to be the sum of two equal but opposite circular motions (Figure 5.28). If we now imagine a magnetic field applied along the z-axis, then these various motions may be modified by it. Firstly, the simple harmonic motion along the z-axis will be unchanged because it lies along the direction of the magnetic field. The simple harmonic motions along the x- and y-axes, however, are cutting across the magnetic field and so will be altered. We may best see how their motion changes by considering the effect upon their circular components. When the magnetic field is applied, the radii of the circular motions remain unchanged, but their

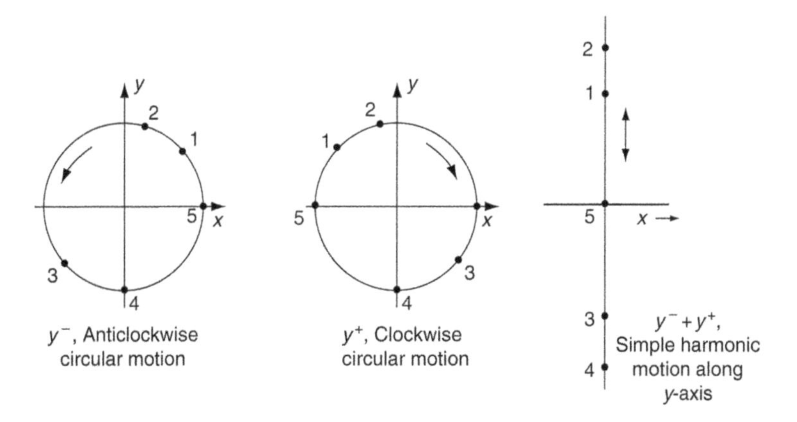

FIGURE 5.28 Resolution of simple harmonic motion along the y-axis into two equal but opposite circular motions.

frequencies alter. If ν is the original frequency of the circular motion and H is the magnetic field strength, then the new frequencies of the two resolved components ν^+ and ν^- are

$$\nu^+ = \nu + \Delta\nu \tag{5.76}$$

$$\nu^- = \nu - \Delta\nu \tag{5.77}$$

where:

$$\Delta\nu = \frac{eH}{4\pi m_e c} \tag{5.78}$$

$$= 1.40 \times 10^{10} H \quad \text{Hz T}^{-1} \tag{5.79}$$

Thus, we may combine the two higher frequency components arising from the x and y simple harmonic motions to give a single elliptical motion in the xy plane at a frequency of $\nu + \Delta\nu$. Similarly, we may combine the lower frequency components to give another elliptical motion in the xy plane at a frequency of $\nu - \Delta\nu$. Thus, the electron's motion may be resolved, when it is in the presence of a magnetic field along the z-axis, into two elliptical motions in the xy plane, plus simple harmonic motion along the z-axis (Figure 5.29), the frequencies being $\nu + \Delta\nu$, $\nu - \Delta\nu$ and ν, respectively. Now if we imagine looking at such a system, then only those components that have some motion across the line of sight will be able to emit light towards the observer because light propagates in a direction perpendicular to its electric vector. Hence, looking along the z-axis (i.e., along the magnetic field lines) only emission from the two elliptical components of the electron's motion will be visible. Since the final spectrum contains contributions from many atoms, these will average out to two circularly polarised emissions, shifted by $\Delta\nu$ from the normal frequency (Figure 5.30), one with clockwise polarisation and the other with anticlockwise polarisation. Lines of sight perpendicular to the magnetic field direction, that is within the xy plane, will in general view the two elliptical motions as two collinear simple harmonic motions, while the z-axis motion will remain as simple harmonic motion orthogonal to the first two motions. Again, the spectrum is the average of many

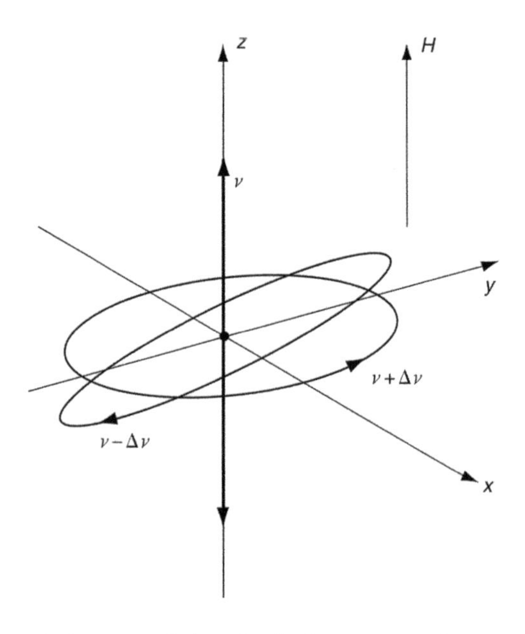

FIGURE 5.29 Components of electron orbital motion in the presence of a magnetic field.

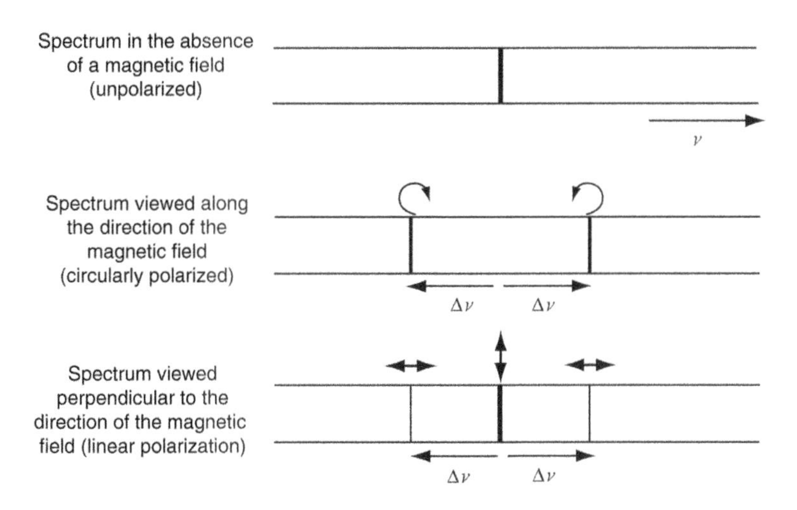

FIGURE 5.30 The normal Zeeman effects.

atoms and so will, therefore, comprise three linearly polarised lines. The first of these is at the normal frequency of the line and is polarised parallel to the field direction. It arises from the z-axis motion. The other two lines are polarised at right angles to the first and are shifted in frequency by $\Delta\nu$ (Figure 5.30) from the normal position of the line. When observing the source along the line of the magnetic field, the two spectrum lines have equal intensities, while when observing perpendicular to the magnetic field, the central line has twice the intensity of either of the other components. Thus, if we imagine the magnetic field progressively reducing, then as the components remix, an unpolarised line results – as one would expect. This pattern of behaviour for a spectrum line originating in a magnetic field is termed the 'normal Zeeman effect'.

In astronomy, absorption lines, rather than emission lines, are the main area of interest and the inverse Zeeman effect describes their behaviour. This, however, is precisely the same as the Zeeman effect except that emission processes are replaced by their inverse absorption processes. This analysis may therefore be equally well applied to describe the behaviour of absorption lines. The one major difference from the emission line case is that the observed radiation *remaining* at the wavelength of one of the lines is preferentially polarised in the *opposite* sense to that of the Zeeman component because the polarised Zeeman component is being *subtracted* from unpolarised radiation.

If the spectrum line does not originate from a transition between singlet levels (i.e., $M_s \neq 0$), then the effect of a magnetic field is more complex. The resulting behaviour is known as the 'anomalous Zeeman effect', but this is something of a misnomer because it is anomalous only in the sense that it does not have a classical explanation. Quantum mechanics describes the effect completely.

The orientation of an atom in a magnetic field is quantised in the following manner. The angular momentum of the atom is given by

$$\left[J\left(J+1\right)\right]\frac{h}{2\pi} \tag{5.80}$$

where J is the inner quantum number and its space quantisation is such that the projection of the angular momentum onto the magnetic field direction must be an integer multiple of $h/2\pi$ when J is an integer, or a half-integer multiple of $h/2\pi$ when J is a half-integer. Thus, there are always $(2J + 1)$ possible quantised states (Figure 5.31). Each state may be described by a total magnetic quantum number, M, which for a given level can take all integer values from $-J$ to $+J$ when J is an integer, or all half-integer values over the same range when J is a half-integer. In the absence of a magnetic field, electrons in these states all have the same energy (i.e., the states are degenerate) and the set

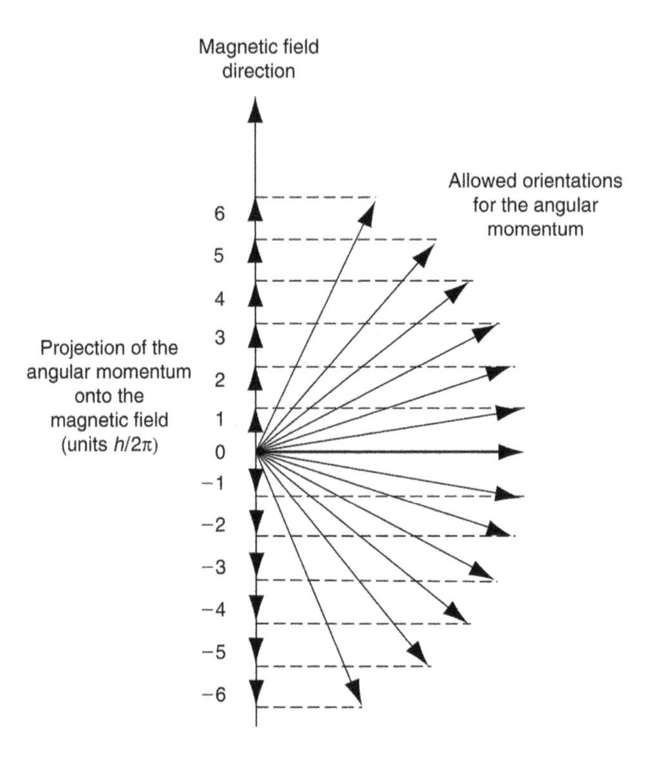

FIGURE 5.31 Space quantisation for an atom with $J = 6$.

of states forms a level. Once a magnetic field is present, however, electrons in different states have different energies, with the energy change from the normal energy of the level, ΔE, being given by

$$\Delta E = \frac{eh}{4\pi m_e c} MgH \tag{5.81}$$

where g is the Landé factor, given by

$$g = 1 + \frac{J(J+1) + M_s(M_s+1) + L(L+1)}{2J(J+1)} \tag{5.82}$$

where L is the total azimuthal quantum number. So, the change in the frequency, $\Delta\nu$, for a transition to or from the state is

$$\Delta \nu = \frac{e}{4\pi m_e c} MgH \tag{5.83}$$

$$= 1.40 \times 10^{10} MgH \qquad \text{Hz}\,\text{T}^{-1} \tag{5.84}$$

Now for transitions between such states we have the selection rule that M can change by 0 or ± 1 *only*. Hence, we may understand the normal Zeeman effect in the quantum mechanical case simply from the allowed transitions (Figure 5.32) plus the fact that the splitting of each level is by the same amount since for the singlet levels

$$M_s = 0 \tag{5.85}$$

so that

$$J = L \tag{5.86}$$

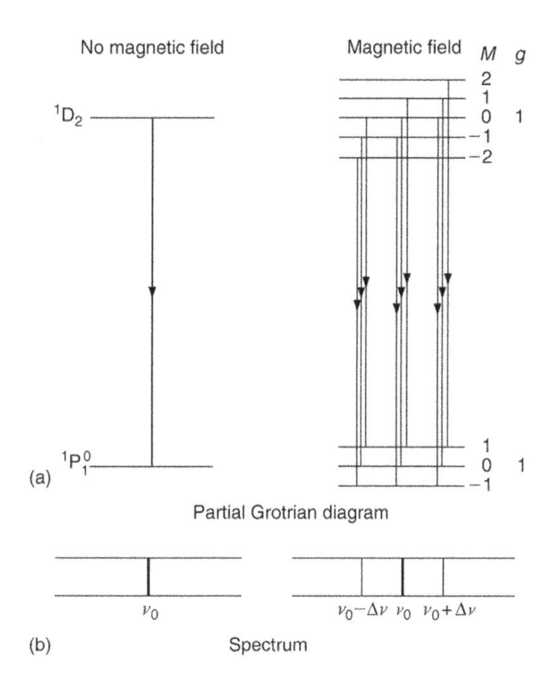

FIGURE 5.32 Quantum mechanical explanation of the normal Zeeman effect: (a) Partial Grotrian diagram and (b) spectrum.

and so

$$g = 1 \tag{5.87}$$

Each of the normal Zeeman components for the example shown in Figure 5.32 is therefore triply degenerate and only three lines result from the nine possible transitions. When M_s is not zero, g will in general be different for the two levels and the degeneracy will cease. All transitions will then produce separate lines, and hence, we get the anomalous Zeeman effect (Figure 5.33). Only one of many possible different patterns is shown in the figure; the details of any individual pattern will depend upon the individual properties of the levels involved.

As the magnetic field strength increases, the pattern changes almost back to that of the normal Zeeman effect. This change is known as the Paschen-Back effect. It arises as the magnetic field becomes strong enough to decouple L and M_s from each other. They then no longer couple together to form J, which then couples with the magnetic field, as just described, but couple separately and independently to the magnetic field. The pattern of spectrum lines is then that of the normal Zeeman effect, but with each component of the pattern formed from a narrow doublet, triplet and so on accordingly as the original transition was between doublet, triplet, and so on levels. Field strengths of around 0.5 T or more are usually necessary for the complete development of the Paschen-Back effect.

At strong magnetic field strengths (>10^3 T), the quadratic Zeeman effect will predominate. This displaces the spectrum lines to higher frequencies by an amount $\Delta\nu$, given by

$$\Delta\nu = \frac{\varepsilon_o h^3}{8\pi^2 m_e^3 c^2 e^2 \mu_o} n^4 \left(1 + M^2\right) H^2 \tag{5.88}$$

$$= 1.489 \times 10^4 n^4 \left(1 + M^2\right) H^2 \qquad \text{Hz} \tag{5.89}$$

where n is the principal quantum number.

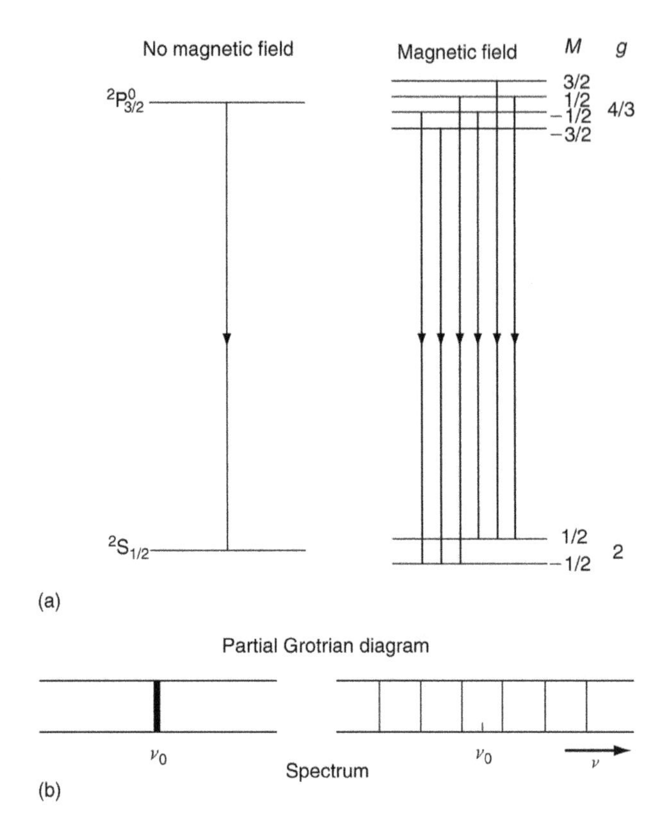

FIGURE 5.33 Quantum mechanical explanation of the anomalous Zeeman effect. (a) the Grotrian diagram and (b) the lines in the spectrum.

5.4.2 Magnetometers

Amongst the direct measuring devices, the commonest type is the fluxgate magnetometer illustrated in Figure 5.34. Two magnetically soft cores have windings coiled around them as shown. Winding A is driven by an alternating current. When there is no external magnetic field, winding B has equal and opposite currents induced in its two coils by the alternating magnetic fields of the cores – and so there is no output. The presence of an external magnetic field introduces an imbalance into the currents, giving rise to a net alternating current output.[19] This may then easily be detected and calibrated in terms of the external field strength. Spacecraft usually carry three such magnetometers oriented orthogonally to each other so that all the components of the external magnetic field may be measured. The four spacecraft of NASA's Magnetospheric Multiscale (MMS) mission (2015– to date), for example, are each equipped with fluxgate magnetometers[20] capable of measuring field strengths to ±5 pT.

Another type of magnetometer that is often used on spacecraft is based upon atoms in a gas oscillating at their Larmor frequency. It is often used for calibrating fluxgate magnetometers in orbit. The vector helium magnetometer operates by detecting the effect of a magnetic field upon

[19] By rearranging the windings in the magnetometer slightly, it is possible to get a DC output in the presence of a steady magnetic field.

[20] N.B. any magnetometer on board a spacecraft usually has to be deployed at the end of a long boom after launch to distance it from the magnetic effects of the spacecraft itself.

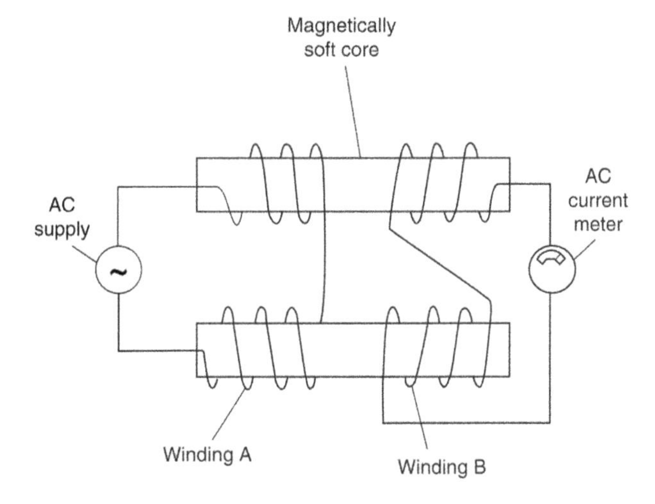

FIGURE 5.34 Fluxgate magnetometer.

the population of a metastable state of helium. A cell filled with helium is illuminated by 1.08-μm radiation which pumps the electrons into the metastable state. The efficiency of the optical pump is affected by the magnetic field and the population of that state is monitored by observing the absorption of the radiation. A variable artificially generated magnetic field is swept through the cell until the external magnetic field is nullified. The strength and direction of the artificial field are then equal and opposite to the external field.

An alternative technique altogether, is to look at the electron flux. Since electron paths are modified by the magnetic field, their distribution can be interpreted in terms of the local magnetic field strength. This enables the magnetic field to be detected with moderate accuracy over a large volume, in comparison with the direct methods that give high accuracies but only in the immediate vicinity of the spacecraft.

Most of the successful indirect work in magnetometry based upon the (inverse) Zeeman effect has been undertaken by Harold and Horace Babcock[21] or uses instruments based upon their designs. For stars, the magnetic field strength along the line of sight may be determined via the longitudinal Zeeman effect. The procedure is made much more difficult than in the laboratory by the widths of the stellar spectrum lines. For a magnetic field of 1 T, the change in the frequency of the line is 1.4×10^{10}Hz (Equation 5.79), which for lines near 500 nm corresponds to a separation for the components in wavelength terms of only 0.02 nm. This is smaller than the normal line width for most stars. Thus, even strong magnetic fields do not cause the spectrum lines actually to split into separate components but merely to become somewhat broader than normal. The technique is saved by the opposite senses of circular polarisation of the components which enables them to be separated from each other.

Babcock's differential analyser, therefore, consists of a quarter-wave plate (Section 5.2) followed by a doubly refracting calcite crystal. This is placed immediately before the entrance aperture of a high dispersion spectroscope (Figure 5.35). The quarter-wave plate converts the two circularly polarised components into two mutually perpendicular linearly polarised components. The calcite is suitably oriented so that one of these components is the ordinary ray and the other the extraordinary ray. The components are therefore separated by the calcite into two beams. These are both accepted

[21] 1882 to 1968 and 1912 to 2003, respectively.

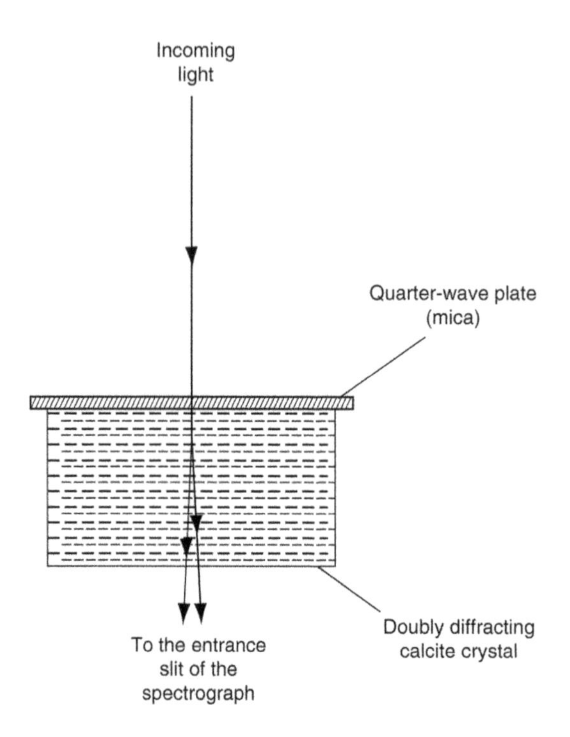

FIGURE 5.35 Babcock's differential analyser.

by the spectroscope slit and so two spectra are formed in close juxtaposition, with the lines of each slightly shifted with respect to the other as a result of the Zeeman splitting. This shift may then be measured and translated back to give the longitudinal magnetic field intensity. A lower limit on the magnetic field intensity of about 0.02 T is detectable by this method, while the strongest stellar fields found have strengths of a few teslas. By convention, the magnetic field is positive when it is directed towards the observer.

The technique of Zeeman-Doppler imaging potentially enables maps of the distribution of magnetic fields over the surfaces of stars to be constructed, though often with considerable ambiguities left in the results. The basis of the technique is best envisaged by imagining a single magnetic region on the surface of a star. The Zeeman effect will induce polarisation into the lines emitted or absorbed from this region, as already discussed. However, when the rotation of the star brings the region first into view, it will be approaching us. The features from the region will therefore be blue-shifted (i.e., Doppler shifted) compared with their 'normal' wavelengths. As the rotation continues to move the region across the disk of the star, the spectral features will change their wavelengths, going through their 'normal' positions to being red-shifted just before they disappear around the opposite limb of the star. If there are several magnetic regions, then each will usually have a different Doppler shift at any given moment and the changing Doppler shifts can be interpreted in terms of the relative positions of the magnetic regions on the star's surface (i.e., a map). The reduction of the data from this type of observations is a complex procedure and MEMs (Section 2.1.1) are usually needed to constrain the range of possible answers. Spectropolarimeters such as SemelPol[22] on the AAT (decommissioned 2018), Echelle SpectroPolarimetric Device for the Observation of Stars (ESPaDons) on the Canada-France-Hawaii Telescope (CFHT) and Neo-Narvalon the Pic du Midi's 2-m Lyot telescope have been used in this manner recently.

[22] This is not an anagram; it is named after its designer, Meir Semel.

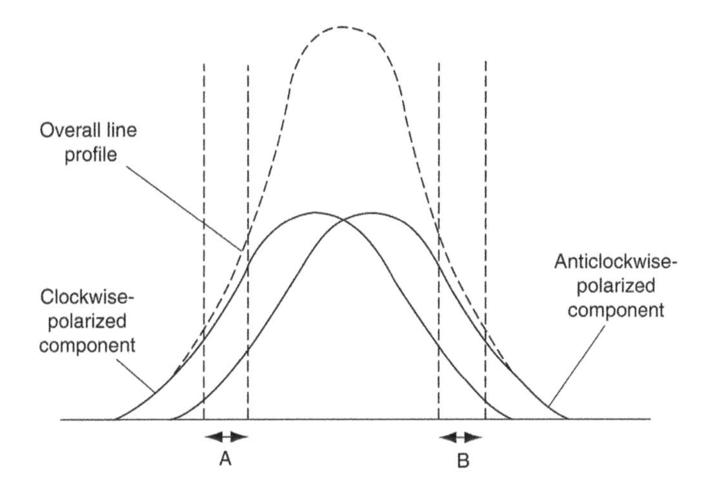

FIGURE 5.36 Longitudinal Zeeman components.

For the Sun, much greater sensitivity is possible and fields as weak as 10^{-5} T can be studied. George Ellery Hale first detected magnetic fields in sunspots in 1908, but most of the modern methods and apparatus are again due to the Babcocks. Their method relies upon the slight shift in the line position between the two circularly polarised components, which causes an exaggerated difference in their relative intensities in the wings of the lines (Figure 5.36). The solar light is passed through a differential analyser as before. This time, however, the quarter-wave plate is composed of ammonium dihydrogen phosphate and this requires an electric potential of some 9 kV across it to make it sufficiently birefringent to work. Furthermore, only one of the beams emerging from the differential analyser is fed to the spectroscope. Originally, a pair of PMTs that accepted the spectral regions labelled A and B in Figure 5.36 detected the line, now CCDs or other array detectors are used. Thus, the wing intensities of the clockwise-polarised component (say) are detected when a positive voltage is applied to the quarter-wave plate, whilst those of the anticlockwise component are detected when a negative voltage is applied. The voltage is switched rapidly between the two states and the outputs are detected in phase with the switching. Since all noise except that in phase with the switching is automatically eliminated and the latter may be reduced by integration and/or the use of a phase-sensitive detector (Section 3.2), the technique is sensitive. The entrance aperture to the apparatus can be scanned across the solar disc so that a magnetogram can be built up with a typical resolution of a few arcseconds and of a few μ*T*. Interline transfer CCD detectors are used which have every alternate row of pixels covered over. The accumulating charges in the pixels are switched between an exposed row and a covered row in phase with the switching between the polarised components. Thus, at the end of the exposure, alternate rows of pixels contain the intensities of the clockwise and anticlockwise components. Similar instruments have been devised for stellar observations but so far are not as sensitive as the previously mentioned system.

Several other types of magnetometer have been devised for solar work. For example, at the higher field strengths, Leighton's method provides an interesting pictorial representation of the magnetic patterns. Two spectroheliograms are obtained in opposite wings of a suitable absorption line and through a differential analyser that is arranged so that it only allows the passage into the system of the stronger circularly polarised component in each case. A composite image is then made of both spectroheliograms, with one as a negative and the other as a positive. Areas with magnetic fields less than about 2 mT are seen as grey, and areas with stronger magnetic fields show up as light or dark according to their polarity. More modern devices use computer imaging but the basic technique remains unaltered. Vector-imaging magnetographs image the Sun over a narrow wavelength range and then rapidly step that image through a magnetically sensitive spectrum line. This enables both

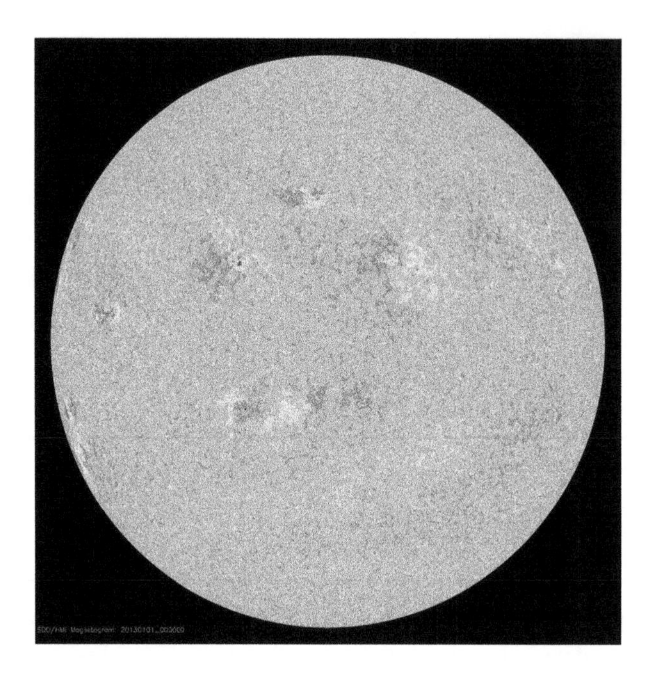

FIGURE 5.37 A line of sight solar magnetogram obtained by the Helioseismic and Magnetic Image (HMI) instrument on the Solar Dynamics Observatory (SDO) spacecraft. (Courtesy of NASA.)

the direction and strength of the magnetic field to be determined over the selected region. Usually, this region is just a small part of the whole solar disk because of the enormous amounts of data that such instruments can generate. The Helioseismic and Magnetic Imager (HMI) on board NASA's Solar Dynamics Observatory (SDO), however, does obtain vector magnetograms of the whole solar disk (Figure 5.37). The Imaging Magnetograph Experiment (IMaX) vector magnetograph on the Sunrise (Section 5.3.2) balloon-borne telescope used the Fe I 525.02-nm line for its measurements, achieving magnetic resolutions of 400 μT and radial velocity resolutions of 5 ms^{-1} at best over a 50-arcsecond field of view.

In the radio region, the Zeeman effect affects lines such as the 1.42-GHz emission from hydrogen (usually better known as the 21-cm line). Their Zeeman splitting is found in a similar manner to Babcock's solar method – by observing in the steepest part of the wings. Interstellar magnetic fields of about 10^{-9} T are detectable in this way.

The strongest fields, up to 10^8 T, are expected to exist in white dwarfs and neutron stars. However, there is usually little spectral structure in such objects from which the fields might be detected. The quadratic Zeeman effect, however, may then be sufficient for circular polarisation of the continuum to be distinguished at points where it is varying rapidly in intensity. Field strengths that are upwards of 1000 T are found in this way.

A few devices have been constructed to detect electric fields via the Stark effect upon the Paschen hydrogen lines. They are basically similar to magnetometers and for solar studies can reach sensitivities of 500 Vm^{-1}.

5.4.3 Data Reduction and Analysis

Once a fluxgate magnetometer has been calibrated, there is little involved in the analysis of its results except to convert them to field strength and direction. However, it may be necessary to correct the readings for influence from the spacecraft and for the effects of the solar wind and the solar or interplanetary magnetic fields. Measurements with accuracies as good as 10^{-11} T are possible when these corrections are reliably known.

With the first of the Babcocks' methods, in which two spectra are produced side by side, the spectra are simply measured as if for radial velocity. The longitudinal magnetic field strength can then be obtained via Equation 5.84. The photoelectric method's data are rather more complex to analyse. No general method can be given because it will depend in detail upon the experimental apparatus and method and upon the properties of the individual line (or continuum) being observed.

With strong solar magnetic fields, such as those found in sunspots, information may be obtainable in addition to the longitudinal field strength. The line in such a case may be split into two or three components and when these are viewed through a circular analyser (i.e., an analyser which passes only one circularly polarised component) the relative intensities of the three lines of the normal Zeeman effect are given by Seare's equations

$$I_V = \frac{1}{4}(1 \pm \cos\theta)^2 I \tag{5.90}$$

$$I_c = \frac{1}{2}(\sin^2\theta) I \tag{5.91}$$

$$I_r = \frac{1}{4}(1 \mp \cos\theta)^2 I \tag{5.92}$$

where I_v, I_c, and I_r are the intensities of the high frequency, central and low frequency components of the triplet, I is the total intensity of all the components and θ is the angle of the magnetic field's axis to the line of sight. Thus, the direction of the magnetic field as well as its total strength may be found.

5.5 EXPERIMENTAL ASTROPHYSICS

Astronomy and astrophysics are, of course, the archetypes for every variety of remote-sensing study – everyone knows that all that astronomers and astrophysicists can do is *look* very hard and very well.

EXCEPT – in the last few decades, solar system astronomers have become accustomed to drilling, digging, sampling, chemically analysing, irradiating and so on actual bits and pieces of planets, asteroids, comets, natural satellites, atmospheres, solar corona and the like – just as though they thought they were chemists or geologists.[23]

Clearly though, it will be a long time before similar work can be undertaken for anything in the far reaches of the Universe.

EXCEPT – things from the far reaches of the Universe, sometimes come to us.

This possibility has seemed much more reasonable since 2017, when 1I/`Oumuamua[24] was spotted and even more so in 2019, when Comet 2I/Borisov was discovered. Both these objects are visitors to the solar system from 'somewhere' well outside the solar system. They were distinguishable from objects that 'belong' to the solar system by their hyperbolic orbits[25] around the Sun. Both objects have been and are being intensively studied by all the normal astronomical remote sensing methods (and so far, do not seem to be different from comparable solar system objects, despite some wild speculations about the possibility of extraterrestrial [ET]spacecraft and the like) – but potentially they could have been visited by (one of our) spacecraft to make direct investigations and even to return samples of them to the Earth – indeed, some moderately serious considerations have been mooted to send missions to them even now.

---

[23] The authors' companion book to this one, *Remote and Robotic Investigations of the Solar System* (2018) discusses these sorts of topics in more detail.

[24] Nomenclature – the 'I' denotes an interstellar object and the '1' that it is the first such object ever discovered.

[25] The type of orbit followed by an object with a speed greater than the escape velocity from the Sun.

With massive and continuous survey programmes like that of the VRO coming on stream at the time of writing, the discovery of interstellar visitors (inanimate – not ETs) is likely to become much more frequent. Furthermore, at least one estimate, suggests that some 10,000 smaller interstellar objects are present inside the orbit of Neptune all the time, and that a few may even be quite close to the Earth. While interstellar objects the size of 1I/`Oumuamua or Comet 2I/Borisov are likely to remain rare, it seems feasible that the more numerous smaller visitors might be discovered in time to be investigated by a suitable ready-and-waiting fast response space mission – ESA and Japan Aerospace Exploration Agency's (JAXA's) proposed Comet Interceptor spacecraft (perhaps to be launched in 2028) might even be able to do the job.

It also seems likely that we already have samples from interstellar space somewhere within our numerous collections of meteorites – but, somehow, we need to be able to distinguish the interstellar meteorites from the homespun solar system meteorites. The fact that there are no interstellar meteorites which are instantly recognisable as such though, itself shows that most of the rest of the Universe is fairly similar to the solar system. A possible interstellar meteor occurred in 2014. Its possible interstellar origin was deduced (via studies in 2019) from its speed of 60 km/s and by its trajectory.

Somewhat less dramatically, we are continuously bombarded by material from beyond the solar system in the form of the primary cosmic rays (Section 1.4) and by the occasional high energy neutrino (Section 1.5).

There are all size-scales of objects between primary cosmic rays' protons and 1I/`Oumuamua, so that we may expect interstellar dust particles to exist as well – and these may already have been found. Some ocean floor samples contain the radioactive isotope Fe-60, which has a half-life of 2.6 million years (so that almost none is left over from before the formation of the Earth). Fe-60, though *is* produced in supernova explosions. The ocean floor samples are suggestive of two supernovae bombarding the Earth with dust particles. The supernovae might have been about 100 pc away and possibly occurred around 1.5 million and 2.3 million years ago. Since supernova ejecta have speeds up to 30,000 km/s, it would have taken the dust particles only a few thousand years to reach the Earth.

5.6 COMPUTERS AND THE INTERNET

5.6.1 Introduction

This section either has to be *much* longer than *all* the rest of the book, or brief – and since all topics under this heading are likely to change radically on timescales of a year or two and so would be out of date by the time the book appears, we will opt for the brief approach!

Information Technology (IT), computers, the internet and their applications are now essential to just about every aspect of astrophysics and indeed to most of the rest of science and to life in general. Many of the uses of computers, such as controlling telescopes, processing data and so on have already been mentioned within other sections of this book. Internet links are now being used quite extensively to transfer data between (say) the individual instruments making up interferometric arrays and in some cases for controlling remote/robotic instruments. There are several programmes aimed at enabling schools to use telescopes (sometimes of quite large sizes) via the internet such as the two 2-m Faulkes robotic optical telescopes and the National Radio Astronomy Observatory's (NRAO's) 20-m radio telescope at Green Bank (part of Skynet – https://skynet.unc.edu/).

There are also bibliographic data bases such as arXiv (http://arxiv.org/) and Astrophysical Data System (ADS; http://adsabs.harvard.edu/index.html) that are essential facilities for astrophysicists needing to keep up-to-date in their specialism. For anyone wishing to keep informed on a wide range of astronomical topics, then Astrobites (http://astrobites.com/) sends out a brief explanation and up-to-date account of selected subjects on a daily basis. Rapid alerts about new transient events are sent out by some observatories. Whilst primarily intended for fellow observers, such alerts will be of interest to many more people. For example, γ-ray bursts (GRBs), high-energy neutrino events and gravitational wave events are all notified within a few minutes of their occurrence via the γ-ray Coordinates Network (GCN) Circulars e-mail system (https://gcn.gsfc.nasa.gov/).

All the major and most of the smaller observatories, institutes, research groups and astronomical societies have their own websites. Many of these have news/update sections and some will send out alerts by e-mail or the social media whenever new items appear. The International Astronomical Union (IAU; www.iau.org/), NASA (www.nasa.gov/), and ESO (www.eso.org/public/) are three such examples, but there are hundreds more to be found. More specialist, but sometimes still quite large, groups have their own sites and discussion groups – such as the American Association of Variable Star Observers (AAVSO; www.aavso.org/).

There are a great many other astronomical resources to be found on the internet and far too little space here to mention even a fraction of them. Furthermore, most come and go or change in other ways rapidly, so readers will have to maintain their own list of favourites. However, there are two additional promising IT applications which deserve sections to themselves. These are the availability of digital sky surveys and the related development of 'virtual' observatories.

5.6.2 Digital Sky Surveys and Catalogues

A major concern of astronomy, dating back to Hipparchus and before, has always been the cataloguing, listing or mapping of objects in the sky, together with their positions and properties. The quantum changes that have occurred in the last few years are that many of these catalogues are now on open access through the internet and that personal computers, affordable by individuals, are now powerful enough to enable them to process some of the vast amounts of data present in those catalogues.

The possibility of conducting real research[26] is therefore open to anyone with the interest and some spare time – no longer just to professional scientists. Of course, there are still problems, the main one being data links. Some of the surveys contain a terabyte (10^{12} bytes) of data and using a 100 Mbs^{-1} broadband connection,[27] this would take about 24 hours to download. Data archives such as Set of Identifications, Measurements and Bibliography for Astronomical Data (SIMBAD; http://simbad.u-strasbg.fr/Simbad/) and Mikulski Archive for Space Telescopes (MAST; http://archive.stsci.edu/ – see below) contain tens of TB at the time of writing and are being added to at a rate of 5 or more TB per year. The cumulative information content within astronomy is now hundreds of TB and one project alone (the voltage controlled oscillator [VCO's] mission) could be generating 5 TB of data *per day* within a few years. Fortunately, for many purposes, it is not necessary to download a whole survey to work with it, a small subset can often be sufficient.

We have already seen (Section 5.1) that the Hipparcos, Tycho and other astrometric catalogues are available via the internet.

We may expect that many of the gigantic surveys, catalogues, and databases that will result from projects now running or soon to commence, such as the James Webb Space Telescope (JWST), Gaia, Square Kilometre Array (SKA) and the VRO, will also quickly become available via the internet. A list of many of the surveys and catalogues available online, which currently contains 20,000 entries, is given by VizieR at the Centre de Données Astronomiques de Strasbourg (CDS)[28]

[26] By data mining the catalogues and data bases (Section 1.4.2.3.1).

[27] Bits and Bytes can easily cause confusion. In information technology a bit is a single binary digit – i.e., a zero or a one. A byte is a collection of bits. Nowadays a byte is almost always made up from eight bits, but there have been variations on this in the past. A byte, therefore, represents a binary number between 0 and 255. A bit is symbolised by the lower case 'b', while a byte is symbolised by the upper case 'B'. The volume of a data set is usually given in bytes, while the capacity of a broadband line is usually given in bits. Thus, a 10 Mbs^{-1} broadband line actually transmits at 1.25 MB s^{-1}.

[28] The CDS also archives bibliographical information, enabling searches for publications relating to objects of interest to be conducted.

(https://cds.u-strasbg.fr/). The CDS also hosts SIMBAD. Other 'catalogues of catalogues' may be found at the National Space Science Data Centre (NSSDC; http://nssdc.gsfc.nasa.gov) and MAST.

5.6.3 Virtual Observatories

Anyone who is interested can easily look up the details on an individual object within the surveys and catalogues discussed in Section 5.5.2 via the internet. However, if more than a few tens objects are needed, then this approach becomes cumbersome and time-consuming, larger projects, therefore, need more powerful search and processing software. Some of the sites provide this on an individual basis, but with little consistency in what is available and how to use it. Recently, therefore, virtual observatories have started to come on-stream providing much greater software and other support, though for some of them their use is limited to accredited scientists.

Virtual observatories are interfaces whereby huge data sets such as the digital sky surveys and collections of observations from individual space missions or observatories and so on, can be handled and data-mined for new information. The virtual observatory is a set of data archives, software tools, hardware and staff that enables data in archives to be found and processed in many different ways. Amongst the type of functions that a virtual observatory can provide there are

- Standardising the formats of differing data sets
- Finding different observations of a given object or a set of objects and sorting through them
- Comparing and combining the information about objects available in various catalogues and other databases
- Comparing and combining observations taken at different times
- Combining archive data with new data obtained from telescopes
- Correlating data from different sources
- Classifying objects
- Performing statistical analyses
- Measuring images
- Image processing.

While the types of scientific study that are possible include

- Multi-wavelength studies
- Large statistical studies
- Searches for changing, moving, rare, unusual or new objects and types of objects
- Combining data in new ways leading to unexpected discoveries.

Virtual observatories started becoming available a decade or two ago and are now proliferating rapidly. Much of the information on virtual observatories is collated by the International Virtual Observatory Alliance (IVOA; www.ivoa.net/). The IVOA attempts to develop universal standards for virtual observatories so that they can work together and currently lists 20 active sites.

Most of the programmes are national but the Euro-VO (www.euro-vo.org/) is pan-European. The US's programme is called the Virtual Astronomical Observatory (VAO) and is to be found at www.usvao.org/. Details and contacts for the others may be found on the IVOA website..

As examples, the Tunka-Rex cosmic ray observatory (Section 1.4.3) is aimed to launch its own virtual observatory shortly providing open access to all its data, whilst for anyone with an interest in the Earth's Moon, NASA's Lunar Mapping, and modelling website (http://lunarscience.nasa.gov/articles/lunar-mapping-and-modeling-project/) is well worth looking at. It is a simple virtual observatory containing data from many lunar missions together with tools to produce 3D visualizations, to overlay images, to determine Sun angles, etc.

Finally, the most widely used virtual observatory of all is GOOGLE-SKY. This forms a part of GOOGLE-EARTH and may be accessed at www.google.com/sky/. GOOGLE-SKY provides a complete map of the sky containing some 100 million stars and 200 million galaxies. The more prominent nebulae and galaxies can be selected for higher resolution images and for notes about the natures of the objects. Best of all – it's available to anyone and it's FREE!

5.6.4 Management of Large Data Samples

Theoretical astronomy, particularly for the simulation of supernovae, binary black hole collisions, galaxy evolution and the like, has long needed the largest of supercomputers – and the progress in those topics has often been limited by the available computing speeds. Observational astronomy, by contrast, has had relatively low power computational requirements. At least, that has been the situation until about the last decade. Now numerous projects are starting to become operational, or will be doing so in a few years, where handling the amount of data produced by the instrument is becoming a major part of designing and affording the instruments. Thus, the VRO may soon be producing 5 TB of data every day, in 2018, the TESS data archive anticipated a download demand rate of 400 TB per day following the spacecraft's first data release, while the SKA's Australian Square Kilometre Array Pathfinder (ASKAP) in 2017 could generate 5.2 TB in one *second* of time[29] – and the instrument then, was only one-third of its finally planned size. Instruments designed to observe transient and/or rapidly varying sources usually operate at high cadences, thus generating large amounts of data in short intervals of time.

Gathering, storing and moving such enormous amounts of data as that produced by ASKAP, is simply beyond our present capabilities. The solution, in the case of ASKAP, is to undertake the preliminary processing of the raw data as soon as it arrives. By storing the *images* produced by ASKAP, but *not* storing the *raw data*, the data load is reduced by a factor of about 200,000 – to something that a fast, domestic broadband connection could handle. BUT – the raw data is gone for good – if better ways of processing the data are developed, or even if something needs checking, then it will no longer be possible to do that.

This is not an unprecedented situation – when astronomers could only make naked-eye observations, the raw data (i.e., what the astronomer saw) disappeared instantly. Only the memory remained: to be written down or sketched after the event – and then perhaps argued over. As far back as Galileo, observers have disagreed about what they could see through the telescope's eyepiece – and one of the great benefits of photography was the permanent physical image that was produced and which could be archived to be checked or reused years, decades, even a century, later. It will probably be an uncomfortable experience for modern astronomers to find themselves back in Galileo's shoes.

Less draconian measures for dealing with the data flood may be possible for less extreme situations than ASKAP's. Thus, for many years, images have been compressed to take up less space in (electronic) storage – sometimes with no loss of data, sometimes with the loss of non-essential data (Section 2.9.2). It may be possible to develop other similar strategies for other data types. Plus, of course, Moore's law suggests that computer performances double every 18 months or so – computer memory improvements may therefore catch up with data demands at some point in the future. Unfortunately, data transmission rates are limited by basic physics. Using e-m radiation, the sampling theorem (Section 2.1.1), puts an upper limit of N/2 bits per second for radiation of frequency N Hz. A 5 TBs^{-1} data production rate could thus be theoretically transmitted by ~100 THz[30] (NIR) radiation,[31] but the present and foreseeable capabilities of fibre-optic transmission cables are way short of that target.

---

[29] At that time, the whole of the internet amounted to about 35 TB/s.

[30] Remember; 1 byte = 8 bits

[31] Of course, a square-wave type signal – such as the '0s' and '1s' of a digital transmission – needs far higher frequencies for reliable representation than the basic signal frequency.

It seems likely, therefore, that some astronomical projects, especially those in space, could soon be limited by their data transmission requirements, rather than by any other parts of their designs.

5.7 ASTRONOMY AND THE REAL WORLD

5.7.1 Introduction

La découverte d' un mets nouveau
fait plus pour le bonheur du genre humain que
la découverte d' une étoile.[32]

(Jean Anthelme Brillat-Savarin, *La Physiologie du Goût*, 1825.)

Fortunately for astronomy, *both* discoveries contribute to many peoples' happinesses. Moreover, many people wish to *participate* in both discoveries and of all the sciences except perhaps botany, astronomy has the longest and proudest record of involving such participations.

For many centuries, it has widely been seen as a part of an astronomers' or astrophysicist's *duty* communicate, in as comprehensible fashion as possible and to as wide a general audience as possible, their discoveries about the wonders of the Universe – then, at least, they are providing some payback for the privilege of being allowed to be an astronomer or astrophysicist,[33] whether amateur or professional.

Participation in astronomy is also a long and worthy tradition at all and every level possible – whether 'just' watching a TV programme on the latest discoveries, visiting a local observatory to look for oneself,[34] buying your own (probably now very sophisticated) telescope or making the life-long commitment required to become a professional astronomer. Most of this book has so far been devoted to the (observational) concerns of professional or fairly advanced amateur astronomers. But participation in the joy of discovering something new about the Universe and contributing significantly to the advance of astronomy is open to many more people than those two groups. This book, therefore, concludes with a brief review of these possible ways for anyone to join in the fun.

5.7.2 Outreach/Education

5.7.2.1 Outreach

Outreach in an astronomical (or more general scientific) context basically means publicity and at some level, giving as many people as possible a piece of the action. Most forms of news media now widely publicise significant astronomical events, such as solar eclipses and efforts to involve people in astronomy directly are proliferating widely.

More technical discoveries will now frequently be publicised by their discoverers via press releases – many large scientific/astronomical institutions or societies, of all types, now have press officers. Occasionally, these items will also be picked up by general news sources. Specialised publications, such as scientific or astronomical magazines, programmes, internet channels, and so on will usually pick up on many more of the more significant (or spectacular) stories.

---

[32] *'The discovery of a new recipe does more for the happiness of human-kind than the discovery of a star'*. Author's (slightly free) translation.

[33] and not just for astronomers – for example; David Hilbert in a 1900 lecture to the International Congress of Mathematicians said *'A mathematical theory cannot be regarded as complete until you have made it so clear that you could explain it to the first man you encounter on the road.'* – Hilbert himself however, attributed this statement to *'an old French mathematician'*. Einstein had similar beliefs, but the statement: *'You do not really understand something until you can explain it to your grandmother"*, which is often attributed to him, is probably apocryphal.

[34] The Author, when Director of a University's observatory, had to limit participants at the regular open evenings to 500 people, many more, though, always applied. See also Wren-Marcario Accessible Telescope (WMAT) – Figure 1.55 and Section 1.1.19.2.

Also, as mentioned previously (Section 5.6.1 and elsewhere), news sites on the internet, provided by observatories and research institutes, allow anyone interested to follow their activities directly. Some observatories send out live feeds from cameras on their sites showing what is happening there in real time – KECKWATCH, for example, may be downloaded from the iTunes App Store. Numerous astronomy-related apps can be downloaded for mobile devices and some provide instant alerts to astronomical developments. Recently, for example, ESASky (https://sci.esa.int/web/astrophysics/-/60099-explore-the-cosmos-with-esasky) has become available, providing access to the whole sky for tablet and mobile users. While 17 robotic telescopes (at the time of writing), widely distributed over the Earth and sending out free and/or paid-for live images over the internet, especially of special events, form the Slooh project (www.slooh.com/guestDashboard).

Huge numbers of people are personally involved in the now-common practice of attaching lists of peoples' names in some way to spacecraft and so sending them off into the Universe (the Parker Solar Probe thus carries a memory card containing more than a million names), or in the naming of telescopes, planetary rovers and the like (the VLT's four telescopes were named Antu, Kueyen, Melipal, and Yepun following an essay contest amongst thousands of school children in Chile while the Martian rovers, Opportunity and Spirit, were named following a similar essay contest sponsored by NASA). It should be noted though, that stars are *not* generally given people's names. The commercial companies that advertise a star-naming service do not have those names recognised or used within astronomy. The few stars that *are* personalised – such as Barnard's star – are named after their discoverer(s) or person(s) who has investigated their peculiarities in great detail.

Awareness of the needs of astronomy are becoming a little wider known through various dark-sky campaigns (Section 1.1.3). Any city dwellers, who make the effort to visit a dark-sky site on a clear night, acquires an instant appreciation of the deprivation that they have previously suffered via the smog and lights of the city – and often then starts taking a serious interest in astronomy more generally.

The IAU now publishes the (free) *Communicating Astronomy with the Public* (CAP) journal. Interested readers may request notification of its issues at www.capjournal.org/subscription.php.

Outreach can, though, be overreached. The massive publication of astronomical events which are trivial, unexciting or in some cases, cannot even be seen, is counterproductive to astronomy, astronomers, science more generally and even to the credibility of news agencies. An example, which has become quite common recently, is the over-hyping of the full moon's appearance when the Moon is near to perihelion in its orbit around the Earth. Often called a 'Super-Moon', the reader of publicity about the event will usually be led to think that they have the *once-in-a-lifetime* opportunity to see the Moon *many* times larger in the sky than normal. In fact, the full moon is only 7.2% angularly larger than average, even at its closest possible perigee[35] – and no one could notice this change, except by using some moderately precise measuring instrument. Furthermore, the full moon occurring when the Moon is closer to the Earth than average, happens five or six times every year.

5.7.2.2 Education

The direct use of, sometimes quite large, telescopes by students in schools and colleges as a part of their education has already been mentioned (Section 1.1.24). The Las Cumbres Observatory, as another example, annually offers to educators 1,000 hours of observational time on its 23 robotic telescopes (which are up to 2 m in aperture, can be used with spectrographs and are located all around the world). Observing time may also be purchased at $300–$400 per hour.

Astronomical topics, though, do tend to be somewhat marginalised in school curricula – perhaps not too surprising given the pressure on schools to give their graduates skills for more likely employment opportunities. There, is however, an option to study for a General Certificate of Secondary

[35] The Moon's perigee distance from the Earth does alter slightly.

Education (GCSE)[36] in Astronomy in UK schools. While science projects at secondary and even primary levels, will often be based upon astronomical subjects. For the educators, 'Discover the Universe' (www.discovertheuniverse.ca/about/) is an online training programme offering workshops and educational material to aid and inspire their astronomy teaching – if there is time for some to be included within their courses.

At tertiary level, courses in/about/containing astronomy are more widespread. Some are intended for future professional astronomers, many more are 'sweeteners' for programmes leading to qualifications in what may be regarded as 'tougher' subjects (like physics or mathematics) – though rude awakenings may then follow in some such cases.

In practice, most students committed to a career in astronomy, read for undergraduate degrees in physics or mathematics, switching to their primary interest for their postgraduate studies.

In 2019, the IAU held a symposium devoted to Astronomy Education and the *Journal of Astronomy and Earth Science Education* specialises in research on the topic.

5.7.3 Pro-Am Collaborations and Citizen Science

It might seem impossible that work undertaken by professional astronomers using gigantic $100 million telescopes or $1,000 million spacecraft could be matched by amateur astronomers with a small telescope in their backyard. BUT – the Universe is a BIG place – and there's room for both. The reason for this situation lies not so much in the telescopes' sizes (though having a 10-m telescope or a spacecraft to hand *is* useful), but in the detectors available.

Photography transformed astronomy when it first started to be used late in the nineteenth century. But even the highly sophisticated photographic emulsions a century later only caught 1% of the photons which passed through them. The CCDs (invented in the late 1960s), though, caught nearly *all* the photons which they received (Section 1.1). Thus, when small 2D CCD arrays became available in the early 1970s, the photographic plate's days were numbered. Almost overnight, amateurs using a 0.2-m telescope and a CCD detector found that they could duplicate the observations of a professional's 2-m telescope which was still relying upon photography for its imaging.

Today, some amateur astronomers have computer-controlled 0.5-m telescopes and 4k × 4k-pixel cooled CCD arrays are (fairly) affordable to a private individual. Thus, amateur astronomers can now undertake observational programmes which, three or four decades ago, would have required the 5-m Hale telescope. However, even if the Hale telescope *were* to use photography now, there would still be plenty of work left for it to do. So, equally, there is still plenty of original research work left for a well-equipped amateur astronomer to undertake.[37]

Some astronomical Pro-Am[38] collaborations are formal arrangements with agreed responsibilities, tasks and programmes, others are more ad hoc and some are just happenstance. Citizen science is sometimes a formal arrangement with science input from the amateur volunteers, at other times it only amounts to computer users allowing some central facility to access their computer during intervals when the main user is not doing so.

The AAVSO, for example, when it was set up in 1911, had the purpose of organising (amateur) astronomer variable star observers' data to pass it on to the (professional) Harvard College Observatory. Today, both amateur and professional variable star observers from more than 100 countries are members of the organisation. To quote from its mission statement;

---

[36] an exam topic usually taken by pupils at about the age of 16.

[37] Embarrassment may also play a factor in providing research opportunities for amateur astronomers. Professional astronomers and/or the funding agencies are likely to be reluctant to be found using a telescope which *could* be observing a 23^m object, Gpc distant from us, for the study of a naked-eye star just a few pc away!

[38] See the footnote in the Preface to this book about the differences, or lack of them, between these two groups of astronomers.

> Professional astronomers have neither the time nor the telescopes needed to gather data on the brightness changes of thousands of variables, and amateurs make a real and useful contribution to science by observing variable stars and submitting their observations to the AAVSO International Database.
>
> **(www.aavso.org/visionmission)**

The search for objects which might collide cataclysmically with the Earth (Near Earth Objects [NEOs]) is currently a research area of great interest (after all, every one of us has a chance of being at the ground-zero of such an impact). Ground-based professional observatories and groups, such as Catalina Sky Survey (CSS), NEAT and Panoramic Survey Telescope and Rapid Response System (PanSTARRS) and space-based missions like NEOWISE[39] have been accumulating data for more than two decades, but the task is still incomplete. The professional results are, therefore, supplemented by amateur observations. For the last 22 years, the Planetary Society has been making grants to advanced amateur observers so that they may upgrade their telescopes and ancillary instruments to professional standards for this purpose. At the time of writing, the society has awarded a total of US$440,000 to more than 60 such observers, who have then made significant contributions to the project.

Rapid transient events may be caught serendipitously by amateur astronomers and then be useful to other, more planned, projects. Thus, during the lunar eclipse of January 2019, three amateur astronomers making video recordings of the event caught the light flash from a meteorite hitting the dark portion of the Moon. NASA's Lunar Impact Monitoring Programme, which has been in operation for more than a decade now, has caught many similar impacts in the past, but this was the first time that one has been detected during a lunar eclipse.

In a totally different fashion, amateur astronomers and/or the just-vaguely-interested general public can now help out professional work. Several projects have recently been crowdfunded (or have attempted to be so). The 2014 outreach programme, 'The Universe in a Box', for example, had the aim of distributing astronomy education kits to underprivileged communities around the world and needed 15,000 € for its 160 kits. In just 31 days, its organisers raised more than €17,000 through a crowd-funding campaign. A campaign to raise $1,000,000 for a space telescope (to be named Arkyd) and on which donors of more than $200 would have had the use of the instrument for 30 minutes of observations, actually raised $1,500,000 from 17,500 donors. In the end, the project had to be cancelled, although for reasons unrelated to the crowdfunding (reportedly the donors did get refunds!). A sophisticated small portable telescope, though, has been developed successfully via crowdfunding. Called the eVscope and starting from 2015, at the time of writing, some $2.1 million has reportedly been raised and more than 2,000 orders for the telescope received.

5.7.4 Citizen Science

Anyone with access to the internet and a computer can join in the work for some surveys. These citizen science projects work in several different ways but generally make use of spare time on individual personal computers.

For a typical citizen science project, some data processing software is downloaded onto the computer via the internet from the host institution's computer, together with some data from a large survey-type project. When the personal computer is on standby, it processes the batch of data that it has been sent and then sends the processed data back to the host computer. A new batch of data is then sent to the personal computer for processing at the next opportunity. Sometimes the work for the citizen science project requires little or no input from owners of the personal computer – they are, in effect, simply donating processing power to the project. Other types of project require much more human input so that the volunteer involved is a part of the project science team and just

[39] An extension of the WISE mission to search for NEOs.

happens to communicate with the rest of the team via the internet. In either case, the distributed network enables projects to have access to far more resources than could be afforded through dedicated facilities.

Amongst the earliest citizen science projects is SETI@home (http://setiathome.berkeley.edu/). It was started in 1999 and currently has more than 5,200,000 participants. It is searching for radio signals from extraterrestrial intelligences, though without any success to date.

Other large projects include Einstein@home (https://einsteinathome.org/ – see also Section 1.6.3.2.1.1.7) with more than 500,000 volunteers and which has been searching data from Advanced Laser Interferometer Gravitational wave Observatory (AdvLIGO; Section 1.6), the Arecibo radio telescope (Section 1.2) and the Fermi γ-ray spacecraft (Section 1.3) for emissions from spinning neutron stars and MilkyWay@home (http://milkyway.cs.rpi.edu/milkyway/) with over 400,000 participants since it started in 2007 and which is analysing data from the SDSS to map the 3D structure of the Milky Way galaxy.

Projects requiring more input from the volunteer include the Galaxy Zoo (www.galaxyzoo.org/) which is classifying galaxies and Planet Hunters and (www.planethunters.org/) which is searching data from the TESS spacecraft for exoplanets.

There are now hundreds of citizen science projects covering most of the sciences and lists of them can be found at the time of writing at the Berkeley Open Infrastructure for Network Computing (BOINC) website (http://boinc.berkeley.edu/), at the Zooniverse website (www.zooniverse.org/?lang=en) and as 'List of Distributed Computing Projects' on Wikipedia, although these sites are not all-inclusive.

Epilogue

In conclusion, I hope that somewhere between the preface and this epilogue you have found what you were searching for, that you discovered the details needed and that the accompanying explanations were adequate.

I also hope, that you may have found things that you were not expecting, or you did not know about, and yet which were still interesting and moreover discovered the enormous range of instruments, techniques, physical ideas, and mathematical processes which have brought about our present understanding of the Universe.

With every new edition of *Astrophysical Techniques*, I find myself amazed at the rate of development of the older approaches to observing the Universe and wonder at the new ideas suddenly burgeoning, often apparently out of nowhere. If anyone (except myself!) still possesses a first edition (written in 1983), they will marvel at the changes that have occurred in observational astronomy in slightly less than decades – but also will probably be astonished at a few items which have not changed at all.

I will end by repeating the opening salutation as valediction:
Clear Skies and Good Observing to you all!

C. R. Kitchin

Bibliography

Some selected journals, books, articles, internet sites, and other sources which provide further reading for various aspects of this book, or from which further information may be sought, are listed below.

RESEARCH JOURNALS (SEE ALSO WIKIPEDIA 'LIST OF ASTRONOMY JOURNALS')

Astronomical Journal
Astronomy and Astrophysics
Astrophysical Journal
Icarus
Messenger
Monthly Notices of the Royal Astronomical Society
Nature
Publications of the Astronomical Society of the Pacific
Science
Solar Physics

POPULAR MAGAZINES

All About Space
Astronomy
Astronomy Now
BBC Sky at Night
Ciel et Espace
Discover
L'Astronomie
Mercury
New Scientist
Scientific American
Sky News
Spaceflight
Sky at Night magazine
Sky and Telescope

OTHER BOOKS BY C. R. KITCHIN

Early Emission Line Stars, 1982 (Adam Hilger), ISBN 0141-1128;8
Stars, Nebulae and the Interstellar Medium, 1987 (IOP Publishing), ISBN 0-85274-580-X
Journeys to the Ends of the Universe, 1990 (Adam Hilger), ISBN 0-7503-0037-X
Optical Astronomical Spectroscopy, 1995 (Institute of Physics Publishing), ISBN 0-7503-0345-X
Photo-Guide to the Constellations, 1997 (Springer), ISBN 3-540-76203-5
Seeing Stars (with R. Forrest), 1997 (Springer), ISBN 3-540-76030-X
Solar Observing Techniques, 2001 (Springer), ISBN 1-85233-035-X
Illustrated Dictionary of Practical Astronomy, 2002 (Springer), ISBN 1-85233-559-8
Galaxies in Turmoil, 2007 (Springer), ISBN, 1-84628-670-0
Exoplanets: Finding, Exploring and Understanding Alien Worlds, 2012 (Springer), ISBN 13-978-1461406433
Telescope & Techniques (3rd Edition), 2013 (Springer), ISBN 13-978-1461448907

Remote and Robotic Investigations of the Solar System, 2019 (CRC Press), ISBN 13-978-0367871666
Understanding Gravitational Waves, 2021 (Springer), In Preparation

EPHEMERIDES, STAR AND OTHER CATALOGUES, ATLASES, SKY GUIDES

Dunlop S., Tirion W., *Guide to the Night Sky*, Published for each year (Collins)

H.M.S.O./U.S. Government Printing Office, *Astronomical Almanac*, Published for each year (H.M.S.O./U.S. Government Printing Office)

Inglis M., *Field Guide to the Deep Sky Objects*, 2011 (Springer), ISBN 13-978-146142656

Philip's Maps, *Philip's Star Chart*, 2018 (Philip's), ISBN 13-978-1849074872

Ridpath I. (Ed), *Norton's Star Atlas 20th Edition*, 2003 (Addison Wesley), ISBN 13-978-0131451643

Sinnott R., *Sky & Telescope Pocket Atlas Jumbo Edition*, 2019 (F & W Publications Inc.), ISBN-13-978-1940038704

Tirion W., *The Cambridge Star Atlas*, 2011 (Cambridge University Press), ISBN 13-978-1858059006

Walker R., *Spectral Atlas for Amateur Astronomers* (Cambridge University Press), ISBN 13-978-1107165908

PRACTICAL ASTRONOMY BOOKS

Bonaque-González S., Miguel-Hernández J., Fernánde Z, *Scientific Reduction of Astronomical Images*, 2017 (Lap Lambert Academic Publishing), ISBN-13-978-3330052574

Burke B. F., Graham-Smith F., Wilkinson P., *An Introduction to Radio Astronomy*, 2019 (Cambridge University Press), ISBN 13-978-1107189416

Fleisch D., *A Student's Guide to the Mathematics of Astronomy*, 2013 (Cambridge University Press), ISBN 13-978-1107610217

Harrison K. M., *Grating Spectroscopes and How to Use Them*, 2012 (Springer), ISBN 13-978-1461413967

Joardar S., *Radio Astronomy: An Introduction*, 2014 (Mercury Learning & Information), ISBN 13-978-1936420353

North G., *Observing Variable Stars, Novae and Supernovae*, 2014 (Cambridge University Press), ISBN 13-978-1107636125

Trypsteen M., Walker R., *Spectroscopy for Amateur Astronomers*, 2017 (Cambridge University Press), ISBN 13-978-1107166189

Warner B., *A Practical Guide to Lightcurve Photometry and Analysis*, 2016 (Springer), ISBN-13-978-3319327495

INTERNET SITES LISTED IN THE TEXT AND OTHERS OF USE

AAVSO – American Association of Variable Star Observers – www.aavso.org/
ADS – Astrophysical Data System – http://adsabs.harvard.edu/index.html
arXiv – http://arxiv.org/
Astrobites – http://astrobites.com/
Astronomical spectroscopy – http://spectroscopy.wordpress.com/
Astropy – www.astropy.org/
BiSON – Birmingham Solar Oscillations Network – http://bison.ph.bham.ac.uk/
BOINC – Berkeley Open Infrastructure for Network Computing – http://boinc.berkeley.edu/
CAP – Communicating Astronomy with the Public – www.capjournal.org/subscription.php
CASA – Common Astronomy Software Applications – https://casa.nrao.edu/
CDS – Centre de Données Astronomiques de Strasbourg – https://cds.u-strasbg.fr/
CERN – Conseil Européen pour la Recherche Nucléaire – https://home.cern/
CRAF – Committee on Radio Astronomy Frequencies – www.craf.eu/
Discover the Universe – www.discovertheuniverse.ca/about/
Einstein@home – https://einsteinathome.org/
ESA – European Space Agency – www.esa.int/
ESASky – https://sci.esa.int/web/astrophysics/-/60099-explore-the-cosmos-with-esasky
ESO – European Southern Observatory – www.eso.org/public/
Euro-VO – www.euro-vo.org/
Galaxy Zoo – www.galaxyzoo.org/
GCN – Gamma-ray Coordinates Network – https://gcn.gsfc.nasa.gov/

GIMP – GNU Image Manipulation Program – www.gimp.org/
GONG – Global Oscillation Network Group – https://gong.nso.edu/
GOOGLE-EARTH – www.google.com/sky/
IAU – International Astronomical Union – www.iau.org/
International Occultation Timing Association – http://occultations.org/
IVOA – International Virtual Observatory Alliance – www.ivoa.net/
Lunar Mapping and Modelling – http://lunarscience.nasa.gov/articles/lunar-mapping-and-modeling-project/
MAST – Mikulski Archive for Space Telescopes – http://archive.stsci.edu/
MilkyWay@home – http://milkyway.cs.rpi.edu/milkyway/
NASA – National Aeronautics and Space Administration – www.nasa.gov/
NSSDC – National Space Science Data Centre – http://nssdc.gsfc.nasa.gov
Planet Hunters – www.planethunters.org/
Public Event Explorer – http://labdpr.cab.cnea.gov.ar/ED-en/index.php
python – www.python.org/
Scintillation and seeing – http://apod.nasa.gov/apod/ap000725.html
SETI@home – http://setiathome.berkeley.edu/
SIMBAD – Set of Identifications, Measurements and Bibliography for Astronomical Data – http://simbad.u-strasbg.fr/Simbad/.
Skynet – https://skynet.unc.edu/
Slooh – www.slooh.com/guestDashboard
TOPCAT – Tool for Operations on Catalogues and Tables – www.star.bris.ac.uk/~mbt/topcat/
VAO – Virtual Astronomical Observatory – www.usvao.org/
VSO – Virtual solar observatory – http://vso.nascom.nasa.gov/cgi-bin/search
Zooniverse – www.zooniverse.org/?lang=en

Index

A

D

G

H

I

J

K

L

M

N

O

P

Q

R

S

T

U

V

W

X

Y

Z

For Product Safety Concerns and Information please contact our EU
representative GPSR@taylorandfrancis.com
Taylor & Francis Verlag GmbH, Kaufingerstraße 24, 80331 München, Germany

www.ingramcontent.com/pod-product-compliance
Ingram Content Group UK Ltd.
Pitfield, Milton Keynes, MK11 3LW, UK
UKHW031042080625
459435UK00013B/557